Methods in Molecular Biology

Series Editor
John M. Walker
School of Life and Medical Sciences
University of Hertfordshire
Hatfield, Hertfordshire, AL10 9AB, UK

For further volumes:
http://www.springer.com/series/7651

Computational Cell Biology

Methods and Protocols

Edited by

Louise von Stechow

NNF Center for Protein Research, University of Copenhagen, Copenhagen, Denmark

Alberto Santos Delgado

Novo Nordisk Foundation Center for Protein Research, University of Copenhagen, Copenhagen, Denmark

Humana Press

Editors
Louise von Stechow
NNF Center for Protein Research
University of Copenhagen
Copenhagen, Denmark

Alberto Santos Delgado
Novo Nordisk Foundation Center for Protein Research
University of Copenhagen
Copenhagen, Denmark

ISSN 1064-3745 ISSN 1940-6029 (electronic)
Methods in Molecular Biology
ISBN 978-1-4939-8617-0 ISBN 978-1-4939-8618-7 (eBook)
https://doi.org/10.1007/978-1-4939-8618-7

Library of Congress Control Number: 2018956725

Cover of the book was designed by Dr. Francesco Russo – Faculty of Health and Medical Sciences, Novo Nordisk Foundation Center for Protein Research, University of Copenhagen, Copenhagen, Denmark

This Humana Press imprint is published by the registered company Springer Science+Business Media, LLC part of Springer Nature.
The registered company address is: 233 Spring Street, New York, NY 10013, U.S.A.

Preface

Technological advances over the past decade have resulted in an explosion of available data, putting an end to researchers' focus on single genes or proteins and promoting system-wide approaches into biomedical research. The so-called big data era brings along the need for ways to extract meaningful information that go beyond manual inspection of large-scale datasets. An expanding toolbox of computational methods is evolving for identification and interpretation of biological phenotypes. Data-driven analyses, gene and protein set enrichment, representation of large-scale data into networks, and mathematical modeling of biological phenotypes are now emerging as means for the sophisticated analysis of the available biological data.

Computational Cell Biology: Methods and Protocols is targeted toward scientists who wish to employ computational techniques for analyses of a wide range of biological contexts, providing a great overview of suitable methods currently used in the field. It is written for a broad audience ranging from researchers who are unfamiliar with computational biology to those with more experience in the field. A number of review-style chapters give an overview of available data resources and analysis methods, while easy-to-follow protocols allow the researcher to apply various computational tools to an array of different data types.

Copenhagen, Denmark

Louise von Stechow
Alberto Santos Delgado

Contents

Contributors

HUSSAH ALOBAID • *Laboratories for Integrative Neuroscience and Endocrinology, School of Clinical Sciences, University of Bristol, Bristol, UK*

ELENA R. ALVAREZ-BUYLLA • *Laboratorio de Genética Molecular, Desarrollo y Evolución de Plantas, México, México; Instituto de Ecología, Centro de Ciencias de la Complejidad (C3), Universidad Nacional Autónoma de México Ciudad Universitaria, México, México*

ANASTASIA BARYSHNIKOVA • *Lewis-Sigler Institute for Integrative Genomics, Princeton University, Princeton, NJ, USA; Calico Life Sciences LLC, South San Francisco, CA, USA*

C. BAUER • *MicroDiscovery GmbH, Berlin, Germany*

KARSTEN M. BORGWARDT • *Machine Learning and Computational Biology Lab, D-BSSE, ETH Zurich, Basel, Switzerland; SIB Swiss Institute of Bioinformatics, Lausanne, Switzerland*

DAVID BOTSTEIN • *Calico Life Sciences, South San Francisco, CA, USA*

AIDAN BUDD • *Earlham Institute, Norwich Research Park, Norwich, UK*

ÁLVARO BUSTOS DELGADO • *Computational Biology Lab, Fundacion Ciencia & Vida, Santiago, Chile*

JOSE LUIS CALDU-PRIMO • *Centro de Ciencias de la Complejidad (C3), Universidad Nacional Autónoma de México, Ciudad Universitaria, México, México*

NATALIE M. CLARK • *Department of Plant and Microbial Biology, North Carolina State University, Raleigh, NC, USA; Biomathematics Graduate Program, North Carolina State University, Raleigh, NC, USA*

FEDERICA CONTE • *Institute for Systems Analysis and Computer Science "A. Ruberti" (IASI), National Research Council (CNR), Rome, Italy*

HELEN V. COOK • *Novo Nordisk Center for Protein Research, University of Copenhagen, Copenhagen, Denmark*

LUCA CSABAI • *Eötvös Loránd University, Budapest, Hungary*

ATTILA CSIKÁSZ-NAGY • *Randall Centre for Cell and Molecular Biophysics and Institute for Mathematical and Molecular Biomedicine, King's College London, London, UK; Faculty of Information Technology and Bionics, Pázmány Péter Catholic University, Budapest, Hungary*

YESID CUESTA-ASTROZ • *Centro de Pesquisas René Rachou (CPqRR), Fundação Oswaldo Cruz (FIOCRUZ), Belo Horizonte, Minas Gerais, Brazil*

THOMAS DANDEKAR • *Department of Bioinformatics, Functional Genomics and Systems Biology Group, Biocenter, Würzburg, Germany; BioComputing Unit, EMBL Heidelberg, Heidelberg, Germany*

JOSE DAVILA-VELDERRAIN • *Centro de Ciencias de la Complejidad (C3), Universidad Nacional Autónoma de México, Ciudad Universitaria, México, México; Departamento de Control Automático, Cinvestav-IPN, Cambridge, México D.F, Mexico; MIT Computer Science and Artificial Intelligence Laboratory, Cambridge, MA, USA; •Broad Institute of MIT and Harvard, Cambridge, MA, USA*

MARIA PAMELA DOBAY • *SIB Swiss Institute of Bioinformatics, Quartier Sorge, Bâtiment Génopode, Lausanne, Switzerland; IQVIA, Basel, Switzerland; Yocto Group Limited, Zurich, Switzerland*

YANNAN FAN • *Institute of Parasitology, McGill University, Sainte Anne de Bellevue, QC, Canada*

DÁVID FAZEKAS • *Eötvös Loránd University, Budapest, Hungary; Earlham Institute, Norwich Research Park, Norwich, UK*

GIULIA FISCON • *Institute for Systems Analysis and Computer Science "A. Ruberti" (IASI), National Research Council (CNR), Rome, Italy*

ADAM P. FISHER • *Department of Plant and Microbial Biology, North Carolina State University, Raleigh, NC, USA*

IGNACIO FUENZALIDA • *Computational Biology Lab, Fundacion Ciencia & Vida, Santiago, Chile*

KATHRYN L. GARNER • *Laboratories for Integrative Neuroscience and Endocrinology, School of Clinical Sciences, University of Bristol, Bristol, UK*

DOMINIK G. GRIMM • *Machine Learning and Computational Biology Lab, D-BSSE, ETH Zurich, Basel, Switzerland; SIB Swiss Institute of Bioinformatics, Lausanne, Switzerland*

ANJA C. GUMPINGER • *Machine Learning and Computational Biology Lab, ETH Zurich, Basel, Switzerland; SIB Swiss Institute of Bioinformatics, Lausanne, Switzerland*

C. HARDT • *Department of Computational Molecular Biology, Max-Planck-Institute for Molecular Genetics, Berlin, Germany*

DAVID G. HENDRICKSON • *Calico Life Sciences, South San Francisco, CA, USA*

ROSA HERNANSAIZ-BALLESTEROS • *Randall Division of Cell and Molecular Biophysics and Institute for Mathematical and Molecular Biomedicine, King's College London, London, UK*

R. HERWIG • *Department of Computational Molecular Biology, Max-Planck-Institute for Molecular Genetics, Berlin, Germany*

KIRSTEN JENKINS • *Randall Division of Cell and Molecular Biophysics and Institute for Mathematical and Molecular Biomedicine, King's College London, London, UK*

LARS JUHL JENSEN • *Novo Nordisk Center for Protein Research, University of Copenhagen, Copenhagen, Denmark*

TAMÁS KORCSMÁROS • *Eötvös Loránd University, Budapest, Hungary; Earlham Institute, Norwich Research Park, Norwich, UK; Quadram Institute, Norwich Research Park, Norwich, UK*

MEIK KUNZ • *Department of Bioinformatics, Functional Genomics and Systems Biology Group, Biocenter, Würzburg, Germany*

ALBERTO J. M. MARTIN • *Computational Biology Lab, Fundacion Ciencia & Vida, Santiago, Chile; Centro Interdisciplinario de Neurociencias de Valparaiso, Valparaiso, Chile; Centro de Genomica y Bioinformatica, Universidad Mayor, Santiago, Chile*

MAURICIO PÉREZ MARTÍNEZ • *Instituto de Investigaciones Biomédicas, Universidad Nacional Autónoma de México, CDMX, Mexico, Mexico*

JUAN CARLOS MARTINEZ-GARCIA • *Departamento de Control Automático, Cinvestav-IPN, México, México*

CRAIG A. MCARDLE • *Laboratories for Integrative Neuroscience and Endocrinology, School of Clinical Sciences, University of Bristol, Bristol, UK*

R. SCOTT MCISAAC • *Calico Life Sciences, South San Francisco, CA, USA*

AKRAM MÉNDEZ • *Instituto de Investigaciones Biomédicas, Universidad Nacional Autónoma de México, CDMX, Mexico, Mexico*

LUIS MENDOZA • *Instituto de Investigaciones Biomédicas, Universidad Nacional Autónoma de México, CDMX, Mexico, Mexico*

MÁRTON ÖLBEI • *Earlham Institute, Norwich Research Park, Norwich, UK; Quadram Institute, Norwich Research Park, Norwich, UK*

GUILHERME OLIVEIRA • *Instituto Tecnológico Vale, Belém, PA, Brazil*
PAOLA PACI • *Institute for Systems Analysis and Computer Science "A. Ruberti" (IASI), National Research Council (CNR), Rome, Italy*
ROHITH PALLI • *Medical Scientist Training Program and Biophysics, Structural & Computational Biology graduate program, Rochester, NY, USA*
MARCO PELLEGRINI • *Institute of Informatics and Telematics (IIT), National Research Council (CNR), Pisa, Italy*
TOMÁS PÉREZ-ACLE • *Computational Biology Lab, Fundacion Ciencia & Vida, Santiago, Chile; Centro Interdisciplinario de Neurociencias de Valparaiso, Valparaiso, Chile*
NATALIA PIETROSEMOLI • *Institut Pasteur, Bioinformatics and Biostatistics Hub, C3BI, USR 3756 CNRS, Paris, France*
ANDREAS PITTROFF • *Department of Bioinformatics, Functional Genomics and Systems Biology Group, Biocenter, Würzburg, Germany*
CARLOS RAMÍREZ • *Instituto de Investigaciones Biomédicas, Universidad Nacional Autónoma de México, CDMX, Mexico, Mexico*
SIMONE RIZZETTO • *School of Medical Sciences, University of New South Wales, Sydney, NSW, Australia; Viral Immunology Systems Program, Kirby Institute for Infection and Immunity, University of New South Wales, Sydney, NSW, Australia*
MILENA RIZZO • *Institute of Clinical Physiology, National Research Council (CNR), Pisa, Italy; Istituto Toscano Tumori (ITT), Firenze, Italy*
DAMIAN ROQUEIRO • *Machine Learning and Computational Biology Lab, D-BSSE, ETH Zurich, Basel, Switzerland; SIB Swiss Institute of Bioinformatics, Lausanne, Switzerland*
FRANCESCO RUSSO • *Novo Nordisk Foundation Center for Protein Research, Faculty of Health and Medical Sciences, University of Copenhagen, Copenhagen, Denmark*
RODRIGO SANTIBÁÑEZ • *Computational Biology Lab, Fundacion Ciencia & Vida, Santiago, Chile; Escuela de Ingeniería, Pontificia Universidad Católica de Chile, Santiago, Chile*
J. SCHUCHHARDT • *MicroDiscovery GmbH, Berlin, Germany*
ILYA SOIFER • *Calico Life Sciences, South San Francisco, CA, USA*
ROSANGELA SOZZANI • *Department of Plant and Microbial Biology, North Carolina State University, Raleigh, NC, USA; Biomathematics Graduate Program, North Carolina State University, Raleigh, NC, USA*
JUILEE THAKAR • *Department of Microbiology and Immunology, University of Rochester Medical Center, Rochester, NY, USA; Department of Biostatistics and Computational Biology, University of Rochester Medical Center, Rochester, NY, USA*
KRASIMIRA TSANEVA-ATANASOVA • *EPSRC Centre for Predictive Modeling in Healthcare, University of Exeter, Exeter, UK; Department of Mathematics and Living Systems Institute, College of Engineering, Mathematics and Physical Sciences, University of Exeter, Exeter, UK*
MARGARITIS VOLIOTIS • *EPSRC Centre for Predictive Modeling in Healthcare, University of Exeter, Exeter, UK; Department of Mathematics and Living Systems Institute, College of Engineering, Mathematics and Physical Sciences, University of Exeter, Exeter, UK*
BERND WRANIK • *Calico Life Sciences, South San Francisco, CA, USA*
JIANGUO XIA • *Institute of Parasitology, McGill University, Sainte Anne de Bellevue, Quebec, Canada; Department of Animal Science, McGill University, Sainte Anne de Bellevue, Quebec, Canada*

Part I

Big Data- and Its Implications in Cell Biology

Chapter 1

Rule-Based Models and Applications in Biology

Álvaro Bustos, Ignacio Fuenzalida, Rodrigo Santibáñez, Tomás Pérez-Acle, and Alberto J. M. Martin

Abstract

Complex systems are governed by dynamic processes whose underlying causal rules are difficult to unravel. However, chemical reactions, molecular interactions, and many other complex systems can be usually represented as concentrations or quantities that vary over time, which provides a framework to study these dynamic relationships. An increasing number of tools use these quantifications to simulate dynamically complex systems to better understand their underlying processes. The application of such methods covers several research areas from biology and chemistry to ecology and even social sciences.

In the following chapter, we introduce the concept of rule-based simulations based on the Stochastic Simulation Algorithm (SSA) as well as other mathematical methods such as Ordinary Differential Equations (ODE) models to describe agent-based systems. Besides, we describe the mathematical framework behind *Kappa* (κ), a rule-based language for the modeling of complex systems, and some extensions for spatial models implemented in PISKaS (Parallel Implementation of a Spatial Kappa Simulator). To facilitate the understanding of these methods, we include examples of how these models can be used to describe population dynamics in a simple predator–prey ecosystem or to simulate circadian rhythm changes.

Key words Stochastic simulation, Rule-based modeling, κ language

1 The Stochastic Simulation Algorithm (SSA)

The SSA, also known as Gillespie's algorithm [10], is the basis of most stochastic simulation tools available. This algorithm and the tools based on it assume there is a homogeneous and "well-stirred" system of particles named agents. Agents can represent any type of entity within a system, i.e., molecules or individuals, and the interactions between agents are determined by a set of rules or equations taking place at certain rates. These rules are ordered and divided into agents to which the rule applies and products (outcome agents). For instance, in a system of chemical reactions described by an equation or rule ($reactants \rightarrow products$), every set of particles matching the left side of the equation (or reactant agents) has an equal probability of being

Louise von Stechow and Alberto Santos Delgado (eds.), *Computational Cell Biology: Methods and Protocols*, Methods in Molecular Biology, vol. 1819, https://doi.org/10.1007/978-1-4939-8618-7_1,

the subject of that rule, that is to undergo the process described by the rule. To clarify, given a reaction of the form $A + B \rightarrow C$ in a system with 1000 particles of type A and 1000 of type B, this "well-stirring" assumption means that every pair of particles $\{A_1, B_1\}, \{A_1, B_2\}, \ldots, \{A_i, B_j\}, \ldots, \{A_{1000}, B_{1000}\}$ has equal probability of interacting to produce a particle of type C. Another important assumption made by the SSA is that the volume or area where the simulation takes place is fixed, and thus, concentrations of agents correspond to the discrete number of agents of each type.

To describe chemical systems, and this can be extended to any other type of system, a specific set of **reactions** is required. Reactions in this algorithm always match the following schema (Eq. (1)):

$$m_1 A_1 + \cdots + m_r A_r \rightarrow n_1 C_1 + \cdots + n_s C_s \quad (1)$$

Whenever a reaction of this type takes place, a set of m reactant particles of types A_i are removed from the simulation (for $i = 1, \ldots, r$) and are in turn replaced by another set of n product particles of types C_j (for $j = 1, \ldots, s$). It should be noted that any of the products C_j could be of the same kind as one of the reactants, and that this schema covers reactions as simple as $A_i \rightarrow A_j$ to more complicated reactions requiring several types of different agents. Other important reactions that follow the same schema are $A_i \rightarrow \emptyset$ to indicate degradation of agents and very similarly, $\emptyset \rightarrow A_i$ to model the addition of a new element in a system. Rules are applied according to reaction rates, which defines different behaviors of the system upon variations in the concentration of its reactants.

The quantity of each type of agent, or state of the system, at a given time t can be represented by a vector of non-negative integers, or **state vector**, in which each entry represents the amount of each agent type. The outcome of a given chemical reaction can also be represented by a **state-change vector**, with the same size as the state-vector at time t. The negative entries in the state-vector depict the consumption of an agent, positive values mean the creation of an agent, and 0 or *null* value indicates no change for a particular agent type. Therefore, if the state vector before a given reaction r is $\vec{X}$ and the associated state-change vector is $\vec{d_r}$, the state of the system changes from $\vec{X}$ to $\vec{X} + \vec{d_r}$ after the reaction occurs. For example, in a medium that has samples of three different chemicals A, B, C, the following vector represents the existence of 1000 molecules of type A, 900 of type B, and 1200 of type C at time t:

$$\vec{X}(t) = \begin{bmatrix} 1000 \\ 900 \\ 1200 \end{bmatrix}$$

In a similar way, the following two chemical reactions r_1 and r_2 correspond, respectively, to the two state-change vectors $\vec{d}_{r_1}$ and $\vec{d}_{r_2}$:

$$2A + B \xrightarrow{r_1} C \qquad\qquad 2C \xrightarrow{r_2} 2A + B + C$$

$$\vec{d}_{r_1} = \begin{bmatrix} -2 \\ -1 \\ 1 \end{bmatrix} \qquad\qquad \vec{d}_{r_2} = \begin{bmatrix} 2 \\ 1 \\ -1 \end{bmatrix}$$

Note that in the second reaction one of the particles of type C acts as a catalyst and the outcome effect of reaction r_2 is the same as of $C \xrightarrow{r_2} 2A + B$ with a $\vec{d}_{r_2}$ state-change vector identical to $-\vec{d}_{r_1}$. However, the relationship between the different probabilities of reactions r_1 and r_2 happening and the amount of particles of type C present lead to different long-term behaviors of the system.

A reaction r is fully specified by the state-change vector $\vec{d}_r$ and a **propensity function** a. This propensity function takes the state vector $\vec{X}$ as argument and calculates the rate of every reaction r in the system; thus, if $a(r_1, \vec{X}) > a(r_2, \vec{X})$ reaction r_1 is more likely to occur than r_2. This is a discrete model; therefore, the function $a(r, \vec{X})$ is combinatorial in nature. For a fixed r, it should, theoretically, be directly proportional to the number of distinct sets of molecules that match the left side of the equation describing the reaction r and the physical properties of the medium being simulated. In this way, a probabilistic mathematical model of any set of reactions can be built given their state-change vectors $\vec{d}_r$ and a propensity function a. The function a reflects the constraints given by the chemical nature of the system being modeled and allows the description, at least indirectly, of the probability distribution of the possible future state of the system, given its initial state and a time lapse:

$$P(\vec{x}, t \mid \vec{x}_0, t_0) := \mathbb{P}(\vec{X}(t) = \vec{x} \mid \vec{X}(t_0) = \vec{x}_0) \qquad (2)$$

Equation (2) is the **Markovian condition** that assumes the future state of the system relies exclusively on the present state $(\vec{x}_0, t_0)$ and the propensity function of each possible reaction. To be precise, to get to state $\vec{x}$ at time $t + dt$ for a dt small enough to ensure that the probability of two reactions occurring in that time interval is negligible, either the state at time t is also $\vec{x}$, or the state at time t is $\vec{x} - \vec{d}_r$, and reaction r takes place during the interval $[t, t + dt]$. Thus, we have the following approximate equality (in which R is the set of all reactions):

$$
\begin{aligned}
&P(\vec{x}, t + dt \mid \vec{x}_0, t_0) \\
&\quad \approx \sum_{r \in R} P(\vec{x} - \vec{d}_r, t \mid \vec{x}_0, t_0)\, \mathbb{P}(\text{reaction } r \text{ happens in } [t, t + dt]) \\
&\qquad + P(\vec{x}, t \mid \vec{x}_0, t_0)\, \mathbb{P}(\text{no reaction happens during } [t, t + dt]) \\
&\quad \approx \sum_{r \in R} P(\vec{x} - \vec{d}_r, t \mid \vec{x}_0, t_0) a(r, \vec{x} - \vec{d}_r)\, dt \\
&\qquad + P(\vec{x}, t \mid \vec{x}_0, t_0) \left(1 - \sum_{r \in R} a(r, \vec{x})\, dt \right)
\end{aligned}
$$

From the last expression (Eq. (1)), moving the term $P(\vec{x}, t \mid \vec{x}_0, t_0)$ to the left side of the equality and dividing by dt, the following identity is obtained as $dt \to 0$:

$$
\begin{aligned}
\frac{d}{dt} P(\vec{x}, t \mid \vec{x}_0, t_0) = \sum_{r \in R} [&P(\vec{x} - \vec{d}_r, t \mid \vec{x}_0, t_0) a(r, \vec{x} - \vec{d}_r) \\
&- P(\vec{x}, t \mid \vec{x}_0, t_0) a(r, \vec{x})]
\end{aligned} \tag{3}
$$

Equation (3), commonly known as the **Chemical Master Equation (CME)**, can be rigorously formalized from the laws of probability and the theory of Markov processes [9], but for simplicity we will use the informal derivation given above. Although the previous equation is theoretically enough to determine the probabilities involved in the simulation at any moment given an initial state (i.e., the function $P(\vec{x}, t \mid \vec{x}_0, t_0)$), determining an explicit form of P analytically from the CME is usually extremely hard. This difficulty is due to Eq. (3) being a system of coupled differential equations with one function for each different state vector , and thus it can potentially have infinite unknown functions. Therefore, using this equation directly as a basis for a simulation is extremely impractical for systems composed of many different types of agents and/or with a large number of rules. However, it is possible to construct accurate numerical Markov simulations that follow the distribution given by the CME [10]. To accomplish this and accurately simulate the future state of a system based on information of the current state, only two questions need to be answered:

- Which reaction will happens next?, and
- How much time will pass from now until it happens?

Thus, for an accurate simulation, we only need information about the conditional probability distribution of the next reaction r and expected time τ. So, we define the function $p(r, \tau \mid \vec{x}, t)$ as follows:

$$
\begin{aligned}
p(r, \tau \mid \vec{x}, t)\, d\tau \approx \mathbb{P}(&\text{the next reaction happens in the interval} \\
&[t + \tau, t + \tau + d\tau] \text{ and is of type } r \mid \vec{X}(t) = \vec{x})
\end{aligned}
$$

If we assume the Markovian memoryless property, this probability should be independent of the current time t; thus, the definition can be simplified slightly by removing references to t:

$$p(r, \tau \mid \vec{x}, t)\, d\tau \approx \mathbb{P}(\text{no reactions during } [0, \tau] \text{ and a reaction of type } r \text{ during } [\tau, \tau + d\tau] \mid \vec{X}(0) = \vec{x}).$$

Assuming that every reaction r takes place independently of all other reactions, the Markovian assumption tells us that the expected time T_r until a reaction of type r is an exponential variable of rate $a(r, \vec{x})$ [17, chapters 6–7]. Thus, the time $T = \min_{r \in R} T_r$ until the next reaction is an exponential variable of rate $a_0(\vec{x}) := \sum_{r \in R} a(r, \vec{x})$, and it is independent of the reaction r chosen. Therefore, an explicit value for the probability density p can be easily determined:

$$p(r, \tau \mid \vec{x}, t) = a(r, \vec{x}) \exp(-\tau a_0(\vec{x})). \quad (4)$$

Equation (4) can be used to generate trajectories that follow the desired distribution, since it implies that the probability of choosing a given reaction r is $a(r, \vec{x})/a_0(\vec{x})$ and it is independent of the expected time T, we get the following simple algorithm for generating valid trajectories given an initial state $\vec{x_0}$:

1. Initialize the state $\vec{X}$ as $\vec{x_0}$ and the current time t to 0.

$$\vec{x_0} = \vec{X}(t = t_0) = \begin{bmatrix} 1000 \\ 900 \\ 1200 \end{bmatrix} \times [\text{A, B, C}] \qquad \begin{aligned} &2\text{A} + \text{B} \xrightarrow{r_1} \text{C} \\ &2\text{C} \xrightarrow{r_2} 2\text{A} + \text{B} + \text{C} \end{aligned}$$

2. Generate two random numbers p_1, p_2 in [0, 1] (uniform distribution), for example:

$$p_1 = 0.18 \text{ and } p_2 = 0.67$$

3. Determine the reactivity of the system as $a_0(\vec{x}) = \sum_{r \in R} a(r, \vec{x})$ and set δt as the value $\ln(1/p_1)/a_0(\vec{X})$. This ensures that the random variable δt has an exponential distribution with rate $a_0(\vec{X})$. For simplicity, each reaction occurs at the same frequency $r = r_1 = r_2 = 1.0\,\text{s}^{-1}$

$$a_0(\vec{x}) = r_1 \times \text{A} \times (\text{A} - 1) \times \text{B} + r_2 \times \text{C} \times (\text{C} - 1)$$

$$a_0(\vec{x}) = r_1 \times 1000 \times (1000 - 1) \times 900 + r_2 \times 1200 \times (1200 - 1)$$

$$\delta t = \ln(1/0.18)/900{,}538{,}800 = 1.90 \times 10^{-9}\,\text{s} = 1.90\,\mu\text{s}$$

4. Suppose the set of reactions is given by $R = \{r_1, \ldots, r_j, \ldots, r_n\}$. The probability to choose any reaction $r \in R$ is $a(r, \vec{X})/a_0(\vec{X})$. We choose the reaction r_j testing the following inequality:

$$\sum_{j=1}^{n-1} \frac{a(r_j, \vec{X})}{a_0(\vec{X})} \leq p_2 < \sum_{j=1}^{n} \frac{a(r_j, \vec{X})}{a_0(\vec{X})}$$

For example, given the two reactions r_1 and r_2, we test the following inequalities:

$$0 \leq 0.67 < \sum_{j=1}^{1} \frac{a(r_j, \vec{X})}{a_0(\vec{X})} = 0.9984$$

$$0.9984 \leq 0.67 < \sum_{j=1}^{2} \frac{a(r_j, \vec{X})}{a_0(\vec{X})} = 1.0000$$

5. Replace the old value of t by the new value $t + \delta t$ and the old value of $\vec{X}$ with $\vec{X} + \vec{d_r}$, where r is the reaction chosen in step (4).

$$\vec{x_1} = \vec{X}(t = t_0 + \delta t) = \begin{bmatrix} 1000 \\ 900 \\ 1200 \end{bmatrix} + \begin{bmatrix} -2 \\ -1 \\ 1 \end{bmatrix} = \begin{bmatrix} 998 \\ 899 \\ 1201 \end{bmatrix}$$

6. Save the new values of t and $\vec{X}$ and go back to step (2) or finish if $a_0(\vec{X}) = 0$.

This is a basic form of the SSA, readers interested in a more in-depth analysis of the model may consult the review by Gillespie in [10]. Common methods for the simulation of rule-based models use adapted versions of this algorithm to generate accurate simulations, each approach making certain assumptions and often requiring a formal language to describe the models. Examples of such SSA-based implementations are BioNetGen [3] and KaSim [13], each with its own formal language (BNGL [7] and Kappa [16], respectively).

2 Introduction to Ordinary Differential Equations Models

Another common approach to the study of the dynamic behavior of complex systems employs ODEs or Partial Differential Equationss (PDEs) based on the empirical **law of mass action** [12, 21]. This law states that the rate of a chemical reaction is proportional to the **activity** of each of its reactants. In order to simplify the model, it is often assumed that such activity values match the concentrations of each reactant. While this is generally not true, for elemental reversible reactions with no intermediate steps, it is a reasonable assumption and an acceptable approximation. For instance, given an elemental reversible reaction such as the following:

$$A + B \rightleftarrows C$$

the rate at which the forward reaction $A + B \rightarrow C$ occurs is proportional to the concentrations of A and B, with a similar remark applying to the backward reaction. This simple reversible equation prompts the following three ODEs systems as a candidate for modeling its evolution or dynamic behavior over time:

$$\frac{d[A]}{dt} = -k_1[A][B] + k_2[C]$$
$$\frac{d[B]}{dt} = -k_1[A][B] + k_2[C]$$
$$\frac{d[C]}{dt} = k_1[A][B] - k_2[C]$$

in which [X] stands for the concentration of the reactant X and k_1 and k_2 are rate constants usually determined from experimental data. The right-hand side of the equation represents that in the forward reaction ($A + B \rightarrow C$), one instance of A and one of B are replaced by one of C, with the opposite happening for the reverse reaction.

This small system of ODEs is usually nonlinear. The model has a very simple structure, and allows both numerical and theoretical analyses. For instance, equilibrium can be calculated assuming that $k_1[A][B] - k_2[C] = 0$, which leads to Eq. (5):

$$K = \frac{k_1}{k_2} = \frac{[C]}{[A][B]} \tag{5}$$

where K is called the **equilibrium constant** of the system and does not depend directly on the concentrations of the reactive substances but only on the rate constants k_1, k_2. K governs the asymptotic behavior of the system as time goes to infinity [19, 20]; more precisely, a system of chemical reactions eventually reaches a situation in which the concentration of each chemical involved remains unchanged, with this value being determined by the constant K [20, chapter 17]. This can be seen mathematically by noticing that the system of equations above has a constant solution whose value depends on K, and any other positive solution converges to this value as $t \rightarrow \infty$ [21].

However, for more complex reactions and systems with more types of agents, the setup of the ODE system and the structure of the resultant reactions become very difficult to simulate using this type of equation [10]. Chemical reactions such as electrolysis, which involves two or more instances of the same reactant, introduce higher-order terms that might induce unexpected and/or difficult-to-explain behavior in numerical simulations. In addition, non-elementary reactions have to be decomposed into a series of elementary reactions, which can greatly increase the number of terms and variables involved in the system. Thus, the ODE approach becomes impractical very quickly in sufficiently complex chemical systems. Another drawback of this approach is that low concentrations or quantifications of agents can lead to unrealistic simulations of the behavior of the system in the long term upon extinction of these agents. This is particularly noticeable in small systems comprised of only hundreds or thousands of agents.

Another characteristic of the ODE-based approach is that it is purely deterministic. Given that in a real chemical system there are random fluctuations and non-deterministic phenomena, a deterministic model might not be able to fully represent all of the possible outcomes of the system. As in the previous paragraph, it is worth mentioning that random fluctuations usually have negligible long-term influence in large systems with sufficiently high concentrations of every species in the system. However, they become much more evident in systems with a lower number of components. In such systems, there are potential alternative outcomes (different from the average behavior simulated by deterministic models) with large quantitative differences and non-negligible probabilities. Hence, taking into account this non-deterministic behavior becomes essential to understand small-scale systems [10].

Lastly, ODE-based models usually carry no spatial information, as the medium is assumed homogeneous and well-stirred, with a uniform distribution of all system components. Here, we describe several biological systems in which those assumptions are invalid. The most straightforward way to create models that take into account spacial information is by replacing the concentration value as a function of simulated time for each entity $[X](t)$ by a spatial density term $\rho_X(t; x, y, z)$, which represents the density of X in a small neighborhood of points in the area or volume comprised by the model. Also, additional terms in the differential equations above are required to model physical phenomena that may affect density. For example, the chemical entities in the simulated system are liquids capable of diffusion; a possible set of equations for the reaction $\mathrm{A} + \mathrm{B} \rightarrow \mathrm{C}$ could be defined as

$$\frac{\partial \rho_\mathrm{A}}{\partial t} = -\Delta \rho_\mathrm{A} - k\rho_\mathrm{A}\rho_\mathrm{B}$$

$$\frac{\partial \rho_\mathrm{B}}{\partial t} = -\Delta \rho_\mathrm{B} - k\rho_\mathrm{A}\rho_\mathrm{B}$$

$$\frac{\partial \rho_\mathrm{C}}{\partial t} = -\Delta \rho_\mathrm{C} + k\rho_\mathrm{A}\rho_\mathrm{B}$$

where each $\Delta\rho$ term corresponds to the following sum of partial derivatives (known as the **Laplacian** or **Laplace operator**):

$$\Delta\rho = \frac{\partial^2 \rho}{\partial x^2} + \frac{\partial^2 \rho}{\partial y^2} + \frac{\partial^2 \rho}{\partial z^2}$$

This conforms to the usual diffusion reaction from physics (as stated in [6, Chapter 2]), $\partial\rho/\partial t + \Delta\rho = 0$ (with constant diffusion rates uniformly equal to 1), after adding the additional terms brought by the law of mass action considering that for a very small neighborhood of a point (x, y, z) the term $[X]$ is proportional to ρ_X and the chemical X may be assumed approximately homogeneous.

3 Parallel Implementation of Spatial κ

3.1 The κ Algorithm

In this section, we introduce a modified version of the SSA to allow more complex simulations in a variety of contexts beyond the standard chemical applications. We do not go in depth into the mathematical formalisms behind the modifications of the SSA introduced here; these details may be consulted in publications about the κ language such as the work of Danos et al. [4, 5]. The discussion below follows the theoretical framework setup by Danos, with a schematic graphical notation whenever possible. For the actual language and syntax used in standard implementations such as KaSim, please consult the KaSim reference manual [14]. Nevertheless, the examples in this and the following sections can be easily implemented in KaSim, which provides all of the standard κ framework. Further examples that involve spatial information are designed to be compatible with PISKaS [18], which is a spatially-enhanced fork of KaSim.

The classical Gillespie's algorithm treats every kind of chemical compound (or, in general, a variation of an agent) as a separate type, no matter how similar it may be to a previously existent type of compound [5]. This becomes problematic when there is a large amount of different compounds that are similar—but not identical—as there is no way to express this similarity properly in the classic SSA framework, even if these cannot participate in the same reactions. Thus, this results in state-vectors with a large number of entries and (usually) several almost-duplicate reactions or rules involving small variants of the same compound, requiring too many computational resources to simulate such systems. Another problem is that the described SSA framework ignores the internal structure of the compounds involved. This is a problem when dealing with complex molecules such as proteins or DNA, since their internal structure can severely influence the outcome of a chemical process out of sheer geometrical positioning, let alone physical or chemical constraints caused by the size of the molecule. Last, biological systems and other complex systems are naturally compartmentalized (cellular compartments), a characteristic difficult to replicate into a model using an algorithm that assumes a homogeneously mixed environment where all reactions take place.

The first two observations made above suggest that a modification of the data structure used to store the current status of the system, as well as a change in the idea of what constitutes an agent, might allow for a more flexible and robust framework. The concerns about information regarding the fixed structure of a compound suggest that an atom is probably a better model than a molecule for the concept of agent. An atom interacts with other atoms in several ways, the covalent bond being among the simplest to understand conceptually: Each atom can form a finite number of preestablished links of a specific type with one or more other atoms, which in turn can also have links between themselves. For instance, an oxygen atom can form two covalent bonds, or in other words, it has two "open places" where other atoms can bind to, while a hydrogen atom can form a single covalent bond. When two hydrogen atoms bind to one oxygen atom by forming two covalent bonds, a water molecule is formed. Similarly, chemical reactions can be expressed as the formation or destruction of links between reactants or agents.

This motivates storing the current state of the system as a **site graph** [5]. This graph corresponds to a network in which the nodes or **agents** have a specific structure that limits the kinds of connections or bonds that can be formed. More specifically,

- Each agent has a **type**. Going with our chemical analogy, this would correspond to the specific element (hydrogen, helium, oxygen...) of each atom.
- Each type has a set of **sites** associated with it. Every site has a set of possible **internal states**. In our example, these sites correspond to the places in the atom where covalent bonds can be formed, while the internal states may correspond to markers of phenomena like partial charges, or differentiators between distinct types of chemical bonds.
- Each link between two nodes (agents) connects exactly one site from one of the agents to one site of the other; reciprocally, a site from an agent can be involved with at most one link with another agent. In our chemical example, this means, for instance, that each of the two "open positions" from an oxygen atom can participate in only one covalent bond with another atom and thus this atom can be bound to at most two other atoms at once.

Reactions or **rules** can be also described via site graphs. A rule r is expressed via a site graph S_r and a set of transformations A_r, which corresponds to the addition or removal of edges between sites of S_r, changing their internal states, or adding or removing agents from S_r.

As a simple example, let us consider the reaction of electrolysis:

$$2H_2O \rightarrow 2H_2 + O_2$$

To describe this reaction, only two types of agents are needed: H and O, each one representing the type of atom. Agents of type H have one site, h_1, while agents of type O have two sites o_1 and o_2. For our current purpose, neither of the sites has a specific internal state. The reactants can be represented by a graph with six nodes (agents), two of type O and the rest of type H; each of the four sites $o_i, i = 1, 2$ is linked to a single h_1 site from one of the H agents. Furthermore, the set of transformations to be applied to this graph are as follows:

- Delete the four $o_i h_1$ links.
- Add a $h_1 h_1$ link to the two H agents corresponding to each water molecule.
- Add two links, $o_1 o_1$ and $o_2 o_2$, between the two O agents.

The effects of these operations on a set of agents can be observed graphically in Fig. 1.

Observe that by the specific combinations of two types of agents, we are able to describe three different species participating in this reaction (H_2O, H_2, O_2). With the same two agents, other chemical species can be easily described. For example, ozone can be described using three O agents and different internal states to represent the hybridization of the chemical bonds involved. Another example is hydrogen peroxide, which uses two H and two O agents, with a pattern of links mimicking the chemical structure of the molecule. By adding only a third type of agent with four sites $c_1, \ldots, c_4$, we can include carbon atoms in our model and thus represent the whole set of hydrocarbon species and other related types of compounds.

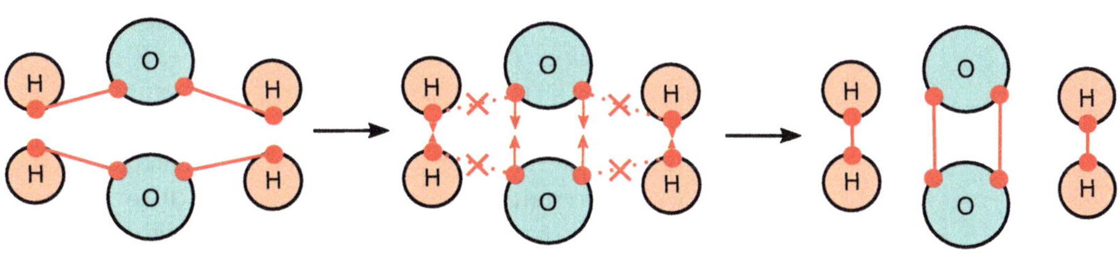

Fig. 1 Rearrangement of agents over the application of a rule corresponding to the reaction of hydrolysis, $2H_2O \rightarrow 2H_2 + O_2$

The κ framework allows declaring certain internal states or site links as "undefined," which allows applying the same rule to similar, but not identical, species. For example, the formation of alcohols from hydrocarbons corresponds to a set of very similar reactions, usually consisting of replacing a single H agent by a two-agent subgraph corresponding to the radical −OH. Therefore, specifying the whole structure of the hydrocarbon involved is usually superfluous.

3.2 Non-chemical Models in κ: A Predator–Prey Ecosystem

While the original motivation for the SSA comes from literal expression of chemical reactions, this framework can be used to model other types of systems where the agents involved do not represent chemical units but instead more complex entities. A simple example of this is the implementation of a predator–prey model, where agents represent a predator species that may consume other agents (prey). In this model, additional agents may also be used to indicate the availability of limited resources, such as plants or edible fruits for the sustenance of a herbivorous prey.

A simple system involving two species A, B (prey and predator, respectively) can be modeled via a set of **Lotka–Volterra equations** (as seen in [15, chapter 7, section B]), which correspond to the following system of differential equations:

$$\frac{dA}{dt} = \alpha A - \beta AB$$

$$\frac{dB}{dt} = \gamma AB - \delta B$$

Here, $\alpha, \beta, \gamma, \delta$ are nonnegative rate constants. The two terms of the first equation are interpreted as follows: αA means that the rate of growth of the prey species is proportional to the number of extant members of the species (i.e., exponential growth); $-\beta AB$ represents the predation rate of members of A by the B species, which assuming a homogeneous population is proportional to the product AB. With respect to the two terms of the second equation, these are interpreted as γAB corresponds to the growth rate of the predator species, which is proportional to the number of extant members of B and A as well as to the number of available resources or the amount of prey population; and the term $-\delta B$ is the rate of extinction of the predator, assumed to be proportional to the current population of B.

Note that the rate of natural death of A is neglected (technically, it can be represented by a diminished value of the constant α) as well as the dependence of A on other resources (for instance, available plants for a herbivorous animal). Moreover, the population density of both species is assumed to be constant. For instance, sexual reproduction is not considered, no age groups are taken into account (which makes this model inaccurate for predator species that target young members of the prey species), and no extinction of any of the involved species can be studied, since the population densities are assumed to be constant.

The simple Lotka–Volterra model can be implemented as a κ model, allowing inclusion of different parameters into the model such as natural dead, variable population densities, sexual reproduction, and age groups. In order to simplify notation, we expressed the model as chemical equations with internal states being represented via parentheses and linked sites via lines whenever necessary.

The rules of reproduction for A and extinction for B have a very simple format:

$$A \xrightarrow{r_1} 2A$$

$$B \xrightarrow{r_2} \varnothing$$

The reproduction rule for B has A as a catalyst, as the frequency at which reproduction of B occurs depends on the A population, as B will attempt to reproduce more often if there are more resources available, but this does not mean they should consume a member of A every time they attempt to reproduce. Thus, the rule is

$$A + B \xrightarrow{r_3} A + 2B$$

Finally, the predation rule has B as a catalyst; for it to occur, a member of A needs to encounter a predator B. In this model, "hunger" or similar states are not considered. Thus, the rule appears as follows:

$$A + B \xrightarrow{r_4} B$$

The rates of each of those rules depend on the values of $\alpha, \beta, \gamma, \delta$ and they can be determined in the same way as if they were chemical reactions. Note that this model does not take into account internal states (e.g., hunger) or links between agents (e.g., two B agents acting together to capture a prey). Next, we will discuss possible improvements of the model by using internal states or links to represent this type of situation.

One simple addition to this model would be the implementation of sexual reproduction. Of course, this will not apply to every type of species, and its effects might be negligible in simple ecological systems; however, for environments with large disparity in sex distribution or acute sexual dimorphism, this approach might provide an accurate model.

To implement sexual reproduction into the model, we can use sites as a property of the agents. Sites are variables that can be used to store a finite set of values or states in the form of qualitative or quantitative descriptors. Thus, we can use a site g in each agent to represent the sex (e.g., ♀, ♂ for male and female, respectively, or ☿ for species with hermaphroditic individuals). Thus, the rules for sexual reproduction are as follows:

$$A(\female) + A(\male) \rightarrow A(\female) + A(\male) + A(\female)$$

$$A(\female) + A(\male) \rightarrow A(\female) + A(\male) + A(\male)$$

$$A(?) + B(\female) + B(\male) \rightarrow A(?) + B(\female) + B(\male) + B(\female)$$

$$A(?) + B(\female) + B(\male) \rightarrow A(?) + B(\female) + B(\male) + B(\male)$$

Note the $A(?)$ term in the left side of the predator reproduction rules. As stated before, we allow for sites or links to be undefined so a single rule can be applied to every combination of internal states of A. In this case, what matters is that there are available resources (i.e., prey) and not the specific sex of the prey animals present. The $A(?)$ term on the right-hand side means that the corresponding term on the left-hand side remains untouched. These rules add new agents of a specific type $(A(\male), A(\female), B(\male), B(\female))$ without affecting the existing ones.

Age information or the stage of maturation of the agents can also be useful to improve the Lotka–Volterra basic model. For instance, we can suppose that predators more often capture young or elderly animals of the prey species due to inexperience, physical weakness, or illness. Similarly, only animals that have reached sexual maturity can reproduce, and in some species elderly animals present diminished fertility.

To incorporate this information into the model, we include an additional internal state d, whose values correspond to the different stages of development of each species, for instance $\{D_c, D_y, D_a, D_e\}$ (child, young or adolescent, adult or sexually mature, elderly or senescent, respectively). We need to define rules of growth that make every agent transit through those internal states sequentially:

$$A(?, D_c) \rightarrow A(?, D_y)$$

For examples of sex- and age-segregated ecological models that served as the inspiration for the set of rules shown here, see *Fundamentals of Mathematical Ecology*, by Mark Kot [15].

Every rule introduced above could be reproduced in a relatively simple way in the usual Gillespie's framework. However, the possibility to link agents through sites has not yet been covered in the κ language in this section. As an example, we will consider the formation of herds in both predator and prey species. A large group of prey animals can fend off a lone predator, while a prey animal can be more easily overwhelmed by a herd of social predators when alone or in a small group.

A potential way to implement this would be to add a few sites through which an agent can be linked to others of the same kind. Those links can represent social relations in the herd, and we can define rules to represent both herd protection and social hunting. For instance, we can add a few "relation sites," e.g., $p^{♀}$, $p^{♂}$, c, m (mother, father, child, mate) to represent a monogamous species with just one offspring per reproduction event and define the following state rules:

$$B(♀, D_a) \overset{m}{—} B(♂, D_a) \rightarrow \overbrace{B(♀, D_a) \underset{c,p^{♀}}{—} B(♀, D_c) \underset{p^{♂},c}{—} B(♂, D_a)}^{m}$$

Here, the notation $B \overset{s,t}{—} B$ means site s from agent B is linked to site t of another agent B. Thus, this rule reads as follows: two agents who are in the "adult" stage of development and are a mating couple (there is a link between the m sites of both agents) engender a third agent (in this case, female) and the c site of each parent agent gets linked to the respective p site of the new $B(♀, D_c)$ agent (the former two agents get marked as parents to the new child agent).

3.3 Spatial Information in κ

Usually, the setup for the simulation algorithm of the κ framework in the standard implementations of κ such as KaSim [5, 14] makes the same physical assumptions as the standard SSA, in particular that the medium is homogeneous and well-stirred, which means that the agents are uniformly distributed in the environment. This simplifies the model defining the probability of two agents interacting as proportional to their respective population. While this assumption is valid for certain systems, e.g., chemical reactions in gases, it might not be applicable to systems that are not homogeneous or have spatial dependences. For instance, the cell membrane provides different chemical and physical properties to the intracellular and extracellular medium. Moreover, the transfer of certain substrates from one medium to the other is in itself a phenomenon of interest, which is entirely ignored by the κ framework. Thus, incorporating spatial information to an implementation of the κ language allows for more realistic models. However, the assumption of homogeneity cannot be completely eliminated, as we usually care only about large-scale tendencies and not individual agent behavior.

In the cell membrane system, there are two different and clearly defined media. Interactions between them consist of transportation of certain agents from one environment to the other through the membrane. We assume that both environments or cellular compartments are homogeneous; hence, the probability of a certain agent approaching the membrane depends on the quantity of that agent, which can be represented using rules.

One way to implement such system of compartments is simply to add a new site w to every agent to represent each medium, which can have two states i and o corresponding to inside and outside the cell. Each rule becomes a set of rules, one per compartment, to ensure that agents only interact with other agents in the same compartment:

$$A + B \to C \text{ becomes } \begin{cases} A(i) + B(i) \to C(i) \\ A(o) + B(o) \to C(o) \end{cases}$$

The transport rules change the states of agents from one compartment to another. In this case, we can represent this via rules like $A(i) \to A(o)$ and $A(o) \to A(i)$, each with a certain rate. In this way, to simulate osmosis through equal volumes, each of those rules should have equal rate, such that the compartment with higher concentration has a higher rate of transportation.

For a more complete system with many more compartments, we can separate the cell space into a series of subspaces that we assume approximately homogeneous. This is analogous to the Riemann sums method to compute an integral:

$$\int_0^1 f(x)\,dx \approx \sum_{k=0}^{n-1} \frac{f(k/n)}{n}$$

To recall, this method allows to approximate the area under a curve given by the function f as a sum of the areas of small rectangles [2]. Here, we divide the interval $[0, 1]$ into n equal subintervals (subspaces) of length $1/n$, and we assume that the value of f in $[k/n, (k+1)/n]$ is approximately $f(k/n)$, i.e, this assumption does not significantly affect the value of f and it is similar to the "approximately homogeneous" assumption from above. For Riemann summations, a large value of n, and, thus, smaller subintervals, gives a better estimation of $\int_0^1 f(x)\,dx$, at least when f is a continuous function. In a similar way, we can suppose that with smaller subspaces we should reach a better approximation.

If the space of agents has some kind of geometrical properties, we can represent them via the transport rules: Compartments that are geometrically adjacent should have a higher rate of transfer reactions. For instance, we could represent a cell as two different compartments, nucleus and cytoplasm (Fig. 2), interconnected by several transport mechanisms, or build a more complex system where the nucleus is represented as a central compartment surrounded by several other compartments representing subspaces of the cytoplasm (Fig. 3).

To represent the geometry of the system, here we assume that there are only transport rules between adjacent compartments,

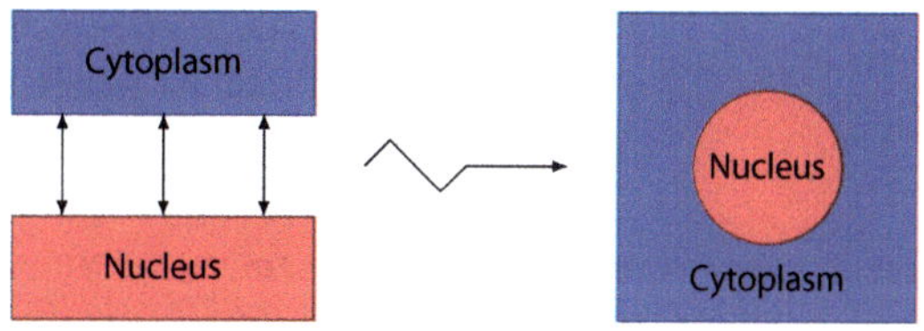

Fig. 2 Internal representation and interpretation of a two-compartment model for a cell. Note that the geometry of the cell is not taken into account and thus the cytoplasm and the nucleus are taken as entirely homogeneous

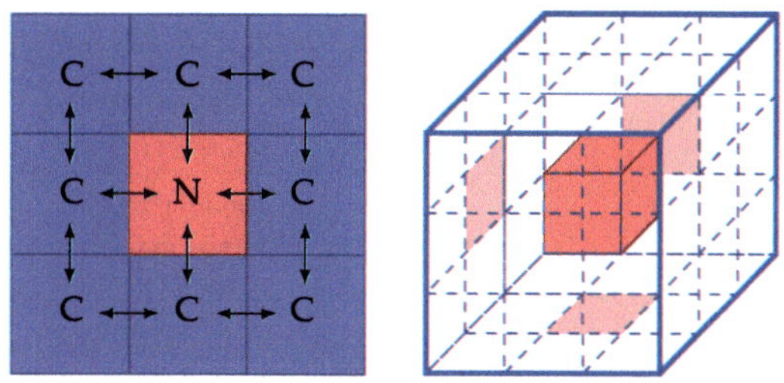

Fig. 3 Potential compartment arrangements to represent the geometry of a cell. (Left) A 2D arrangement of nine compartments with eight representing the cytoplasm and the central one representing the nucleus. The arrows represent possible transport rules. (Right) A potential 3D version of the former 2D arrangement. The marked (central) compartment corresponds to the nucleus, while the other 26 represent the cytoplasm; there are transport rules among compartments that share a face

automatically considering that the rate of any transport rule between nonadjacent compartments is zero.

As stated before, this does not require a special implementation of κ and can be done in the usual implementations such as KaSim by using a special site whose states correspond to the different compartments. However, since we need a copy of a rule for each compartment, this may result in several redundant rules that may severely impact execution time. Thus, an implementation allowing explicit declaration of compartments can reduce the total number of rules needed and improve the performance of the simulation. In what follows, we will discuss one such implementation, namely Parallel Implementation of a Spatial Kappa Simulator (PISKaS) [18], which is a forked branch of KaSim. PISKaS focuses on independent, parallel simulation of each compartment; however, most remarks below should apply to any spatial κ implementation.

3.4 Metapopulation Dynamics in a Predator–Prey Model with Explicit Spatial Information

In the following section, we will introduce a model of population dynamics that attempts to reproduce the experimental results obtained in the work of Holyoak and Lawler [11]. The study subject here was the evolution of the population of a set of predator and prey species in an environment that allows spatial migration. The goal is to verify whether a prey species who is prone to extinction by predation can survive in a medium that allows for migration. The end result of the simulation appears to conform to the experimental results; further details of an implementation of this simulation can be consulted in [8].

The experiment in [11] studied two species of microorganisms, *Didinium nasutum* (predator) and *Colpidium striatum* (prey), which inhabit an environment consisting of several bottles (compartments) linked by four-way connectors in a specific configuration. Given that the prey species shows logistic behavior (i.e., short-term exponential growth, which eventually is stunted due to lack of resources, resulting in a stable population) in the absence of predators, it becomes natural to suggest a Lotka–Volterra model to represent the interaction between those two species, with rules similar to the ones introduced in Sect. 3.2.

This model incorporates transport rules to represent microorganisms from both species moving through the different bottles. A coarse geometry has to be introduced, representing the spatial arrangement of the system of bottles and connections. Each bottle can be assumed to be a homogeneous medium and, thus, can be taken as a compartment. In the original experiment, the configurations consisted of square grids of bottles joined by four-way connectors that linked each bottle to the ones adjacent horizontally, vertically, and diagonally, as seen in Fig. 4.

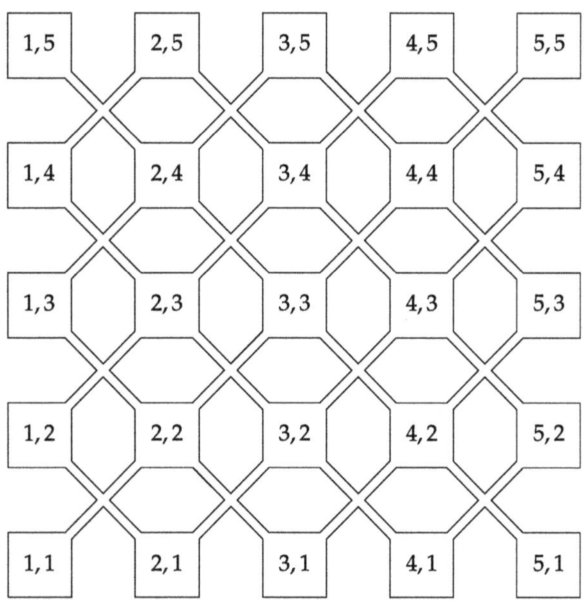

Fig. 4 Representation of the bottle arrangement and the four-way connections linking each bottle to its neighbors

Thus, this simulation is performed in an arrangement of 25 compartments, where each one represents a single bottle. The bottles have labels that represent their position in a square matrix of 5 rows and 5 columns; namely, a bottle labeled (i, j) is placed in the i-th column, j-th row. For example, the third bottle from the second row (from bottom to top) is labeled $(3, 2)$. Transport rules allow the microorganisms from a bottle to move to other bottles but only to adjacent ones; for instance, a microorganism in bottle $(3, 4)$ can move only to the bottles with labels $(2, 3), (2, 4), (2, 5), (3, 3), (3, 5), (5, 3), (5, 4)$, or $(5, 5)$.

The rate of movement of the microorganisms between bottles is determined by analysis of the physical characteristics of the system. If no further information regarding the physical properties of the system is provided and there are no external factors affecting inter-bottle transportation, it is reasonable to assume that the movement of the prey species through the bottles corresponds to simple diffusion. In this way, diffusion happens at the same rate through each connection the bottles have, which is linked to the physical capacity of those connections (assumed equal in all directions). The predator species follows similar rules. This is summarized by the following set of rules, with equal rate λ_T for every non-boundary compartment:

$$\text{transport } A : (i, j) \to (i - 1, j - 1)$$
$$\text{transport } A : (i, j) \to (i, j - 1)$$
$$\text{transport } A : (i, j) \to (i + 1, j - 1)$$
$$\vdots$$
$$\text{transport } A : (i, j) \to (i + 1, j + 1)$$

with eight rules in total, one for every adjacent bottle. For bottles located in the border, the same reasoning applies, but with less rules (e.g., the bottle with label $(1, 2)$ should have five neighbors instead of eight, so there are five rules in that case). Given that some bottles may have more than one connection with another bottle (e.g., to get from $(2, 2)$ to $(3, 2)$ one can go up right first and then down right or down right first and then up right), we do not really need to assume all transport rates are equal, and may adjust them accordingly if needed.

After stating the transport rules, we specify the behavior of both species. First of all, the agents will be of two types, prey (A) and predator (B); the prey species reproduces by simple mitosis, while the predator requires a certain minimal mass before it can undergo this process [11]. To implement this distinction, we need to introduce a way to count how many agents of the prey species have been consumed by a specific predator agent, and only allow "sated" predators to reproduce. We implement this by introducing agents with the following specifications:

- **Prey:** The associated agents have two sites, i and s. We use the second site to store the status of the prey agent (in this case, either alive, ℓ or dead, d), while the first site is used to link the prey agent to other agents to set up the aforementioned "counter."
- **Predators:** The corresponding agents have a single site a that is used to create links to a prey agent and store the state of the predator agent (h, f, d corresponding to hungry, sated or fed, and dead, respectively).

The rules are as follows:

- Prey species reproduction (mitosis):

$$A(?, \ell) \rightarrow A(?, \ell) + A(?, \ell)$$

- Predator feeding:

$$B(h) + A(?, \ell) \rightarrow B(h) \overset{a,i}{\text{———}} A(?, d)$$

$$? \overset{?,i}{\text{———}} A(?, d) + A(?, \ell) \rightarrow ? \overset{?,i}{\text{———}} A(?, d) \overset{s,i}{\text{———}} A(?, d)$$

The first rule is a rather straightforward statement of the situation that happens when a hungry predator meets an alive prey animal that then is consumed. The link formed between the B agent and the dead A agent indicates that the latter is incorporated into the mass of the predator.

The second rule, although less intuitive in formulation, corresponds to the exact same statement; if an alive prey agent meets a dead prey agent that has its i site linked to some other agent, then the dead agent must be part of the mass of a predator, and thus this means the alive prey has encountered a predator. Once again, when this rule is applied, the alive prey is killed and consumed. Since both rules represent the same phenomenon, they should be assigned equal rates.

- Predator satiation:

$$B(h) \overset{a,i}{\text{———}} A(?,d) \overset{s,i}{\text{———}} A(?,d) \overset{s,i}{\text{———}} \ldots \overset{s,i}{\text{———}} A(?,d) \rightarrow B(f)$$

This rule specifies that a $B(h)$ agent with a sufficiently long chain of $A(?,d)$ agents linked to it (with "sufficiently long" corresponding to a specific number that represents the average amount of prey eaten before a predator undergoes mitosis) becomes a $B(f)$ agent (i.e., not hungry but sated) and the $A(?,d)$ agents are discarded and eliminated from the simulation.

To ensure that the $A(?,d)$ agents are eliminated from the simulation immediately, so that a predator does not continue feeding after reaching the satiation level, we give this rule a rate of ∞.

- Predator mitosis:

$$B(f) \rightarrow B(h) + B(h)$$

This is similar to prey mitosis, with the only difference being that we only allow $B(f)$ agents (i.e., predators that have gathered enough mass) to undergo this process. Since the agents representing the progeny of the parent agents will only have a fraction of the mass of their parent, they need to gather additional mass themselves before they undergo reproduction, thus starting in the h (hungry) state.

- Predator unfed:

$$? \overset{?,i}{\text{———}} A(?,d) \overset{s,i}{\text{———}} A(?,d) \rightarrow ? \overset{?,i}{\text{———}} A(?,d)$$

This rule is the opposite of the second feeding rule. It states that a predator that has not been able to consume prey for long stretches of time has to utilize some of the mass it has already consumed and stockpiled for sustenance.

- Prey death:

$$A(?,\ell) \rightarrow \varnothing$$

- Predator death:

$$B(h) \rightarrow \varnothing$$

$$B(f) \rightarrow \varnothing$$

$$B(h) \overset{a,?}{\text{———}} ? \rightarrow \varnothing$$

The reason to have three distinct rules for predator death is to give different rates to each agent depending on how "hungry" they are. A $B(h)$ agent with no links does not have food reserves and is "starving" (the end result of a long period unfed), thus it has a higher rate of death than a $B(h)$ agent linked to others, i.e a predator that has been able to feed recently. Similarly, a $B(f)$ agent has a big reserve of mass and thus is not susceptible to starvation, hence having a lesser rate of death than other B agents.

- Prey cleanup:

$$A(?, d) \rightarrow \varnothing$$

This eliminates any $A(?, d)$ agents that do not have their i site linked to any other agent. Those $A(?, d)$ agents appear whenever a predator $B(h)$ linked to one or more $A(?, d)$ dies before reaching satiation, and since they serve no purpose anymore they are removed from the simulation. As before, to ensure those agents are removed immediately we assign a rate of ∞ to this rule.

With those rules, together with the transport scheme outlined above, we have set up a description for the predator–prey compartmentalized system as set up in the experiment by Holyoak and Lawler [11]. For a comparison between the results of the simulation and the experimental output, the reader may consult Fuenzalida et al. [8].

An equivalent ODE-based model applied in this specific example provides a fixed output: either both species always go extinct or both species manage to survive. This depends only on the parameters (rates) of the model, and thus it is impossible to see the effect that random fluctuations have on the output. The experiment by Holyoak and Lawler [11] shows that these fluctuations are actually of importance in the current situation, as some iterations of the experiment showed total extinction while others result in survival under the same initial conditions. This can be interpreted as a result of the way in which the population of prey species migrates and distributes along the bottles (which can be seen as random dispersal) and how the formation or dispersion of aggregates determines its survival, and by extension, the survival of the predator species as well. The usage of stochastic models in place of a deterministic ODE-based set of equations allows us to observe the effect that those random events have on the outcome of the simulations.

This model also allows us to observe the importance of spatial characterization in a simulation (*see* Fig. 5). In this framework, we can easily observe how an isolated bottle usually reaches extinction events very quickly, which is analogous to the situation of a homo-

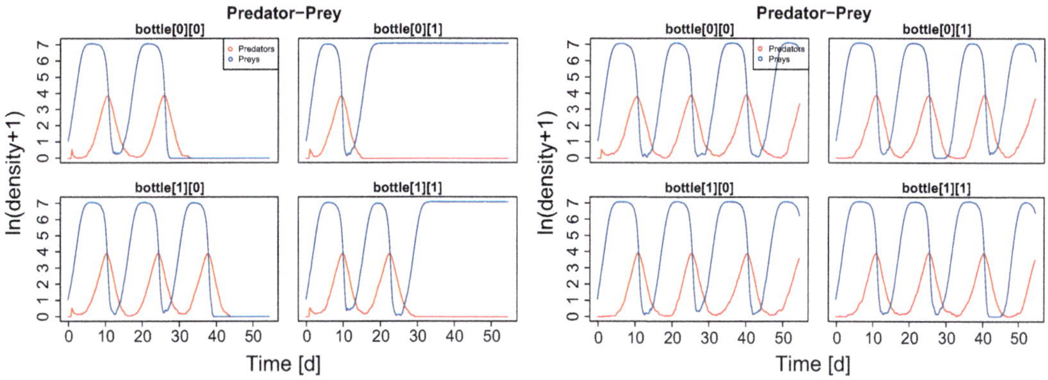

Fig. 5 Population behavior of the species in a bottle arrangement of four cells, showing the influence of spatial configuration, in particular, isolation on the left and migration on the right and extinction events

geneous medium with uniform population densities of the prey and predator species, while the complete system of linked bottles shows a much higher likelihood of survival of both species, also exhibiting the oscillatory behavior of both populations associated with the prey species favoring bottles with less population densities.

3.5 A Circadian Clock Model

In this section, we discuss a simplified model of the mammalian circadian clock. Our goal is to represent how the day–night cycle affects the transcription processes inside the cell, resulting in a 24-h periodic behavior regarding the concentration of proteins, transcription factors, mRNA, and others. This is the second sample model described in [8] as an example of the usage of PISKaS for simulations, and it is based on a certain system of transcription factors regulated by the presence of sunlight and the molecular interactions and feedback loops initiated by them, as described in [1].

The system modeled here consists of two compartments, corresponding to the nucleus and cytosol of the cell. Additional complexity can be added by dividing the cytosol into a set of compartments to represent the heterogeneity of the environment. For instance, the cell may be represented by a cube of $3 \times 3 \times 3 = 27$ compartments as shown in the right panel of Fig. 3, with the central compartment representing the nucleus and the remaining 26 the cytosol, reflecting its geometric structure. However, for the sake of simplicity, this example uses the least complex two-compartment model.

We modeled the periodic behavior of five different genes in this model, *PER1*, *PER2*, *CRY1*, *CRY2*, and *NR1D1*, each coding a protein. In addition, there are two transcription factors (CLOCK and BMAL1) and a phosphatase (CKI), all of which are assumed to be at constant concentrations within the cell. The model also considers an agent for each mRNA and their transportation to the

Table 1
Protein components of the circadian clock model and their respective UniProt IDs

Type	Name	UniProt Id
Genes	*PER1*	O15534
	PER2	O15055
	CRY1	Q16526
	CRY2	Q49AN0
	NR1D1	P20393
Transcription factors	CLOCK	O15516
	BMAL1	O00327
Phosphatase	CKI	Q06486 or P49674

cytoplasm. All protein components of this model are described in Table 1.

Genes and corresponding messenger RNA are represented by agents $G(i, s_1, \ldots, s_5)$ and $R(i)$, respectively, where the i site is used as identifier of the encoded protein and, in the case of the gene agents, the $s_1, \ldots, s_5$ sites allow linking to other agents which represent transcription factors. Those agents, T_e and T_r, will have two sites, one to allow binding to DNA, while the other allows binding to certain proteins.

The proteins will be actually represented by different types of agents instead of a single one with an "identifier" site. The reason for this is different proteins interact in specific ways with the transcription factors and with each other, thus resulting in different numbers of binding sites and reactions. In this case, our protein agents will be declared as follows:

- $PER(i, p_1, p_2, s_{cry}, s_{cki})$: here, i is an identifier taking the values 1 or 2 (as we include two types of proteins that form similar kinds of bonds), p_1 and p_2 are phosphorylation sites (with states p phosphorylated and d not), and the remaining sites allow linking to other proteins and agents.
- $CRY(i, s_{per}, s_{clk})$: same as before, the i site is an identifier with two possible values, and the s sites are for interaction with other agents.
- $NR1D1(r)$: for this agent, we only include one site, which allows linking to a transcription factor T_r.

The rules of the processes of phosphorylation and dephosphorylation that involve PER protein agents are handled by the phosphatase agents CKI.

Transport rules are to be limited to certain kinds of agents, since, for instance, we do not allow DNA to "leak" to the cytosol. Unlike the previous example, the linking between compartments is not symmetrical and has specified directions for each transport rule. For instance, we allow mRNA to move from the nucleus, where it is synthesized, to the cytosol where protein production occurs. However, there is no transport of mRNA from the cytosol back to the nucleus. In contrast, we allow certain PER and $NR1D1$ proteins to move in both directions depending of the status of certain binding sites. A sample of those rules is as follows:

$$\text{transport } R : \text{nucleus} \rightarrow \text{cytosol}$$

$$\text{transport } \quad PER \overset{s_{cry}, s_{per}}{\text{———}} CRY \quad : \text{nucleus} \rightarrow \text{cytosol}$$

$$\text{transport } PER(?, p, u) : \text{cytosol} \rightarrow \text{nucleus}$$

$$\text{transport } NR1D1 : \text{cytosol} \rightarrow \text{nucleus}$$

R agents are only allowed to go in one direction, while PER agents are allowed to move in both directions, but only in certain arrangements (e.g., intranuclear transport is only allowed during a specific state of phosphorylation, while movement toward the cytosol is affected by the interaction between PER and CRY proteins).

The remaining rules are also different in the two compartments. For example, since there is no DNA in the cytosol, there is no need to process rules pertaining to DNA transcription in the corresponding compartment, and similarly, there is no translation in the nucleous. PISKaS and similar rule-based compartmentalized software usually allow to declare some rules as exclusive for a subset of compartments. In this case, we have several rules to represent the phases of the encoding and expression process:

- **Translation rules:** Those rules govern the production of proteins with the information encoded in the corresponding messenger RNA, and thus are cytosol-exclusive. They take a very simple form, e.g.,

$$R(per_1) \rightarrow R(per_1) + PER(1, u, u, s_{cry}, s_{cki})$$

in which the identifier site of the mRNA agent R takes the value per_1 to indicate that it encodes a $PER(1, \ldots)$ protein.

- **Transcription rules:** This kind of rule is nucleus-exclusive and manages the production of mRNA ($R(i)$ agents) from the corresponding DNA in the nucleus ($G(i, \ldots)$ gene agents) in the presence of adequate transcription factors. Those rules are stated in ways similar to this example:

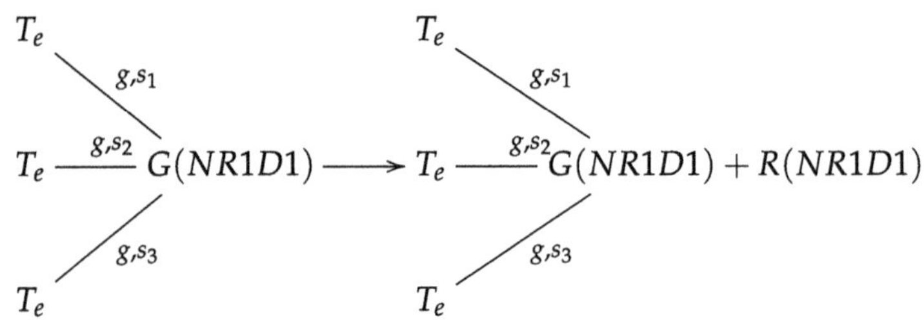

  in which a $NR1D1$-encoding gene bonded through its first three sites to adequate transcription factors produces a $R(NR1D1)$ agent, representing the corresponding messenger RNA. This agent is afterward moved to the cytosol compartment using the respective transport rules, where it induces the creation of a $NR1D1(r)$ agent, representing the phenomenon of protein translation. Note that, as it is to be expected, $G(NR1D1)$ agents act only as catalysts with no modifications either to themselves or to the associated transcription factors.

- **RNA and protein degradation:** Rules to represent degradation, usage, or elimination of mRNA and proteins are included, in a similar way as the death rules in the previous example:

$$R(?) \rightarrow \varnothing$$

  RNA degradation rules are deemed exclusive to the cytosol, while protein degradation rules are applied in both compartments.

- **Phosphorylation rules:** These are limited to PER agents and managed by CKI agents:

$$CKI \overset{\ell, s_{cki}}{\text{———}} PER(1, u, u) \rightarrow CKI \overset{\ell, s_{cki}}{\text{———}} PER(1, p, u)$$

  Different rates can be given to other configurations involving distinct kinds of proteins or with other states of phosphorylation.

- **Protein reactions:** A set of rules to determine the interaction between proteins and transcription factors both inside and outside the nucleus. These rules are either exclusive to the nucleus or to the cytosol. For example, the interaction of phosphatase agents with PER proteins and their further linking to CRY proteins only happens in the cytosol:

$$PER(?) + CKI \rightarrow \quad PER \overset{s_{cki}, \ell}{\text{———}} CKI$$

$$PER(?, p) + CRY \rightarrow \quad PER(?, p) \overset{s_{cry}, s_{per}}{\text{———}} CRY$$

  Note that the second reaction has to follow the first as phosphorylation of the PER agent has to be achieved.

- **Light-dependent transcription:** To express the dependence of this phenomena to the day–night cycle, we incorporate an additional set of transcription rules, which do not depend on T_e or T_r agents. Those rules have a variable rate, with a much higher activity during the "daytime"; otherwise, they are similar to previously shown transcription rules [1, 8]:

$$G(per_1) \rightarrow G(per_1) + R(per_1)$$

  Only $PER(1)$ and $PER(2)$ agents are generated by those rules, as these are the proteins intended to be dependent on the circadian clock.

3.6 Perturbations

In this section, we discuss a final feature of PISKaS and other rule-based simulation environments: the possibility to incorporate **perturbations**. Perturbations correspond to changes in the status of the system attributed to external factors; in κ, they can be manifested as variations on the rate of certain rules, for instance.

In the current example, perturbations are implemented as if-then-else statements, which verify the timer of the simulation and assign values to certain variables accordingly:

if $0 \leq T \bmod 24 < 12$ **then**
  set $\lambda_L \leftarrow 0.2$
else if $12 \leq T \bmod 24 < 24$ **then**
  set $\lambda_L \leftarrow 0.01$

Here, T is the timer of the simulation, thus "$T \mod 24$" represents the value shown in a 24-h clock. Daylight corresponds to the time when the clock is between 0 and 12, while nighttime corresponds to the remaining values in this time frame. The variable λ_L is, then, assigned as a rate to the rules from the Light-dependent transcription section above. Other kinds of perturbation rules exist and may be applied to different contexts.

This model is a very simple example of an application of rule-based simulation of a biological process without explicit reference to the underlying rules of chemistry. It also shows a simulation in which there is a natural requirement for two compartments (as the nucleus and cytosol show vastly different phenomena) and how having an explicit framework for those compartments separation benefits the description of the model, as it allows for simpler expressions of the rules and other factors involved. It also shows how other features of the simulation environment, i.e., how the possibility of incorporating perturbations, allow us to describe many phenomena of interest.

As stated above, it is also possible to generate more complex models of this phenomenon that incorporate, together with the mentioned two-environment situation, more fine-grained spatial information regarding the cytosol. This allows for a much richer simulation but does not notably affect the expression of the rules as we only need to distinguish nucleus and non-nucleus compartments. Thus, this gives us a model with good readability as the rules match closely the observed phenomena and we do not need to restate rules for each compartment. In addition, this model also allows faster execution in a practical situation as the number of rules impacts the execution time and, for standard κ environments, this number scales at least linearly with the quantity of compartments involved. Analysis performed and described in [8] shows that the results in an environment that allows for explicit compartment declaration such as PISKaS do not suffer from severe losses of accuracy when the synchronization time is small, and thus follow the behavior of the observed biological system.

4 Conclusions and Outlook

In this chapter, we have described and introduced rule-based stochastic simulation. We highlighted the characteristics of this type of modeling and compared it with deterministic approaches widely employed in the modeling literature. In doing so, we explained several example models which will help the reader to understand the strengths and limitations of this approach. Simulation engines such as KaSim and its fork PISKaS to allow explicit spatial declaration are freely available in public repositories (KaSim can be obtained at https://github.com/Kappa-Dev/KaSim and PISKaS at https://github.com/DLab/PISKaS). The models describing the predator–prey ecosystem with explicit spatial information and the circadian clock for both simulation engines can be obtained at https://github.com/DLab/models.

Nomenclature

Term	Definition
Agent	Abstract representations of entities on a system. An agent can bind each other's agents through its *sites*. Optionally, each *site* could harbor a *state*, a label that recapitulates a feature of the mentioned *site*, or a numeric property of the agent.
Bond	A representation of binding between two *sites* of two different agents.
Compartment	Declaration that represents a physical or logical space or volume which is part of a system.
Rule	Chemical equations that represent elemental reactions where reactants are *agents* with a set of features necessary and sufficient for a transformation to occur (*Left-Hand Side*) and the resulting pattern for each participating *agent* (*Right-Hand Side*). κ *rules* declare reactions that change the value of a *site*, create or destroy *bonds* between *agents*, and create or remove *agents* on the modeled system.
Site	Abstract representation of a physical or logical interface where an *agent* binds another *agent* or where different *states* are declared.
Specie	Each of the individual instances of an *agent*.
State	Abstract representation of a qualitative or quantitative characteristic that recapitulates a feature of the declared *site*.
Transport	Declaration that states the *link* that uses an *agent* to travel from one compartment to another. Additionally, it may declare the frequency and time employed to move the agent between compartments.

Acknowledgements

The authors would like to kindly acknowledge the financial support received from FONDECYT Inicio 11140342 and award numbers FA9550-16-1-0111 and FA9550-16-1-0384 of the USA Air Force Office of Scientific Research. This research was partially supported by the supercomputing infrastructure of the Chilean NLHPC [ECM-02]. Basal Funding Program from CONICYT PFB-16 to Fundacion Ciencia & Vida and Instituto Milenio Centro Interdisciplinario de Neurociencia de Valparaiso CINV ICM-Economia [P09-022-F].

References

1. Agostino PV, Golombek DA, Meck WH (2011) Unwinding the molecular basis of interval and circadian timing. Front Integr Neurosci 5:64
2. Apostol TM (1980) Calculus. Volumes 1 and 2, Wiley Eastern
3. Chylek LA, Harris LA, Faeder JR, Hlavacek WS (2015) Modeling for (physical) biologists: an introduction to the rule-based approach. Phys Biol 12(4):45007
4. Danos V, Feret J, Fontana W, Harmer R, Krivine J (2007) Rule-based modelling of cellular signalling, invited paper. In: CONCUR 2007 – concurrency theory. Lecture notes in computer science, vol 4703. Springer, Berlin, pp 17–41
5. Danos V, Fontana W, Krivine J (2007) Scalable simulation of cellular signaling networks. In Programming languages and systems. Lecture notes in computer science, vol 4807. Springer, Berlin, pp 139–157
6. Evans LC (2010) Partial differential equations. Graduate studies in mathematics. American Mathematical Society, Providence
7. Faeder JR, Blinov ML, Hlavacek WS (2009) Rule-based modeling of biochemical systems with BioNetGen. Methods Mol Biol (Clifton, NJ) 500:113–67
8. Fuenzalida I, Martin AJM, Perez-Acle T (2015) PISKa: a parallel implementation of spatial kappa. F1000Research
9. Gillespie DT (1992) A rigorous derivation of the chemical master equation. Phys A Stat Mech Appl 188(1–3):404–425
10. Gillespie DT (2007) Stochastic simulation of chemical kinetics. Annu Rev Phys Chem 58(1):35–55
11. Holyoak M, Lawler SP (1996) Persistence of an extinction-prone predator–prey interaction through metapopulation dynamics. Ecology 77(6):1867–1879
12. Jost J (2013) Partial differential equations. Graduate texts in mathematics, vol. 214. Springer, New York
13. Kappa Language (2017). http://www.kappalanguage.org
14. KaSim Reference Manual http://dev.executableknowledge.org/docs/KaSim-manual-master/KaSim_manual.htm
15. Kot M (2001) Elements of mathematical ecology. Cambridge University Press, Cambridge
16. Murphy E, Vincent D, Feret J, Krivine J, Harmer R (2010) Rule based modeling and model refinement. In: Elements of computational systems biology, chap. 4. Wiley, Hoboken, pp 83–114
17. Pardoux E (2008) Markov processes and applications: algorithms, networks, genome and finance. Wiley series in probability and statistics. Wiley, New York
18. Perez-Acle T, Fuenzalida I, Martin AJM, Santibañez R, Avaria R, Bernardin A, Bustos AM, Garrido D, Dushoff J, Liu JH (2017) Stochastic simulation of multiscale complex systems with PISKaS: a rule-based approach. Biochem Biophys Res Commun 498:342–351
19. Renner T (2007) Quantities, units and symbols in physical chemistry. The Royal Society of Chemistry, Cambridge
20. Silberberg MS (2006) Chemistry. Molecular nature of matter and change, 4th international edn. McGraw Hill, Boston
21. Tu PNV (1994) Review of ordinary differential equations. Springer, Berlin, pp 5–38

Chapter 2

Optimized Protein–Protein Interaction Network Usage with Context Filtering

Natalia Pietrosemoli and Maria Pamela Dobay

Abstract

Protein–protein interaction networks (PPIs) collect information on physical—and in some cases–functional interactions between proteins. Most PPIs are annotated with *confidence scores*, which reflect the probability that a reported interaction is a true interaction. These scores, however, do not allow users to isolate interactions relevant in a particular biological context. Here, we describe solutions for performing context filtering on PPIs to allow biological data interpretation and functional inference in two publicly available PPIs resources (HIPPIE and STRING) and in the proprietary pathway analysis tool and knowledge base Ingenuity Pathway Analysis.

Key words Protein–protein interaction networks, Context filtering, Orthogonal text mining resources

1 Introduction

Protein–protein interaction networks (PPIs) are typically used in biological interpretation to explain differences among the observed phenotypes, particularly in enrichment analyses and in functional inference via guilt-by-association methods [1]. PPIs can be grouped into *physical* interaction networks and *functional* PPIs. Tools which provide physical interaction networks include HIPPIE, which integrates various direct interactions from public and curated databases [2] and STRING, which provides both physical interactions between proteins and functional interactions, such as in signaling cascades [3, 4]. While PPIs built using HIPPIE are derived from experimental results, STRING also incorporates other interactions sources such as text mining, coexpression evidence, co-occurrence evidence, genomic neighborhood (i.e., genes that colocalize in the genome), gene fusion, as well as information from other PPIs databases to infer interactions (Table 1) [3]. Other PPIs that focus on experimentally validated physical interaction information include the Human Protein Reference

Louise von Stechow and Alberto Santos Delgado (eds.), *Computational Cell Biology: Methods and Protocols*, Methods in Molecular Biology, vol. 1819, https://doi.org/10.1007/978-1-4939-8618-7_2,

Database (HPRD, [5]), Molecular INTeraction database (MINT, [6]), IntAct [7], and the Biological General Repository for Interaction Datasets (BioGRID, [8]). Both STRING and HIPPIE consolidate information from all these other PPIs and can thus be considered "meta"-PPI resources [9].

With the exception of HIPPIE, which provides options for filtering PPIs by biological context [2, 10], most interaction network tools filter PPIs using *confidence scores*, which measure the reliability of an interaction as a function of the amount and kind of experimental evidence that supports it.

Other proprietary resources such as Ingenuity Pathway Analysis (IPA, http://www.ingenuity.com) permit creating networks linking query genes. IPA does not present confidence scores, but instead allows users to choose among three different confidence levels: experimentally observed interactions, predicted high and predicted moderate interactions. To start building networks, IPA queries the Ingenuity Knowledge Base (KB) for both interquery interactions and then for interactions with all other objects stored in the KB. Similar to STRING, the KB of IPA is a repository of curated biological interactions and functional annotations created from individually modeled relationships between proteins, genes, complexes, cells, tissues, drugs, and diseases gathered from both public and private biomedical databases. A main extension provided in IPA is the integration of small molecule data and links with databases on diseases and disease biomarkers. A main issue regarding PPI or KB use in biological interpretation is *degeneracy* or the lack of concordance among the first-degree neighboring protein *vertices* (i.e., proteins with a direct connection to the queries used to build the network) retrieved from using different PPI tools, and even different versions of the same PPI tool [9] (*see* **Note 1**).

As an example, we show the differences in the network neighborhoods of 22 influenza A virus (IAV) entry factors (Table 1) in different PPI tools (Fig. 1a). We see that not only are the derived network topologies obtained from using the default settings quite diverse among the different tools, but the sizes of their network neighborhoods and, thus, their extent of overlap, calculated by dividing the total number of overlapping components (edges of vertices) by the number of components of the smaller of two graphs G_1 and G_2 being compared, are highly different too (Fig. 1b, c). Finally, if we extract the abstracts that are used as evidence for an interaction from PubMed and check the frequency of biological keywords associated with these (*see* **Notes 2** and **3**), we see that majority of the interactions are linked to cancer (STRING, Fig. 2a) or signaling in general (STRING and IPA, Fig. 2a, b), and only a small proportion of supporting evidence from literature have been specifically reported in virus infection-related processes.

Table 1
Comparison of content, filtering options and evidence sources of PPIs

PPI	Content	Available filters[a]	Primary evidence sources	Other PPIs from which PPI draws information	URL
STRING	Functional and physical protein-protein interactions	Organism; confidence score	Other PPIs, KEGG pathways, textmining, coexpression experiments, curated experimental data, genomic co-occurrence, genomic neighborhood, gene fusions	BIND, HPRD, MINT, DIP, IntAct, BioGRID	http://string-db.org
HIPPIE	Physical protein–protein interactions	Confidence score; interaction type; tissue expression; functional annotation (GO/MeSH)	Other PPIs	BIND, MINT, DIP, IntAct	http://cbdm-01.zdv.uni-mainz.de/~mschaefer/hippie/index.php
MINT	Functional and physical protein–protein interactions	Organism; interaction type; detection methods; publication	Experimental data, obtained through textmining, then manually curated	Merged with IntAct database	http://mint.bio.uniroma2.it
IntAct	Functional and physical protein–protein interactions; compound–protein relationships	Interactor types: Proteins, complexes, compounds, nucleic acids and genes	Experimental data, obtained through textmining, then manually curated	Merged with MINT database	http://www.ebi.ac.uk/intact/
BioGRID	Functional and physical protein–protein interactions, including genetic interactions, compound–protein interactions, and posttranslational modifications	Organism; publication; interaction type (e.g., raw interactions, posttranslational modifications)	Experimental data, obtained through textmining, then manually curated	–	https://thebiogrid.org

[a]Online version

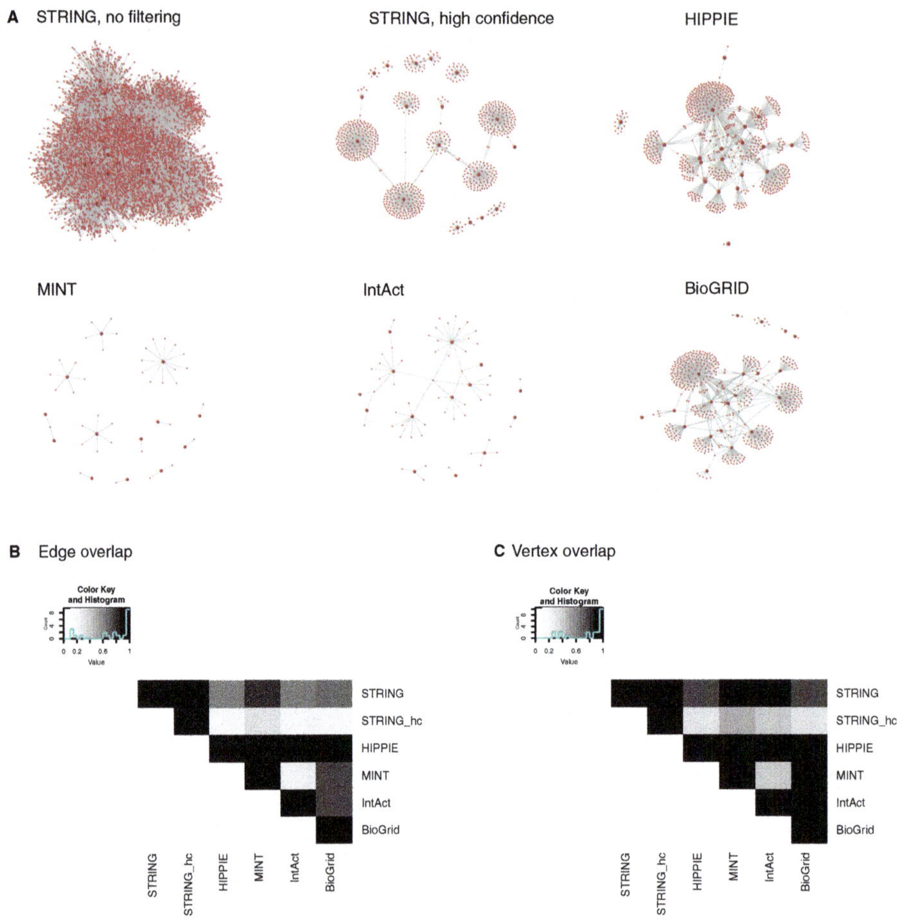

Fig. 1 Network neighborhoods of influenza A virus (IAV) entry factors in different protein–protein interaction networks and IPA (**a**). Query vertices from the entry screen are shown in dark red. Pairwise network neighborhood overlaps for edges (**b**) and vertices (**c**) of all tools except IPA. Overlaps are calculated as the number of vertices or edges in common between the graphs divided by the number of vertices or edges of the smaller graph. A filtered version of STRING restricted to high-confidence edges (STRING_hc, confidence score > 800, A)—calculated as the fraction of overlapping edges or vertices with respect to the smaller graph—indicates that smaller network neighborhoods are not always completely included in the larger ones

This observation evidences an important bias in the supporting literature [9].

Thus, to maximize the utility of PPI tools, it is necessary to isolate interactions, which are relevant in specific biological contexts. Here we describe how to use context filters in two meta-PPI resources, STRING and HIPPIE, and the proprietary IPA tool. Additionally, we show how to use the R package **rentrez** (https://cran.r-project.org/web/packages/rentrez/index.html) to check if the retrieved edges are indeed implicated in the process of interest (i.e., molecules involved in the interactions—*incident vertices*—

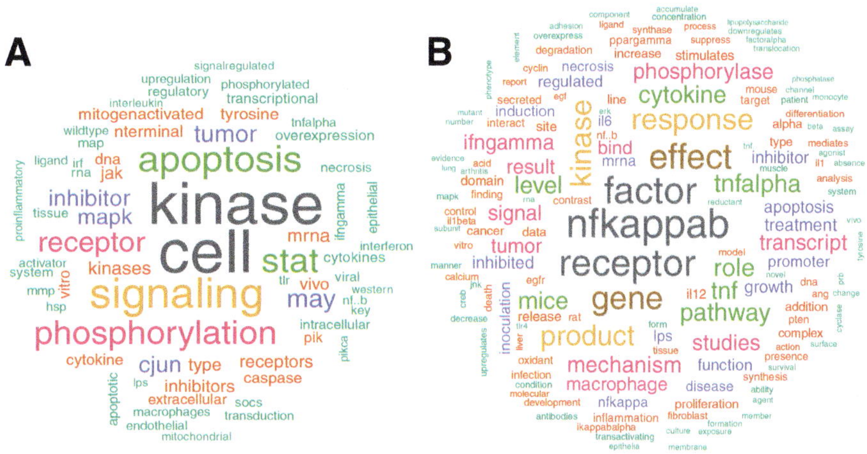

Fig. 2 Most frequently used words from textmining evidence of the entry network neighborhood of the unfiltered STRING network (**a**) and IPA network (**b**). Evidence for the STRING network neighborhood is characterized by the predominance of signaling-related terms, and mainly implicates STAT, MAPK, and caspase signaling. For IPA, the implicated signaling pathways terms include the NF-kappa B, TNF, and interferon signaling pathways. Note that influenza A hijacks various pathways, including NF-kappa B, PI3K/Akt, MAPK, PKC/PKR, TLR/RIG-I, mTOR, EGFR, and ERK signaling [20], with early signaling being predominantly linked to PKC/PKR [21] and EGFR [22]; nonetheless, information directly linking interactants to mechanical effectors of entry, such as clathrin or caveolin, are not immediately evident from the network neighborhood. Details on how the wordclouds are generated are described in [9]

are comentioned together with the process of interest in a record abstract, *see* also Subheading 5, **Note 2**).

2 Materials

2.1 Data

The following versions of PPIs were used in this analysis:

STRING: v.10, online version and STRINGdb R package (v. 1.14.0).

HIPPIE: online version; for R-based manipulations, we used HIPPIE v.1.8.

IntAct: (ftp://ftp.ebi.ac.uk/pub/databases/intact/current/all.zip, downloaded in December 2015).

HPRD: Release9_062910.

MINT: 2012-10-29 (last release).

IAV entry factors can be found under the following link: https://github.com/pampernickel/flu_ppi/blob/master/data/annotations/hgnc.csv

Mappings between HUGO gene nomenclature committe (HGNC) symbols [11] and entrez IDs can be found in this file:

https://raw.githubusercontent.com/pampernickel/flu_ppi/master/data/annotations/hgnc.csv

2.2 Software Packages

All analyses were performed using the R language, version 3.2.3. Analyses were also tested and confirmed to run on R 3.2.2 and 3.3.3.

2.3 Code Repositories

Critical code and functions required to reproduce results shown in this book chapter are deposited in: https://github.com/pampernickel/flu_ppi/blob/master/sample_codes/sampleScripts.r. sampleScripts.r is provided to illustrate vertex- and edge-based filtering. All custom function dependencies are likewise provided in the same repository: https://github.com/pampernickel/flu_ppi/tree/master/sample_codes/functions

3 Methods

3.1 First-Degree Network Neighborhood Construction and Context Filtering in HIPPIE

Context filtering has been introduced in HIPPIE in the form of tissue expression, cellular compartment (cc) and biological process (bp) annotations. The following steps describe how a context-filtered first-degree network neighborhood for your query protein(s) in HIPPIE is created:

1. For a single query, input the query protein name on the default HIPPIE tab ("protein query"); for multiple queries, select the network tab and input a list of query proteins (Fig. 3a). Note that in HIPPIE, both the single query and multiple query modes yield the first-degree neighborhoods of the query proteins.
2. Select the output type; the default output is a visualization of the network on the web browser ("show in browser—visualization"). If you wish to import the results into an analysis platform (e.g., R software environment), a tab-delimited format (HIPPIE TAB or PSI-MI TAB formats) is recommended.
3. Check all relevant filters (Fig. 3b), which include the interaction type restrictions (interaction type filter), tissue expression localization (tissue filter); and Gene ontology (GO) or Medical Subject Heading (MeSH) annotations (functional filter). GO terms, which are standardized terms ("ontologies") that describe gene function and their relationships [12], are further categorized into biological processes or cellular compartments in HIPPIE. MeSH headings [13] are likewise standardized terms that facilitate indexing of journal articles in the life sciences—and by extension, their content—according to its subject. HIPPIE MeSH terms are restricted to disease-associated terms. Note that all functional filters are menus (delineated by a "+" symbol), and clicking on the "+" sign expands the choices to more specific GO or MeSH terms.
4. After selecting filters and output options, click on the "SEARCH" button, located just below the query box.

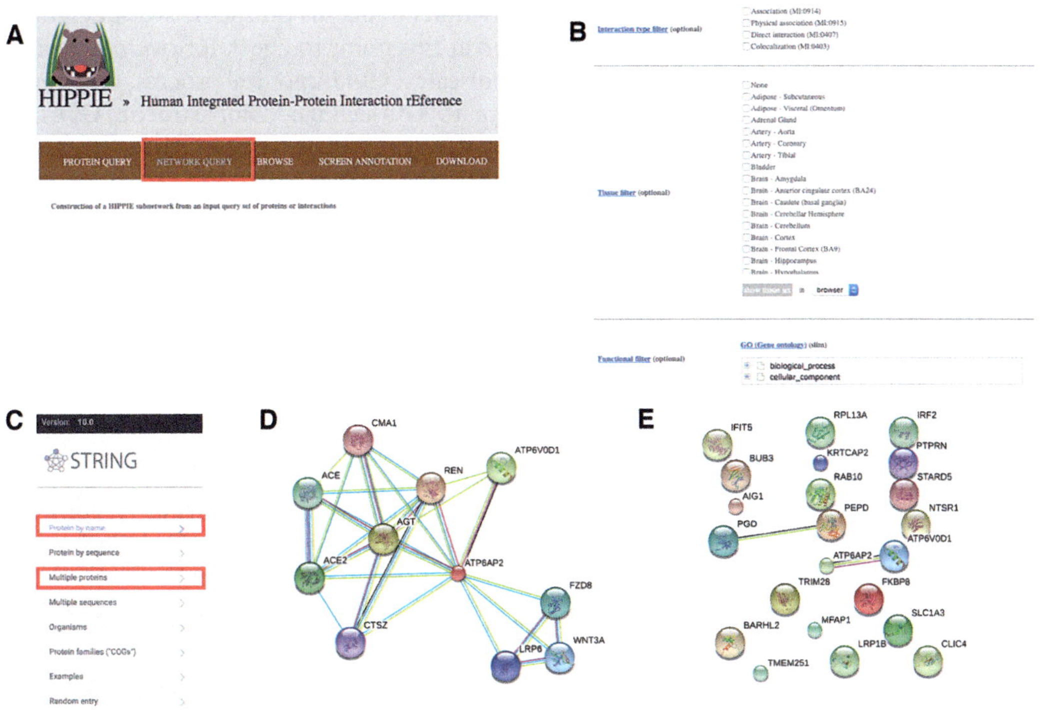

Fig. 3 Context filters in HIPPIE and STRING. Context filtering options in HIPPIE can be accessed in the network query mode (red box, **a**). The menu of filtering options (**b**) allows the user to impose various restrictions on the retrieved network neighborhood based on the interaction type, the tissue(s), cellular context ("biological_process") or location ("cellular_component") in which the interaction has been reported, or the diseases in which the interaction has been implicated ("MeSH," not shown). STRING allows searching with default parameters using a single protein name or multiple protein names as input parameters (red boxes, **c**). Note that querying with a single name retrieves the first-level interactors of the query protein (**d**), while querying with multiple names retrieves the interaction network *between* the query proteins, without the inclusion of first-level interactors (**e**)

5. If you selected a tab-delimited format as an output, results will be automatically downloaded; if you selected the default option, the interaction network is shown on the same window.
   (a) In the case of more than ten queries, or queries involving signaling pathway molecules (e.g., MAPK or EGFR), where the number of neighboring vertices is expected to exceed more than 40, the HIPPIE visualization result is not informative. An option to improve the visualization is to download the network neighborhood as a JSON object or tab-delimited text, which can be imported into Cytoscape (*see* also **Note 4**), a platform for network visualization and analysis [14].

(b) Options for enrichment analysis on the network neighborhood are offered at this point to check for overrepresented genes linked to diseases, GO biological process, GO molecular functions or to GO cellular compartment. GO term enrichment calculations are done with the PANTHER overrepresentation test, which is based on the binomial distribution, rather than the hypergeometric distribution [15].

3.2 First-Degree Network Neighborhood Construction and Context Checking in STRING (Web Interface)

STRING allows the submission of single or multiple protein names as queries, and with a medium confidence filter (Fig. 3c). When the query is comprised of a single protein, it returns the first-degree network (Fig. 3d), whereas a query with multiple proteins yields the interactions between these proteins (Fig. 3e). This is different from how single and multiple queries are processed in HIPPIE.

The following steps describe how to check the interactions for a single protein in STRING (for users of the web interface, the procedure is very limited in terms of performance)

1. In the main STRING page, select "Proteins by name" and enter the name of your query. Indicate the species for which you want to retrieve the interactions; alternately, leave the auto-detect option on. Click "search."
2. If the species was specified, the network neighborhood for the query protein is displayed as an image, together with options for data exploration and download; if the species was not specified, you will be first redirected to a page where you can select the [auto-detected] species.
3. To access the different types of evidence linked to the network neighborhood, click on the "evidence" button below the image. This reveals a list of evidence sources used to build the neighborhood. To specifically check the context of the text mining evidence associated with the network, click on the "textmining" button, which shows excerpts from the abstract or full text where the interacting proteins are comentioned.
4. Alternately, if you want to check the text mining evidence specifically linked to an edge, click on the green lines linking two protein vertices. This opens a popup window that shows the evidence types available for that specific interaction. Click "show" on the text mining evidence box. This redirects you to a page listing the text mining reference(s) specifically associated with that edge.
5. As in HIPPIE, it is possible to run enrichment analyses on the network neighborhood of your query. Click on the "analysis" button below the graphical network to reveal enriched GO terms (biological processes, molecular functions and cellular components), functional pathways (KEGG) protein domains (PFAM and INTERPRO) and enriched interactions. By click-

ing on a selected enriched term, it is possible to visualize all proteins in the network annotated with this term. You may change the background of the enrichment analysis from the whole genome to just the druggable genome or the kinome.

3.3 First-Degree Network Neighborhood Construction and Context Filtering in STRING (STRINGdb R Package)

Context filtering is currently not available in the web versions of STRING, but with the release of STRING for R (STRINGdb, [4]), it is possible to implement script-based workarounds, which are described in detail below. While the methods are recommended for users with basic knowledge of the R language [16], we provide several running examples that walk the user through the minimum steps in R (and the **STRINGdb** and **igraph** R packages) required to extract a context-filtered network (*see* **Note 3**, https://github.com/pampernickel/flu_ppi/blob/master/sample_codes/sampleScripts.r). Note that all the STRING tables and STRING output in R are converted, where possible, to igraph objects, which allow the direct application of graph algorithms on STRING content.

3.3.1 Network Neighborhood Extraction in STRINGdb for R

1. Create a vector of gene names for which you want to get the network neighborhood; to ensure higher chances of identification, use HGNC symbols.
2. Load STRINGdb to a variable (**string.db**) using the **STRINGdb$new** command. Specify the version, organism, and confidence score threshold that you wish to work with. In our example, we use version 10 for *Homo sapiens* (Taxonomy identifier: 9606), with a score threshold of 0 to include the full network.
3. Load the full STRING graph to your workspace as an igraph object with the **getGraph** command (**string.db$getGraph**).
4. Map your vector of gene names to STRING identifiers using the map command (**string.db$map**). Note that the vector of gene names is passed to a custom function, **prepareMap** to process it in the format that STRING requires. This step returns a data frame (**vertexMap** in our example) consisting of your original vector of gene names and the corresponding STRING ids.
5. Extract the first-degree neighborhood using the **getNeighbors** function applied to the STRING identifiers in **vertexMap** (from Subheading 3.3.1, **step 4**). Use the STRING graph from Subheading 3.3.1, **step 3** as the query graph. Convert the final object (a list of all neighbors per query), using the **constructGraph** function to an **igraph** object.
6. Get a reverse mapping of the vertex names (currently in the form of STRING identifiers) to HGNC symbols using the **string.db$get_aliases** function; select the subset of aliases that are HGNC symbols.

7. The resulting form of the graph is now ready for GO biological process annotation.

3.3.2 Vertex-Based Filtering (GO Biological Process)

A first step to use a similar vertex filter as in HIPPIE entails the annotation of network components with GO terms in R. In our example, we use the **getgos** and the **nodeToGO** functions, which are dependent on the org.Hs.eg.**db** package (**v.3.1.2**). For usage, see **sampleScripts.r**. Note that this method is applicable to any network without GO annotations.

1. Create a vector of GO terms of interest. Given that GO terms can vary widely in specificity, choose a GO term at a level that is specific enough to be informative. In **sampleScripts.r,** we use the GO bp terms (vesicle-mediated transport) and GO:0060627 (regulation of vesicle-mediated transport) and selected descendants of this GO term (i.e., excluding terms linked to synaptic vesicle transport) for filtering (Fig. 4a).
2. Use the **GOBPCHILDREN** mapping from the **GO.db** package to include the descendants of the GO terms of interest; check the names of the descendant terms using the **getGOnames** function to remove GO daughter terms that fall under the main GO term query, but are not of interest (e.g., GO:1903421, regulation of synaptic vesicle recycling).
3. Using the **getNodeAttribute** function and a combination of R base functions, you can check which vertices linked to a query vertex are annotated with a GO term of interest. Only edges connecting query vertices and vertices annotated with a GO term of interest are retained (Fig. 4b).

3.3.3 Edge-Based Filtering (Keyword Filtering of Textmining Evidence)

Most edges in STRING are supported by textmining evidence, which give some contextual information regarding the interactions. The textmining evidence was obtained from the application of a text retrieval and analysis algorithm of STRING run on PubMed abstracts and when available, on full article texts.

1. Create the entry network neighborhood following **steps 1–5** of Subheading 3.3.1 (i.e., up to step where an igraph object is created with STRING identifiers, and not HGNC identifiers).
2. Convert the igraph object in Subheading 3.3.3, **step 1**. to a data frame, then extract all references linked to each edge using the **string.db$get_pubmed_interaction** function. This returns a vector of PubMed and Online Mendelian Inheritance in Man (OMIM) identifiers, when applicable; OMIM identifiers indicate if a protein has been implicated in a genetic phenotype [17].
3. Extract the abstracts linked to these PubMed IDs using **entrez_fetch**; use XML as the return type; use functions

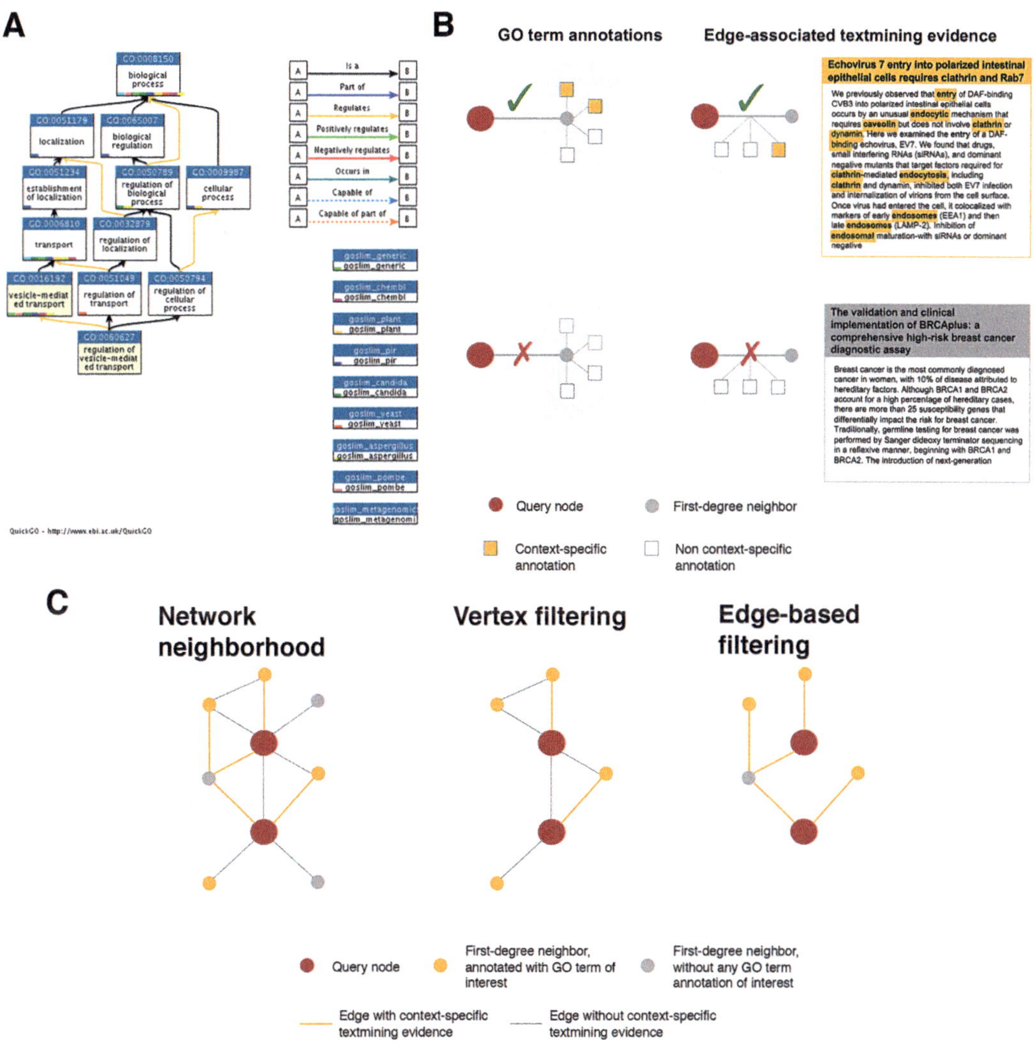

Fig. 4 Vertex- and edge-based context filtering. Examples of entry-related terms (in yellow) in the GO biological process (bp) hierarchy for the GO term, regulation of vesicle-mediated transport (GO:0060627, http://www.ebi.ac.uk/QuickGO) (**a**). Note that term associations are nested, with some terms being less specific than others (e.g., "localization" encompasses more genes than "vesicle-mediated transport"). Context filtering using GO annotations (**b**) results in the retention of edges with relevant GO term annotations (**b**, top left) connected to a query vertex; analogously, edges supported by textmining evidence mentioning keywords of interest are retained (**b**, top right). Note that context filtering can still result in degenerate solutions (**c**), but all retained vertices and edges are at least implicated in the process of interest

xmlTreeParse and the **getAbstract** to retrieve the abstracts as a character vector.

4. Create a vector of keywords to search for in the retrieved abstracts, then check the frequency of occurrence of each of the keywords in the abstracts. Note that the keywords can be patterns rather than full keywords (e.g., "endosom"

instead of "endosome," which matches both "endosome" and "endosomal"). Retain edges that are supported by abstract(s) with a minimum number of keywords (Fig. 4b).

Note that vertex- and edge-based filtering may still result in degenerate solutions (Fig. 4c). Nonetheless, unlike in the case of confidence score filtering, it is at least certain that retained elements would be restricted to those linked to the process of interest.

Alternatively, a recently released Cytoscape application, stringApp (http://apps.cytoscape.org/apps/stringapp, compatible with Cytoscape 3.3) also allows the direct import of STRING data into Cytoscape [18]. The application allows three main query modes, namely protein names (as described in Subheading 3.2), disease, and PubMed query. The PubMed query option mirrors results that can be obtained in Subheading 3.3.3. The disease mode yields the top proteins linked to a disease based on information from the DISEASE database [19]. Finally, if the information is available for the organism from which a query protein is from, information regarding the subcellular localization and tissue expression are also indicated and can be used to contextualize the network.

3.4 First-Degree Network Neighborhood Construction and Context Filtering in IPA

Ingenuity Pathway Analysis is a proprietary software that allows modeling and analysis of biological systems. There are several ways of building a network in IPA (Stand alone version: Version 28,820,210; Building krikkit Date 2016-09-24), but the general workflow consists of the following steps:

1. Format your input list of molecules according to the IPA standards. All formatting is done outside IPA in a spreadsheet program (e.g., Excel) to produce the input file.
2. The header must consist of only one row, and there should be no empty cells. Unlike in HIPPIE or STRING where the only data required are protein names, IPA allows also associating measurement values (up to three) for each molecule (fold changes and p-values for expression data, variant loss/gain values, phosphorylation ratio, differential metabolomics data, etc.).
3. Mixed identifier types (Gene Symbol HGNC, Ensembl, Entrez, Uniprot, Unigene, etc.) are allowed and mapped to a common identifier
4. Upload your molecule list file as a flat file (.cvs, .txt, .tsv) or an excel file (.xls) using the file menu: "File"; "Upload dataset"; "Formatted data file". Define the upload settings, including the "file format" (Flexible Format), the "column header" (Yes) and the "identifier type," which is usually automatically detected (e.g., HGNC gene symbol). If using data from microarrays, the platform details need to be specified

for identifier mapping, otherwise the platform is set to "Not specified/applicable."

5. Select columns of the input file to use in the analysis: "ID column," and optionally, the observation column(s) (including potential threshold columns such as *p*-value and expression value) and their corresponding measurement type.
6. Choose "Save the dataset," as all datasets are automatically annotated when they are uploaded into IPA. It is important to verify the mapping of the dataset in order to identify mapped and unmapped identifiers (IDs). The annotated dataset comprises several columns describing the uploaded molecules that provide specific details such as gene ID, *Symbol*, *Entrez Gene Name*, a unique subcellular location (column *Location*; which can be: cytoplasm, nucleus, plasma membrane or other), one functional gene family (column *Type(s)*; which can be: enzyme, G-protein coupled receptor, ion channel, kinase, ligand-dependent nuclear receptor, peptidase, transcription regulator, transporter or other) and association with drugs (column *Drug(s)*).
7. Run the core analysis on the annotated dataset by choosing "Analyze/Filter Dataset"; "Core Analysis."
8. Define all the filter settings (most of them have default values which may be kept):
9. "General settings Reference set": Population of genes to consider as a reference or background set to be used for ranking the statistical significance (i.e., *p*-value calculation) in the enrichment analyses. Options from the IPA KB include the genes or proteins only (default), metabolites only, or a combination of genes and metabolites. The user can also specify the background based on the microarray platforms or a user-uploaded dataset.
10. Type of relationships to consider (that affect networks and upstream regulator analysis), which could be direct (default) and indirect. Direct relationships imply a physical interaction between two molecules (e.g., a kinase and its known substrate, or a drug and its protein target). Indirect relationships may occur through intermediates (e.g., the relationship between a chemokine receptor and the gene whose expression it induces via downstream signaling).
11. Parameters for network construction: option to include (default) or exclude endogenous chemicals, number of molecules per network (35 by default), and number of networks per analysis (25 by default).
12. Specify the vertex types to be included: this parameter allows to build or filter the network based on the specific types of molecules selected (e.g., biological drug, chemical

(kinase inhibitor, chemical drug, protease inhibitor), cytokines, enzymes, growth factors, microRNAs, transcription regulators, transmemebrane receptors).

13. Specify the data sources to be used: allows contextual data analysis filtering. The list includes primary PPI databases (e.g., BioGRID, IntAct, *see* Table 1), ontological annotations (e.g., GO, OMIM), and other sources (COSMIC database for mutations, DrugBank) from which data in network reconstruction should be derived.

    Other contextual filtering parameters include:

    (a) "Confidence": this parameter determines the confidence level on the reported interactions. It allows three different types: experimentally observed (default); high or moderate confidence.

    (b) "Species": by selecting items in this filter, you are specifying to consider only ortholog genes of the selected species. Broad categories are mammal (human, mouse, rat) and uncategorized. This parameter includes an additional filter: "stringent" or "relaxed." The first option will return only those molecules and relationships relevant to the selected species. The relaxed filter will match molecules that contain an ortholog that includes the selected species. This option filters on the orthologs, but not the relationships between the orthologs.

    (c) "Tissues and Cell Lines": filter for genes expressed in a particular tissue or cell line. Currently, IPA includes data from 21 different cell lines, mainly cancer cell lines. Again, there are two filtering choices: "Stringent" and "Relaxed" which are analogous to those for the "Species" filter.

    (d) "Mutations" findings from the Ingenuity Knowledge Base are now available for inclusion in the generated networks. Such findings involve a mutant form of at least one of the genes. You can choose to exclude mutant findings using this filter—the default is to include relationships that involve wild type and mutant forms of genes.

The results of the analysis consist in all the molecules that interact with other molecules in the Ingenuity Knowledge Base, which are identified as "Network Eligible" molecules and serve as "seeds" for generating networks. Network Eligible molecules are combined into networks that maximize their specific connectivity (i.e, their interconnectedness with each other relative to all molecules they are connected to in the Ingenuity KB). Additional molecules from the Ingenuity KB are used to connect two or more smaller networks by merging them into a larger one.

Results can be exported in three tab-delimited files (references, molecules, relationships). The relationships file details all interactions (experimental or a combination of experimental and predicted), as well as the nature of interactions (activation, inhibition, etc.). The references file contains IDs linked to various data sources (selected in **step 6e**) from which relationships were derived. Unlike in STRING, references *cannot* be associated one-to-one with the edges.

4 Summary and Outlook

PPIs contain a wealth of information that may be used in functional inference. Given the varied contexts from which the information is derived, however, it is useful to be able to filter PPIs according to the context in which each interaction is reported. The context can be defined in terms of common ontological annotations of the vertices, including cellular localization and biological function. Here, we show how to take advantage of information – in particular, the supporting literature associated with each edge— to isolate subnetworks whose elements have been reported in a set of biological contexts of interest.

5 Notes

1. We have noted that disparities in network-neighborhoods can occur between different *versions* of the same PPI [9]. It is clearly good practice to check the consistency of your results whenever a new version of a PPI is released by checking, for instance, if the change occurred at the level of one of the primary databases, or (in the case of STRING) if there are differences in the textmining sources used as edge evidence.
2. One way of evaluating the relevance of retrieved network neighborhoods without performing preliminary experiments is to check for articles that either mention the vertices of a network neighborhood together with keywords of interest (i.e., the vertices have been implicated in the biological context of interest, but the interaction in the PPI might be novel, as could be the case in vertex-based filters), or check for articles that comention the incident vertices together with keywords specific to a biological process of interest (i.e., the interaction has been documented in the biological context of interest). In the GO-filtered STRING network for virus entry, which is comprised only of vertices and edges that have been previously annotated with a GO term specific for entry (Fig. 4a) we

have checked the frequency at which each retained vertex is comentioned with one of eight keywords of interest that were not used in filtering. To efficiently perform such checks, one can use PubMed e-utilities (https://www.ncbi.nlm.nih.gov/books/NBK25501/) or wrappers, which are a set of functions that simplify exploration and access of NCBI databases, such as rentrez. A typical search in e-utilities is defined by the database and the query term; it is also convenient to set the maximum number of retrieved terms. This is done by specifying the following fields in your query on https://eutils.ncbi.nlm.nih.gov/entrez/eutils/esearch.fcgi?:

(a) **db**: NCBI database on which to perform the search; to search for abstracts, **pubmed** should be specified.

(b) **term**: search term

(c) **retmax**: number of retrieved matches that should be returned; note that in the case of PubMed abstracts, hits are ordered both by time (i.e., more recent articles first) and relevance.

Typing the following on your web browser, for instance, yields the PubMed IDs of the 1000 most recent articles on influenza from PubMed in an XML file: https://eutils.ncbi.nlm.nih.gov/entrez/eutils/esearch.fcgi?db=pubmed&term=influenza&retmax=1000.

3. For users with experience in R programming, using rentrez is a more flexible option. To retrieve the same results as above, the following command yields an entrez list object:

entrez_search(db="pubmed", term="influenza", retmax=1000) -> res

This object has the following fields:

(a) **ids**: PubMed IDs of the relevant results, limited to the number specified in retmax; can be obtained by typing **res$ids**

(b) **count**: total number of records in PubMed that match the query term

(c) **retmax**: number of retrieved matches that should be returned.

(d) **Query translation**: in case the search term matches any ontological term in PubMed, the search is expanded to include all these terms. In this example, the query "influenza" is translated into "influenza, human"[MeSH Terms] OR ("influenza"[All Fields] and "human"[All Fields]) OR "human influenza["All Fields] OR "influenza" [All Fields]

Additional examples of the use of **rentrez** to check the availability of information regarding pairs of proteins reported in an interaction network are shown in Example 3 of **sampleCodes.r**.

4. There are various open-source tools that can be used for network visualization and analyses, including Cytoscape (http://www.cytoscape.org), Graphviz (http://www.graphviz.org), and Gephi (https://gephi.org), which have graphical user interfaces. For users with more experience in programming, igraph (http://igraph.org) is an efficient analysis tool that is available in R, Python, and C/C++.

Acknowledgments

This work was supported by the Swiss Initiative in Systems Biology, SystemsX, through a fellowship (2013/137) provided to M.P.D.

References

1. Rual JF, Venkatesan K, Hao T, Hirozane-Kishikawa T, Dricot A, Li N et al (2005) Towards a proteome-scale map of the human protein-protein interaction network. Nature 437:1173–1178
2. Schaefer MH, Fontaine JF, Vinayagam A, Porras P, Wanker EE, Andrade-Navarro MA (2012) HIPPIE: integrating protein interaction networks with experiment based quality scores. PLoS One 7:e31826
3. von Mering C, Jensen LJ, Snel B, Hooper SD, Krupp M, Foglierini M et al (2005) STRING: known and predicted protein-protein associations, integrated and transferred across organisms. Nucleic Acids Res 33:D433–D437
4. Franceschini A, Szklarczyk D, Frankild S, Kuhn M, Simonovic M, Roth A et al (2013) STRING v9.1: protein-protein interaction networks, with increased coverage and integration. Nucleic Acids Res 41:D808–D815
5. Peri S, Navarro JD, Amanchy R, Kristiansen TZ, Jonnalagadda CK, Surendranath V et al (2003) Development of human protein reference database as an initial platform for approaching systems biology in humans. Genome Res 13:2363–2371
6. Zanzoni A, Montecchi-Palazzi L, Quondam M, Ausiello G, Helmer-Citterich M, Cesareni G (2002) MINT: a molecular INTeraction database. FEBS Lett 513:135–140
7. Hermjakob H, Montecchi-Palazzi L, Lewington C, Mudali S, Kerrien S, Orchard S et al (2004) IntAct: an open source molecular interaction database. Nucleic Acids Res 32:D452–D455
8. Stark C, Breitkreutz BJ, Reguly T, Boucher L, Breitkreutz A, Tyers M (2006) BioGRID: a general repository for interaction datasets. Nucleic Acids Res 34:D535–D539
9. Dobay MP, Stertz S, Delorenzi M (2017) Context-based retrieval of functional modules in protein-protein interaction networks. Brief Bioinform
10. Schaefer MH, Lopes TJ, Mah N, Shoemaker JE, Matsuoka Y, Fontaine JF et al (2013) Adding protein context to the human protein-protein interaction network to reveal meaningful interactions. PLoS Comput Biol 9:e1002860
11. Gray KA, Yates B, Seal RL, Wright MW, Bruford EA (2015) Genenames.org: the HGNC resources in 2015. Nucleic Acids Res 43:D1079–D1085
12. Ashburner M, Ball CA, Blake JA, Botstein D, Butler H, Cherry JM et al (2000) Gene ontology: tool for the unification of biology. The gene ontology consortium. Nat Genet 25:25–29
13. Beyerly E (1962) New medical subject heading lists: a comparative review of American and soviet works. Bull Med Libr Assoc 50:196–202
14. Shannon P, Markiel A, Ozier O, Baliga NS, Wang JT, Ramage D et al (2003) Cytoscape:

a software environment for integrated models of biomolecular interaction networks. Genome Res 13:2498–2504
15. Mi H, Muruganujan A, Casagrande JT, Thomas PD (2013) Large-scale gene function analysis with the PANTHER classification system. Nat Protoc 8:1551–1566
16. R Development Core Team (2014) R: a language and environment for statistical computing. the R Foundation for Statistical Computing, Vienna
17. McKusick VA (2007) Mendelian inheritance in man and its online version, OMIM. Am J Hum Genet 80:588–604
18. Szklarczyk D, Morris JH, Cook H, Kuhn M, Wyder S, Simonovic M et al (2017) The STRING database in 2017: quality-controlled protein-protein association networks, made broadly accessible. Nucleic Acids Res 45:D362–D368
19. Pletscher-Frankild S, Palleja A, Tsafou K, Binder JX, Jensen LJ (2015) DISEASES: text mining and data integration of disease-gene associations. Methods 74:83–89
20. Gaur P, Munjhal A, Lal SK (2011) Influenza virus and cell signaling pathways. Med Sci Monit 17:RA148–RA154
21. Ludwig S, Planz O, Pleschka S, Wolff T (2003) Influenza-virus-induced signaling cascades: targets for antiviral therapy? Trends Mol Med 9:46–52
22. Eierhoff T, Hrincius ER, Rescher U, Ludwig S, Ehrhardt C (2010) The epidermal growth factor receptor (EGFR) promotes uptake of influenza a viruses (IAV) into host cells. PLoS Pathog 6:e1001099

Part II

Data-driven Analyses of High-throughput Datasets

Chapter 3

SignaLink: Multilayered Regulatory Networks

Luca Csabai, Márton Ölbei, Aidan Budd, Tamás Korcsmáros, and Dávid Fazekas

Abstract

Biological networks are graphs used to represent the inner workings of a biological system. Networks describe the relationships of the elements of biological systems using edges and nodes. However, the resulting representation of the system can sometimes be too simplistic to usefully model reality. By combining several different interaction types within one larger multilayered biological network, tools such as SignaLink provide a more nuanced view than those relying on single-layer networks (where edges only describe one kind of interaction). Multilayered networks display connections between multiple networks (i.e., protein–protein interactions and their transcriptional and posttranscriptional regulators), each one of them describing a specific set of connections. Multilayered networks also allow us to depict cross talk between cellular systems, which is a more realistic way of describing molecular interactions. They can be used to collate networks from different sources into one multilayered structure, which makes them useful as an analytic tool as well.

Key words Molecular network, Multilayered network, Signal transduction, Signaling, Network resources, Network format, Data integration, Workflow

1 Introduction

1.1 From Signaling Pathways to Networks

Intracellular signaling pathways play a crucial role in regulating physiological and pathological cell functions [1]. Cellular signals are often transmitted by subtle chemical changes and mediate a variety of molecular processes. These signals are a vital component of the cellular decision-making machinery. Deregulated signal transmission can cause many different diseases, making the analysis of signaling pathways important for medical and basic research [2].

From a functional viewpoint, we can differentiate independent pathways that represent individual biological processes. However, seen from the systems biology perspective, if we want to analyze elements of signaling processes, we cannot separate the networks [3] that represent each pathway, since there are many cross talks between them, which create intertwined, complex networks. This

Louise von Stechow and Alberto Santos Delgado (eds.), *Computational Cell Biology: Methods and Protocols*, Methods in Molecular Biology, vol. 1819, https://doi.org/10.1007/978-1-4939-8618-7_3,

highlights the importance of studying and understanding signaling networks as a large network of connected pathways. In this chapter, we refer to pathways as the individual functional biological processes in an organism, while we define a signaling network as an entity, which contains elements of pathways and their connections to each other.

Until relatively recently, signaling pathways were typically thought of as linear chains of chemical reactions. It was common for researchers to focus on building understanding of one or a few components of a single pathway [4–6]. There has been a paradigm shift in the past decade. Instead of a collection of functionally distinct pathways, the consensus has shifted toward viewing biological processes in terms of a single signaling network that comprises all the intertwined pathways formerly thought of as separate entities [7]. However, this paradigm shift does not obviate the need and value of more focused, reductionist analyses concentrating on individual signaling molecules, or pathways. Such analyses remain essential for addressing more specific nuanced questions about specific interactions. There is a significant amount of overlap among biological pathways, i.e., multiple members actively participate in more than one signaling pathway.

So far signaling pathway databases have usually been created via integrating high-throughput interactions screens, text mining, and most effectively manual collection of interactions from the relevant literature [8]. A major problem that curated databases such as Signor [9], Reactome [10], and SignaLink [11] have in common is that the curation criteria, and the definition of which entities constitute a biological pathway, depend on the curator. Many resources fail to emphasize the importance of multipathway proteins, which function in different signaling pathways, and these proteins must therefore be very carefully and precisely regulated so that they can contribute appropriately to signal transduction in multiple different processes [8].

SignaLink is a database that has been constructed with the previously mentioned concerns in mind (representing signaling networks as connected functional pathways, including multipathway proteins and connections between pathways) [11]. The primary goal of SignaLink (http://signalink.org) is to provide a map of global signaling pathways, and to serve as a valuable tool for systems-level studies of cellular signaling. The curation process is specifically aimed at reducing manual errors, while maintaining a high level of quality.

The database lists signaling proteins and directed signaling interactions between pairs of proteins within healthy (i.e., nondiseased) cells of three species: *Homo sapiens, Drosophila melanogaster, and Caenorhabditis elegans.* SignaLink currently does not distinguish between cell types, and recommend tissue based filtering of the signaling pathways based on expression datasets. Besides integrating data from already existing databases, full texts of peer-reviewed papers were used to characterize pathways. Despite the inherent risk of curation processes (whether manual or automatic) to introduce biases toward certain datasets, through the application of clearly defined, published rules for annotations, SignaLink ensures that it is possible to explore and test the assumptions that have been used to assign individual annotations [12].

With SignaLink 2, the database was expanded to a multilayered data structure. The layers, which are essentially subsections of the signaling network, contain the core signaling pathways, their regulators and modifier enzymes, as well as transcriptional and posttranscriptional regulators of these components [11].

1.2 Multilayered Networks

Most systems can be described and analyzed by representing them as networks. To understand multilayered networks, we must first get a glimpse of what makes up single layered networks. Networks are graphs in which nodes indicate entities (e.g., genes, proteins), and edges represent relations between nodes. If two nodes are connected by an edge, then these nodes are adjacent. However, these single-layer networks can often be too simple to provide useful models of more complex networks (such as biological systems) [2]. It is more precise to describe complex systems as a network of multiple networks in connection with one another. In these multilayered networks many different types of interactions between two entities can be represented. For instance, in biological systems, interaction of proteins can be a physical interaction that depends on the colocalization of those proteins. If we were to illustrate these physical interactions in single layered graphs, we may neglect other type of interactions regulating physical interactions [13].

In multilayered networks, next to nodes and edges, we also should consider layers as structural units of the network. Nodes can be present in one or more layers. In general, edges in multilayered networks describe pairwise connections between any pair of nodes. If this connection is between two nodes in the same layer, then that interaction is an intralayer link, while if the edge connects nodes in different layers, it is considered an interlayer link [14] (*see* Fig. 1).

As data curation technology advances, more and more data are collected that need to be interpreted in order to further our understanding of biological systems. As a result, more and more pathway resources are created [8]. However, most of these pathway resources employ the same concept of presenting information about the molecules involved in a specific type of interaction pathway. In reality these pathways' molecules communicate and interact with each other. Biological processes rely on multiple types of interacting molecules and networks. These mechanisms can be better understood if we examine the biological process as a single complex network, rather than as a collection of smaller separate networks.

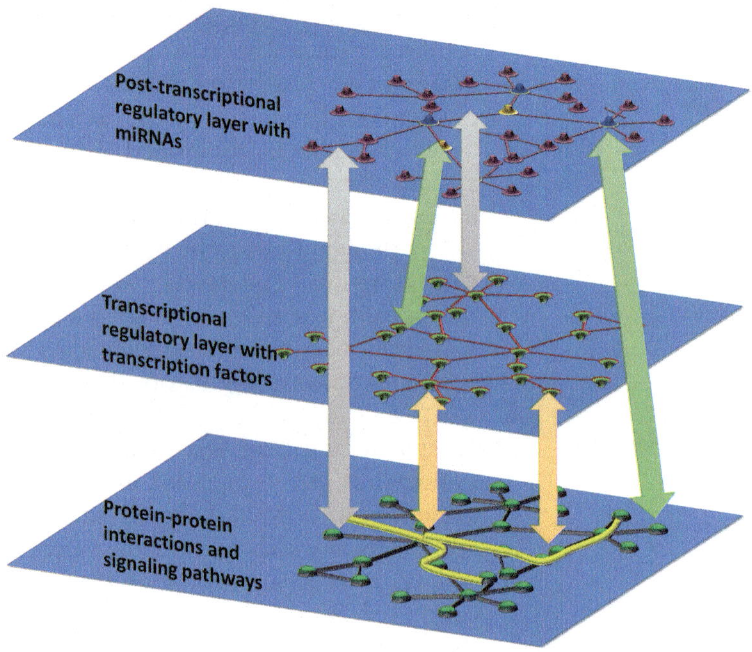

Fig. 1 Schematic representation of a multilayered biological network. The multiple layers build on the base network of PPIs, adding further information with each level, therefore giving us a more complex understanding of the network. Nodes represent different type of molecules (miRNAs, transcription factors, signaling proteins), while arrows on the different layers represent different type of connections (miRNA regulatory connections, transcriptional regulation, and enzymatic reactions)

There are many excellent databases containing information about molecular interactions in model organisms (e.g., BioGRID [15] and ENCODE [16]), however these can be only used for those specific organisms, and do not provide data as integrated multilayer networks. This means that they cannot be used directly to gain an overview of how multiple intertwining pathways cooperate with each other and create complicated multilayered networks. Therefore biological pathways could be further understood, when examining all aspects of a signaling network together, rather than individually, layer by layer.

With SignaLink, we are focusing on protein–protein interaction (PPI) networks while also considering transcriptional, post-transcriptional, and posttranslational regulations. We thereby hope to create even more accurate biological networks, while providing a simple, easy-to-use interface so researchers can access the data.

To achieve this, we have to curate interaction data from literature and from specialized data sources (PPI databases, transcription factor-target gene association databases and miRNA–mRNA interaction databases), and then integrate those levels of information

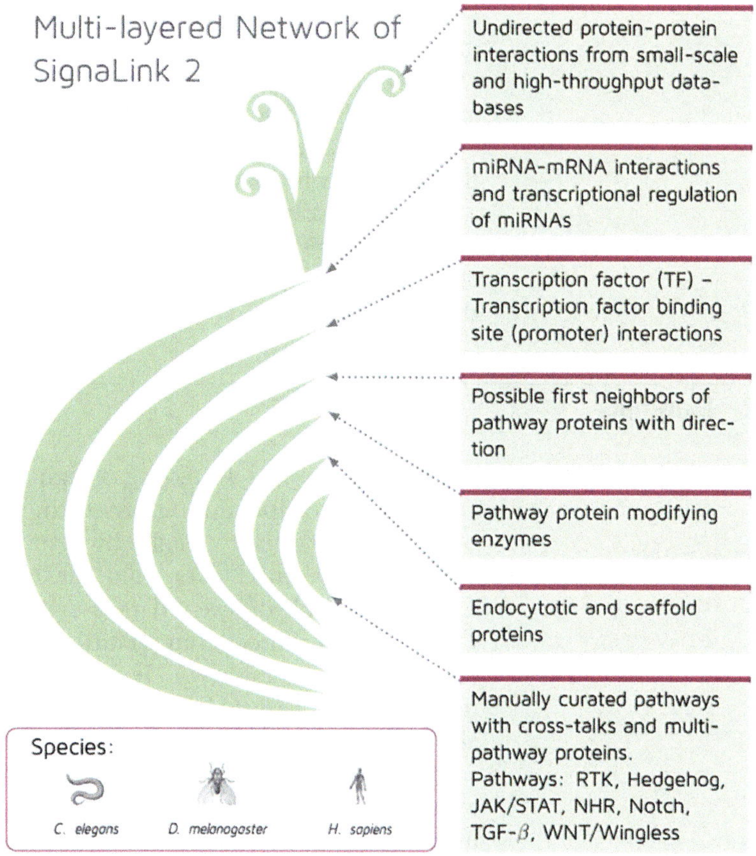

Fig. 2 The onion-like structure of the SignaLink multilayered network database. There are six additional layers on top of the manually curated core pathways. They contain additional connections of various biological importance, which adds further information to the pathways

into one multilayered database. Since not all data sources use the same data format, assigned molecular identifiers have to be mapped to a widely used standard format—UniProt in case of proteins and miRBase in case of miRNAs—in order to make biological network data universal and the nodes within identical. Besides the problems described above, we have to overcome multiple obstacles while creating a multilayered biological database, which will be further presented in this chapter [17] (*see* Fig. 2).

1.3 Open Data, Open Science

The scientific method in general is largely built upon the "collection, analysis, publication, reanalysis, critique and reuse" of data [18]. This process allows us to confirm or refute our hypotheses and allows others to test our findings, and build upon them. Currently, not all scientific data are available in a way that could be beneficial to most researchers. Many factors block the free flow of information: paywalls, restrictions on the specific usage of published data, poor formats, lack of annotations or widely applied data standards [18].

The Open Science movement tries to overcome those problems [19, 20]. They urge scientists to strive for a higher level of transparency, reproducibility, and accessibility. This is something that has been more prevalent in the software development world with many great projects, like the open source Linux operating system [21]. Making information publicly available reduces the chances of fraud, increases the number of citations, and speeds up the research process.

2 Materials

2.1 Data Sources

When creating and analyzing a signaling pathway, a quick and effective way of integrating preexisting human expert knowledge on the pathway is to begin by structuring the interactions of the pathway around data from the literature. This data can be found in multiple different databases such as Signor, SignaLink [8]. Different data sources contain different types of data with multiple levels of detail (Table 1). It is often difficult to decide for an unexperienced user which of these resources should be used either alone or in combination during a pathway modeling process [8].

The sources that we used for SignaLink have been identified from the available public resources containing data on signaling interactions in model organisms or in humans [11]. Some of the utilized sources contain information about causal interactions (when the sign of the interactions (activation/inhibition) is known), biochemical reactions (enzymatic modifications), and undirected interactions (PPIs), to name a few. In more detail, directed interactions involve processes in which one interaction partner affects the other in a specific directed way, e.g., a protein kinase (interactor A) activates a kinase substrate (interactor B) by phosphorylating it—in this case, the direction of the phosphorylation interaction is from kinase (A) to substrate (B), and following this reaction the activation of substrate (B) will increase. In some cases the information about this direction is not available, or the interaction is inherently undirected. The analyzed data for each interaction are uniformly represented alongside with references, directionality, sign (if available), and optionally added details (i.e., localization, mechanistic details).

To demonstrate the importance of integrating different manually curated data sources, we created the OmniPath resource (http://omnipathdb.org; [8]). By integrating high-confident manually curated signaling databases, we found that OmniPath covers approximately three times more proteins (7984) and four times more interactions (36,557) than the largest resource it contains [8].

Table 1
Collection of currently available, widely used interaction databases. Relevant for working with biological networks and integrating their data into our own database

Resource name	Description	Reference	URL
ACSN	The original sources of information come from review articles in high-impact journals. The information is extracted from these papers and represented in the form of biochemical interactions. This map is enriched with data from recent discoveries. It is necessary for the represented biochemical process to have evidences from more than two studies.	[44, 45]	https://acsn.curie.fr
Alz Pathway	From AD accessible PubMed articles, an AD pathway map is created. Molecules, reactions and cell types are all distinguished by multiple types. All reactions have PubMed ID references which are accessible only from XML files.	[46, 47]	http://alzpathway.org/AlzPathway.html
ARN	Signaling proteins and interactions are listed from reviews and are completed by additional signaling interaction data. The interactions are manually researched multiple times.	[48]	http://autophagyregulation.org/
BioCarta	Large number of pathways curated by experts. The downloadable files do not contain references and some pathways may be outdated.	[49]	http://www.biocarta.com/
BioGrid	Human hippocampal CA1 region neuron signaling network. Key components of pathways are curated from published research papers demonstrating direct interactions that are supported by biochemical or physical effect. The included interactions are directed and up to date. Accessible in tabular format with UniProt IDs and PubMed references.	[27, 42, 50]	http://thebiogrid.org/
Consensus PathDB	Interaction data containing data of physical interactions, biochemical reactions and gene regulations. The source databases for the prior two are supplied and permits linking the original sources.	[51, 52]	http://cpdb.molgen.mpg.de/CPDB

(continued)

Table 1
(continued)

Resource name	Description	Reference	URL
dbPTM	11 PTM related biological databases are integrated in this source. Includes MS/MS-identified peptides in association with PTMs. Mapped PTM sites are supplied with PubMed IDs	[53, 54]	http://dbptm.mbc.nctu.edu.tw/
Death Domain	Contains information about multiple DD superfamily proteins and physical binding between them. Data can only be accessed by HTML parsing.	[55]	http://deathdomain.org/
DEPOD	Manually curated human dephosphorylation database with PubMed references in MITAB format with UniProt IDs. Includes information about protein and nonprotein substrates, dephosphorylation sites and involved pathways verified by experimental data. It also supplies references to kinase databases.	[56]	http://www.koehn.embl.de/depod/index.php
DIP	Database of high-throughput interactions of proteins. Annotated with PubMed IDs, evidences and mechanisms. Available in MITAB format.	[57, 58]	http://dip.doe-mbi.ucla.edu/dip/Main.cgi
HPRD	Human protein reference database, containing PPIs and posttranslational modifications curated from experiments supplied with their types annotated. It also contains information about upstream enzymes responsible for protein modifications and alternative subcellular localization of the described proteins.	[59, 60]	http://www.hprd.org/
Human Signaling Network	Aims to combine different manually curated networks. Description of the used sources and methods are not provided. Also, the data file does not have a header or key, which makes the database less efficient.	[61]	http://www.bri.nrc.ca/wang/

(continued)

Table 1
(continued)

Resource name	Description	Reference	URL
HuPho	Dynamically updated database of human phosphate portal which provides proteome-wide data of the phosphate interactome. This makes it possible to browse experimental evidences from many scientific articles about experiments backing protein interactions in which at least one of the participating proteins is a phosphatase. Interactions are supported by multiple evidences.	[62]	http://hupho.uniroma2.it/
InnateDB	Manually curated binary protein interaction resource. A platform designed to represent systems-level mammalian innate immune response analyzes. It is a comprehensive database of human, mouse and bovine molecular pathways and interactions curated from public molecular interaction databases, and is not only limited to innate immunity-relevant data. Data is supplied with UniProt IDs, PubMed references and experimental evidences.	[63, 64]	http://www.innatedb.com/
IntAct	Database mostly containing PPI data, annotated by IMEx standards, containing detailed description of experimental conditions of the interactions.	[26, 65]	http://www.ebi.ac.uk/intact/
KEGG	Kyoto encyclopedia of genes and genomes. Data can be accessed in KGML files, which contain binary interactions mostly between large complexes. This resource does not supply references.	[66]	http://www.genome.jp/kegg/
MatrixDB	Contains data about intracellular and membrane proteins with intention to create a detailed network of the partners of extracellular molecules. Protein data is supplied with UniProtKB/SwissProt accession numbers and is annotated with GO terms 'extracellular region' and 'extracellular space' (used for proteins found in biological fluids). Interactions are associated to the human protein's accession numbers.	[67–69]	http://matrixdb.univ-lyon1.fr/

(continued)

Table 1 (continued)

Resource name	Description	Reference	URL
MINT	Molecular interaction database.	[70]	http://mint.bio.uniroma2.it/mint/Welcome.do
MPPI	The MIPS mammalian PPI database. This resource favors quality over completeness, therefore it includes strictly published data from individual experiments (not large-scale surveys). Data is supplied with UniProt IDs and PubMed references.	[71]	http://mips.helmholtz-muenchen.de/proj/ppi/
Negatome	Includes data from only mammalian species' proteins. To keep a high standard of reliability, data from high-throughput experiments were excluded.	[72]	http://mips.helmholtz-muenchen.de/proj/ppi/negatome/
NetPath	In the tab delimited format pathway memberships of genes, PubMed references can be found. There is no data of interaction partners. During curation, data is studied and reviewed in multiple stages by different researchers and experts.	[73]	http://netpath.org/
PathwayCommons	Includes public pathway information from multiple organisms, and provides researchers with access to a through collection of biological pathways from different sources for gene and metabolic pathway analysis.	[74]	http://www.pathwaycommons.org/
Phospho.ELM	Database, which includes manually curated and experimentally verified phosphorylation sites. It is developed as a part of the ELM resource. Information can be found here about substrate proteins with the exact positions of residues known to be phosphorylated by cellular kinases. It also supplies literature references, disuse distribution and other additional data.	[75, 76]	http://phospho.elm.eu.org/
PhosphoSite	Combination of low-, and high-throughput data sources of phosphorylation sites in human, mouse and other species.	[77, 78]	http://www.phosphosite.org/homeAction.do

(continued)

Table 1 (continued)

Resource name	Description	Reference	URL
Reactome	It is not possible to extract binary information programmatically from its dataset. The curation method doesn't include binary interactions, the available lists are based on automatic expansion of reactions and complexes, which makes the data unreliable. Also, it doesn't make it possible to assign references to interactions.	[25, 79, 80]	http://reactome.org/
SignaLink	A database assigning proteins to signaling pathways using the full texts of pathway reviews. Compared to most signaling resources, SignaLink uses more than 20 review papers per pathway on average. It aims at reducing data and curation errors during the curation process.	[11, 12]	http://signalink.org/
Signor	Signaling network open resource aims to store and organize signaling information published in scientific literature in a structured format, as binary causative relationships between biological entities. Relationships are provided with literature reporting experimental evidence. The contained data is mapped to the human proteome.	[9]	http://signor.uniroma2.it/
SPIKE	Contains data on relationships between entities from different sources. The interactions represented in the database are assigned quality values between 1 and 4. Those relationships derived from biochemical studies are given a high quality (1 or 2), and those from high-throughput experiments a lower quality.	[81, 82]	http://www.cs.tau.ac.il/~spike/
TRIP	Mammalian transient receptor potential channel-interacting protein database. TRP channel proteins with binary interactions. Nonstandard protein names are used, which makes bioinformatical use of the data practically impossible.	[83–85]	http://www.trpchannel.org

(continued)

Table 1 (continued)

Resource name	Description	Reference	URL
WikiPathways	Aims to collect data about biological pathways in a way that is accessible and understandable for both human and computational analysis. Only interactions are available in BioPAX format, with no references.	[86]	http://www.wikipathways.org/index.php/WikiPathways

2.2 Other Sources

The main differences between different signaling databases can be broken down into a few key elements: they usually either differ in the content they aggregate, the way they are presented, or the way they are accessible to the public. Based on content, we differentiate between primary (manually curated), secondary (aggregated, collated from different sources), and hybrid databases, which are a combination of the two. Hybrid databases contain both manually curated and aggregated data from outside sources [22].

Manual curation is still the most accepted way of integrating data, although it is not without fault [23]. In particular, in contrast to automatic curation, its reliability depends on the applied curation protocol and the expertise of the curator, and as such human biases can occur. Collecting pathway data requires intensive work and a deep knowledge of the specific field. Because of this the level of quality can be quite heterogeneous. This can be corrected by having pathways reviewed by experts of the given field. Despite all problems, manual curation is still the most accepted method of data curation, used by various popular databases [24, 25].

Most pathway databases attach links to the literature references used to annotate a given portion of the network. This helps the users to get familiar with the details of the signaling process at hand, and also allows constant verifications. While nodes of pathways are practically always referenced, the interactions themselves are not. To overcome this issue, PPI (protein–protein interaction) databases, such as BioGrid or IntAct list interaction data between proteins with unique identification numbers [26, 27]. It is worth mentioning, that some pathway databases, such as KEGG or Reactome already include this feature [28]. Based on accessibility, databases can be divided into commercially available and public, academic databases [22] (*see* Fig. 3).

2.3 Most Common Interaction Data Formats: BioPax, SBML, PSI-MI

When building biological databases, the format in which the data is stored can often differ from the format the end user receives. If the database is collated from multiple sources, and/ or built up in multiple steps, there is a need for an inner data structure to act as a common denominator.

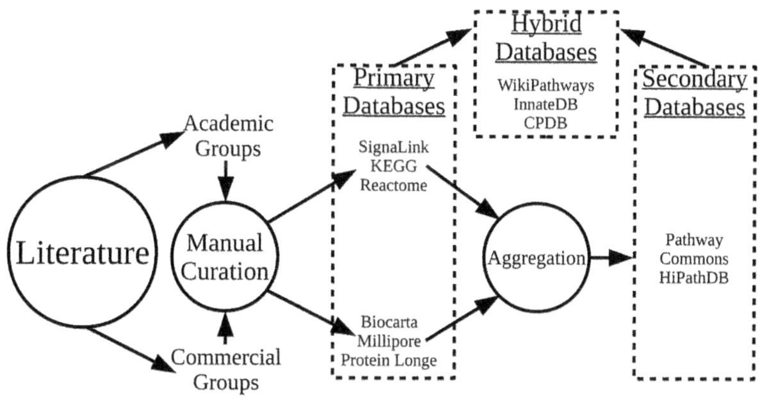

Fig. 3 The general workflow of database development. Scientific literature is used as the source of information that can be processed and aggregated in various ways. Based on the collected content we differentiate between manually curated primary databases, secondary databases that are collated from different sources, and hybrid databases that are a combination of the two

Alternatively, rather than developing a new internal format for this purpose, one can use one of the many commonly used standardized formats. One of these standard languages is BioPax [29], which is meant to describe biological pathways on multiple (cellular or molecular) levels. The main goal of the BioPax project is to make pathway data easier to collect. BioPax is designed to be able to incorporate a large amount of very diverse information about many specific details of an interaction network. However, if the user does not have enough information about the interactions, this flexibility and power can be problematic, as it puts a significant overhead on working with BioPax for no benefit, as the high level of specificity it provides is not needed [29].

Another frequently used language is SBML, which stands for Systems Biology Markup Language [30]. Similar to the BioPax format, SBML is based on XML (Extensible Markup Language), a human and machine-readable standard format. The aim of SBML is to provide a machine-readable format for modeling biological systems [http://sbml.org/Basic_Introduction_to_SBML]. The individual pathway elements are broken down into multiple components (Compartment, Species, Reaction, Parameter, Unit definition, Rule), which collectively make up a model [30].

When building up SignaLink, the PSI-MITAB format was used. Developed by the HUPO (Human Proteome Organization) Proteomics Standards Initiative (PSI), it acts as a community standard for data representation. MITAB stands for Molecular Interactions in a tab delimited exchange format. It is both machine and human readable. Since PSI-MITAB is tab delimited, it is especially easy for humans to handle and work with. While its first 1.0 version only focused on protein–protein interactions, the capabilities of the format have been notably extended in the past years [31]. Besides adding a variety of interactors to the system, the interactions between each molecule can now be detailed by the description of the interacting domains and the kinetic parameters of the reaction [31].

3 Methods

3.1 Data Integration

In this section we describe several important issues to have in mind when building a new biological database. First, the dataset incorporated in the database should be unique and valuable for the scientific community (measured by number of downloads, page visits, citations). The curation process—especially when done by hand from primary resources such as journals—requires a lot of time and dedication. Once collected, the information should be available in several widely accepted, standardized formats, with both manual and automatic ways of accessing it. The data should be kept up to date and the database should continue running, since future research might depend on its availability [32].

While building SignaLink, the first step was to integrate signaling proteins and their interactions from review articles (and from WormBook in *C. elegans*) and to add further signaling interactions to them from resources like WormBase [33]. Only those genetic interactions were added from the literature, which could manually be proven to be experimentally validated direct interactions. A similar method was used while collecting data for the *Drosophila melanogaster* pathways. Only those genetic interactions were included from FlyBase [34] that were validated by at least one yeast-2-hybrid experiment or other approach demonstrating the physical binding. In the case of human data, the directions and reliability of the protein–protein interactions were checked with two relevant search engines, iHOP and Chilibot [35, 36].

3.2 ID Mapping

The use of different molecule types in simulations of biological systems creates the problem of converting molecular information [2]. If done incorrectly, this may result in a loss of precision, as well as subsequent loss of data. Posttranscriptional processes may occur, which modify a protein's sequence compared to that of its coding gene. Therefore, it is important that we differentiate the two. Additionally, acquiring information about specific protein sequences (through mass spectrometry) is a more difficult process than DNA sequencing, thus it is less accessible [37].

Another common problem with creating molecule-based networks is the array of different identifiers that have been assigned to them. Various data sources supply proteins with IDs varying in length and character types (as some sources use identifiers only containing letters or numbers, while others use a combination of both). This creates a complex system of intertwining definitions of molecules, and referencing these entities in any network layer difficult. For this reason during the development of a new biological database it could increase the interoperability and ease of maintenance for a resource if the molecules are not defined with internally created, new IDs, rather defining them with already existing identifiers (or mapping the internal identifiers to existing ones) both within the database and as an output. While this has no effect on the user, as for the user it does not matter which identifiers are used internally, when creating a multilayered network resource with integrating different sources and data types, using existing identifiers could decrease the complexity of the development and maintenance of the resource. The external ID should be from a commonly accepted, accurate and wide-ranging database (such as UniProt and miRbase). However, when using multilayered networks, for different layers, different data sources may be needed. Although there are many ID conversion tools, including an ID mapper within the structure of a new resource is preferred. Creating a network that can be accessed by widely used IDs makes the process of searching, and using features of that network easier for researchers, and creates a uniform identification system. An issue may arise if the network uses molecules that have not been previously recorded in another database. In this case, the creation of an internal ID cannot be avoided [38, 39].

3.3 Workflow

In the course of developing our network the utilized data is processed in multiple steps before establishing a finalized database. After an initial phase of collecting relevant information from various resources, the data is conformed into an internal structure. This step is necessary, since different sources may use separate data structures. We use a standard internal structure (PSI) [37], which is accessible and simple to modify. During this process there is no change in data quality. Subsequently, internal IDs are mapped to widely accepted universal identifiers (Uniprot [40] and miRbase [41]) to make the network more practical for research purposes. Note that structure modification and ID mapping are done in different steps, compared to many other workflows combining these processes. Then the acquired material is compressed (i.e., multiple sources containing the same information will be considered as one) during the course of data integration. The established database will have the same inner data structure created in the second step, with common IDs and ubiquitous and compact data. Finally, data is exported from this database; however, it is always possible to revert back to the information stored here.

In comparison, many other databases do not use these methods of data integration. BioGRID [15] is an interaction database focusing on biomedical relevance and human disease-associated interaction networks. BioGRID offers a variety of different downloadable formats (including PSI-MI), which makes the data accessible to many researchers, as opposed to only using one universal internal structure throughout the entire database. However, providing the data in multiple formats does not contribute to making biological interaction databases interoperable, thus eliminating problems with integrating data from different sources with various structures, which is one of the key aims of SignaLink [11]. An important component of the SignaLink database is the conversion of all external data structures into a standard format. This contributes to the accessibility of interaction data, since combination and comparison between other sources using the same format becomes easier. Another major difference is that while SignaLink maps identifiers from external sources to more commonly used IDs, BioGRID does not. This makes referencing and researching certain molecules less effective. Similar to SignaLink, in BioGRID, the curated information is regularly reviewed, and annotations are frequently updated in order to compress entries with the same information into a single search result [42].

4 Working with SignaLink

Users have two options to access and work with SignaLink data.

Firstly, they can access the data via the SignaLink website (http://signalink.org). This focuses on providing users with an answer to the question "Which signaling molecules are described as interacting with my signaling molecule of interest?". To address this question, the user queries SignaLink via the text query box at the top of the home page with an identifier describing their molecule of interest. The user is then taken to a page describing a list of all signaling molecules described as interacting with the query molecule in the SignaLink multilayer network. By clicking on the links associated with the names of these molecules, the user can explore the list of molecules described as interacting with them.

Secondly, the user can download the full (or a selected part of the) SignaLink network, from the dedicated Download page. This provides the data in a range of different formats (BioPAX, SBML, PSI-MI, CSV). Users can then import this data into other tools, such as Cytoscape [43], to carry out further visualizations and analyses.

On the right side of any page in the SignaLink website a Feedback tab allows the user to ask for help or submit a problem to the active SignaLink user community.

5 Summary and Outlook

SignaLink has been developed over more than 10 years and is accessible to the scientific community interested in high-confidence multilayer networks of interactions in several model organisms.

SignaLink is currently being reimplemented using new back- and front-end technologies, updating the manually curated interactions, integrating new data types (Long noncoding RNAs: lncRNAs, subcellular and tissue localizations) and new external data resources. However, the principles described in this book chapter will continue to be applied to this new version of the resource (SignaLink 3.0). The advantage of a multilayered and well-designed network database structure is that it is inherently scalable if new data or interaction types become available. In conclusion, multilayered networks, despite increasing the complexity of the stored data, they help the interpretation of primary biological data.

Acknowledgment

The authors are grateful for the past and present developers and coauthors of SignaLink, and also to the members of the Netbiol, LINK, and Korcsmaros groups. This work was supported by a fellowship to TK in computational biology at the Earlham Institute (Norwich, UK) in partnership with the Quadram Institute (Norwich, UK), and strategically supported by the Biotechnological and Biosciences Research Council, UK (BB/J004529/1 and BB/P016774/1).

References

1. Pires-daSilva A, Sommer RJ (2003) The evolution of signalling pathways in animal development. Nat Rev Genet 4:39–49. https://doi.org/10.1038/nrg977
2. Csermely P, Korcsmáros T, Kiss HJM et al (2013) Structure and dynamics of molecular networks: a novel paradigm of drug discovery: a comprehensive review. Pharmacol Ther 138:333–408. https://doi.org/10.1016/j.pharmthera.2013.01.016
3. Valdespino-Gómez VM, Valdespino-Castillo PM, Valdespino-Castillo VE (2015) Cell signalling pathways interaction in cellular proliferation: potential target for therapeutic interventionism. Cir Cir 83:165–174. https://doi.org/10.1016/j.circen.2015.08.015
4. Hansson EM, Lendahl U, Chapman G (2004) Notch signaling in development and disease. Semin Cancer Biol 14:320–328. https://doi.org/10.1016/j.semcancer.2004.04.011
5. Ingham PW, Kim HR (2005) Hedgehog signalling and the specification of muscle cell identity in the zebrafish embryo. Exp Cell Res 306:336–342. https://doi.org/10.1016/j.yexcr.2005.03.019
6. Nayak L, Bhattacharyya NP, De RK (2016) Wnt signal transduction pathways: modules, development and evolution. BMC Syst Biol 10(Suppl 2):44. https://doi.org/10.1186/s12918-016-0299-7
7. Xia Y, Yu H, Jansen R et al (2004) Analyzing cellular biochemistry in terms of molecular networks. Annu Rev Biochem 73:1051–1087. https://doi.org/10.1146/annurev.biochem.73.011303.073950

8. Türei D, Korcsmáros T, Saez-Rodriguez J (2016) OmniPath: guidelines and gateway for literature-curated signaling pathway resources. Nat Methods 13:966–967. https://doi.org/10.1038/nmeth.4077
9. Perfetto L, Briganti L, Calderone A et al (2016) SIGNOR: a database of causal relationships between biological entities. Nucleic Acids Res 44:D548–D554. https://doi.org/10.1093/nar/gkv1048
10. Fabregat A, Sidiropoulos K, Garapati P et al (2016) The Reactome pathway knowledgebase. Nucleic Acids Res 44:D481–D487. https://doi.org/10.1093/nar/gkv1351
11. Fazekas D, Koltai M, Türei D et al (2013) SignaLink 2–a signaling pathway resource with multi-layered regulatory networks. BMC Syst Biol 7:7. https://doi.org/10.1186/1752-0509-7-7
12. Korcsmáros T, Farkas IJ, Szalay MS et al (2010) Uniformly curated signaling pathways reveal tissue-specific cross-talks and support drug target discovery. Bioinformatics 26:2042–2050. https://doi.org/ 10.1093/bioinformatics/btq310
13. De Domenico M, Nicosia V, Arenas A, Latora V (2015) Structural reducibility of multilayer networks. Nat Commun 6:6864. https://doi.org/10.1038/ncomms7864
14. Kivelä M, Arenas A, Barthelemy M et al (2014) Multilayer networks. J Complex Netw 2:203–271. https://doi.org/10.1093/comnet/cnu016
15. Chatr-Aryamontri A, Breitkreutz B-J, Heinicke S et al (2013) The BioGRID interaction database: 2013 update. Nucleic Acids Res 41:D816–D823. https://doi.org/10.1093/nar/gks1158
16. ENCODE Project Consortium (2012) An integrated encyclopedia of DNA elements in the human genome. Nature 489:57–74. https://doi.org/10.1038/nature11247
17. Santra T, Kolch W, Kholodenko BN (2014) Navigating the multilayered organization of eukaryotic signaling: a new trend in data integration. PLoS Comput Biol 10:e1003385. https://doi.org/10.1371/journal.pcbi.1003385
18. Molloy JC (2011) The open knowledge foundation: open data means better science. PLoS Biol 9:e1001195. https://doi.org/10.1371/journal.pbio.1001195
19. Vizcaíno JA, Deutsch EW, Wang R et al (2014) ProteomeXchange provides globally coordinated proteomics data submission and dissemination. Nat Biotechnol 32:223–226. https://doi.org/10.1038/nbt.2839
20. Omenn GS, States DJ, Adamski M et al (2005) Overview of the HUPO plasma proteome project: results from the pilot phase with 35 collaborating laboratories and multiple analytical groups, generating a core dataset of 3020 proteins and a publicly-available database. Proteomics 5:3226–3245. https://doi.org/10.1002/pmic.200500358
21. Woelfle M, Olliaro P, Todd MH (2011) Open science is a research accelerator. Nat Chem 3:745–748. https://doi.org/10.1038/nchem.1149
22. Chowdhury S, Sarkar RR (2015) Comparison of human cell signaling pathway databases–evolution, drawbacks and challenges. Database (Oxford) 2015:bau126. https://doi.org/10.1093/database/bau126
23. Cusick ME, Yu H, Smolyar A et al (2009) Literature-curated protein interaction datasets. Nat Methods 6:39–46. https://doi.org/10.1038/nmeth.1284
24. Pico AR, Kelder T, van Iersel MP et al (2008) WikiPathways: pathway editing for the people. PLoS Biol 6:e184. https://doi.org/10.1371/journal.pbio.0060184
25. Croft D, Mundo AF, Haw R et al (2014) The Reactome pathway knowledgebase. Nucleic Acids Res 42:D472–D477. https://doi.org/10.1093/nar/gkt1102
26. Orchard S, Ammari M, Aranda B et al (2014) The MIntAct project–IntAct as a common curation platform for 11 molecular interaction databases. Nucleic Acids Res 42:D358–D363. https://doi.org/10.1093/nar/gkt1115
27. Stark C, Breitkreutz B-J, Chatr-Aryamontri A et al (2011) The BioGRID interaction database: 2011 update. Nucleic Acids Res 39:D698–D704. https://doi.org/10.1093/nar/gkq1116
28. Kanehisa M, Furumichi M, Tanabe M et al (2017) KEGG: new perspectives on genomes, pathways, diseases and drugs. Nucleic Acids Res 45:D353–D361. https://doi.org/10.1093/nar/gkw1092
29. Demir E, Cary MP, Paley S et al (2010) The BioPAX community standard for pathway data sharing. Nat Biotechnol 28:935–942. https://doi.org/10.1038/nbt.1666
30. Hucka M, Finney A, Sauro HM et al (2003) The systems biology markup language (SBML): a medium for representation

and exchange of biochemical network models. Bioinformatics 19:524–531. https://doi.org/10.1093/bioinformatics/btg015
31. Kerrien S, Orchard S, Montecchi-Palazzi L et al (2007) Broadening the horizon–level 2.5 of the HUPO-PSI format for molecular interactions. BMC Biol 5:44. https://doi.org/10.1186/1741-7007-5-44
32. Helmy M, Crits-Christoph A, Bader GD (2016) Ten simple rules for developing public biological databases. PLoS Comput Biol 12:e1005128. https://doi.org/10.1371/journal.pcbi.1005128
33. Harris TW, Antoshechkin I, Bieri T et al (2010) WormBase: a comprehensive resource for nematode research. Nucleic Acids Res 38:D463–D467. https://doi.org/10.1093/nar/gkp952
34. Tweedie S, Ashburner M, Falls K et al (2009) FlyBase: enhancing drosophila gene ontology annotations. Nucleic Acids Res 37:D555–D559. https://doi.org/10.1093/nar/gkn788
35. Chen H, Sharp BM (2004) Content-rich biological network constructed by mining PubMed abstracts. BMC Bioinformatics 5:147. https://doi.org/10.1186/1471-2105-5-147
36. Hoffmann R, Valencia A (2005) Implementing the iHOP concept for navigation of biomedical literature. Bioinformatics 21(Suppl 2):ii252–ii258. https://doi.org/10.1093/ bioinformatics/bti1142
37. Hermjakob H, Montecchi-Palazzi L, Bader G et al (2004) The HUPO PSI's molecular interaction format–a community standard for the representation of protein interaction data. Nat Biotechnol 22:177–183. https://doi.org/10.1038/nbt926
38. Ling F, Kang B, Sun X-H (2014) Id proteins: small molecules, mighty regulators. Curr Top Dev Biol 110:189–216. https://doi.org/10.1016/B978-0-12-405943-6.00005-1
39. Jamil HM (2015) Improving integration effectiveness of ID mapping based biological record linkage. IEEE/ACM Trans Comput Biol Bioinform 12:473–486. https://doi.org/10.1109/TCBB.2014.2355213
40. The UniProt Consortium (2017) UniProt: the universal protein knowledgebase. Nucleic Acids Res 45:D158–D169. https://doi.org/10.1093/nar/gkw1099
41. Kozomara A, Griffiths-Jones S (2014) miRBase: annotating high confidence microRNAs using deep sequencing data. Nucleic Acids Res 42:D68–D73. https://doi.org/10.1093/nar/gkt1181
42. Stark C, Breitkreutz B-J, Reguly T et al (2006) BioGRID: a general repository for interaction datasets. Nucleic Acids Res 34:D535–D539. https://doi.org/10.1093/nar/gkj109
43. Shannon P, Markiel A, Ozier O et al (2003) Cytoscape: a software environment for integrated models of biomolecular interaction networks. Genome Res 13:2498–2504. https://doi.org/10.1101/gr.1239303
44. Kuperstein I, Bonnet E, Nguyen HA et al (2015) Atlas of cancer signalling network: a systems biology resource for integrative analysis of cancer data with Google maps. Oncogene 4:e160. https://doi.org/10.1038/oncsis.2015.19
45. Calzone L, Gelay A, Zinovyev A et al (2008) A comprehensive modular map of molecular interactions in RB/E2F pathway. Mol Syst Biol 4:173. https://doi.org/10.1038/msb.2008.7
46. Mizuno S, Iijima R, Ogishima S et al (2012) AlzPathway: a comprehensive map of signaling pathways of Alzheimer's disease. BMC Syst Biol 6:52. https://doi.org/10.1186/1752-0509-6-52
47. Ogishima S, Mizuno S, Kikuchi M et al (2016) Alzpathway, an updated map of curated signaling pathways: towards deciphering alzheimer's disease pathogenesis. Methods Mol Biol 1303:423–432. https://doi.org/10.1007/978-1-4939-2627-5_25
48. Türei D, Földvári-Nagy L, Fazekas D et al (2015) Autophagy regulatory network–a systems-level bioinformatics resource for studying the mechanism and regulation of autophagy. Autophagy 11:155–165. https://doi.org/10.4161/15548627.2014.994346
49. Nishimura D (2001) BioCarta. Biotech Software Internet Report 2:117–120. https://doi.org/10.1089/152791601750294344
50. Breitkreutz B-J, Stark C, Reguly T et al (2008) The BioGRID interaction database: 2008 update. Nucleic Acids Res 36:D637–D640. https://doi.org/10.1093/nar/gkm1001
51. Kamburov A, Wierling C, Lehrach H, Herwig R (2009) ConsensusPathDB–a database for integrating human functional interaction networks. Nucleic Acids Res 37:D623–D628. https://doi.org/10.1093/nar/gkn698
52. Kamburov A, Pentchev K, Galicka H et al (2011) ConsensusPathDB: toward a more complete picture of cell biology.

Nucleic Acids Res 39:D712–D717. https://doi.org/10.1093/nar/gkq1156
53. Lu C-T, Huang K-Y, Su M-G et al (2013) DbPTM 3.0: an informative resource for investigating substrate site specificity and functional association of protein post-translational modifications. Nucleic Acids Res 41:D295–D305. https://doi.org/10.1093/nar/gks1229
54. Lee T-Y, Huang H-D, Hung J-H et al (2006) dbPTM: an information repository of protein post-translational modification. Nucleic Acids Res 34:D622–D627. https://doi.org/10.1093/nar/gkj083
55. Kwon D, Yoon JH, Shin S-Y et al (2012) A comprehensive manually curated protein-protein interaction database for the death domain superfamily. Nucleic Acids Res 40:D331–D336. https://doi.org/10.1093/nar/gkr1149
56. Duan G, Li X, Köhn M (2015) The human DEPhOsphorylation database DEPOD: a 2015 update. Nucleic Acids Res 43:D531–D535. https://doi.org/10.1093/nar/gku1009
57. Xenarios I, Rice DW, Salwinski L et al (2000) DIP: the database of interacting proteins. Nucleic Acids Res 28:289–291
58. Xenarios I, Salwínski L, Duan XJ et al (2002) DIP, the database of interacting proteins: a research tool for studying cellular networks of protein interactions. Nucleic Acids Res 30: 303–305
59. Peri S, Navarro JD, Amanchy R et al (2003) Development of human protein reference database as an initial platform for approaching systems biology in humans. Genome Res 13:2363–2371. https://doi.org/10.1101/gr.1680803
60. Keshava Prasad TS, Goel R, Kandasamy K et al (2009) Human protein reference database–2009 update. Nucleic Acids Res 37:D767–D772. https://doi.org/10.1093/nar/gkn892
61. Gao Y, Qi G, Guo L, Sun Y (2016) Bioinformatics analyses of differentially expressed genes associated with acute myocardial infarction. Cardiovasc Ther 34:67–75. https://doi.org/10.1111/1755-5922.12171
62. Liberti S, Sacco F, Calderone A et al (2013) HuPho: the human phosphatase portal. FEBS J 280:379–387. https://doi.org/10.1111/j.1742-4658.2012.08712.x
63. Breuer K, Foroushani AK, Laird MR et al (2013) InnateDB: systems biology of innate immunity and beyond–recent updates and continuing curation. Nucleic Acids Res 41:D1228–D1233. https://doi.org/10.1093/nar/gks1147
64. Lynn DJ, Winsor GL, Chan C et al (2008) InnateDB: facilitating systems-level analyses of the mammalian innate immune response. Mol Syst Biol 4:218. https://doi.org/10.1038/msb.2008.55
65. Kerrien S, Alam-Faruque Y, Aranda B et al (2007) IntAct–open source resource for molecular interaction data. Nucleic Acids Res 35:D561–D565. https://doi.org/10.1093/nar/gkl958
66. Ogata H, Goto S, Sato K et al (1999) KEGG: Kyoto encyclopedia of genes and genomes. Nucleic Acids Res 27:29–34. https://doi.org/10.1093/nar/28.1.27
67. Chautard E, Ballut L, Thierry-Mieg N, Ricard-Blum S (2009) MatrixDB, a database focused on extracellular protein-protein and protein-carbohydrate interactions. Bioinformatics 25:690–691. https://doi.org/10.1093/bioinformatics/btp025
68. Chautard E, Fatoux-Ardore M, Ballut L et al (2011) MatrixDB, the extracellular matrix interaction database. Nucleic Acids Res 39:D235–D240. https://doi.org/10.1093/nar/gkq830
69. Launay G, Salza R, Multedo D et al (2015) MatrixDB, the extracellular matrix interaction database: updated content, a new navigator and expanded functionalities. Nucleic Acids Res 43:D321–D327. https://doi.org/10.1093/nar/gku1091
70. Chatr-aryamontri A, Ceol A, Palazzi LM et al (2007) MINT: the molecular INTeraction database. Nucleic Acids Res 35:D572–D574. https://doi.org/10.1093/nar/gkl950
71. Pagel P, Kovac S, Oesterheld M et al (2005) The MIPS mammalian protein-protein interaction database. Bioinformatics 21:832–834. https://doi.org/ 10.1093/bioinformatics/bti115
72. Blohm P, Frishman G, Smialowski P et al (2014) Negatome 2.0: a database of non-interacting proteins derived by literature mining, manual annotation and protein structure analysis. Nucleic Acids Res 42:D396–D400. https://doi.org/10.1093/nar/gkt1079
73. Kandasamy K, Mohan SS, Raju R et al (2010) NetPath: a public resource of curated signal transduction pathways. Genome Biol 11:R3. https://doi.org/10.1186/gb-2010-11-1-r3

74. Cerami EG, Gross BE, Demir E et al (2011) Pathway commons, a web resource for biological pathway data. Nucleic Acids Res 39:D685–D690. https://doi.org/10.1093/nar/gkq1039
75. Diella F, Cameron S, Gemünd C et al (2004) Phospho.ELM: a database of experimentally verified phosphorylation sites in eukaryotic proteins. BMC Bioinformatics 5:79. https://doi.org/10.1186/1471-2105-5-79
76. Dinkel H, Chica C, Via A et al (2011) Phospho.ELM: a database of phosphorylation sites–update 2011. Nucleic Acids Res 39:D261–D267. https://doi.org/10.1093/nar/gkq1104
77. Hornbeck PV, Chabra I, Kornhauser JM et al (2004) PhosphoSite: a bioinformatics resource dedicated to physiological protein phosphorylation. Proteomics 4:1551–1561. https://doi.org/10.1002/pmic.200300772
78. Hornbeck PV, Kornhauser JM, Tkachev S et al (2012) PhosphoSitePlus: a comprehensive resource for investigating the structure and function of experimentally determined post-translational modifications in man and mouse. Nucleic Acids Res 40:D261–D270. https://doi.org/10.1093/nar/gkr1122
79. Matthews L, Gopinath G, Gillespie M et al (2009) Reactome knowledgebase of human biological pathways and processes. Nucleic Acids Res 37:D619–D622. https://doi.org/10.1093/nar/gkn863
80. Haw R, Hermjakob H, D'Eustachio P, Stein L (2011) Reactome pathway analysis to enrich biological discovery in proteomics data sets. Proteomics 11:3598–3613. https://doi.org/10.1002/pmic.201100066
81. Elkon R, Vesterman R, Amit N et al (2008) SPIKE–a database, visualization and analysis tool of cellular signaling pathways. BMC Bioinformatics 9:110. https://doi.org/10.1186/1471-2105-9-110
82. Paz A, Brownstein Z, Ber Y et al (2011) SPIKE: a database of highly curated human signaling pathways. Nucleic Acids Res 39:D793–D799. https://doi.org/10.1093/nar/gkq1167
83. Shin Y-C, Shin S-Y, So I et al (2011) TRIP database: a manually curated database of protein-protein interactions for mammalian TRP channels. Nucleic Acids Res 39:D356–D361. https://doi.org/10.1093/nar/gkq814
84. Shin Y-C, Shin S-Y, Chun JN et al (2012) TRIP database 2.0: a manually curated information hub for accessing TRP channel interaction network. PLoS One 7:e47165. https://doi.org/10.1371/journal.pone.0047165
85. Chun JN, Lim JM, Kang Y et al (2014) A network perspective on unraveling the role of TRP channels in biology and disease. Pflugers Arch 466:173–182. https://doi.org/10.1007/s00424-013-1292-2
86. Kelder T, van Iersel MP, Hanspers K et al (2012) WikiPathways: building research communities on biological pathways. Nucleic Acids Res 40:D1301–D1307. https://doi.org/10.1093/nar/gkr1074

Chapter 4

Interplay Between Long Noncoding RNAs and MicroRNAs in Cancer

Francesco Russo, Giulia Fiscon, Federica Conte, Milena Rizzo, Paola Paci, and Marco Pellegrini

Abstract

In the last decade noncoding RNAs (ncRNAs) have been extensively studied in several biological processes and human diseases including cancer. microRNAs (miRNAs) are the best-known class of ncRNAs. miRNAs are small ncRNAs of around 20–22 nucleotides (nt) and are crucial posttranscriptional regulators of protein coding genes. Recently, new classes of ncRNAs, longer than miRNAs have been discovered. Those include intergenic noncoding RNAs (lincRNAs) and circular RNAs (circRNAs). These novel types of ncRNAs opened a very exciting field in biology, leading researchers to discover new relationships between miRNAs and long noncoding RNAs (lncRNAs), which act together to control protein coding gene expression. One of these new discoveries led to the formulation of the "competing endogenous RNA (ceRNA) hypothesis." This hypothesis suggests that an lncRNA acts as a sponge for miRNAs reducing their expression and causing the upregulation of miRNA targets. In this chapter we first discuss some recent discoveries in this field showing the mutual regulation of miRNAs, lncRNAs, and protein-coding genes in cancer. We then discuss the general approaches for the study of ceRNAs and present in more detail a recent computational approach to explore the ability of lncRNAs to act as ceRNAs in human breast cancer that has been shown to be, among the others, the most precise and promising.

Key words MicroRNAs, Long noncoding RNAs, Competing endogenous RNAs, Sponge, Cancer, Long noncoding RNA-derived microRNAs, Host genes

1 Introduction

1.1 MicroRNAs in Cancer

miRNAs are intensively studied small ncRNA of 20–22 nt length, which have been recognized over the past decade as key controllers of gene expression by targeting messenger RNAs (mRNAs). miRNAs are involved in several biological processes [1–3]. The miRNA molecules recognize their targets by base-pairing to partially complementary sequences in the 3′- and 5′-untranslated regions

Francesco Russo, Giulia Fiscon, and Federica Conte contributed equally to this work.

Louise von Stechow and Alberto Santos Delgado (eds.), *Computational Cell Biology: Methods and Protocols*, Methods in Molecular Biology, vol. 1819, https://doi.org/10.1007/978-1-4939-8618-7_4,

(3′-UTR and 5′-UTR respectively) or in the open reading frames of the targets. The current version of miRBase (http://www.mirbase.org/), the miRNA registry [4], contains 1881 precursors and 2588 mature human miRNAs. Those numbers represent the set of annotated miRNAs. Yet only a subset of those miRNAs has been extensively characterized at a functional level.

miRNAs are frequently deregulated in cancer. For instance, in leukemia cells the miRNA genes encoding for miR-15a and miR-16 are deleted [5]. Several studies have shown that these two miRNAs are tumor suppressors and their deletion or downregulation leads to the upregulation of antiapoptotic proteins such as BCL2 [6, 7]. Conversely, miRNAs can also be amplified in cancer. An example for this is the miR-17-92 cluster, which targets, among others, Pten, Ppp2r5e, Prkaa1, and Bim [8, 9].

1.2 Long Noncoding RNAs in Cancer

The recently acknowledged long noncoding RNAs (lncRNAs) are non-protein coding transcripts longer than 200 nucleotides, which lack extended open reading frames [10–28].

In cancer lncRNAs can be deleted or amplified, similar to miRNAs. For instance, extra copies of the chromosomal region 8q24.21 have been shown to be common in many human cancers and are associated with poor prognosis [29]. This region not only contains the well-known oncogene MYC but also the lncRNA PVT1. It has been shown that the copy number alteration of MYC is correlated to the increase of PVT1 copies [29], which stabilizes MYC protein and potentiates its activity [30].

Copy number alterations and mutations can alter the transcriptional regulation of lncRNAs. Furthermore, specific single-nucleotide polymorphisms (SNPs) can be associated with an increased or decreased risk of specific diseases. It has been shown that the genetic variant rs7763881 in the lncRNA HULC may contribute to the decreased risk of developing Hepatitis B virus-related Hepatocellular carcinoma [31].

1.3 Competing Endogenous RNAs

Recent findings show that coding genes are not the only targets of miRNAs. In fact, it has been reported that different noncoding/coding RNAs compete for the same miRNA enabling the reduction of the amount of miRNAs available for interaction via the binding of the miRNA recognition/response elements (MREs) [32–40].

These RNA transcripts act as competing endogenous RNAs (ceRNAs), also known as miRNA “decoy” or miRNA “sponges” and appear to be involved in many disease conditions, including cancer development and progression [35, 36, 39, 41–43].

One of the possible ways of functioning of ceRNA mechanism was proposed by Poliseno et al. [44], where ceRNAs recruit the miRNAs and thus effectively de-repress other targets of that miRNA: the more copies of a ceRNA that acts as a sponge for

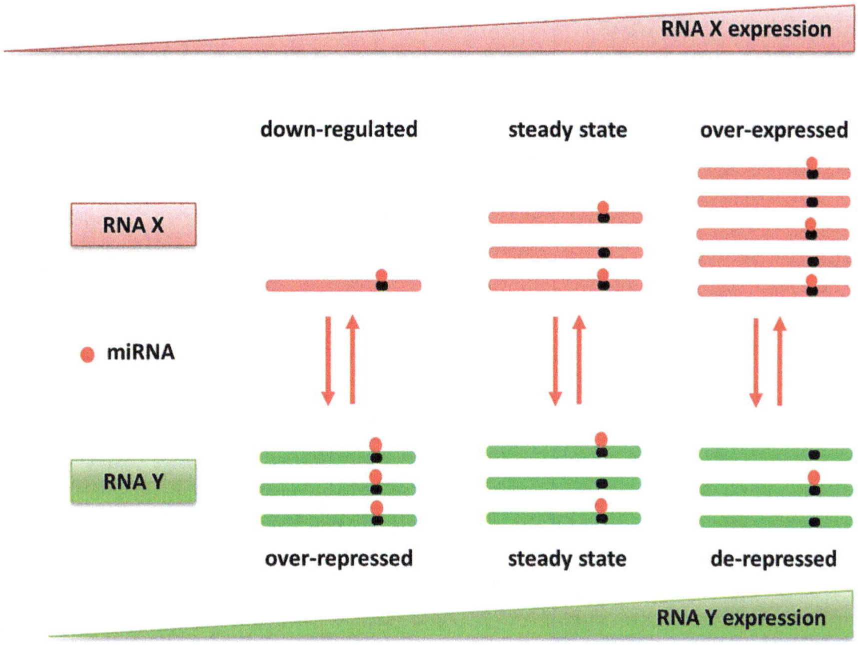

Fig. 1 CeRNA mechanism proposed by Poliseno et al. [44] to explain the experimentally observed highly positive correlation between competing endogenous RNAs. X and Y are two RNA transcripts that compete for binding the same miRNA(s). In the steady state (middle), the microRNA molecules and their targets X and Y are in equilibrium and the microRNA will be equally distributed between its targets. In a downregulation condition of the RNA X (left), the availability of microRNA molecules to bind the RNA Y increased determining the decrease of RNA Y expression. On the contrary, in an overexpression condition of the RNA X (right), less miRNA molecules are free to bind the RNA Y, and thus the RNA Y abundance increases. Legend. Red dots: microRNA molecules; light red boxes: RNA X; green boxes: RNA Y

a specific miRNA are expressed, the more copies of the target mRNAs should be present (Fig. 1). As a consequence, ceRNAs and "canonical" miRNA targeted RNA have highly correlated expression profiles [44].

Such a mechanism of regulation of miRNA activity was first discovered in plants and was called "target mimicry" process [45]. So far, researchers all over the world have focused on the study of this mechanism as evidenced by the increased number of publications on ceRNAs in the past few years [32–40].

Moreover, publicly available databases of miRNA sponge interactions have emerged, listing both predicted [46–48] and experimentally confirmed [46, 49] interactions.

In this chapter, we show different levels of regulation between lncRNAs and miRNAs. Then, we describe step by step the recent approach proposed by Paci et al. [50] for the discovery of ceRNA–miRNA interactions.

2 LncRNA-Derived MiRNAs

An interesting aspect of miRNA and lncRNA research is the evidence that several lncRNAs are host genes of miRNAs, that is, many overlapping transcripts exist (Fig. 2). The lncRNA-derived miRNAs are an example of a complex gene regulatory network [51]. A reciprocal regulation of these two types of RNAs can lead to strong cellular effects, in terms of posttranscriptional regulation and protein expression.

Recent works showed that lncRNA-derived miRNAs are involved in development but also in human diseases such as cancer [51, 52].

Mangiavacchi et al. [51] showed that the lncRNA linc-223 had a crucial role in acute myeloid leukemia (AML). The authors discovered that the alternative production of miR-223 and linc-223 is finely regulated during monocytic differentiation. They furthermore demonstrated that endogenous linc-223 localizes in the cytoplasm and acts as a competing endogenous RNA for miR-125-5p, an oncogenic miRNA in leukemia. In particular, they showed that linc-223 directly binds to miR-125-5p and its knockdown increases the repressing activity of miR-125-5p. This effect, resulted in the downregulation of miR-125-5p target interferon regulatory factor 4 (IRF4), involved in the inhibition of the oncogenic activity of miR-125-5p in vivo. Furthermore, data from primary AML samples showed significant downregulation of linc-223 in different AML subtypes. These findings indicate that the newly identified lncRNA linc-223 may have an important role in myeloid differentiation and leukemogenesis by cross talk with IRF4 mRNA and miR-125-5p.

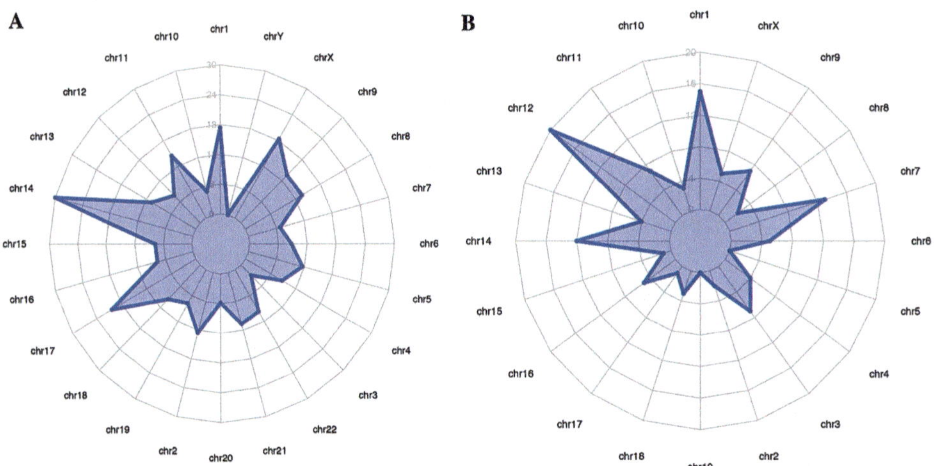

Fig. 2 Radar plot of chromosome distribution. Chromosome distribution of miRNAs within lncRNAs for humans (**a**) and mice (**b**). The length of the lines indicates the number of miRNAs within lncRNAs for each chromosome

For the purpose of this chapter we mapped the chromosomal locations of human and murine miRNAs to show the landscape of lncRNA-derived miRNAs. We retrieved the chromosomal locations of miRNAs from miRBase (http://www.mirbase.org/) [4] and the coordinates of lncRNAs from Gencode (http://www.gencodegenes.org/) [53]. We considered all miRNAs located within lncRNAs loci. Taking into consideration the strand specificity, we obtained a total number of 180 lncRNAs and 256 miRNAs for human. For mouse, we obtained 69 lncRNAs and 113 miRNAs. When we look at the chromosomal distribution of these overlaps, we can see that some chromosomes have a high number of overlaps for both human and mouse. Looking at the specific genomic positions, we find that this high number is related to the presence of miRNA clusters. Some published examples of these miRNA clusters are the imprinted genomic regions in the chromosome 12qF1 in mouse and 14q32 in human. These regions contain a large number of imprinted miRNAs that are conserved in mammals, seem to be involved in development and are highly expressed in the placenta and the embryo, whereas in the adult the expression is limited to the brain [54].

Another example of an imprinted genomic region containing miRNAs, as well as the H19 lncRNA, is the H19-miR-675 axis. In mice, these genes are located at the chromosomal position 7qF5, while in human they are within 11p15. The imprinted region H19-miR-675 has been reported to be deregulated in pediatric and adult cancer [52] and it has been shown that ncRNAs in this region can act as tumor suppressor or oncogene. These results underline the importance of the epigenetic control in cancer and at the same time highlight the potential diagnostic impact of these genomic regions.

3 LncRNAs as ceRNAs

Recent studies have shown that lncRNAs may have a key regulatory role linked not only to their secondary structure [55–64] but also to their primary structure (i.e., their nucleotide sequence). Indeed, increasing experimental evidence supports the hypothesis that lncRNAs may exploit ceRNA activity [41, 65–75].

The first experimental evidence of lncRNAs acting as ceRNAs in mammalian cells has been found in a wide variety of cancers by Poliseno et al. [44]. The authors investigated the functioning of pseudogenes (i.e., degenerate copies of genes that mostly originate from DNA duplication or retrotransposition of cellular RNAs) as miRNA sponges of their ancestral genes: the pseudogene PTENP1 competes with its homologous gene PTEN for shared miRNAs (i.e., miR-17, miR-19b, miR-20a, miR-21, miR-26, and miR-214 family), influencing the PI3K signaling cascade and subsequently acting on cell proliferation in prostate cancer [44]. In addition, the authors found that pairs FOXO3B/FOXO3 and KRASIP/KRAS of pseudogene/ancestral gene functioning as a miR-182- and miR-143/let-7-sponge, respectively [42].

Other lncRNAs functioning as ceRNAs can be also observed in human and mouse muscle cells [76], where the long intergenic noncoding RNA (lincRNA) linc-MD1 controls muscle differentiation by targeting miR-133 and miR-135 to regulate the expression of MAML1 and MEF2C. Wang et al. [77] found that the linc-RoR acts as a miR-145-sponge. miR-145 affects the expression of core transcription factors such as NANOG, OCT4, and SOX2. Those factors play a key role in cell pluripotency and self-renewing of human embryonic stem cells. Moreover, Fan et al. [43] observed that the thyroid-specific lncRNA PTCSC3 can act as ceRNA by targeting miR-574-5p in human thyroid cancer. Kallen et al. [78] demonstrated that the H19 lncRNA modulates the let-7 miRNAs family availability by acting as a molecular sponge and causing precocious muscle differentiation.

4 circRNAs as ceRNAs

Most recently, also the newly appreciated circular RNAs (circRNAs) were found to act as miRNA sponges [40, 79, 80]. circRNAs are a class of noncoding RNAs derived mostly from a noncanonical form of alternative splicing, whereby the exon ends are joined to form a continuous loop [81–83].

circRNAs are much more stable than linear transcripts as they are more resistant to exonuclease [79]. In view of their higher stability with respect to that of linear transcripts, circRNAs enable a more efficient suppression of miRNA activity.

The first circRNA was discovered over two decades ago and it is encoded the testis-specific cirRNA Sry (sex-determining region Y) [84]. Recently, circRNAs have gained a great interest as a number of recent studies demonstrated their widespread and abundant expression in eukaryotes [38, 79, 80, 82, 85].

Although the general function of most circRNAs remains unknown, until now three circRNAs have been experimentally shown to act as miRNA sponges in mammals [40]: the already mentioned testis-specific cirRNA Sry, which serves as sponge for miR-138 in mouse when it is overexpressed [84]; the circular CDR1as transcript (also known as cIRS-7), which has been identified as a miR-7 sponge in the central nervous system [86]; the transcript circITCH that controls the level of itchy E3 ubiquitin protein ligase (ITCH) by sponging miR-7, miR-17, and miR-214 in esophageal squamous cell carcinoma (ESCC) [87].

5 Databases of ceRNA–miRNA Interactions

In view of the increasing interest in miRNA sponge interactions, several databases collecting these interactions, both experimentally validated and computationally predicted, were developed. We next review the most common databases of ceRNA–miRNA.

The first miRNA sponge interaction database was ceRDB [46]. ceRDB allows users to predict miRNA sponges for a specific mRNA target by evaluating the co-occurrence of miRNA response elements in the 3′UTR sequence of mRNAs. However, the putative interactions may not be very reliable since the co-occurrence database used, TargetScan v5.2, is outdated.

Another database of miRNA sponge interactions is starBAse [47] that utilizes large-scale CLIP-Seq data (HITS-CLIP, PAR-CLIP, iCLIP)—where CLIP refers to a method to purify protein-RNA complexes with the use of ultraviolet cross-linking and immunoprecipitation [88]—of 108 datasets from 37 studies and experimental results providing physical binding information of miRNA–mRNA, miRNA–lncRNA, miRNA–circRNA, miRNA–pseudogene, and miRNA–sncRNA binding. In order to evaluate if a miRNA sponge pair shares significant common mRNAs, starBAse uses a hypergeometric test.

The database lnCeDB [48] contains human lncRNAs which can potentially act as miRNA sponges. The miRNA–mRNA interactions in lnCeDB are predicted by using TargetScan [89] while the miRNA–lncRNA interactions are either retrieved from miR-code [90], or predicted by lnCeDB's own algorithm. lnCeDB not only allows users to browse lncRNA–mRNA pairs sharing the same miRNAs but also compares the expression data of that pair in 22 human tissues.

LncACTdb (lncRNA-associated competing triplets database) is a database containing 5119 experimentally supported and over 530,000 computationally predicted lncRNA–miRNA–gene interactions, which are obtained by integrating heterogeneous data from many in silico target prediction studies, Argonaute-CLIP experiments, and RNA-seq expression profiles [91].

HumanViCe [92] is a comprehensive database that contains a vast number of coding and noncoding RNAs acting as potential miRNA sponges in virus-infected human cells, where the putative human miRNA targets on human protein-coding transcripts were predicted by existing miRNA-target prediction databases.

The first experimentally validated ceRNA–miRNA interaction database is miRSponge [49], which collects 185 unique miRNA sponge interactions in 11 species by manually curating scientific literature. This database is extremely useful tool to verify computational predictions.

Finally, a freely accessible repository that greatly assists the miRNA research community is miRWalk2.0, a comprehensive archive of predicted and experimentally verified miRNA–target interactions [93]. In particular, miRWalk2.0 combines the information of miRNA binding sites within the complete sequence of a gene with the results of existing miRNA-target prediction databases such as DIANA-microT [94], miRanda [95], PicTar [96], PITA [97], and also provides experimentally verified miRNA–target interactions obtained via an automated text-mining search and data from existing resources such as miRTarBase [98].

Obviously, on a case-by-case basis, users can choose to consider a single independent database or combine multiple databases to identify candidate miRNA sponge interactions. Notably, the user can benefit by integrating all the information contained in different databases, conveniently incorporating missing data. However, in order to take more control over the prediction/validated interactions data, the user has to take into account that retrieving and integrating information from multiple sources can be affected by redundancy and requires general familiarity with the database contents and structure, with their query language, and more.

6 Computational Approaches for Identifying ceRNA–miRNA Interactions

In order to analyze and predict the behaviors of the ceRNA regulatory mechanism, different computational approaches have been developed. Such methods can be classified into: pair-wise correlation-based methods, partial association methods, and mathematical modeling approaches [99].

Pairwise correlation-based methods are based on the principle that the expression levels of pairs of RNAs that compete for the same miRNA are positively correlated [33, 44]. Such a principle stems from the observation of the titration mechanism, which states that the increase (decrease) of the competing RNA concentrations of a miRNA target decreases (increases) the availability of the miRNA, thus relieving the miRNA repression on its target RNA (that in turn acts as a ceRNA). As a result, the expression levels of the two ceRNAs rise or decrease together, showing a positive correlation (Fig. 1).

Methods belonging to this class share the same procedure: first they search for all pairs of RNAs that share the same MREs, then they perform a hypergeometric test to calculate the significance of sharing miRNAs, and finally they predict the positively correlated pairs as miRNA sponges [100–104]. In particular, Zhou et al. [100] built the miRNA sponge interaction network in human breast cancer using matched miRNA and gene expression data, and, by performing a survival analysis, they found that the hub nodes (i.e., nodes with the number of incoming and outgoing edges exceeding 5 [105]) are good candidate biomarker in breast cancer. Similarly, Xu et al. [101] inferred the miRNA sponge interactions landscape across 20 cancer types, identifying both cancer-specific and pancancer interactions. Shao et al. [102] identified dysregulated ceRNA–miRNA interactions in lung adenocarcinoma by integrating ceRNA expression levels and miRNA–target interactions. Finally, Chiu et al. [104] investigated the optimal conditions of the miRNA sponge regulation mechanism in various cancer types (e.g., glioblastoma, ovarian, and lung carcinoma) by combining ceRNA expression profiles and putative miRNA–target interactions.

Partial association methods take into account both miRNAs and ceRNA expression levels computing either the mutual information [103, 106] or the partial correlation [50]. In particular, Sumazin et al. [106] investigated ceRNAs activity in human glioblastoma, driven by the a priori information on putative/validated pairs of RNAs sharing a statistically significant number of common miRNAs. Specifically, they combined expression data of RNA–RNA pairs sharing a significant overlap of common miRNAs with predicted miRNA–target regulatory interactions. Then, they estimated the difference between the mutual information and conditional mutual information to identify the RNA–miRNA–RNA triplets. Chiu et al. [103] proposed a prediction method, which they tested on breast cancer data, that allows to simultaneously identify both miRNA–target and miRNA-mediated sponge interaction networks. Paci et al. [50] developed a purely data-driven approach focused on the identification of new putative lncRNAs acting as ceRNAs by using expression data of breast invasive carcinoma available at The Cancer Genome Atlas (TCGA) [107, 108].

Mathematical modeling approaches exploit deterministic or stochastic models to analyze and predict the behavior of ceRNA regulatory networks [109]. Deterministic models exploit the network connectivity information and make use of the kinetic parameters characterizing the biochemical reactions in order to determine how the system changes in time and space under external stimulation. Each biological network is affected by stochastic components. However, when the number of involved molecules of each species is quite large, the law of mass action can be used to accurately calculate the change in concentrations, and little or no stochastic effect is observable. Conversely, when the number of molecules is small, significant stochastic effects may be seen and then it is preferable to choose a stochastic model. Examples of mathematical modeling approaches aiming to quantitatively understand ceRNA–miRNA interactions networks can be found in [110, 111]. Here, a mass-action model is used to determine the optimal conditions of miRNA sponge activity in silico. In [112] the authors proposed a stochastic model to analyze the equilibrium and out-of-equilibrium properties of a network of M miRNAs interacting with N mRNA targets in terms of a titration mechanism. More recently, Yuan et al. [113] performed a model-quantitative analysis for a miRNA sponge interaction system and validated their computational results by using synthetic gene circuits in human embryonic kidney 293 cells.

7 Case Study: Algorithm for Identifying ceRNA–miRNA Interactions in Breast Cancer Data

A recent review [99] reported a comparison study of the widespread computational methods for identifying ceRNA–miRNA interactions. Among these methods the algorithm proposed by Paci et al. [50] resulted as the best one in terms of the percentage of discovered miRNA sponge interactions associated with breast invasive carcinoma. We here present in detail such a computational analysis that aims to identify putative lncRNAs acting as miRNAs sponge in breast cancer.

In this study, the authors used normalized level 3 RNA- and miRNA-sequencing expression data of breast invasive carcinoma from IlluminaHiSeq platform that were retrieved from TCGA [107, 108].

The study comprised 72 samples for which the complete sets of tumor and matched healthy profiles (for both RNA-seq and miRNA-seq data) were available. Entries with more than 10% of missing values were filtered out. Coding versus noncoding RNAs were separated based on entrez gene identifiers and human annotation obtained from NCBI.

The analysis was restricted to those mRNAs with an available 3′-UTR sequence at least equal to 500 nt in the curated UTRdb database [114]. Altogether, a total of 10,492 mRNAs, 311 miRNAs, and 833 lncRNAs, were analyzed in [50].

The computational model developed to analyze these data is based on three hypotheses:

1. *The RNAs competing for the same miRNA are characterized by a highly positive Pearson correlation*
   The top-correlated mRNA–lncRNA pairs in healthy tissue and cancer data sets were selected by setting the correlation threshold to the 99th percentile of the corresponding overall Pearson correlation distribution in both cases.
2. *The interaction between the RNAs competing for the same miRNA is indirect, i.e., mediated by miRNA.*
   To investigate the scenario in which specific miRNAs may mediate the interactions of the top-correlated mRNA–lncRNA pairs, the authors applied a well-established tool of multivariate analysis (namely, the partial correlation) to each selected

mRNA/lncRNA pair with respect to each miRNA in their dataset. In general, the partial correlation measures the extent to which an observed correlation between two variables X and Y (here, the expression profiles of a mRNA and a lncRNA) relies on the presence of a third controlling variable Z (here, the expression profile of a miRNA) and it is computed as:

$$\rho_{XY|Z} = \frac{\rho_{XY} - \rho_{XZ}\rho_{ZY}}{\sqrt{1-\rho_{XZ}^2}\sqrt{1-\rho_{ZY}^2}}$$

where ρ_{XY} is the Pearson's correlation. Then, the *sensitivity correlation* S was defined as:

$$S = \rho_{XY} - \rho_{XY|Z}$$

The XYZ triplets with $S > 0.3$, corresponding to a drop of about the 30% in the correlation between XY when Z is removed, were selected. The sensitivity distribution of the top-correlated mRNA–lncRNA pairs (XY) is plotted removing one miRNA (Z) molecule at time.

3. *The RNAs competing for the same miRNA harbor one or more MREs for the miRNA that they sponge.*
   A seed match analysis was performed in order to select only those mRNA/lncRNA/miRNA triplets that are enriched in binding sites of the shared miRNA (hypergeometric test with p-value <0.01, after adjustment for multiple testing performed with the Benjamini–Hochberg method).

   The minimal pairing requirement to predict an miRNA target recognition is a perfect match to positions 2 to 7 (6-mer miRNA seed) at the 5′-end of the mature miRNA sequence [89]. The miRNA seed sequences were obtained by mapping TCGA miRNA identifiers to miRBase [4]. Complementary DNA (cDNA) sequences (i.e., without introns) for lncRNAs were obtained querying the Ensembl data portal through its R/Bio-conductor interface provided by the package biomaRt [4, 115] and by using Entrez gene identifiers. For each 3′-UTR sequence included in the dataset analyzed in [50], all the occurrences matching the reverse-complement of the 6-mer seed for the miRNAs analyzed were recorded. Similarly, for each lncRNA included in the dataset, analyzed in [50], all occurrences of short sites matching the reverse-complement of a miRNA seed in the entire transcript sequence were stored. The lists of coding and noncoding RNA identifiers used to retrieve corresponding sequences were built based on gene annotations obtained from NCBI.

Integrating the results of multivariate statistical analysis and seed match analysis, the so-called miRNA-mediated interactions network (MMI-network), was built both in healthy and cancer tissues [50]. Nodes in the networks represent mRNAs and lncRNAs with highly correlated expression profiles while edges represent miRNAs mediating their interactions. Concretely, linked nodes are required to meet three conditions: (1) matching high values of the Pearson correlation coefficients between their expression profiles (i.e., $\rho > 0.7$ in normal and $\rho > 0.4$ in cancer); (2) matching high values of the sensitivity correlation values ($S > 0.3$); (3) sharing binding sites for miRNAs (6-mer miRNA seed match).

The study in [50] revealed the existence of a complex regulatory network in healthy tissue samples that appears to be missing in tumor samples (and vice versa). In particular, the normal MMI-network (1738 nodes and 32,375 edges) is marked by a clear segregation into two internally well connected components: a larger one (1354 nodes and 31,417 edges) mainly dominated by the mir-200 family and a smaller one (378 nodes and 954 edges) mainly controlled by mir-452. Notably, in the whole normal MMI-network, the lncRNA PVT1 with its 2169 edges represents the first hub, it is connected to 753 different mRNAs (about 50% of total mRNAs in the network) and the mir-200 family members mediate over 80% of these interactions. Also for the similarly constructed cancer MMI-network (415 nodes and 1103 edges), a clear segregation into two components was observed, yet marked by fewer interactions and nodes compared to the normal case. Indeed, the larger subnetwork (mainly controlled by mir-150) here is composed by 383 nodes and 1070 edges whereas the smaller one is composed of only 20 nodes and 26 edges. Two lncRNAs—MEG3 (Maternally Expressed Gene 3) and KIAA0125—compete for the role of the first hub and regulate the expression of the almost totality of the mRNAs in the cancer-MMI-network, by antagonizing mir-379 and mir-150, respectively.

To summarize, the computational analysis proposed by Paci et al. [50] highlighted a marked rewiring in the ceRNA program documented by its "on/off" switch from healthy to cancerous breast tissues, and vice versa (i.e., RNA transcripts acting as ceRNA in healthy tissues were not found as ceRNA in cancer tissues and vice versa). This mutually exclusive activation confers an interesting character to ceRNAs as potential oncosuppressive, or oncogenic, protagonists in human cancer. At the heart of this phenomenon is the oncogene PVT1, which has received a great amount of attention in recent studies [25, 29, 116–133]. PVT1 switches from being the first of the hubs in the normal MMI-network to fall outside the list of nodes of the cancer network. In the healthy network, PVT1 revealed a net binding preference toward the miR-200 family [50], which antagonizes to regulate the expression of hundreds of mRNAs that are known to be related to cancer development and progression (e.g., GATA3 [134], TP53, TP63, and TP73 [135]).

8 Conclusions

In this chapter, we present recent advances in the field of ncRNAs and in particular the cross talk between lncRNAs and miRNAs. A particular emphasis is given to ceRNA–miRNA interactions, presenting a recent algorithm proposed by Paci and colleagues [50]. They computationally showed a remarkable rewiring in the ceRNA program between normal and pathological breast tissue, supported by the "on/off" switch from normal to cancer condition (and vice versa) of the ceRNA regulatory networks. The rationale behind this rewiring and the specific conditions required for a ceRNA–miRNA interaction to occur are still unknown, but the following hypotheses can be formulated: (1) an exon skipping mechanism, i.e., the presence of alternative transcription start sites causes the skipping of exons where the MREs reside, which could lead to a preferential expression in tumor tissue of some isoforms that lack the binding sites required for a given miRNA sponge; (2) a mechanism of titration, i.e., large variations in the ceRNA expression levels can overcome, or relieve, the repression of miRNA on its competitors, or similarly, overexpression of a miRNA can abolish the competition between the two transcripts. Recently, several studies [33, 136] have stressed the importance of the relative concentration of RNA molecules that participate in the sponge mechanism but further works are needed to explore this result, and to understand in deep the ceRNA–miRNA interactions and their role in cancer and other diseases.

References

1. Bartel DP (2004) MicroRNAs: genomics, biogenesis, mechanism, and function. Cell 116(2):281–297
2. Bartel DP (2009) MicroRNAs: target recognition and regulatory functions. Cell 136(2):215–233
3. Filipowicz W, Bhattacharyya SN, Sonenberg N (2008) Mechanisms of post-transcriptional regulation by microRNAs: are the answers in sight? Nat Rev Genet 9(2):102–114
4. Kozomara A, Griffiths-Jones S (2014) miRBase: annotating high confidence microRNAs using deep sequencing data. Nucleic Acids Res 42(Database issue):D68–D73
5. Calin GA et al (2002) Frequent deletions and down-regulation of micro- RNA genes miR15 and miR16 at 13q14 in chronic lymphocytic leukemia. Proc Natl Acad Sci U S A 99(24):15524–15529
6. Cimmino A et al (2005) miR-15 and miR-16 induce apoptosis by targeting BCL2. Proc Natl Acad Sci U S A 102(39): 13944–13949
7. Calin GA et al (2008) MiR-15a and miR-16-1 cluster functions in human leukemia. Proc Natl Acad Sci U S A 105(13):5166–5171
8. Hayashita Y et al (2005) A polycistronic microRNA cluster, miR-17-92, is overex-

pressed in human lung cancers and enhances cell proliferation. Cancer Res 65(21):9628–9632
9. Mavrakis KJ et al (2010) Genome-wide RNA-mediated interference screen identifies miR-19 targets in Notch-induced T-cell acute lymphoblastic leukaemia. Nat Cell Biol 12(4):372–379
10. Qureshi IA, Mattick JS, Mehler MF (2010) Long non-coding RNAs in nervous system function and disease. Brain Res 1338:20–35
11. Nagano T, Fraser P (2011) No-nonsense functions for long noncoding RNAs. Cell 145(2):178–181
12. Clark MB, Mattick JS (2011) Long noncoding RNAs in cell biology. Semin Cell Dev Biol 22(4):366–376
13. Wang KC, Chang HY (2011) Molecular mechanisms of long noncoding RNAs. Mol Cell 43(6):904–914
14. Gibb EA et al (2011) The functional role of long non-coding RNA in human carcinomas. Mol Cancer 10(1):38–55
15. Prensner JR, Chinnaiyan AM (2011) The emergence of lncRNAs in cancer biology. Cancer Discov 1(5):391–407
16. Moran VA, Perera RJ, Khalil AM (2012) Emerging functional and mechanistic paradigms of mammalian long non-coding RNAs. Nucleic Acids Res 40(14):6391–6400
17. Tano K, Akimitsu N (2012) Long non-coding RNAs in cancer progression. Front Genet 3:219
18. Tang J-Y et al (2013) Long Noncoding RNAs-related diseases, cancers, and drugs. Scientific World Journal 2013:943539
19. Li X et al (2014) LncRNAs: insights into their function and mechanics in underlying disorders. Mut Res 762:1–21
20. Fatica A, Bozzoni I (2014) Long non-coding RNAs: new players in cell differentiation and development. Nat Rev Genet 15(1):7–21
21. Dey BK, Mueller AC, Dutta A (2014) Long non-coding rnas as emerging regulators of differentiation, development, and disease. Transcription 5(4):e944014
22. Yang G, Lu X, Yuan L (2014) LncRNA: a link between RNA and cancer. Biochim Biophys Acta 1839(11):1097–1109
23. Morlando M et al (2014) The role of long noncoding RNAs in the epigenetic control of gene expression. ChemMedChem 9(3):505–510
24. Hansji H et al (2014) Keeping abreast with long non-coding RNAs in mammary gland development and breast cancer. Front Genet 5:379
25. Iden M et al (2016) The lncRNA PVT1 contributes to the cervical cancer phenotype and associates with poor patient prognosis. PLoS One 11(5):e0156274
26. Parasramka MA et al (2016) Long non-coding RNAs as novel targets for therapy in Hepatocellular Carcinoma. Pharmacol Ther 161:67–78
27. Shi Q, Yang X (2016) Circulating microRNA and long noncoding RNA as biomarkers of cardiovascular diseases. J Cell Physiol 231(4):751–755
28. Liu F-T et al (2016) Long noncoding RNA ANRIL: a potential novel prognostic marker in cancer A meta-analysis. Minerva Med 107(2):77–83
29. Tseng Y-Y et al (2014) PVT1 dependence in cancer with MYC copy-number increase. Nature 512:82
30. Tseng YY, Bagchi A (2015) The PVT1-MYC duet in cancer. Mol Cell Oncol 2(2):e974467
31. Liu Y et al (2012) A genetic variant in long non-coding RNA HULC contributes to risk of HBV-related hepatocellular carcinoma in a Chinese population. PLoS One 7(4):e35145
32. Ebert MS, Neilson JR, Sharp PA (2007) MicroRNA sponges: competitive inhibitors of small RNAs in mammalian cells. Nat Methods 4(9):721–726
33. Salmena L et al (2011) A ceRNA hypothesis: the Rosetta Stone of a hidden RNA language? Cell 146(3):353–358
34. Tay Y, Rinn J, Pandolfi PP (2014) The multilayered complexity of ceRNA crosstalk and competition. Nature 505(7483):344–352
35. Ergun S, Oztuzcu S (2015) Oncocers: ceRNA-mediated cross-talk by sponging miRNAs in oncogenic pathways. Tumor Biol 36(5):3129–3136
36. Qi X et al (2015) ceRNA in cancer: possible functions and clinical implications. J Med Genet 52(10):710–718
37. Kagami H et al (2015) Determining associations between human diseases and non-coding RNAs with critical roles in network control. Sci Rep 5:14577
38. Guo L-L et al (2015) Competing endogenous RNA networks and gastric cancer. World J Gastroenterol 21(41):11680–11687

39. Yang C et al (2016) Competing endogenous RNA networks in human cancer: hypothesis, validation, and perspectives. Oncotarget 7(12):13479–13490
40. Thomson DW, Dinger ME (2016) Endogenous microRNA sponges: evidence and controversy. Nat Rev Genet 17(5):272–283
41. Wang J et al (2010) CREB up-regulates long non-coding RNA, HULC expression through interaction with microRNA-372 in liver cancer. Nucleic Acids Res 38(16):5366–5383
42. Poliseno L, Pandolfi PP (2015) PTEN ceRNA networks in human cancer. Methods 77:41–50
43. Fan M et al (2013) A long non-coding RNA, PTCSC3, as a tumor suppressor and a target of miRNAs in thyroid cancer cells. Exp Ther Med 5(4):1143–1146
44. Poliseno L et al (2010) A coding-independent function of gene and pseudogene mRNAs regulates tumour biology. Nature 465(7301):1033–1038
45. Franco-Zorrilla JE et al (2007) Target mimicry provides a new mechanism for regulation of microRNA activity. Nat Genet 39(8):1033–1037
46. Sarver AL, Subramanian S (2012) Competing endogenous RNA database. Bioinformation 8(15):731–733
47. Li J-H et al (2014) starBase v2. 0: decoding miRNA-ceRNA, miRNA-ncRNA and protein–RNA interaction networks from large-scale CLIP-Seq data. Nucleic Acids Res 42:D92
48. Das S et al (2014) ln Ce DB: database of human long noncoding RNA acting as competing endogenous RNA. PLoS One 9(6):e98965
49. Wang P et al (2015) MiRSponge: a manually curated database for experimentally supported miRNA sponges and ceRNAs. Database 2015:pii: bav098
50. Paci P, Colombo T, Farina L (2014) Computational analysis identifies a sponge interaction network between long non-coding RNAs and messenger RNAs in human breast cancer. BMC Syst Biol 8:83
51. Mangiavacchi A et al (2016) The miR-223 host non-coding transcript linc-223 induces IRF4 expression in acute myeloid leukemia by acting as a competing endogenous RNA. Oncotarget 7:60155
52. Matouk IJ et al (2015) The non-coding RNAs of the H19-IGF2 imprinted loci: a focus on biological roles and therapeutic potential in lung cancer. J Transl Med 13:113
53. Harrow J et al (2012) GENCODE: the reference human genome annotation for The ENCODE Project. Genome Res 22(9):1760–1774
54. Seitz H et al (2004) A large imprinted microRNA gene cluster at the mouse Dlk1-Gtl2 domain. Genome Res 14(9): 1741–1748
55. Engreitz JM, Ollikainen N, Guttman M (2016) Long non-coding RNAs: spatial amplifiers that control nuclear structure and gene expression. Nat Rev Mol Cell Biol 17(12):756–770
56. Zhang X et al (2010) Maternally expressed gene 3 (MEG3) noncoding ribonucleic acid: isoform structure, expression, and functions. Endocrinology 151(3):939–947
57. Liang JC, Bloom RJ, Smolke CD (2011) Engineering biological systems with synthetic RNA molecules. Mol Cell 43(6):915–926
58. Saxena A, Carninci P (2011) Long non-coding RNA modifies chromatin. Bioessays 33(11):830–839
59. Novikova IV, Hennelly SP, Sanbonmatsu KY (2012) Structural architecture of the human long non-coding RNA, steroid receptor RNA activator. Nucleic Acids Res 40(11):5034–5051
60. Mortimer SA, Kidwell MA, Doudna JA (2014) Insights into RNA structure and func tion from genome-wide studies. Nat Rev Genet 15(7):469–479
61. Mercer TR, Mattick JS (2013) Structure and function of long noncoding RNAs in epigenetic regulation. Nat Struct Mol Biol 20(3):300–307
62. Somarowthu S et al (2015) HOTAIR forms an intricate and modular secondary structure. Mol Cell 58(2):353–361
63. Fiscon G et al (2015) A new procedure to analyze RNA Non-branching Structures. BSP Curr Bioinformatics 9(5):242–258
64. Fiscon G, Iannello G, Paci P (2016) A perspective on the algorithms predicting and evaluating the RNA secondary structure. J Genet Genome Res 3:023
65. Tay Y et al (2011) Coding-independent regulation of the tumor suppressor PTEN by competing endogenous mRNAs. Cell 147(2):344–357
66. Karreth FA et al (2011) In vivo identification of tumor-suppressive PTEN ceRNAs in an

oncogenic BRAF-induced mouse model of melanoma. Cell 147(2):382–395
67. Wang L et al (2013) Pseudogene OCT4-pg4 functions as a natural micro RNA sponge to regulate OCT4 expression by competing for miR-145 in hepatocellular carcinoma. Carcinogenesis 34(8):1773–1781
68. Huarte M (2015) The emerging role of lncRNAs in cancer. Nat Med 21(11): 1253–1261
69. Marques AC et al (2012) Evidence for conserved post-transcriptional roles of unitary pseudogenes and for frequent bifunctionality of mRNAs. Genome Biol 13(11):1
70. Johnsson P et al (2013) A pseudogene long-noncoding-RNA network regulates PTEN transcription and translation in human cells. Nat Struct Mol Biol 20(4):440–446
71. Liu Q et al (2013) LncRNA loc285194 is a p53-regulated tumor suppressor. Nucleic Acids Res 41(9):4976–4987
72. Yu G et al (2014) Pseudogene PTENP1 functions as a competing endogenous RNA to suppress clear-cell renal cell carcinoma progression. Mol Cancer Ther 13(12):3086–3097
73. Xie J et al (2015) Microarray analysis of lncRNAs and mRNAs co-expression network and lncRNA function as cerna in papillary thyroid carcinoma. J Biomater Tissue Eng 5(11):872–880
74. Zhou X et al (2015) The interaction between MiR-141 and IncRNA-H19 in regulating cell proliferation and migration in gastric cancer. Cell Physiol Biochem 36(4):1440–1452
75. Zheng L et al (2015) The 3′ UTR of the pseudogene CYP4Z2P promotes tumor angiogenesis in breast cancer by acting as a ceRNA for CYP4Z1. Breast Cancer Res Treat 150(1):105–118
76. Cesana M et al (2011) A long noncoding RNA controls muscle differentiation by functioning as a competing endogenous RNA. Cell 147(2):358–369
77. Wang Y et al (2013) Endogenous miRNA sponge lincRNA-RoR regulates Oct4, Nanog, and Sox2 in human embryonic stem cell self-renewal. Dev Cell 25(1):69–80
78. Kallen AN et al (2013) The imprinted H19 lncRNA antagonizes let-7 microRNAs. Mol Cell 52(1):101–112
79. Jeck WR, Sharpless NE (2014) Detecting and characterizing circular RNAs. Nat Biotechnol 32(5):453
80. Memczak S et al (2013) Circular RNAs are a large class of animal RNAs with regulatory potency. Nature 495(7441):333–338
81. Zlotorynski E (2015) Non-coding RNA: circular RNAs promote transcription. Nat Rev Mol Cell Biol 16:206
82. Rybak-Wolf A et al (2015) Circular RNAs in the mammalian brain are highly abundant, conserved, and dynamically expressed. Mol Cell 58(5):870–885
83. Memczak S et al (2015) Identification and characterization of circular RNAs as a new class of putative biomarkers in human blood. PLoS One 10(10):e0141214
84. Capel B et al (1993) Circular transcripts of the testis-determining gene Sry in adult mouse testis. Cell 73(5):1019–1030
85. Hansen TB et al (2013) Natural RNA circles function as efficient microRNA sponges. Nature 495(7441):384–388
86. Hansen TB (2013) J.o. Kjems, rgen, and C.K. Damgaard, Circular RNA and miR-7 in cancer. Cancer Res 73(18):5609–5612
87. Li F et al (2015) Circular RNA ITCH has inhibitory effect on ESCC by suppressing the Wnt/beta-catenin pathway. Oncotarget 6(8):6001–6013
88. Ule J et al (2005) CLIP: a method for identifying protein-RNA interaction sites in living cells. Methods 37(4):376–386
89. Lewis BP, Burge CB, Bartel DP (2005) Conserved seed pairing, often flanked by adenosines, indicates that thousands of human genes are microRNA targets. Cell 120(1):15–20
90. Jeggari A, Marks DS, Larsson E (2012) miRcode: a map of putative microRNA target sites in the long non-coding transcriptome. Bioinformatics 28(15):2062–2063
91. Wang P et al (2015) Identification of lncRNA-associated competing triplets reveals global patterns and prognostic markers for cancer. Nucleic Acids Res 43(7):3478–3489
92. Ghosal S et al (2014) HumanViCe: host ceRNA network in virus infected cells in human. Front Genet 5:249
93. Dweep H, Gretz N (2015) miRWalk2.0: a comprehensive atlas of microRNA-target interactions. Nat Methods 12(8):697
94. Paraskevopoulou MD et al (2013) DIANA-microT web server v5.0: service integration into miRNA functional analysis workflows. Nucleic Acids Res 41(Web Server issue):W169–W173

95. Betel D et al (2010) Comprehensive modeling of microRNA targets predicts functional non-conserved and non-canonical sites. Genome Biol 11(8):R90
96. Krek A et al (2005) Combinatorial microRNA target predictions. Nat Genet 37(5):495–500
97. Kertesz M et al (2007) The role of site accessibility in microRNA target recognition. Nat Genet 39(10):1278–1284
98. Chou CH et al (2016) miRTarBase 2016: updates to the experimentally validated miRNA-target interactions database. Nucleic Acids Res 44(D1):D239–D247
99. Le TD et al (2017) Computational methods for identifying miRNA sponge interactions. Brief Bioinform 18:577
100. Zhou X, Liu J, Wang W (2014) Construction and investigation of breast-cancer-specific ceRNA network based on the mRNA and miRNA expression data. IET Syst Biol 8(3):96–103
101. Xu J et al (2015) The mRNA related ceRNA-ceRNA landscape and significance across 20 major cancer types. Nucleic Acids Res 43(17):8169–8182
102. Shao T et al (2015) Identification of module biomarkers from the dysregulated ceRNA-ceRNA interaction network in lung adenocarcinoma. Mol Biosyst 11(11):3048–3058
103. Chiu H-S et al (2015) Cupid: simultaneous reconstruction of microRNA-target and ceRNA networks. Genome Res 25(2):257–267
104. Chiu Y-C et al (2015) Parameter optimization for constructing competing endogenous RNA regulatory network in glioblastoma multiforme and other cancers. BMC Genomics 16(4):1
105. Han JD et al (2004) Evidence for dynamically organized modularity in the yeast protein-protein interaction network. Nature 430(6995):88–93
106. Sumazin P et al (2011) An extensive microRNA-mediated network of RNA-RNA interactions regulates established oncogenic pathways in glioblastoma. Cell 147(2):370–381
107. Network CGAR et al (2013) The Cancer Genome Atlas Pan-Cancer analysis project. Nat Genet 45(10):1113–1120
108. Tomczak K et al (2015) The Cancer Genome Atlas (TCGA): an immeasurable source of knowledge. Contemp Oncol (Pozn) 19(1A):A68–A77
109. Karlebach G, Shamir R (2008) Modelling and analysis of gene regulatory networks. Nat Rev Mol Cell Biol 9(10):770–780
110. Figliuzzi M, Marinari E, De Martino A (2013) MicroRNAs as a selective channel of communication between competing RNAs: a steady-state theory. Biophys J 104(5):1203–1213
111. Ala U et al (2013) Integrated transcriptional and competitive endogenous RNA networks are cross-regulated in permissive molecular environments. Proc Natl Acad Sci 110(18):7154–7159
112. Bosia C, Pagnani A, Zecchina R (2013) Modelling competing endogenous RNA networks. PLoS One 8(6):e66609
113. Yuan Y et al (2015) Model-guided quantitative analysis of microRNA-mediated regulation on competing endogenous RNAs using a synthetic gene circuit. Proc Natl Acad Sci 112(10):3158–3163
114. Grillo G et al (2010) UTRdb and UTRsite (RELEASE 2010): a collection of sequences and regulatory motifs of the untranslated regions of eukaryotic mRNAs. Nucleic Acids Res 38(suppl 1):D75–D80
115. Kinsella RJ et al (2011) Ensembl BioMarts: a hub for data retrieval across taxonomic space. Database 2011:bar030
116. Colombo T et al (2015) PVT1: a rising star among oncogenic long noncoding RNAs. Biomed Res Int 2015:304208
117. Huppi K et al (1990) Pvt-1 transcripts are found in normal tissues and are altered by reciprocal (6; 15) translocations in mouse plasmacytomas. Proc Natl Acad Sci 87(18):6964–6968
118. Guan Y et al (2007) Amplification of PVT1 contributes to the pathophysiology of ovarian and breast cancer. Clin Cancer Res 13(19):5745–5755
119. Brooksbank C et al (2014) The European Bioinformatics Institute's data resources 2014. Nucleic Acids Res 42(Database issue):D18–D25
120. Huppi K et al (2008) The identification of microRNAs in a genomically unstable region of human chromosome 8q24. Mol Cancer Res 6(2):212–221
121. Gerstein MB et al (2007) What is a gene, post-ENCODE? History and updated definition. Genome Res 17(6):669–681
122. Lemay G, Jolicoeur P (1984) Rearrangement of a DNA sequence homologous to a cell-virus

junction fragment in several Moloney murine leukemia virus-induced rat thymomas. Proc Natl Acad Sci 81(1):38–42
123. Graham M, Adams JM, Cory S (1984) Murine T lymphomas with retroviral inserts in the chromosomal 15 locus for plasmacytoma variant translocations. Nature 314(6013): 740–743
124. Villeneuve L et al (1986) Proviral integration site Mis-1 in rat thymomas corresponds to the pvt-1 translocation breakpoint in murine plasmacytomas. Mol Cell Biol 6(5):1834–1837
125. Graham M, Adams JM (1986) Chromosome 8 breakpoint far 3′ of the c-myc oncogene in a Burkitt's lymphoma 2; 8 variant translocation is equivalent to the murine pvt-1 locus. EMBO J 5(11):2845
126. Huppi K, Siwarski D (1994) Chimeric transcripts with an open reading frame are generated as a result of translocation to the Pvt-1 region in mouse B-cell tumors. Int J Cancer 59(6):848–851
127. Hodgson G et al (2001) Genome scanning with array CGH delineates regional alterations in mouse islet carcinomas. Nat Genet 29(4):459–464
128. Meyer KB et al (2011) A functional variant at a prostate cancer predisposition locus at 8q24 is associated with PVT1 expression. PLoS Genet 7(7):e1002165
129. Chapman MH et al (2012) Whole genome RNA expression profiling of endoscopic biliary brushings provides data suitable for biomarker discovery in cholangiocarcinoma. J Hepatol 56(4):877–885
130. Wang F et al (2014) Oncofetal long noncoding RNA PVT1 promotes proliferation and stem cell-like property of hepatocellular carcinoma cells by stabilizing NOP2. Hepatology 60(4):1278–1290
131. Zhuang C et al (2015) Tetracycline-inducible shRNA targeting long non-coding RNA PVT1 inhibits cell growth and induces apoptosis in bladder cancer cells. Oncotarget 6(38):41194–41203
132. Zhou Q et al (2016) Long noncoding RNA PVT1 modulates thyroid cancer cell proliferation by recruiting EZH2 and regulating thyroid-stimulating hormone receptor (TSHR). Tumor Biol 37(3):3105–3113
133. Cui D et al (2016) Long non-coding RNA PVT1 as a novel biomarker for diagnosis and prognosis of non-small cell lung cancer. Tumor Biol 37(3):4127–4134
134. Asselin-Labat ML et al (2011) Gata-3 negatively regulates the tumor-initiating capacity of mammary luminal progenitor cells and targets the putative tumor suppressor caspase-14. Mol Cell Biol 31(22):4609–4622
135. Fridman JS, Lowe SW (2003) Control of apoptosis by p53. Oncogene 22(56):9030–9040
136. Conte F et al (2017) Role of the long noncoding RNA PVT1 in the dysregulation of the ceRNA-ceRNA network in human breast cancer. PLoS One 12(2):e0171661

Chapter 5

Methods and Tools in Genome-wide Association Studies

Anja C. Gumpinger, Damian Roqueiro, Dominik G. Grimm, and Karsten M. Borgwardt

Abstract

Many traits, such as height, the response to a given drug, or the susceptibility to certain diseases are presumably co-determined by genetics. Especially in the field of medicine, it is of major interest to identify genetic aberrations that alter an individual's risk to develop a certain phenotypic trait. Addressing this question requires the availability of comprehensive, high-quality genetic datasets. The technological advancements and the decreasing cost of genotyping in the last decade led to an increase in such datasets. Parallel to and in line with this technological progress, an analysis framework under the name of *genome-wide association studies* was developed to properly collect and analyze these data. Genome-wide association studies aim at finding statistical dependencies—or associations—between a trait of interest and point-mutations in the DNA. The statistical models used to detect such associations are diverse, spanning the whole range from the frequentist to the Bayesian setting.

Since genetic datasets are inherently high-dimensional, the search for associations poses not only a statistical but also a computational challenge. As a result, a variety of toolboxes and software packages have been developed, each implementing different statistical methods while using various optimizations and mathematical techniques to enhance the computations.

This chapter is devoted to the discussion of widely used methods and tools in genome-wide association studies. We present the different statistical models and the assumptions on which they are based, explain peculiarities of the data that have to be accounted for and, most importantly, introduce commonly used tools and software packages for the different tasks in a genome-wide association study, complemented with examples for their application.

Key words Genome-wide association studies, Missing heritability, Linkage disequilibrium, Phenotypes, Univariate mapping, Population structure correction, Genomic inflation, Multilocus mapping, Multiple hypothesis correction, Meta-analysis, GWAS tools

1 Introduction

1.1 Genome-wide Association Studies: An Overview

Genome-wide association studies (GWAS) have become a valuable tool to identify associations between genetic variants in a group of individuals and a phenotype present in these individuals. The phenotype in question can be a trait such as height, the presence or absence of a disease, the response to a drug treatment, or any

Louise von Stechow and Alberto Santos Delgado (eds.), *Computational Cell Biology: Methods and Protocols*, Methods in Molecular Biology, vol. 1819, https://doi.org/10.1007/978-1-4939-8618-7_5,

other phenotype of interest. The genetic variants used in GWAS are primarily single-nucleotide polymorphisms (SNPs), which correspond to single base-pairs in the DNA that are known to vary between individuals. The goal of a genome-wide association study is to determine which SNPs are associated with the phenotype in a statistically significant manner.

Historically, and prior to the emergence of GWAS, *linkage mapping* was a technique used to detect genetic markers that segregated within families affected by rare diseases. Linkage mapping was successful for rare and Mendelian diseases, for example, at identifying loci associated to Huntington's disease [1] and at detecting mutations in the Cystic Fibrosis Transmembrane Conductance Regulator (CFTR) gene in patients with cystic fibrosis [2]. Nevertheless, for complex diseases such as type II diabetes and schizophrenia, it is the cumulative effect of dozens or hundreds of variants throughout the entire genome—each with a small effect on the phenotype—that confer a greater risk of developing the disease. This is the reason why linkage mapping was less successful in the realm of complex diseases [3, 4], which in turn allowed GWAS to rise to prominence as a tool to identify associations between a trait and genetic variants of smaller effects. There are numerous reasons why GWAS, in contrast to linkage mapping, are better equipped to detect these associations: (a) in GWAS, hundreds of thousands to millions of SNPs are surveyed, (b) GWAS do not necessarily rely on pedigree information, and (c) GWAS have larger sample sizes than linkage mapping studies and, thus, have more power to detect associations [5].

Whether one intends to perform GWAS in plants to increase crop yields, or in livestock to identify genes associated with economically important traits such as fertility, or in humans to find SNPs associated with common diseases, this chapter serves as a guide to all the theoretical aspects of GWAS. The chapter also contains detailed protocols on how to conduct GWAS with different tools and how to overcome potential pitfalls in the analysis.

1.2 Performing GWAS

The haploid human genome comprises approximately three billion base-pairs, 3% of which show variation among individuals (estimate based on the 84.7 million reported SNPs by the 1000 Genomes project [6]). These base-pairs that vary across a population represent the single-nucleotide polymorphisms mentioned in the previous section. Endeavors such as the original HapMap project [7], and the more recent 1000 Genomes project [6] have aimed at genotyping a multitude of human genomes to detect and annotate genetic variation among individuals.

For performing a genome-wide association study, we assume that two types of data are readily available for all individuals in the study: their phenotype and genotypes. The latter can be obtained through state-of-the-art sequencing technologies [8], or through a genotyping array [9].

In a traditional (univariate) genome-wide association study, a measure of association or statistical dependence between each individual SNP and the phenotype is computed. Then a *p*-value is derived for each association score, which represents the probability of observing an association signal of the same strength or stronger under the null hypothesis of no association between the SNP and the phenotype. If the *p*-value falls below a predefined significance threshold α, commonly 0.01 or 0.05, the null hypothesis is rejected, which means that there is an association between the SNP and the phenotype. Despite the strong evidence against the null hypothesis in this case, there remains a chance of $\alpha * 100\%$ that the low *p*-value is purely due to random chance and that the detected association is therefore a false positive result. Avoiding false positive findings is among the major challenges in GWAS.

1.3 Challenges in GWAS

1.3.1 Avoiding False Positive Findings

In GWAS, typically hundreds of thousands to millions of SNPs are tested simultaneously for association with the phenotype. Since each of these hypotheses are rejected at a significance level α—typically, $\alpha = 0.05$—this multiplicity of tests can lead to the reporting of spurious associations if no correction for multiple hypothesis testing is performed. There is rich statistical literature on various methods to correct for multiple hypothesis testing [10] and the most prevalent ones, namely **family wise error rate** and **false discovery rate**, are discussed in detail later in this chapter.

Another possible source of false positive findings is the presence of confounders, such as environmental factors, population or family structure, cryptic relatedness, age, and gender. Autoimmune diseases in humans are a good example of how gender can be a confounder as approximately 80% of patients affected by an autoimmune disease are women [11]. Population structure, on the other hand, refers to differences in allele frequencies between groups of individuals in a study due to systematic ancestry differences [12], such as geographic proximity or individuals sharing the same ethnicity. Not correcting for these confounding factors may lead to false positive findings: SNPs that seem to be associated to the phenotype while they are actually associated with the confounding factor, e.g., with population structure.

Similar to the importance of avoiding false positive findings in GWAS, it is equally crucial to avoid missing true associations, the so-called false negatives. False negatives occur when the statistical signal of the marker is not strong enough to reach genome-wide significance. Possible reasons for this are (a) little evidence in the data to support the statistical association, e.g., because of a small sample size, or (b) the significance threshold being too low. In general, there is a trade-off between the number of false positives and false negatives and we defer this discussion to Subheading 2.7.1 when we address the topic of multiple-hypothesis testing correction.

1.3.2 Missing Heritability

Over the last decade, GWAS have been successfully performed on different organisms, such as *Arabidopsis thaliana* [13–15], rice [16], fruit flies [17], mice [18], and humans [19–21]. In humans, particularly in the field of autoimmune and metabolic diseases, GWAS have revealed important insights into the genetic mechanisms of disease development and progression [4]. Despite these and many other successes, the association signals detected in univariate GWAS often explain only a small fraction of the total phenotypic variability. This phenomenon has been referred to as **missing heritability** [4, 22–24].

In the literature, different strategies have been proposed to discover the missing heritability. One class of approaches aims at changing the hypothesis underlying GWAS by considering the joint effects of multiple SNPs [25]. Examples covered in this chapter are (a) the search for nonlinear SNP-SNP interactions [26–28], also known as epistasis, (b) the joint analysis of SNPs overlapping with genes [29–32], and (c) the interaction of SNPs when superimposed on a biological network [33–36].

Another class of approaches tries to alleviate the burden of multiple hypothesis testing by attempting to increase the per-hypothesis significance threshold while decreasing the number of hypothesis to be tested. Examples of these approaches that are also covered in this chapter are (a) gene-based approaches [29–32] as well as (b) methods analyzing intervals of genetic markers [37], or (c) clustering with subsequent hierarchical testing [38].

1.4 Outline of the Chapter

This chapter provides both a theoretical and practical guide to GWAS. In Subheading 2, *Methods and Definitions*, we summarize important theoretical concepts of GWAS. We present formal definitions and discuss different types of GWAS as well as the evaluation of their results. In Subheading 3, *Tools and Software*, we introduce different software tools and packages that are commonly used to perform GWAS. The presentation of each tool is accompanied by a tutorial on how to use it. We conclude the chapter by reviewing the development of GWAS over the last decade, and highlighting current challenges and future directions of research.

In addition to the contents of this chapter, we provide a virtual machine (VM) with preinstalled tools and all the necessary scripts to run the examples in Subheading 3. The blue boxes in Subheading 3 provide step-by-step protocols on how to run the scripts on the VM. The VM and sample scripts can be accessed online at: https://www.bsse.ethz.ch/mlcb/gwas. The link also provides additional details on how to install the VM and how to use the sample scripts.

2 Methods and Definitions

Before delving into the methodological aspects of GWAS, we first introduce the concept of linkage disequilibrium followed by a discussion on how genomic data are frequently encoded and preprocessed. We then cover the main statistical models for univariate association testing, i.e., when each SNP is analyzed separately. We proceed to complement the ideas presented in univariate methods by discussing more sophisticated scenarios in which methods that rely on gene-based analyses and interactions between genetic markers are considered. We conclude this theoretical section by providing details on how the results of GWAS can be evaluated and combined.

2.1 Linkage Disequilibrium

When working with genetic data, a concept of utmost importance that needs to be taken into consideration is that of **linkage disequilibrium** (LD). For a given population, two SNPs are said to be in LD when their genotypes are correlated in a way that will not arise by chance. From a biological perspective, LD between loci is determined by local recombination rates. As a result of this, older populations that went through a higher number of recombination events show different LD patterns than younger populations do. In humans, for example, LD patterns vary between different ethnic groups [7]. One of the key technological achievements that greatly facilitated GWAS was the development of SNP arrays [7, 39]. SNP arrays do not survey the entirety of SNPs in a genome, but instead rely on information about regions of SNPs in high LD. From each of these regions, a tag SNP is selected and ultimately genotyped under the assumption that, if the tag SNP is associated to the phenotype, then the tag SNP is in high LD with the presumably causal SNP [3].

2.2 Data Encoding and Preprocessing Steps

2.2.1 Encoding of Genotypes and Phenotypes

Genotype data can have different formats, depending on the genotyping platform that was used. In what is normally considered to be *raw* data, a SNP for a given individual is represented by its two alleles (for diploid cells) in the letters A, C, G, or T (*see* Fig. 1a) for each of the four nucleotides. Many of the statistical models described in this section require that the raw allele information be encoded as a numerical value. This is referred to as the **encoding** of the genotype. There exist different models in which genotypes can be encoded and each of them reflects prior assumptions about the underlying mode of inheritance, such as the additive, the dominant and the recessive encoding (*see* Fig. 1b). Of these models, the most commonly used one is the additive model, in which each SNP is represented by the number of its minor alleles, i.e., 0, 1, or 2 [40].

The phenotype can be either qualitative or quantitative. The former refers to phenotypes that have discrete class labels for each

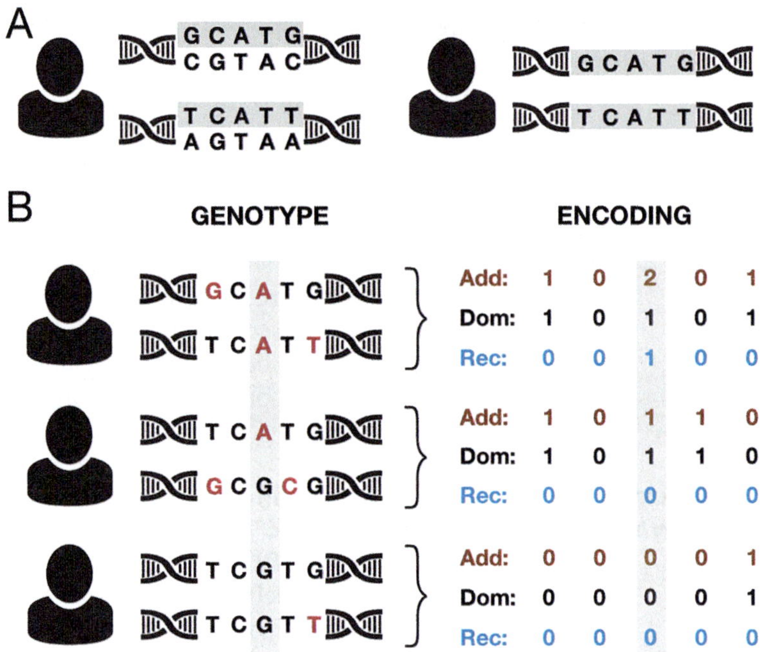

Fig. 1 Data representation in GWAS for a diploid individual. (**a**) For an individual, the genotype is represented by the alleles on two strands of DNA, one from each chromosome. (**b**) Three different encoding schemes (additive, dominant, and recessive) for SNP data. The alleles highlighted in red are the minor alleles. The additive encoding counts the number of minor alleles. The dominant encoding maps the homozygous major alleles to 0 and all other genotypes to 1. The recessive encoding maps the homozygous minor allele to 1, and all other genotypes to 0

individual, for example case/control. Quantitative phenotypes, on the other hand, are represented by a real number. Typical examples of this type of phenotype are height, body-mass index, or blood pressure [41]. The phenotype being qualitative or quantitative is an important criterion for the choice of the statistical method in a genome-wide association study.

2.2.2 Data Preprocessing and Quality Control

Before we can start testing for associations between SNPs and the phenotype of interest, certain preprocessing tasks are commonly applied to minimize the chances of reporting false results.

Transformation of Phenotypes

Some statistical methods make a particular assumption about the distribution of the phenotype and its noise. For example, when applying linear regression in GWAS, one of the assumptions of the method is that, given the genotype, the phenotype follows a Gaussian distribution. In practice, however, this assumption rarely holds. Therefore, a common preprocessing task is to transform or normalize the phenotypic data such that it follows the expected distribution of the statistical model.

Filtering Using Hardy–Weinberg Equilibrium

The Hardy–Weinberg equilibrium (HWE) is a model that allows for the prediction of genotype frequencies from one generation to the next. SNPs in GWAS are expected to be in HWE and any deviation can be assessed through a statistical test with a common threshold on the *p*-value of 1e−06 [42] (this threshold is commonly used in humans, for other organisms the consensus threshold may be different). If an SNP is found not to be in HWE, this is normally because of a sampling or a pure genotyping error. When this occurs, the SNP is removed from the analysis in a preprocessing step.

Filtering by Minor Allele Frequency

The minor allele frequency (MAF) is the frequency of the less common allele at a biallelic locus [3]. SNPs that have low MAF with values smaller than 0.05 or 0.01, the so-called rare variants, are commonly excluded from standard GWAS [5]. Unless sample sizes are very large or the effects of the rare variants are high, standard GWAS techniques are underpowered to detect associations with rare variants [5, 43]. There is an entire class of methods collectively known as burden tests that are well-suited for rare variants [44–46]. In addition to burden tests, there are other methods such as the C-alpha test [47] or SKAT [43] which have proven to be effective at analyzing rare variants. This chapter focuses on methods that utilize common variants and the reader is referred to refs. [43–47] for more details about the analysis of rare variants.

Filtering by Missing Genotypes

Another important quality control step that needs to be performed before conducting GWAS is to exclude from the analysis: (a) individuals with a large number of missing genotypes, and (b) SNPs with a high rate of missing genotypes across all individuals [42]. The first case is normally a consequence of poor DNA quality or low DNA concentration. This affects the accuracy of SNP calling algorithms, which then report large numbers of missing SNPs for the individual. In the second case, when a SNP has a low call rate, i.e., a high missing rate, it is considered of low quality and is excluded from the analysis to avoid false positives [48]. There are well-defined protocols to impute the values of missing SNPs, and we refer the interested reader to [49] for more details.

2.3 Concepts in Univariate GWAS

After quality control and preprocessing we are in a position to start the analysis of the data. We herein describe the methods that are most commonly used to test SNPs for association to the phenotype. Since all SNPs are tested independently, these methods belong to the class of single-locus or univariate GWAS [3].

2.3.1 Two Sample Tests

When analyzing a qualitative phenotype, as it is done in a case/control study, a common way to test an SNP for its association to the phenotype is to create a contingency table by counting the allele frequencies in cases and controls. This contingency table can then be used to test for association using a discrete test statistic, such as Fisher's exact test [50] or a χ^2 test [51].

2.3.2 Linear Models

Linear models are often used to test a single SNP for its association with a phenotype [42]. The underlying assumption in linear models is that the phenotype can be modeled as an additive (linear) combination of the genotype values of the SNP and (possibly) covariates, such as age or gender, each of them with certain effect sizes. Moreover, linear models contain a residual term that captures noise in the data. Let us consider a dataset with n individuals, where for each individual we have: (a) a phenotype value, (b) a genotype value for a given SNP, and (c) p covariates. Then, a linear model can be described as follows:

$$\boldsymbol{y} = \beta_0 \vec{1} + \beta_1 \boldsymbol{x}_{\mathrm{G}} + \boldsymbol{X}_{\mathrm{C}} \boldsymbol{\beta}_2^T + \boldsymbol{\epsilon} \tag{1}$$

where $\boldsymbol{y}$ represents the $n \times 1$ vector of phenotypes, $\vec{1}$ is an $n \times 1$ vector of ones, $\boldsymbol{x}_{\mathrm{G}}$ is the $n \times 1$ vector of the genotypes for a given SNP, $\boldsymbol{X}_{\mathrm{C}}$ is the $n \times p$ matrix of covariates and $\beta_0 \in \mathbb{R}$, $\beta_1 \in \mathbb{R}$, $\boldsymbol{\beta}_2 \in \mathbb{R}^{1 \times p}$ are the offset, the genotype and the covariate effects, respectively. Additionally, $\boldsymbol{\epsilon} \in \mathbb{R}^{n \times 1}$ is the vector of residuals, which are assumed to follow a known probability distribution, allowing the model parameters β_0, β_1, and $\boldsymbol{\beta}_2$ to be estimated using probabilistic techniques such as maximum likelihood estimation (MLE) [52]. The inclusion of the covariates $\boldsymbol{X}_{\mathrm{C}}$ into the model accounts for known factors that might have an influence on the phenotype, such as age or gender. It is common to reformulate Eq. 1 as:

$$\boldsymbol{y} = \boldsymbol{X} \boldsymbol{\beta}^T + \boldsymbol{\epsilon} \tag{2}$$

where $\boldsymbol{X} = \left[\vec{1}, \boldsymbol{x}_{\mathrm{G}}, \boldsymbol{X}_{\mathrm{C}}\right] \in \mathbb{R}^{n \times (2+p)}$ and $\boldsymbol{\beta} = [\beta_0, \beta_1, \boldsymbol{\beta}_2] \in \mathbb{R}^{1 \times (2+p)}$. The matrix $\boldsymbol{X}$ is referred to as the design matrix. When applying linear models to test SNPs for their association with the phenotype, an individual linear model has to be fitted for each SNP. Depending on the number of SNPs, this can be computationally expensive.

Linear Regression

A linear regression model assumes that the residuals, and therefore also the phenotype $\boldsymbol{y}$, are normally distributed given the genotype $\boldsymbol{x}_{\mathrm{G}}$, making it applicable to quantitative phenotypes. Using methods such as MLE, the parameters β_0, β_1, and $\boldsymbol{\beta}_2$ in the model Eq. 1 can be estimated from the data. Once the model is fitted, the parameter β_1 describes the effect the SNP has on the phenotype. The parameter's deviation from zero is an indicator of the effect size: the larger its absolute value is, the higher the contribution of x_G to the phenotype is. With a statistical test, it is possible to assess if the deviation from zero is statistically significant at a predefined significance level.

Logistic Regression

Logistic regression is a special case of linear models when the phenotype is qualitative, as in a case/control study [52].

While in linear regression the phenotype is modeled through a linear combination of genetic and covariate effects, in logistic regression the logarithm of the odds is modeled as:

$$\log\left(\frac{\boldsymbol{P}(y=1)}{1-\boldsymbol{P}(y=1)}\right)=\beta_0\overrightarrow{1}+\beta_1\boldsymbol{x}_{\mathrm{G}}+\boldsymbol{X}_{\mathrm{C}}\boldsymbol{\beta}_2^T \tag{3}$$

where all parameters are defined as in Eq. 1. Logistic regression models are commonly used to model the probability of a target variable to take a specific value (in the case of GWAS, the phenotype). For a more extensive coverage on logistic regression, the reader is referred to [52].

Linear Mixed Models

A further variation of linear regression is the linear mixed model (LMM). Over the last few years LMMs have gained popularity in the field of GWAS [53]. They constitute a flexible framework for the analysis of genetic data, with a linear combination of fixed and random effects accounting for the phenotypic variation. While in linear regression the parameters of the model are fixed, in LMMs some parameters are assumed to follow a Gaussian distribution, the so called random effects. In many applications, the genetic variant and the covariates are modeled as fixed effects, while the genetic similarity among samples is modeled as a random effect, adding an additional term to the model in Eq. 2:

$$\boldsymbol{y}=\boldsymbol{X}\boldsymbol{\beta}^T+\boldsymbol{\gamma}+\epsilon \tag{4}$$

where $\boldsymbol{\gamma}=\boldsymbol{X}_{\mathrm{S}}\boldsymbol{\beta}_3^T$ is called a random effect, with each parameter in the vector $\boldsymbol{\beta}_3\in\mathbb{R}^{1\times w}$ drawn from the same Gaussian distribution, and $\boldsymbol{X}_{\mathrm{S}}$ is the $n\times w$ design matrix of the random effect. This is equivalent to $\boldsymbol{\gamma}$ being drawn from a multivariate Gaussian distribution with a covariance matrix proportional to $\boldsymbol{K}=\boldsymbol{X}_{\mathrm{S}}\boldsymbol{X}_{\mathrm{S}}^T\in\mathbb{R}^{n\times n}$ [54]. As with linear regression, this results in a phenotype with a Gaussian distribution and therefore in a model applicable to quantitative traits.

Bayesian Mixed Models

From a Bayesian point of view, the Gaussian distribution of $\boldsymbol{\beta}_3$ in the LMM of Eq. 4 can be interpreted as a prior on the parameter vector $\boldsymbol{\beta}_3$. This prior can be replaced by other distributions, thereby incorporating different prior assumptions into the model and giving rise to Bayesian mixed models. The choice of the prior distribution has implications on the estimation of the model parameters. For example in BOLT-LMM [54], SNP effects are modeled as random effects, and the Gaussian prior on $\boldsymbol{\beta}_3$ in Eq. 4 is replaced with a mixture of two Gaussians, keeping the parameter estimation simple due to the convenient mathematical representation of the Gaussian.

2.4 Population Structure Correction

As mentioned in the introduction, the presence of population structure in the data can lead to false positive associations. A common way to assess the degree of population structure in the data is to compute the genomic inflation factor λ_{GC} [55]. It describes the deviation of the median of the observed test statistics from the median of the expected test statistics. Since the distribution of the test statistics under the null hypothesis is known (either because there is a closed form representation, or because it has been derived empirically using permutation testing), so is its median. Considering the assumption that most of the SNPs are not associated with the trait, λ_{GC} should be close to one if no confounder is present [56]. If λ_{GC} is inflated (larger than 1) or deflated (lower than 1), a correction for population structure is recommended to avoid false positives or false negatives, respectively. The inflation or deflation of *p*-values can be easily visualized in a Q-Q plot (refer to Subheading 2.7.2 for more details about these plots). The three most common correction approaches are based on (a) the principal components (PCs) of a genetic similarity matrix [12], (b) the genomic inflation factor λ_{GC} [57], and (c) a combination of an LMM with the genetic similarity matrix as random effect [58, 59].

2.4.1 Principal Components of the Genetic Similarity Matrix

For a genetic dataset with n individuals, the **genetic similarity matrix** corresponds to the $n \times n$ matrix that captures the similarity between the individuals based on their SNP information (genotypes). To compute this matrix, assume $\boldsymbol{X}_{\mathbf{G}}$ to be the $n \times d$ matrix containing d SNPs for n individuals. Prior to the computation of the similarity matrix, each SNP in $\boldsymbol{X}_{\mathbf{G}}$ is commonly centered around its mean and normalized by its allele-frequencies [12], resulting in the matrix $\widehat{\boldsymbol{X}}_{\mathbf{G}}$. Then, the similarity matrix is computed as:

$$\boldsymbol{K} = \frac{1}{n}\widehat{\boldsymbol{X}}_{\mathbf{G}}\widehat{\boldsymbol{X}}_{\mathbf{G}}^{T} \in \mathbb{R}^{n \times n} \tag{5}$$

It has been shown that including the leading PCs of the genetic similarity matrix $\boldsymbol{K}$ as covariates in a linear model corrects for population structure [12]. Nevertheless, it is not a priori clear how many PCs should be included as covariates. Commonly, the association analysis is repeated multiple times, with an increasing number of leading PCs included as covariates. For each run, the

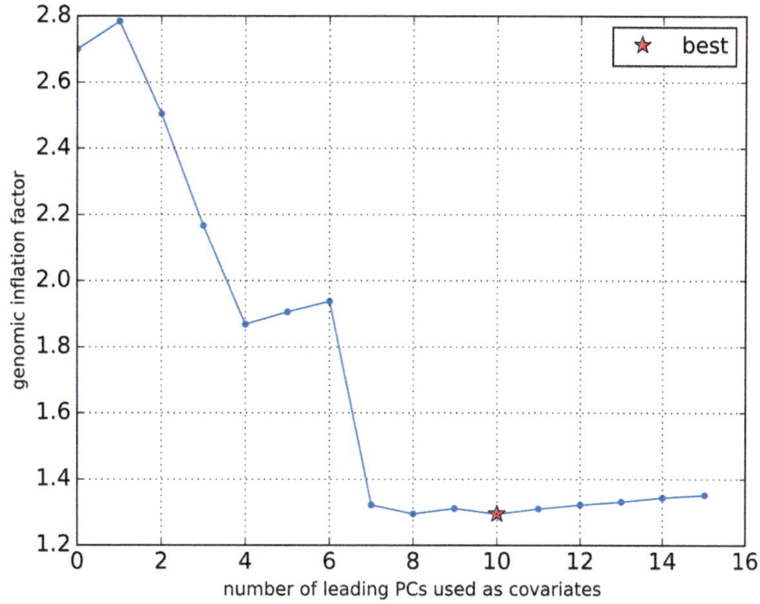

Fig. 2 The genomic inflation factor, when different numbers of leading PCs are included as covariates into a linear regression model

genomic inflation is computed and plotted against the number of PCs in the model (*see* Fig. 2). Finally, the number of leading PCs that are used is the one that yields a genomic inflation factor λ_{GC} close to 1.

2.4.2 Adjusting Test Statistics for Genomic Inflation

Devlin et al. [55, 57] have suggested to adjust the raw test statistics with the genomic inflation factor λ_{GC}. Dividing each test statistic by λ_{GC} and subsequently computing the p-value from the adjusted test statistic reduces the genomic inflation:

$$T_{\text{adjusted}} = \frac{T_{\text{raw}}}{\lambda_{GC}}.$$

2.4.3 Correction for Population Structure in LMMs

LMMs have been successful at correcting for population structure, family relatedness and cryptic relatedness in GWAS [53]. The correction is achieved by explicitly accounting for structure among individuals in the model, e.g., by including a random effect with covariance proportional to the genetic similarity matrix defined in Eq. 5. Depending on the type of relatedness between individuals, coupling more than one genetic similarity matrix, and if necessary, additionally including PCs as fixed effects into the model, can increase the power of GWAS [59]. An often-encountered downside of LMMs for GWAS is the computationally demanding task of obtaining the genetic similarity matrix. This problem has been addressed in different approaches, such as [58, 60–62], which are based on approximations of the matrix, or on computationally efficient exact methods.

2.5 Gene-Based Approaches

In contrast to the univariate methods previously described, gene-based approaches aim at deriving p-values of association, not for single SNPs but for genes. Gene-based approaches rely on first mapping SNPs to genes, followed by specific methods to test the resulting genes for association. Both of these topics are discussed below.

2.5.1 Mapping SNPs to Genes

In gene-based GWAS, a gene is represented by the set of SNPs that overlaps with it. Moreover, it is a common practice to also include SNPs in close proximity to the gene (between 20 and 50 kb, upstream and downstream) [25, 63, 64]. These SNPs are theorized to have regulatory effects on gene expression, thereby affecting the phenotype. Nevertheless, special care has to be taken when expanding the width of this margin as it increases the chance of assigning the same SNP to multiple genes. This results in a violation of the assumption that genes are independent, and might lead to inflated association signals [65].

2.5.2 Association Testing of Genes

There exist two main approaches for association testing of a gene: (a) two-step approaches, which consist of computing univariate test statistics for each SNP in the gene and, subsequently, collapsing them into one test statistic for the whole gene, and (b) one-step approaches, in which all SNPs in the gene are used simultaneously to derive a test statistic [64, 66].

Two-Step Methods

The first step consists of computing the p-values, $p_{g,1}, \ldots, p_{g,m}$, of the m SNPs that overlap with the gene g, derived from the univariate test statistics $s_{g,1}, \ldots, s_{g,m}$. In the second step, the test-statistics are combined using one of the following approaches:

(a) Minimum p-value [32, 64]: The gene p-value is set to the minimum of the SNP p-values overlapping with the gene, i.e., $p_g = \min(p_{g,1}, \ldots, p_{g,m})$.

(b) Sum of test statistic [29, 32]: If the test statistics $s_{g,1}, \ldots, s_{g,m}$ come from a χ^2 test with one degree of freedom, the gene test statistic s_g can be computed as $s_g = \sum_{i=1}^{m} s_{g,i}$, which follows a χ^2-distribution with m degrees of freedom, assuming independence between the statistics. In reality, this assumption of independence does not hold due to LD, which requires the development of methods taking LD into account.

(c) Average test statistic [29, 67]: Computation of the average over the k most significant test statistics, i.e., the gene test statistic corresponds to $s_g = \frac{1}{k} \sum_{i=1}^{k} s_{g,i}$. Since the null-distribution of this statistic cannot be analytically derived, a p-value of association is obtained by performing permutation testing.

One-Step Methods

Alternatively, all the SNPs overlapping with a gene can be tested simultaneously in one test by using the linear mixed model framework. For this purpose, the model in Eq. 1 has to be adapted such that the genotype vector $x_G \in \mathbb{R}^{n\times 1}$ is replaced by a genotype matrix $X_G \in \mathbb{R}^{n\times m}$, and the m columns in the matrix correspond to the genotypes of the m SNPs mapped to the gene [68].

Both one- and two-step approaches have their respective advantages. The one-step approaches, in contrast to the two-step approaches, make no prior assumptions about the direction of effects of the SNPs harbored in the gene. An advantage of two-step approaches is that they do not require access to the original genotype data, but it is sufficient to use the summary statistics of a univariate analysis for each SNP. We refer the reader to [64] for a more comprehensive discussion about these methods.

2.6 Detection of Interactions Between Genetic Loci

As mentioned in Subheading 1.3.2, missing heritability refers to the gap between the heritability of a trait and the phenotypic variation that can be accounted for by association signals from univariate GWAS. One of the possible sources of missing heritability is hypothesized to be the fact that univariate GWAS do not account for nonlinear interactions between loci [69]. Thus, extensions that take into account interactions between SNPs or genes have been proposed to complement the existing univariate approaches. Below we discuss two of these extensions.

2.6.1 Epistasis

There exist various definitions of epistasis, with the most commonly used one referring to the deviation from additivity in a linear model [26]. This can be captured in a statistical model as follows:

$$y = \beta_0 \vec{1} + \beta_1 x_1 + \beta_2 x_2 + \beta_{12} x_1 x_2 + \epsilon \tag{6}$$

In model Eq. 6, x_1 and x_2 correspond to the genotype at SNP_1 and SNP_2, respectively. The multiplicative term $x_1 x_2$ is the element-wise product of x_1 and x_2 and corresponds to the epistatic interaction between the two loci. The parameter β_{12} is the effect size that will be tested for deviation from zero to obtain a p-value of the epistatic interaction. The statistical test is analogous to the one in the linear models explained earlier. For a more comprehensive review of different test statistics used in the analysis of genome-wide interactions, the reader is referred to [70].

2.6.2 Network-Based Approaches

To complement the analysis of epistasis, one can explore higher-order interactions between SNPs in the hope that a group of them can be found to act in concert to disrupt a certain biological process. Testing all possible combinations of SNPs at a genome-wide level for association is computationally infeasible. Therefore, the class of methods known as **network-based GWAS** is frequently used to minimize the search space by incorporating prior knowledge in the form of biological networks and testing only interactions between markers in these networks [25].

Commonly used networks are for example protein–protein interaction networks [71–74] in which a node corresponds to a gene (or its product, a protein) and an edge represents any type of interaction between the nodes at both ends. The interactions can be derived from different sources, which renders some interactions as having higher confidence than others. While some interactions are derived from repositories of experiments, others are predicted or manually curated from the literature [72, 73]. In general, these interactions are not context specific and hold for a variety of tissues and cell types. Exploring and mining these networks with respect to a phenotype is the goal of network-based GWAS. Most methods that were developed to include network information aim at finding dense subgraphs (called modules) within the network that are associated to the phenotype [33–36]. Since these networks are based on genes/proteins, a mapping between genes/proteins and SNPs needs to take place before testing for association.

2.7 Evaluation of GWAS

Once a genome-wide association study has been performed, there are additional post-processing steps that are normally performed on the association signals. These steps include, but are not limited to: (a) correcting *p*-values for multiple hypothesis testing, (b) visualizing the results, (c) (potentially) merging results obtained in an individual genome-wide association study with a larger meta-analysis of the same phenotype, and (d) searching for a biological interpretation of the final results. Each of these steps is described in the subsections below.

2.7.1 Multiple Hypothesis Testing

The analysis of large numbers of SNPs results in the simultaneous execution of equally large numbers of tests. This gives rise to a problem known as **multiple hypothesis testing** (MHT) [75]. As an example, a univariate genome-wide association study with d SNPs will result in d tests, each of them tested at a significance level α. On average, $d \times \alpha$ tests will be false positives and as d is normally in the order of 10^5 or 10^6, it is indispensable to apply a correction to the *p*-values to avoid reporting large numbers of false positives. The two most prominent approaches to correct for multiple hypothesis testing are controlling the family-wise error rate (FWER) and the false discovery rate (FDR).

Controlling the Family-Wise Error Rate (FWER)

The FWER is defined as the probability of obtaining one or more false positives. The most common method for controlling the FWER at level α is the Bonferroni correction [76], which guarantees the FWER to be smaller than α, with the per-test threshold $\delta = \alpha / d$.

Controlling the False Discovery Rate (FDR)

The FDR is defined as the expectation of the false discovery proportion, which in turn is the proportion of false associations among all significant associations. The most common method for controlling the FDR is the Benjamini–Hochberg procedure [77], although there exist other approaches as well, such as Benjamini–Yekutieli [78] and Storey–Tibshirani [79].

Comparison of FWER and FDR

The difference between the FWER and the FDR lies in the number of false positives one is willing to accept in the outcome. Controlling the FWER at a 5% significance level means that there is at most a 5% chance that one or more of the significant hits are false positives. In order to achieve this, only loci with strong association signals are deemed significant by the Bonferroni threshold. While this reduces the number of false positives, the number of false negatives tends to increase. When using the FDR to control for MHT, it means that on average up to 5% of the significant results might actually be false positives. While this approach is less conservative than the Bonferroni correction, the number of false positives might be higher and with more loci being considered significant, the number of false negatives tends to decrease.

2.7.2 Visualization of GWAS Results

Manhattan Plots

In a Manhattan plot [42], each dot corresponds to one genetic marker (SNP). Its genetic location is indicated on the *x*-axis, and the $-\log_{10}$ of its *p*-value on the *y*-axis (see plots in second column of Fig. 3a–d). This transformation results in SNPs with low *p*-values (and therefore stronger association) to have high values on the *y*-axis. Due to LD, SNPs in close proximity to each other show similar association to the phenotype, resulting in spikes in close proximity to SNPs with low *p*-values. This resembles the Manhattan skyline and gave rise to the term Manhattan plot.

Quantile-Quantile Plots

Under the null-hypothesis of no association, *p*-values follow a uniform distribution. Quantile-quantile (Q-Q) plots illustrate this expected distribution of *p*-values compared to the observed distribution. In a Q-Q plot, each dot corresponds to one SNP, and its position corresponds to its expected *p*-value (*x*-axis) against its observed *p*-value (*y*-axis) in $-\log_{10}$ space (see plots in first column of Fig. 3).

Under the general assumption in GWAS that only a small portion of the SNPs are associated to the phenotype [56], the majority of the expected and observed values should coincide (i.e., lie on the bisecting line of the plot). Deviations for a high number of markers, especially in the range of intermediate to high *p*-values, indicate the presence of confounders that artificially inflate the *p*-values (e.g., Fig. 3a). This inflation can be caused, for example, by population structure or cryptic relatedness among the individuals. As mentioned in Subheadings 1.3.1 and 3.4—for

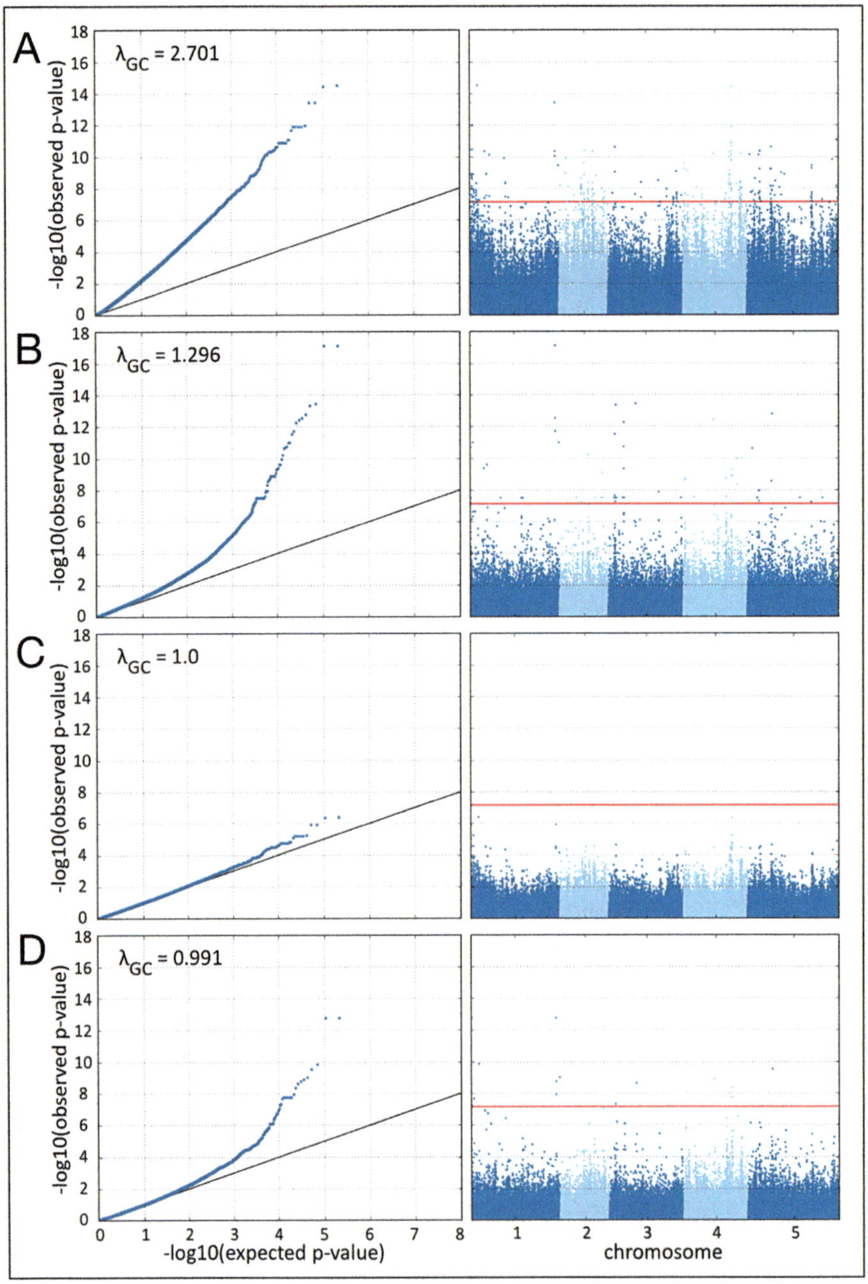

Fig. 3 Visualization of results for population structure correction with Q-Q plots (left) and Manhattan plots (right) for the *A. thaliana* dataset with the "FT Field" phenotype. The red line in the Manhattan plots represents the Bonferroni threshold. (**a**) Baseline: *p*-values derived with linear regression without any correction. (**b**) Using the ten leading PCs of the genetic similarity matrix as covariates in the linear regression. (**c**) *p*-values after correction with the genomic inflation factor λ_{GC}. (**d**) *p*-values generated with FaST-LMM

population structure—a correction needs to be performed in order to avoid reporting false positive results.

2.7.3 Meta-Analysis

Historically, the meta-analysis was developed as a tool to combine results from similar clinical trials [80]. After the advent of GWAS, meta-analysis has proved to be a robust methodology to combine results obtained from different studies. As each individual genome-wide association study has normally a modest sample size, a meta-analysis of multiple GWAS has the ability to increase the overall power and to reduce false positives [81].

It is a common practice in a meta-analysis of GWAS to pool the association signals detected in different studies without explicitly using the underlying genetic data. This is another aspect that makes meta-analysis such an appealing method, as access to genotype data is frequently regulated by strict privacy protections. Individuals that join a study may consent for their genetic data to be used in that specific study, but only allow for summary statistics to be disseminated among other research groups. These summary statistics are the association signals—there is no genotype information—that are aggregated in a meta-analysis.

There exist different ways to integrate signals from different studies. The most commonly used techniques are Fisher's method, Stouffer's method [82], and Stouffer's weighted method, as well as approaches based on fixed and random effect models [83]. The decision of which method is most appropriate to combine GWAS results heavily depends on the underlying assumptions of the studies at hand. The reader is referred to [81] for a comprehensive review of meta-analysis in GWAS.

2.7.4 Implications of Significant Findings

In the case of finding a truly significant SNP there are two different association outcomes: the SNP can either be causal and the association is called a **direct association**, or the SNP is in high LD with the causal variant, in which case one speaks of an **indirect association** [84]. Indirect associations occur when the causal SNP is not genotyped in the study, and the statistical test picks up the signal of the tag SNP marking the LD pattern that includes the causal variant. It is important to notice that without further experiments direct and indirect associations cannot be distinguished.

Another aspect of finding a significant variant is its biological implication. In many cases, markers that are deemed significant in a GWAS do not overlap with coding regions of a gene, but lie in intergenic regions [66]. It is common practice to map the significant SNP to genes that lie in close proximity, and assess if the gene is known to be involved in phenotype-specific pathways or functions.

3 Tools and Software

The previous section presented models, methods and equations that constitute the theoretical foundations of GWAS. There is a plethora of toolboxes and software packages that have been developed for all the different facets of GWAS previously described. While some of these tools are flexible and allow the user to conduct GWAS using a variety of models, others are more specialized and tailored for a specific subset of methods.

Here we start by summarizing the different tools and software in an attempt to provide a high-level view of the functional groups to which they belong. The remainder of this section presents details of a selected subset of tools, including clear protocols on how to run them. The vast majority of these tools are installed in the virtual machine (VM) that accompanies this book chapter and we encourage the reader to run the protocols we present here on the VM. The VM also contains a wiki page that facilitates the navigation through all sample scripts with their respective output and plots. Please refer to Subheading 1.4 for more details about the VM.

PLINK [67] is one of the most widely used software packages for GWAS. It allows the user to perform different kinds of analysis on SNP data, including univariate GWAS using two-sample tests (Subheading 2.3.1) and linear regression models (Subheading 2.3.2), as well as set-based tests (Subheading 2.5) and epistasis screenings (Subheading 2.6.1). Another flexible framework for deciphering the architecture of complex traits is GCTA [85], which started as a project to estimate phenotypic variation from SNP data and has subsequently been extended to accommodate more functionality, including GWAS using linear mixed models (Subheading 2.3.2). In addition to GCTA there are various toolboxes and software packages that implement different approaches to association testing with linear mixed models [53], among them FaST-LMM [58], EMMAX [86], GEMMA [62], and GRAMMAR-Gamma [87] as well as an extension to the Bayesian setting (Subheading 2.3.2) called BOLT-LMM [54].

Moving from the SNP to the gene level, there exist a variety of tools implementing different gene-based tests (Subheading 2.5). Examples of these are VEGAS [29] and PASCAL [32], which follow a two-step approach (Subheading 2.5.2), or MAGMA [88] and FaST-LMM-set [68] that are based on linear models (Subheading 2.3.2).

Another branch of software tools—those that implement network-based approaches (Subheading 2.6.2)—allow for the joint test of multiple variants by including prior knowledge in the form of biological networks. Some examples of such tools are SConES [35], dmGWAS [33, 36] and DAPPLE [34]. Other methods, that also analyze sets of SNPs but that were not covered in Subheading 2, have the particularity of defining the sets to test on the fly. Two examples of these methods are FAIS [37] and hierGWAS [38].

Different tools and software packages not only distinguish themselves by the association methods and tests they provide, but also by their computational efficiency. In conducting univariate GWAS, for example, a tool can choose to test SNPs sequentially or to perform many association tests in parallel as there is no impediment to test multiple SNPs simultaneously. There are levels of parallelization that the user can easily implement, e.g., run a univariate genome-wide association study of human data on each of the 22 chromosomes separately (most GWAS in humans are performed on autosomal chromosomes and exclude chromosomes X, Y, and SNPs from the mitochondrion). But it is the level of parallelization offered by the tool that can clearly improve its runtime and computational efficiency. Some tools allow multi-threading on CPUs while others were designed to run on graphics processing units (GPUs) to achieve maximal parallelization of tasks. Good examples of the latter are the tools that provide efficient implementations of methods to search for epistasis (Subheading 2.6.1) such as EPIBLASTER [27] and GLIDE [28].

Our summary of tools and software would not be complete if we did not mention the GWAS workbenches that are available as online resources. Notable examples are EMMA [60], DGRP2 [17], Matapax [89], GWAPP [90] and easyGWAS [91]. In essence, they allow the user to perform GWAS, analyze and (in certain cases) annotate the results, all within a web server. The functionality provided by these web tools differs, but their main advantage consists in abstracting the user from the tedious work of having to conduct the analyses in their local installation.

In the following subsections, we give a descriptive introduction to commonly used tools for GWAS. Additionally, we provide examples that illustrate how the different types of GWAS introduced in Subheading 2 can be performed. These examples are presented as short snippets of code. Most of these examples require the specification of input data files by giving the complete path to the file. To facilitate the understanding of our examples, we assume a hypothetical file called mydata.txt. In referring to this file we use the following convention:

- Data directory or path: /home/gwasuser/data, the directory where the file is stored. In code snippets, the path will be stored in a variable, for example, $path.
- Full path: /home/gwasuser/data/mydata.txt, is the fully qualified name of the file in the file system, obtained simply by joining the data directory and filename.
- Extension of the filename: .txt, which normally is used to indicate the type of file. In this example, .txt means ASCII text.

Table 1
Execution times, on the VM, of all the scripts presented in Subheading 3

Script name	Execution time (hh:mm:ss)
example_3.1_binary2plain.sh	0:00:01
example_3.1_plain2binary.sh	0:00:02
example_3.2_preprocessing.sh	0:00:01
example_3.3.1_linear.sh	0:00:21
example_3.3.1_model.sh	0:00:01
example_3.3.1_assoc.sh	0:00:12
example_3.3.1_logistic.sh	0:00:20
example_3.3.2_lmm.sh	0:01:55
example_3.4.1.1_compute_pcs_plink.sh	0:00:02
example_3.4.1.2_compute_pcs_eigensoft.sh	0:00:25
example_3.4.1_correction_with_pcs.sh	0:05:20
example_3.4_generate_plots.sh	0:03:27
example_3.5.1.1_set_flag.sh	0:09:00
example_3.5.1.1_make_set_flag.sh	0:10:37
example_3.5.1.2_vegas.sh	No runtime
example_3.6.1.2_epistasis.sh	29:40:01
example_3.7.1_dmgwas.sh	0:10:25

The times are in hours:minutes:seconds and represent the actual elapsed time by the script (obtained with the Linux command time, field real)

- Prefix of the filename: mydata, obtained by removing the extension from the filename.

Some of the examples in this section will have combinations of the items above, such as $path/mydata.txt (to specify the full path) or $path/mydata (as it is the case in many tools that exclude the extension and only require the path and prefix of the file). In addition to these code snippets, we provide sample scripts that can be executed on the VM. Table 1 shows the running times on the VM of all the scripts presented in Subheadings 3.1 through 3.7.

3.1 Data Formats in GWAS

The file format that has become a *de facto* standard for GWAS is the file format used by PLINK [67]. This is commonly referred to as **PLINK format**. Genotype data, in PLINK format, can be stored as two different types of files: plain text and binary. Text files, normally, are delimited by a white space or tab and have the extensions .ped and .map (Fig. 4a, b). Each line in a .ped file corresponds to one individual. The first six columns of the file contain the family ID, individual ID, paternal ID, maternal ID, gender, and phenotype, in that order. The remaining columns contain the genotype information. To complement the .ped file, the .map file contains one line for each SNP, indicating the SNP location and its identifier (if given). The ordering of the SNPs in the .ped file (as columns) matches the ordering of the SNPs in the .map file (as rows). Storing genotype data as plain text implies that the .ped file is both large in size (in the order of dozens of GB) and cumbersome to process. This is the reason why the second type of PLINK file, the binary file, is the one that is most commonly used. The PLINK binary file has the extension .bed and contains, essentially, the same genotype information as the .ped file, albeit in binary format. Full datasets for GWAS, in PLINK binary format, consist of three files with identical prefix but with different extensions: .bed, .bim, and .fam (Fig. 4c, d). The .bim and .fam files are delimited by a white space. A .bim file contains information about the SNPs in the study (similarly to a .map file). The .fam file has the same information stored in the first six columns as a .ped file.

Example of Data Conversion on the VM (using PLINK)

To run an example of data conversion in which a PLINK file in binary format is converted to text format, navigate to the directory containing the sample code by typing at the prompt "cd $EXAMPLES/3.1_data_formats". Execute the script by typing "./example_3.1_binary2plain.sh". Alternatively, if the conversion is from plain text to binary, you can execute another of the sample scripts by typing "./example_3.1_plain2binary.sh". These scripts will perform data conversion on a sample dataset of the plant *Arabidopsis thaliana* (abbreviated as *A. thaliana*) included in the VM.

3.2 Data Preprocessing with PLINK

Most of the data preprocessing in GWAS can be easily performed with PLINK. Depending on the filtering criteria to use (*see* Subheading 2.2.2) different flags must be specified. As an example, the command:

```
plink --bfile $path/mydata --make-bed
      --out $path/mydata_filtered --maf 0.01
      --hwe 1e-6
```

processes the binary genotype data stored in mydata.bed (with its accompanying files mydata.bim and mydata.fam), and generates a filtered output file mydata_filtered.bed (again, with the accompanying files mydata_filtered.bim and mydata_filtered.fam). These new files contain genotype data that passed two filtering criteria: (a)

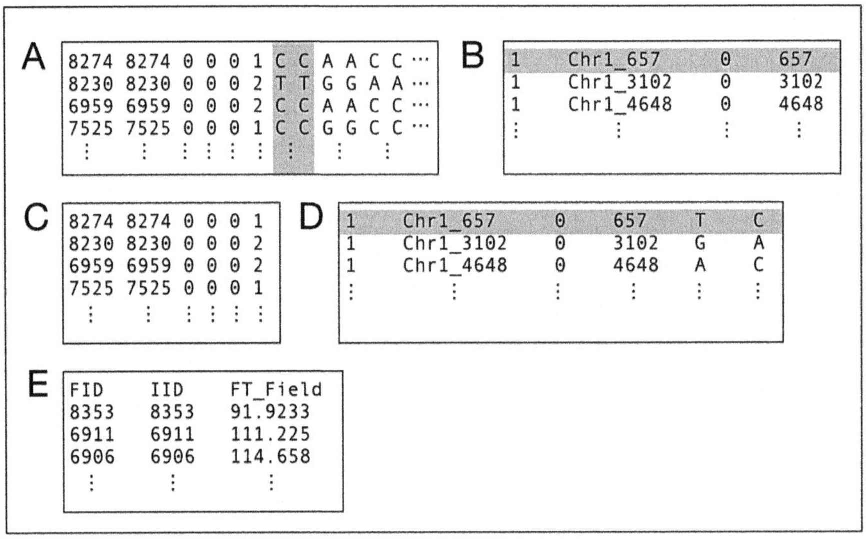

Fig. 4 The different types of PLINK files for the *A. thaliana* dataset. (**a**) .ped file: Each line corresponds to one sample. The first six columns are the family ID, individual ID, paternal ID, maternal ID, sex, and phenotype, respectively. The remaining columns are the SNP data, with two columns per SNP, representing to the two alleles (data for one SNP is highlighted in grey). (**b**) The .map file, in which each line represents one SNP. The columns correspond to the chromosome, the SNP identifier, the genetic distance in morgans (optional), and the base-pair position, respectively. The SNP highlighted in grey corresponds to the one highlighted in (**a**). (**c**) The .fam file corresponds to the first six columns of the .ped file. (**d**) Example of the .bim file. It has the same format as the .map file plus two additional columns indicating the two alleles of the SNP. (**e**) Example of the phenotype file with the "FT Field" phenotype. The columns correspond to the family ID, the individual ID and the phenotype value

Table 2
Flags to invoke different data filtering in PLINK and their commonly used values

Flag	Description	Standard value
--maf {x}	minor allele frequency (MAF); SNPs with lower MAF than {x} are excluded from the analysis	0.01 or 0.05
--hwe {x}	*p*-value threshold of HWE below which SNPs are excluded	1e–6
--mind {x}	Samples with more than {x} * 100% missing genotypes are excluded	0.1
--geno {x}	SNPs with more than {x} * 100% missing values are excluded	0.1

SNPs with minor allele frequencies larger than 1% and (b) SNPs in Hardy–Weinberg equilibrium at a 1e−6 significance threshold. *See* Table 2 for additional filtering flags.

Example of Data Preprocessing on the VM (using PLINK)

To run an example of data preprocessing on the *A. thaliana* dataset, navigate to the directory containing the code by typing at the prompt "cd $EXAMPLES/3.2_data_preprocessing". Execute the script by typing "./example_3.2_preprocessing.sh".

Table 3
Flags to invoke commonly used association tests in PLINK and the phenotypes for which the tests are appropriate

Flag	Description	Phenotype
--assoc	1-degree-of-freedom χ^2 test	case/control
--model	1df dominant, 1df recessive, 2df genotypic, Cochran-Armitage trend	case/control
--logistic	Logistic regression	case/control
--linear	Linear regression	quantitative

3.3 Univariate Association Studies

As described in Subheading 2.3, a univariate association study tests each SNP separately for association with the phenotype. Below we present in more detail two tools to conduct univariate association studies: (a) PLINK, which implements a variety of methods, and (b) FaST-LMM, designed to support different linear mixed model approaches.

3.3.1 Univariate GWAS Using PLINK

A standard case/control association study can be run with PLINK by using the following command:

Basic Analysis

```
plink --bfile $path/mydata_filtered
      --out $path/mydata_filtered --assoc
```

The `--assoc` flag calls a 1-degree-of-freedom χ^2 test on the SNPs in the binary dataset mydata_filtered.bed (with its .bim and .fam files) and writes the output to mydata_filtered.assoc. Other than the χ^2 test, PLINK can be used to find associations with models such as linear or logistic regression. Table 3 lists commonly used flags for univariate testing implemented in PLINK.

Additional Options

PLINK offers a variety of additional flags. Explaining all of them is beyond the scope of this chapter, so we focus on the ones that are of specific use in standard GWAS.

A useful flag that can be added is `--adjust`. It will generate an additional output file with the suffix .adjusted in its filename. This file contains the raw *p*-values as well as the *p*-values after correction for multiple hypothesis testing.

Upon availability, including covariates such as age, gender or sex into a genome-wide association study reduces the amount of false positives. When a linear model is chosen, i.e., a linear or logistic regression, covariates specified in a covariate file can be included using the `--covar` flag, followed by the full path filename of the covariate file (*see* Fig. 4e). The `--covar-number` and `--covar-name` flags allow the user to select a subset of all the covariates in the file. These flags must be followed by the indices of the columns or the names of the chosen covariates. In case the name of a covariate is used, it must match the name of a column in the header of the file.

In addition to the *p*-values computed from the theoretical null-distributions of the test, empirical *p*-values can be obtained via permutation testing. PLINK offers two options for this, namely the `perm` method which performs adaptive Monte Carlo permutation testing, or the `mperm={number of permutations}` method, that computes a max(T) permutation. Both methods can be called with PLINK by adding them directly after the model in the command line, for example:

```
plink --bfile $path/mydata_filtered
      --out $path/mydata_filtered
      --assoc perm plink
      --bfile $path/mydata_filtered
      --out $path/mydata_filtered
      --assoc mperm=1000
```

Both calls result in output files with the additional extensions perm and mperm, respectively.

By default, PLINK uses the phenotype that is specified in the .fam file. Nevertheless, a different phenotype can be specified by using the `--pheno` flag followed by the full path to a text file containing the phenotypes of interest, *see* Fig. 4e for the file format. In case `--pheno` is given, the phenotype specified in the .fam file will be ignored.

3.3.2 Univariate GWAS Using FaST-LMM

FaST-LMM (short for Factored Spectrally Transformed Linear Mixed Models) [58] is a method and software package for association studies using LMMs. FaST-LMM computes a genetic similarity matrix and includes it in the linear model via a random effect, thereby correcting for structure among the samples that potentially could cause false positives. By default, and in order to avoid proximal contamination, FaST-LMM uses a **leave-out-one-chromosome** approach. This means that the genetic similarity matrix is computed from SNPs on all chromosomes, except for those on the chromosome containing the SNP to be tested for association. Here we explain how to work with the Python implementation of FaST-LMM.

Basic Analysis

To run a basic association analysis with FaST-LMM, a .bed file containing the SNPs to be tested is needed, as well as a separate text file with the phenotypes. FaST-LMM is then invoked from Python with the `single_snp` function from the fastlmm package (https://pypi.python.org/pypi/fastlmm):

```
single_snp(test_snps=$path/mydata,
    pheno=$path/mypheno.txt,
    output_file_name=$path/myoutput.txt)
```

This will perform the association study for the genotype data in mydata.bed and the phenotype specified in mypheno.txt. The output will be written to myoutput.txt. FaST-LMM can be applied to both, quantitative and qualitative phenotypes without having to specify this in the function call.

Additional Options

As mentioned above, FaST-LMM always includes at least one genetic similarity matrix via a random effect in the model. By setting the argument `K0` as a .bed file, these data will be used to compute the first genetic similarity matrix. If `K0` is not specified, the data in `test_snps` will be used for its computation.

In addition to the first genetic similarity matrix, a second one can be included by adding the parameter `K1` followed by a .bed file containing the genotypes for its computation. Similar to `K0`, `K1` will also be added as a random effect to the model.

Furthermore, FaST-LMM allows for the inclusion of covariates as fixed effects, using the `covar` argument. The format of the covariate file is the standard PLINK covariate file format (*see* Fig. 4e). In contrast to PLINK it is not possible to specify which covariates to use, but all covariates in the file will be included.

3.3.3 Other Tools

Besides PLINK and FaST-LMM there exist many different tools for performing univariate GWAS that implement different approaches and methodologies. Among the most widely used ones are the following:

(a) GCTA (Genome-wide Complex Trait Analysis) [85] is a flexible toolbox for the analysis of complex traits that comprises a high number of functionalities, such as estimation of genetic relationships, estimation of phenotypic variance, data transformation, and GWAS with linear mixed models.

(b) BOLT-LMM [54] is a method for GWAS using Bayesian mixed models with a mixture-of-Gaussians prior.

(c) EMMAX (Efficient Mixed-Model Association eXpedited) [86] constitutes a toolbox for GWAS with linear mixed models.

Examples of Univariate GWAS

To run the different approaches of univariate GWAS on the *A. thaliana* dataset, navigate to the directory containing the code by typing at the prompt "cd $EXAMPLES/3.3_univariate_gwas". There are four scripts to perform a univariate analysis with PLINK, each with a different model, and one script to perform an analysis with FaST-LMM. To run any of them, type "./example_3.3.1_{plink_model}.sh" or "./example_3.3.2_lmm.sh".

3.4 Population Structure Correction

The three methods to correct for population structure introduced in Subheading 2.4 are based on (a) the principal components (PCs) of the genetic similarity matrix, (b) the genomic inflation factor, and (c) the application of LMMs.

3.4.1 Correction Using Principal Components of the Genetic Similarity Matrix

The main task in this approach to correct for population structure is the computation of the leading principal components of the genetic similarity matrix. Once the PCs have been computed, they can be included in any statistical model for GWAS that allows for covariates, *see* Subheading 3.3 for details. Here we present how the PCs can be computed using PLINK and EIGENSTRAT [12]. The main difference between the two methods is that EIGENSTRAT allows for automatic outlier removal in addition to providing an efficient approximation for very large datasets. In the examples shown below, the PCs obtained from both tools are comparable.

Computation of the Principal Components Using PLINK

The following PLINK command generates the leading `num_pcs` principal components of the genetic similarity matrix, computed from the genotype data in mydata.bed:

```
plink --bfile $path/mydata

      --out $path/mypcs --pca num_pcs header
```

It generates the two output files mypcs.eigenvec and mypcs.eigenval, containing the leading eigenvectors (corresponding to the PCs) and the associated eigenvalues, respectively. The `header` modifier adds a header line to the eigenvector file, making it directly usable for tools such as PLINK or FaST-LMM.

Computation of the Principal Components Using EIGENSTRAT

Another tool for the computation of the principal components of the similarity matrix described in Eq. 5 is EIGENSTRAT, implemented in EIGENSOFT [92]. It contains the function smartpca.perl, which can be called from the command line as follows:

```
smartpca.perl -i $path/mydata
    -a $path/mydata -b $path/mydata -k {num_pcs}
    -o $path/myoutput -p $path/myplot
    -l $path/mylog -e $path/myeigenvalues
```

The -i, -a, and -b flags expect the full path filenames of the .bed, .bim and .fam files, respectively. With the -k flag, the default number of ten PCs can be changed. With the -o flag, the output prefix is specified; -p is followed by the filename of an output plot displaying the data along the first two PCs. The -l flag specifies the filename of the log-file, and the -e flag corresponds to the filename to which the eigenvalues should be written. In order to use the PCs as covariates with PLINK or FaST-LMM, they have to be brought into the PLINK phenotype file format (Fig. 4e). We provide a function for this in our allgwas module on the VM located under /home/gwasuser/tools/allgwas.

3.4.2 Correction Using the Genomic Inflation Factor

Computing *p*-values that are corrected for population structure by using the genomic inflation can be done with PLINK by adding the --adjust flag to the GWAS command, for example:

```
plink --bfile $path/mydata
      --out $path/myoutput --assoc --adjust
```

In addition to the standard output file myoutput.assoc, this generates a second output file myoutput.assoc.adjusted that contains the raw *p*-values and different types of adjusted *p*-values. Among these, the *p*-values adjusted for population structure with the genomic inflation factor are indicated in the header line by "GC".

3.4.3 Correction Using LMMs

Due to the success of LMMs in GWAS, many toolboxes that implement this method have been developed. One of them is FaST-LMM which, as discussed in Subheading 3.3.1, automatically computes the genetic similarity matrix using a **leave-out-one-chromosome** technique and includes it as the covariance matrix of a random effect into the model.

3.4.4 Comparison of the Different Approaches for population Structure Correction

To illustrate the different correction methods for population structure, we apply them to the *A. thaliana* dataset, together with the quantitative phenotype "FT Field" which measures the number of days to flowering of plants grown in the field [11]. Figure 3a displays the baseline, i.e., the *p*-values obtained from a linear regression without any form of correction for population structure. It is obvious from the Q-Q plot and Manhattan plot that the *p*-values are inflated. In the Q-Q plot this inflation manifests itself by the deviation of the *p*-values from the bisecting line, whereas in the Manhattan plot one can recognize the inflation by the large number of significant SNPs. This contradicts the prior assumption that only few loci are associated to the phenotype.

We compute the PCs of the genetic similarity matrix and subsequently use them as covariates in the linear regression model. The genomic inflation factor λ_{GC} shows the least deviation from one when using the leading ten PCs as covariates (*see* Fig. 2). Including them as covariates in the linear regression model results in the *p*-values shown in Fig. 3b. Although the inflation measured by λ_{GC} drops closer to one, the *p*-values remain inflated and still a large number of the SNPs are significant after Bonferroni correction.

When correcting for population structure using λ_{GC}, the inflation decreases substantially, leading to no genome-wide significant results (*see* Fig. 3c). The problem with this approach is that, although the inflation vanishes, actually there is no correction. The ranking of the SNPs with respect to their *p*-value remains unchanged. This means that the SNPs with the lowest *p*-values prior to the correction will remain the ones with lowest *p*-values after correction, despite the fact that they might actually be associated to population structure.

Figure 3d shows the *p*-values obtained from the LMM approach implemented in FaST-LMM. In comparison to the baseline and the PC approach, this results in the least inflation.

Examples of population structure correction

To compute the principal components of the genetic similarity matrix using PLINK or EIGENSOFT for the *A. thaliana* dataset, navigate to the directory containing the code by typing at the prompt "cd $EXAMPLES/3.4_population_structure_correction" and perform the computation by typing "./example_3.4.1.1_compute_pcs_plink.sh" or "./example_3.4.1.2_compute_pcs_eigensoft.sh" at the command prompt. When executing "./example_3.4.1.2_compute_pcs_eigensoft.sh", the inflation factor for different numbers of leading PCs is computed, and Fig. 2 is generated. The script example_3.4.4_generate_plots.sh generates the individual plots in Fig. 3. To run it, type "./example_3.4.4_generate_plots.sh".

3.5 Gene-Based Tests

In Subheading 2.5 two different approaches for gene-based testing were introduced, namely the two-step approaches that combine univariate SNP test-statistics into test-statistics for genes, and the one-step approaches that consider all SNPs mapped to a gene jointly to derive a *p*-value for that gene.

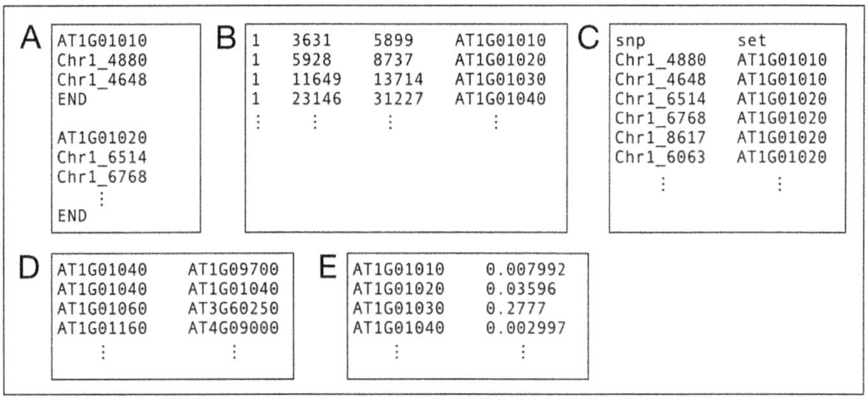

Fig. 5 Examples of file types for gene-based and network-based GWAS for the *A. thaliana* dataset. (**a**) PLINK's *set* file. Each gene is indicated by the gene-name, followed by the SNPs in the gene. The end of the gene-set is marked by END. (**b**) Gene annotation file: each line describes a gene, the columns correspond to the chromosome, the starting and ending position of the gene, as well as the name of the gene. (**c**) Gene file required for FaST-LMM-set: each line corresponds to an SNP and the gene, the SNP is mapped to. (**d**) dmGWAS network interaction file: each row corresponds to an interaction between the two genes in the first and second column. (**e**) dmGWAS gene *p*-value file: each row corresponds to one gene, the first column is the gene name, the second column is the *p*-value of the gene

3.5.1 Two-Step Approaches for Gene-Based Testing

Gene-Based Testing Using PLINK

PLINK computes a gene *p*-value using the *average test statistic* approach. By default, it will use the five most significant univariate test statistics for this average, after omitting SNPs in high LD (default >0.5). Since the distribution under the null hypothesis is not known, the method relies on permutation testing. The number of permutations can be adjusted using the `--mperm` flag, with a default value of 1000. The basic command to run a gene-based test with PLINK is the following:

```
plink --bfile $path/mydata
      --out $path/gene_results
      --set-test --assoc --mperm 1000
```

In addition to these flags, the information of the SNP-to-gene assignment has to be included. There are two different ways to do this:

(a) The first option is to generate a set file that assigns the SNPs to the genes (*see* Fig. 5a), and to pass its full path filename to PLINK with the `--set` flag, i.e.

```
plink --bfile $path/mydata
      --out $path/gene_results
      --set-test --assoc --mperm 1000
      --set $path/myset.set
```

(b) The second option is to use gene-range lists, and include them with the `--make-set` flag (*see* Fig. 5b). This allows the specification of borders around the genes, and SNPs lying within these borders will also be assigned to the gene. The flag `--make-set-borders` lets the user define the size of the borders in kilobases (kb):

```
plink --bfile $path/mydata
      --out $path/gene_results
      --set-test --assoc --mperm 1000
      --make-set $path/hg_built.txt
      --make-set-borders 20
```

For both options (a) and (b), the user can change the maximum number of SNPs used for the average by including the flag `--set-max`. The flag `--set-p` determines the maximum univariate *p*-value an SNP can have to be considered in the average. The `--set-p` flag overrules the `--set-max` flag in the sense that SNPs with *p*-values above the set-p threshold will never be included, even if this implies including fewer than set-max SNPs in the average.

Gene-Based Testing Using VEGAS

VEGAS (Versatile Gene-based Association Study) [29] is a command line tool to compute gene-based *p*-values using the *sum of test statistics* approach. As explained in Subheading 2.5.2, SNPs in close proximity commonly show high LD, violating the assumption of independence of the test-statistics in the sum. In order to obtain the null distribution while taking the correlation between SNPs into account and thus obtain valid *p*-values, permutation testing is indispensable, but it comes at the cost of a high computational effort. VEGAS circumvents this computational burden by replacing the permutations with draws from a *p*-value distribution that accounts for the observed LD pattern (Monte Carlo simulations).

VEGAS is applicable to human data only, since it expects SNP-identifiers to follow the *rs#* naming convention. The tool can be used either online or as a downloadable command line tool. Here, we explain briefly how the downloadable version of VEGAS is used. The following command invokes a basic run:

```
vegas $path/mydata.txt -pop {population}
    -out $path/myoutput
```

where mydata.txt is a two-column file delimited by a white space, containing the univariate *p*-values for the SNPs, and `{population}` corresponds to the name of a reference population from which LD should be estimated. The output will be saved to myoutput.out.

As an alternative to the `-pop` flag, the user can invoke VEGAS with the `-custom` flag, pointing to a .bed file. When this flag is set, LD is estimated from the genotypes in the .bed file.

3.5.2 One-Step Approaches for Gene-Based Testing

Gene-Based Testing Using FaST-LMM-Set

In FaST-LMM-set, sets of genetic variants are tested for their association with a phenotype by implementing the one-step method introduced in Subheading 2.5.2. Defining each gene (plus a border region) as a set, FaST-LMM-set fits an LMM with the genotype-matrix as the covariance matrix of a random effect. In contrast to the univariate implementation (Subheading 3.4.2), a random effect to model the relatedness among the individuals is optional and will not be added by default to the model. Upon availability, covariates can be included as a fixed effect. Once a model for a gene is fitted, a likelihood ratio or score test is applied to test the gene effect.

The Python implementation of FaST-LMM-set requires a file containing the mapping of SNPs to genes (*see* Fig. 5c). FaST-LMM-set can be invoked using the Python function `snp_set` from the fastlmm package with the following command in Python:

```
snp_set(test_snps=$path/mydata,
    set_list=$path/set_file.txt,
    pheno=$path/mypheno.txt,
    output_file_name=$path/myoutput.txt)
```

where the genotype data is stored in mydata.bed, the phenotype information in the file mypheno.txt, and the SNP-set information in set_file.txt. This will generate an output file myoutput.txt, which contains the statistics and *p*-values of the genes.

The user has the option to include covariates with the `covar` argument. Furthermore, an additional random effect can be included to model the relatedness among the samples. To do so, the `G0` argument has to be set to point to a .bed file that contains the SNPs from which genetic similarity should be estimated. Both arguments `covar` and `G0` help reduce the bias of confounders on the association results. The statistical test used can be changed using the test argument, with two settings available: the score test, invoked with `test='sc_davies'`, or the likelihood ratio test (default), invoked with `test='lrt'`.

Examples for Gene-based Testing

To perform gene-based testing on the *A. thaliana* dataset, navigate to the directory containing the code by typing "cd $EXAMPLES/3.5_gene_based_testing" at the prompt. The two different PLINK approaches can be executed by typing "./example_3.5.1.1_make_set_flag.sh" or "./example_3.5.1.1_set_flag.sh". Typing "./example_3.5.2.1_fastlmm_set.sh" runs the FaST-LMM-set method.

In addition, we provide a script to run VEGAS on a PLINK toy-dataset. The script does not execute because it requires supplementary files that do not fit in the VM. Nevertheless, the script is available as an example of how to invoke VEGAS after running a univariate genome-wide association study with PLINK. It can be found at "$EXAMPLES/example_3.5.1.2_vegas.sh".

3.6 Epistasis Search

The search for epistatic interactions is computationally more challenging than the search for single associations. In this search, every locus will be tested against every other locus, which squares the number of models to fit compared to univariate GWAS. In order to alleviate this burden, it is common practice to first run a univariate analysis and only use high ranking SNPs for an epistasis screening. However, this approach might not detect interactions between variants with moderate to low effect sizes. Another way to deal with the computational complexity of the problem is to rely on parallel programming, e.g., on graphics processing units (GPUs). Here, we describe how a search for epistatic interactions can be performed with PLINK and briefly mention two other approaches based on GPU-computing.

3.6.1 Two-Locus Epistasis Search Using PLINK

PLINK provides two functions to run an epistasis screening, namely the fast-epistasis method and the epistasis method.

The fast-epistasis constitutes a fast scan for epistatic interactions only with a qualitative phenotype, e.g., in a case/control study. By default, every two loci are first collapsed from a 3×3 contingency table into a 2×2 table separately for cases and controls. Then, for the two classes, the odds ratios are computed based on the tables, and their difference is tested for significance. While this is a fast method, it does not take the individual effect of each locus into account, and should therefore rather be used for a first screening before carrying out a proper epistasis study.

The second option, epistasis, is based on the linear model described in Eq. 6. As opposed to fast-epistasis, it works with qualitative and quantitative phenotypes. It requires fitting a model for each combination of SNPs and is thus much more time consuming than the fast-epistasis method. *See* Table 1 for a comparison of the execution times between the two methods.

Epistasis Screening Using Fast-Epistasis

A fast-epistasis run can be invoked with the following command:

```
plink --bfile $path/mydata

        --out $path/myoutput --fast-epistasis
```

where the genotype data is contained in mydata.bed (with its corresponding .bim and .fam files). This will produce two output files, myoutput.epi.cc and myoutput.epi.cc.summary. The former is a text file where each line corresponds to one SNP-pair with an epistasis p-value smaller than 0.0001 (this threshold can be modified using the `--epi1` flag; for small datasets with a small number of pairs it can be set to 1). In the file myoutput.epi.cc.summary each line corresponds to one SNP. It contains information on how often the SNP occurred in an epistatic interaction with p-value lower than 0.01 (column with header "N_SIG", this threshold can be adapted with `--epi2`) and the interaction partner with which the lowest epistasis p-value was reached.

Epistasis Screening Using Epistasis

Analogously to fast-epistasis, one can run a full epistasis screen based on linear/logistic regression with the following command:

```
plink --bfile $path/mydata
      --out $path/myoutput --epistasis
```

This generates two output files: myoutput.epi.qt and myoutput.epi.qt.summary for a quantitative trait, or myoutput.epi.cc and myoutput.epi.cc.summary for a case/control phenotype. These output files report the same statistics as in the fast-epistasis case, and the flags `--epi1` and `--epi2` behave in the same way.

3.6.2 Other Tools

Apart from PLINK, other commonly used tools for the detection of epistatic interactions are:

(a) EPIBLASTER [27]: GPU-based tool to detect epistatic interactions in case/control datasets. The detection of two-locus interactions is based on a two-stage approach, where in the first stage SNP-SNP pairs are selected based on the difference in correlation to the phenotype in cases versus controls. In the second stage, selected pairs are tested using logistic regression with a likelihood ratio test.

(b) GLIDE [28]: GPU-based epistasis tool based on linear regression and a 4-degree-of-freedom t-test. It is applicable to quantitative and qualitative phenotypes and has high performance due to an efficient GPU implementation.

Examples for Epistasis Searches

To perform an epistasis search on the *A. thaliana* dataset, navigate to the directory containing the code by typing at the prompt "cd $EXAMPLES/3.6_epistasis". The two different PLINK-approaches can be run by typing "./example_3.6.1.1_fast_epistasis.sh" or "example_3.6.1.2_epistasis.sh". Please note that performing epistasis search is time-consuming.

3.7 Network-Based GWAS

3.7.1 Dense Module Searching for Genome-Wide Association Studies (dmGWAS)

dmGWAS [33, 36] is a method implemented in an R package to find modules within a biological interaction network that are enriched with low *p*-value genes. The search for these modules follows a greedy strategy [93], i.e., the algorithm iteratively adds the gene with the lowest *p*-value to the current module. This corresponds to making the locally optimal decision at each step, which might not necessarily lead to a globally optimal decision. While it allows for an efficient exploration of large networks, it comes with the downside of not analyzing all possible modules.

As input, dmGWAS requires two files: the first one, edge_file.txt, is a two-column, tab-delimited text file containing the network. Each line corresponds to one edge, represented by its two adjacent nodes (*see* Fig. 5d). The other file, pvalue_file.txt, is also a two-column, tab-delimited text file, containing the *p*-values assigned to each node (first column is the node and the second column the *p*-value) (*see* Fig. 5e). *P*-values can be obtained, for example, by implementing one of the methods described in Subheading 3.5. The following commands in R convert the text files to R-readable tables that can be used with dmGWAS:

```
network <- read.delim($path/edge_file.txt,
    as.is=T, header=F)
pvalues <- read.delim($path/pvalue_file.txt,
    as.is=T, header=F)
```

In case the files contain a header in the first line, the header flag should be set to `header=T`. With the data in this format, dmGWAS can be invoked with the following command:

```
res.list <- dms(network, pvalues,
    expr1=NULL, expr2=NULL, d=2, r=0.1)
```

The arguments `expr1` and `expr2` correspond to files containing the gene-expression for cases and controls, respectively, and are used to compute edge-weights. We will not use this functionality here and refer the interested reader to [36]. There are two more parameters, `d` and `r`, that affect the behavior of the greedy search algorithm. Following the recommendation of the authors [33, 36], these values should be set to `d=2` and `r=0.01`.

3.7.2 Other Methods for Network-Based GWAS

(a) SConES (Selecting Connected Explanatory SNPs) [35] is a method implemented in MATLAB to efficiently detect sets of SNPs that are connected in an underlying network and have maximal association to a phenotype, while meeting sparsity and connectivity constraints.

(b) DAPPLE (Disease Association Protein-Protein Link Evaluator) [34] is based on testing the hypothesis that modules of associated genes are more densely connected than expected by random chance. In order to test this hypothesis, permutation testing is applied.

Examples for Network-based GWAS

To perform a network-based GWAS on the *A. thaliana* dataset, navigate to the directory containing the code by typing at the prompt "cd $EXAMPLES/3.7_network_based". It contains a script to run dmGWAS which can be executed by typing "./example_3.7.1_dmgwas.sh" at the prompt.

3.8 Other Types of GWAS

3.8.1 Hierarchical GWAS (hierGWAS)

Buzdugan et al. [38] presented a hierarchical testing procedure for SNP data using a linear regression coupled with a multiple sample splitting procedure for fitting. By clustering SNPs according to a predefined criterion, and hierarchically testing those clusters from largest (all SNPs) to smallest (single SNPs), they developed a framework for testing the cluster-tree and correcting for MHT. Their approach is implemented in the R package *hierGWAS*.

3.8.2 Combinatorial Association Mapping

The goal of combinatorial association mapping is to determine the statistical association of higher-order interactions between genetic markers and a phenotype of interest. When making a naive attempt to test all higher-order interactions between SNPs one faces two main obstacles: firstly, the vast number of association tests that need to be performed creates a computational bottleneck and, secondly, there is a sharp decrease in statistical power due to corrections for MHT (e.g., Bonferroni correction). Newly developed algorithms based on significant pattern mining provide an alternative to mitigate these computational and statistical challenges. These algorithms can be used to perform region-based association studies [94] and to detect higher-order epistatic interactions of genetic variants [95], while allowing for the correction of categorical covariates such as age or gender. These approaches are implemented in the Python and R package CASMAP.

3.9 Online Tools

Apart from downloadable tools, some online tools have been developed to allow the user to perform GWAS on web-servers. Since these web-servers provide user-friendly graphical interfaces, they are usually more intuitive to use than command line tools or Python/R packages, which require the user to have some basic knowledge in computer science. Due to current legislation however, users are not allowed to upload and analyze human genetic data with those tools.

3.9.1 GWAPP

GWAPP [90] is a web application for performing GWAS in the model organism *A. thaliana*. There are a variety of accessions available on the web-server that can be analyzed with a user-specified phenotype. GWAPP implements three methods to test for associations, namely linear regression, an accelerated mixed model and a Wilcoxon rank sum test [96]. Correction for MHT is done using the Benjamini–Yekutieli method to control the FDR [78]. Interactive Manhattan plots, LD structure plots, and different statistics (depending on the model) are generated as final analysis results.

3.9.2 easyGWAS

easyGWAS [91] is a cloud platform for computing, sharing and comparing the results of GWAS on user-specific datasets and phenotypes, as well as publicly available genotype and phenotype data from model organisms, such as *Arabidopsis thaliana* [13, 15, 97–99] and *Pristionchus pacificus* [100]. Based on the phenotype and genotype, the user can choose between various state-of-the-art GWAS methods, including the Wilcoxon rank-sum test, linear/logistic regression and FaST-LMM [58]. easyGWAS allows the user to apply different genotype encodings of the data (*see* Fig. 1b) and phenotype transformations. Furthermore, the user has the possibility to also perform meta-analysis on already computed GWAS or compare its results using a novel intersection analysis. In addition, the user can choose between four different multiple hypothesis correction methods: Bonferroni [76], Benjamini–Hochberg [77], Benjamini–Yekutieli [78], and Storey–Tibshirani [79]. easyGWAS offers dynamic visualizations to facilitate the analysis of GWAS results. Some examples are: interactive Manhattan plots with clickable SNPs, LD plots with gene annotations and chord diagrams to illustrate associated SNPs that are shared between different GWAS.

4 Remarks and Conclusions

Genome-wide association studies have been instrumental in identifying common genetic variants that correlate with specific traits or diseases. The early successes in GWAS, in particular at identifying new susceptibility loci for different types of cancer [101], led some companies to produce and market risk tests that would allow an individual to assess their risk level of developing certain diseases. This was cause for alarm at the time, especially because most of the risk alleles in the reported SNPs only conferred very small increments in risk, between 1.1 and 1.5-fold [102]. As a result, some of these risk tests, such as one that was produced in 2008 for breast cancer, drew intense criticism [103]. Essentially, the test was based on seven SNPs, none of them located on the genes BRCA1 and BRCA2 whose mutations account for up to 40% of familial breast cancer. Although the seven SNPs had been reported in GWAS as being associated to breast cancer, the scientific consensus was that in order to effectively assess risk in a patient, more associated variants needed to be discovered. Events like this, coupled with the fact that the SNPs reported in early GWAS had small effects and explained a small proportion of the heritability [23], launched a debate over the ability and efficacy of GWAS to elucidate the genetic architecture of complex diseases [4]. Nevertheless, and despite the early criticisms, GWAS in humans have invariably advanced our understanding of numerous complex diseases and traits [104–106]. The sheer amount of studies in the *GWAS Catalog*—a resource that contains all published studies for humans [107]—attests to the success of GWAS. At the time of writing this chapter, the catalog contained more than 2700 studies on 2000 traits with more than 30,000 associations reported between SNPs and traits [108].

After more than a decade of research in this area, however, the fact remains that most human SNPs reported in GWAS have small effect-sizes and explain a small proportion of the total heritability [109]. In attempting to address one of the potential reasons of the missing heritability, namely the limited sample size, researchers have conducted GWAS on increasingly larger sample sizes. This was achieved either by increasing the number of individuals in a single genome-wide association study, or by combining results of different GWAS through meta-analysis. As a result, the prediction of "larger sample sizes would yield stronger association signals" [4] finally materialized. A case in point are the studies of schizophrenia [110]. The gene ZNF804A was deemed borderline significant in a study of 479 cases and 2937 controls [111], but the association signal of the gene reached statistical significance in a meta-analysis of 21,274 cases and 38,675 controls [112]. Moreover, the number of loci known to be associated to schizophrenia jumped from 30 to 108 after a large collaborative effort based on a meta-analysis of 36,989 cases and 113,075 controls [104]. This has led to new ambitious initiatives like the one at the Centers for Common Disease Genomics (news release about funding: https://www.genome.gov/27563570/) which aims at sequencing up to 200,000 individuals in hope of elucidating the genomic architecture of complex diseases (with an initial focus on autism and cardiovascular diseases).

The slow pace at which results obtained from GWAS have been applied in clinical settings still remains a point of contention and represents an important aspect in need of future improvement. A possible explanation for this delay is related to structural hurdles in the healthcare system. As an example, in the USA there is an average delay of 17 years between the discovery and adoption of best care practices in the clinic [113]. However, there are cases in which GWAS findings could make their way to the clinic, for example: (a) using GWAS results for risk prediction of type 1 diabetes, especially since early detection can avoid the irreversible depletion of β-cells [114], (b) the repurposing of approved drugs that target genes identified by GWAS, in particular given the costly and lengthy pipeline required to develop new drug targets [115], and (c) the screening of patients, to avoid adverse drug reactions, by using SNPs from GWAS conducted in the field of pharmacogenomics [116].

Although the first GWAS were conducted in humans, the last decade has shown a steady increase of GWAS in model organisms such as *D. melanogaster* [117], *A. thaliana* [15], *C. elegans* [118], *M. musculus* [119], and *Canis lupus familiaris* [120]. The latter in particular, illustrates the success of GWAS in organisms with limited genetic diversity such as dog breeds. The studies of canine diseases whose phenotypes closely resemble those in humans, as it is the case for obsessive-compulsive disorder (OCD), yielded insights into the genetic underpinnings of such diseases in humans. In OCD, for example, statistically significant associations reported in dogs, have functional associations to top SNPs—that did not reach genome-wide significance—in humans [121]. The importance of these developments cannot be overstated, especially because, as opposed to human genetics, model organisms (a) interact with an environment in which the researcher can exert maximum control, and (b) are the ideal vehicle to perform functional validation of candidate genes identified in GWAS. This type of validation, in turn, paves the way for the discovery of where the missing heritability of a trait may lie [122].

In addition to model organisms, GWAS have been widely adopted in crops and livestock. For crops, prominent examples are maize [123], rice [16], tomato [124], and grapevine [125]. In these cases, GWAS have been used in concert with other genetic tools to increase crop yield by making breeding more efficient [126]. Similarly, GWAS in livestock have been instrumental in identifying variants that are associated to economically important traits such as meat quality, milk yield, egg production, and others [127].

Despite a number of successes mentioned above, various challenges in GWAS still lie ahead. One important challenge is, and will continue to be, disentangling the missing heritability in humans. A promising line of research, and one that can potentially lead to a reduction of the missing heritability, is the search for association in high-order SNP interactions. In a genome-wide association study with d SNPs the number of subsets of SNPs that can potentially exhibit association to the phenotype is 2^d. Current approaches limit this vast combinatorial search space either by looking at 2-locus interactions (epistasis) or by superimposing biological networks (network-based approaches). To expand the search space, algorithms that perform pattern mining coupled with sophisticated methods of correction for multiple hypothesis testing have been proposed [128–130]. These methods are based on item-set mining and do not require a priori information about which items can interact with each other. This, in turn, allows them to potentially consider any number of interacting items. It is important to note that an item, in the context of a genome-wide association study, corresponds to a SNP with a minor allele genotype. In the realm of GWAS, another important aspect to consider is the presence of confounders, such as population structure, that can be prevalent in the data. Pattern mining approaches that can account and correct for confounders are essential for identifying SNPs that are jointly associated to the phenotype while reducing the number of false positives [95]. Future research in this area should focus on (a) enabling the abovementioned methods to handle data that must not necessarily be binary, and (b) scaling the methods to make them applicable to GWAS with large sample sizes and where *d* is in the order of hundreds of thousands or even millions.

Genome-wide association studies, coupled with the ever-decreasing costs in DNA sequencing and subsequent genotyping are painting a bright future ahead of us. Due to their pervasive of their pervasive use in a wide variety of organisms and traits, understanding the methods, assumptions, and statistical models behind GWAS has become a necessary skill of every researcher working with genomic data.

References

1. MacDonald ME, Novelletto A, Lin C et al (1992) The Huntington's disease candidate region exhibits many different haplotypes. Nat Genet 1:99–103
2. Kerem B-S (1989) Identification of the cystic fibrosis gene: genetic analysis. Trends Genet 5:363
3. Bush WS, Moore JH (2012) Chapter 11: Genome-wide association studies. PLoS Comput Biol 8:e1002822
4. Visscher PM, Brown MA, McCarthy MI et al (2012) Five years of GWAS discovery. Am J Hum Genet 90:7–24
5. Altshuler D, Daly MJ, Lander ES (2008) Genetic mapping in human disease. Science 322:881–888
6. The 1000 Genomes Project Consortium (2015) A global reference for human genetic variation. Nature 526:68–74
7. Gibbs RA, Belmont JW, Hardenbol P et al (2003) The international HapMap project. Nature 426:789–796
8. Davey JW, Hohenlohe PA, Etter PD et al (2011) Genome-wide genetic marker discovery and genotyping using next-generation sequencing. Nat Rev Genet 12:499–510
9. Fan J-B, Chee MS, Gunderson KL (2006) Highly parallel genomic assays. Nat Rev Genet 7:632–644
10. Dudoit S, van der Laan MJ (2008) Multiple hypothesis testing. In: Multiple testing procedures with applications to genomics. Springer, New York, NY, pp 1–47
11. Fairweather D, Frisancho-Kiss S, Rose NR (2008) Sex differences in autoimmune disease from a pathological perspective. Am J Pathol 173:600–609
12. Price AL, Patterson NJ, Plenge RM et al (2006) Principal components analysis corrects for stratification in genome-wide association studies. Nat Genet 38:904–909
13. Atwell S, Huang YS, Vilhjálmsson BJ et al (2010) Genome-wide association study of 107 phenotypes in *Arabidopsis thaliana* inbred lines. Nature 465:627–631
14. Meijón M, Satbhai SB, Tsuchimatsu T et al (2014) Genome-wide association study using cellular traits identifies a new regulator of root development in Arabidopsis. Nat Genet 46:77–81
15. Alonso-Blanco C, Andrade J, Becker C et al (2016) 1,135 Genomes reveal the global pat-

tern of polymorphism in *Arabidopsis thaliana*. Cell 166:481–491
16. Zhao K, Tung C-W, Eizenga GC et al (2011) Genome-wide association mapping reveals a rich genetic architecture of complex traits in Oryza sativa. Nat Commun 2:467
17. Mackay TFC, Richards S, Stone EA et al (2012) The *Drosophila melanogaster* genetic reference panel. Nature 482:173–178
18. Kirby A, Kang HM, Wade CM et al (2010) Fine mapping in 94 inbred mouse strains using a high-density haplotype resource. Genetics 185:1081–1095
19. Scott LJ, Mohlke KL, Bonnycastle LL et al (2007) A genome-wide association study of type 2 diabetes in Finns detects multiple susceptibility variants. Science 316: 1341–1345
20. Chasman DI, Schürks M, Anttila V et al (2011) Genome-wide association study reveals three susceptibility loci for common migraine in the general population. Nat Genet 43:695–698
21. Freilinger T, Anttila V, de Vries B et al (2012) Genome-wide assoiation analysis identifies susceptibility loci for migraine without aura. Nat Genet 44:777–782
22. Visscher PM, Hill WG, Wray NR (2008) Heritability in the genomics era — concepts and misconceptions. Nat Rev Genet 9:255–266
23. Manolio TA, Collins FS, Cox NJ et al (2009) Finding the missing heritability of complex diseases. Nature 461:747–753
24. Lee SH, Wray NR, Goddard ME et al (2011) Estimating missing heritability for disease from genome-wide association studies. Am J Hum Genet 88:294–305
25. Pedroso I, Breen G (2011) Gene set analysis and network analysis for genome-wide association studies. Cold Spring Harb Protoc 2011:pdb.top065581
26. Cordell HJ (2002) Epistasis: what it means, what it doesn't mean, and statistical methods to detect it in humans. Hum Mol Genet 11:2463–2468
27. Kam-Thong T, Czamara D, Tsuda K et al (2011) EPIBLASTER-fast exhaustive two-locus epistasis detection strategy using graphical processing units. Eur J Hum Genet 19:465–471
28. Kam-Thong T, Azencott C-A, Cayton L et al (2012) GLIDE: GPU-based linear regression for detection of epistasis. Hum Hered 73:220–236
29. Liu JZ, Mcrae AF, Nyholt DR et al (2010) A versatile gene-based test for genome-wide association studies. Am J Hum Genet 87:139–145
30. Listgarten J, Lippert C, Heckerman D (2013) FaST-LMM-select for addressing confounding from spatial structure and rare variants. Nat Genet 45:470–471
31. Lippert C, Xiang J, Horta D et al (2014) Greater power and computational efficiency for kernel-based association testing of sets of genetic variants. Bioinformatics 30:3206–3214
32. Lamparter D, Marbach D, Rueedi R et al (2016) Fast and rigorous computation of gene and pathway scores from SNP-based summary statistics. PLoS Comput Biol 12:e1004714
33. Jia P, Zheng S, Long J et al (2011) dmGWAS: dense module searching for genome-wide association studies in protein-protein interaction networks. Bioinformatics (Oxford, England) 27:95–102
34. Rossin EJ, Lage K, Raychaudhuri S et al (2011) Proteins encoded in genomic regions associated with immune-mediated disease physically interact and suggest underlying biology. PLoS Genet 7:e1001273
35. Azencott C-A, Grimm D, Sugiyama M et al (2013) Efficient network-guided multi-locus association mapping with graph cuts. Bioinformatics 29:i171–i179
36. Wang Q, Yu H, Zhao Z et al (2015) EW_dmGWAS: edge-weighted dense module search for genome-wide association studies and gene expression profiles. Bioinformatics (Oxford, England). 31:2591–2594
37. Llinares-López F, Grimm DG, Bodenham DA et al (2015) Genome-wide detection of intervals of genetic heterogeneity associated with complex traits. Bioinformatics 31:i240–i249
38. Buzdugan L, Kalisch M, Navarro A et al (2016) Assessing statistical significance in multivariable genome wide association analysis. Bioinformatics 32:1990–2000
39. Matsuzaki H, Dong S, Loi H et al (2004) Genotyping over 100,000 SNPs on a pair of oligonucleotide arrays. Nat Methods 1:109–111
40. Clarke GM, Anderson CA, Pettersson FH et al (2011) Basic statistical analysis in genetic case-control studies. Nat Protoc 6:121–133
41. Plomin R, Haworth CMA, Davis OSP (2009) Common disorders are quantitative traits. Nat Rev Genet 10:872–878

42. Power RA, Parkhill J, de Oliveira T (2017) Microbial genome-wide association studies: lessons from human GWAS. Nat Rev Genet 18:41–50
43. Wu MC, Lee S, Cai T et al (2011) Rare-variant association testing for sequencing data with the sequence kernel association test. Am J Hum Genet 89:82–93
44. Morgenthaler S, Thilly WG (2007) A strategy to discover genes that carry multi-allelic or mono-allelic risk for common diseases: a cohort allelic sums test (CAST). Mutat Res 615:28–56
45. Madsen BE, Browning SR (2009) A groupwise association test for rare mutations using a weighted sum statistic. PLoS Genet 5:e1000384
46. Morris AP, Zeggini E (2010) An evaluation of statistical approaches to rare variant analysis in genetic association studies. Genet Epidemiol 34:188–193
47. Neale BM, Rivas MA, Voight BF et al (2011) Testing for an unusual distribution of rare variants. PLoS Genet 7:e1001322
48. Anderson CA, Pettersson FH, Clarke GM et al (2010) Data quality control in genetic case-control association studies. Nat Protoc 5:1564–1573
49. Marchini J, Howie B (2010) Genotype imputation for genome-wide association studies. Nat Rev Genet 11:499–511
50. Fisher RA (1925) Statistical methods for research workers. Genesis Publishing Pvt Ltd., Edinburgh
51. Pearson K (1900) X. On the criterion that a given system of deviations from the probable in the case of a correlated system of variables is such that it can be reasonably supposed to have arisen from random sampling. Philos Mag Ser 5(50):157–175
52. Fahrmeir L, Kneib T, Lang S et al (2013) Regression: models, methods and applications. Springer Science & Business Media, New York, NY
53. Yang J, Zaitlen NA, Goddard ME et al (2014) Advantages and pitfalls in the application of mixed-model association methods. Nat Genet 46:100–106
54. Loh P-R, Tucker G, Bulik-Sullivan BK et al (2015) Efficient Bayesian mixed-model analysis increases association power in large cohorts. Nat Genet 47:284–290
55. Devlin B, Roeder K (1999) Genomic control for association studies. Biometrics 55:997–1004
56. Yang J, Weedon MN, Purcell S et al (2011) Genomic inflation factors under polygenic inheritance. Eur J Hum Genet 19:807–812
57. Devlin B, Roeder K, Wasserman L (2001) Genomic control, a new approach to genetic-based association studies. Theor Popul Biol 60:155–166
58. Lippert C, Listgarten J, Liu Y et al (2011) FaST linear mixed models for genome-wide association studies. Nat Methods 8:833–835
59. Widmer C, Lippert C, Weissbrod O et al (2014) Further improvements to linear mixed models for genome-wide association studies. Sci Rep 4:6874
60. Kang HM, Zaitlen NA, Wade CM et al (2008) Efficient control of population structure in model organism association mapping. Genetics 178:1709–1723
61. Zhang Z, Ersoz E, Lai C-Q et al (2010) Mixed linear model approach adapted for genome-wide association studies. Nat Genet 42:355–360
62. Zhou X, Stephens M (2012) Genome-wide efficient mixed-model analysis for association studies. Nat Genet 44:821–824
63. Veyrieras J-B, Kudaravalli S, Kim SY et al (2008) High-resolution mapping of expression-QTLs yields insight into human gene regulation. PLoS Genet 4:e1000214
64. Mooney MA, Nigg JT, McWeeney SK et al (2014) Functional and genomic context in pathway analysis of GWAS data. Trends Genet 30:390–400
65. Sedeño-Cortés AE, Pavlidis P (2014) Pitfalls in the application of gene-set analysis to genetics studies. Trends Genet 30:513–514
66. Ballard DH, Cho J, Zhao H (2010) Comparisons of multi-marker association methods to detect association between a candidate region and disease. Genet Epidemiol 34:201–212
67. Purcell S, Neale B, Todd-Brown K et al (2007) PLINK: a tool set for whole-genome association and population-based linkage analyses. Am J Hum Genet 81:559–575
68. Listgarten J, Lippert C, Kang EY et al (2013) A powerful and efficient set test for genetic markers that handles confounders. Bioinformatics 29:1526–1533

69. Zuk O, Hechter E, Sunyaev SR et al (2012) The mystery of missing heritability: genetic interactions create phantom heritability. Proc Natl Acad Sci 109:1193–1198
70. Ueki M, Cordell HJ (2012) Improved statistics for genome-wide interaction analysis. PLoS Genet 8:e1002625
71. Cowley MJ, Pinese M, Kassahn KS et al (2012) PINA v2.0: mining interactome modules. Nucleic Acids Res 40:D862
72. Szklarczyk D, Franceschini A, Kuhn M et al (2011) The STRING database in 2011: functional interaction networks of proteins, globally integrated and scored. Nucleic Acids Res 39:D561–D568
73. Franceschini A, Szklarczyk D, Frankild S et al (2013) STRING v9.1: protein-protein interaction networks, with increased coverage and integration. Nucleic Acids Res 41:D808–D815
74. Li T, Wernersson R, Hansen RB et al (2017) A scored human protein-protein interaction network to catalyze genomic interpretation. Nat Methods 14:61–64
75. Johnson RC, Nelson GW, Troyer JL et al (2010) Accounting for multiple comparisons in a genome-wide association study (GWAS). BMC Genomics 11:724
76. Bonferroni CE (1936) Teoria statistica delle classi e calcolo delle probabilita. Libreria internazionale Seeber, Firenze
77. Benjamini Y, Hochberg Y (1995) Controlling the false discovery rate: a practical and powerful approach to multiple testing. J R Stat Soc B Methodol 57:289–300
78. Benjamini Y, Yekutieli D (2001) The control of the false discovery rate in multiple testing under dependency. Ann Stat 29:1165–1188
79. Storey JD, Tibshirani R (2003) Statistical significance for genomewide studies. Proc Natl Acad Sci 100:9440–9445
80. Thompson JR, Attia J, Minelli C (2011) The meta-analysis of genome-wide association studies. Brief Bioinform 12:259–269
81. Evangelou E, Ioannidis JPA (2013) Meta-analysis methods for genome-wide association studies and beyond. Nat Rev Genet 14:379–389
82. Stouffer SA, Suchman EA, DeVinney LC et al (1949) The American soldier: adjustment during army life. In: Studies in social psychology in World War II, vol 1. Princeton University Press, Princeton, NJ
83. Borenstein M, Hedges LV, Higgins JPT et al (2010) A basic introduction to fixed-effect and random-effects models for meta-analysis. Res Syn Methods 1:97–111
84. Hirschhorn JN, Daly MJ (2005) Genome-wide association studies for common diseases and complex traits. Nat Rev Genet 6:95–108
85. Yang J, Lee SH, Goddard ME et al (2011) GCTA: a tool for genome-wide complex trait analysis. Am J Hum Genet 88:76–82
86. Kang HM, Sul JH, S.K. Service et al (2010) Variance component model to account for sample structure in genome-wide association studies. Nat Genet 42:348–354
87. Svishcheva GR, Axenovich TI, Belonogova NM et al (2012) Rapid variance components-based method for whole-genome association analysis. Nat Genet 44:1166–1170
88. de Leeuw CA, Mooij JM, Heskes T et al (2015) MAGMA: generalized gene-set analysis of GWAS data. PLoS Comput Biol 11:e1004219
89. Childs LH, Lisec J, Walther D (2012) Matapax: an online high-throughput genome-wide association study pipeline. Plant Physiol 158:1534–1541
90. Seren Ü, Vilhjálmsson BJ, Horton MW et al (2012) GWAPP: a web application for genome-wide association mapping in Arabidopsis. Plant Cell 24:4793–4805
91. Grimm DG, Roqueiro D, Salome P et al (2017) easyGWAS: a cloud-based platform for comparing the results of genome-wide association studies. Plant Cell 29:5
92. Galinsky KJ, Bhatia G, Loh P-R et al (2016) Fast principal-component analysis reveals convergent evolution of ADH1B in Europe and East Asia. Am J Hum Genet 98:456–472
93. Cormen TH, Leiserson CE, Rivest RL et al (2009) Introduction to algorithms. MIT Press, Cambridge, MA
94. Llinares-López, Papaxanthos L, Bodenham D, Roqueiro D (2017) COPDGene Investigators, Karsten Borgwardt; Genome-wide genetic heterogeneity discovery with categorical covariates. Bioinformatics 33(2): 1820–1828
95. Papaxanthos L, Llinares-Lopez F, Bodenham D et al (2016) Finding significant combinations of features in the presence of categorical covariates. In: Lee DD, Sugiyama M, Luxburg UV et al (eds) Advances in

neural information processing systems, vol 29. Curran Associates, Inc, Red Hook, NY, pp 2271–2279
96. Wilcoxon F (1945) Individual comparisons by ranking methods. Biometrics 1:80–83
97. Horton MW, Hancock AM, Huang YS et al (2012) Genome-wide patterns of genetic variation in worldwide *Arabidopsis thaliana* accessions from the RegMap panel. Nat Genet 44:212–216
98. Seymour DK, Chae E, Grimm DG et al (2016) Genetic architecture of nonadditive inheritance in *Arabidopsis thaliana* hybrids. Proc Natl Acad Sci 113:E7317–E7326
99. Seren Ü, Grimm D, Fitz J et al (2017) AraPheno: a public database for *Arabidopsis thaliana* phenotypes. Nucleic Acids Res 45:D1054–D1059
100. McGaughran A, Rödelsperger C, Grimm DG et al (2016) Genomic profiles of diversification and genotype-phenotype association in Island nematode lineages. Mol Biol Evol 33:2257–2272
101. Easton DF, Eeles RA (2008) Genome-wide association studies in cancer. Hum Mol Genet 17:R109–R115
102. Kraft P, Hunter DJ (2009) Genetic risk prediction–are we there yet? N Engl J Med 360:1701–1703
103. Couzin J (2008) DNA test for breast cancer risk draws criticism. Science 322:357–357
104. Schizophrenia Working Group of the Psychiatric Genomics Consortium (2014) Biological insights from 108 schizophrenia-associated genetic loci. Nature 511:421–427
105. Wood AR, Esko T, Yang J et al (2014) Defining the role of common variation in the genomic and biological architecture of adult human height. Nat Genet 46:1173–1186
106. Fuchsberger C, Flannick J, Teslovich TM et al (2016) The genetic architecture of type 2 diabetes. Nature 536:41–47
107. Welter D, MacArthur J, Morales J et al (2014) The NHGRI GWAS Catalog, a curated resource of SNP-trait associations. Nucleic Acids Res 42:D1001–D1006
108. T. Burdett, P.N. Hall, E. Hastings, et al. The NHGRI-EBI catalog of published genome-wide association studies. www.ebi.ac.uk/gwas.
109. Gusev A, Bhatia G, Zaitlen N et al (2013) Quantifying missing heritability at known GWAS loci. PLoS Genet 9:e1003993
110. Bergen SE, Petryshen TL (2012) Genome-wide association studies (GWAS) of schizophrenia: does bigger lead to better results? Curr Opin Psychiatry 25:76–82
111. O'Donovan MC, Craddock N, Norton N et al (2008) Identification of loci associated with schizophrenia by genome-wide association and follow-up. Nat Genet 40:1053–1055
112. Williams HJ, Norton N, Dwyer S et al (2011) Fine mapping of ZNF804A and genome wide significant evidence for its involvement in schizophrenia and bipolar disorder. Mol Psychiatry 16:429–441
113. Richardson WC, Berwick DM, Bisgard J et al (2001) Crossing the quality chasm: a new health system for the 21st century. Institute of Medicine, National Academy Press, Washington, DC
114. Manolio TA (2013) Bringing genome-wide association findings into clinical use. Nat Rev Genet 14:549–558
115. Lencz T, Malhotra AK (2015) Targeting the schizophrenia genome: a fast track strategy from GWAS to clinic. Mol Psychiatry 20:820–826
116. Chan SL, Jin S, Loh M et al (2015) Progress in understanding the genomic basis for adverse drug reactions: a comprehensive review and focus on the role of ethnicity. Pharmacogenomics 16:1161–1178
117. Huang W, Massouras A, Inoue Y et al (2014) Natural variation in genome architecture among 205 *Drosophila melanogaster* genetic reference panel lines. Genome Res 24:1193–1208
118. Andersen EC, Gerke JP, Shapiro JA et al (2012) Chromosome-scale selective sweeps shape *Caenorhabditis elegans* genomic diversity. Nat Genet 44:285–290
119. Farber CR, Bennett BJ, Orozco L et al (2011) Mouse genome-wide association and systems genetics identify Asxl2 as a regulator of bone mineral density and osteoclastogenesis. PLoS Genet 7:e1002038
120. Hayward JJ, Castelhano MG, Oliveira KC et al (2016) Complex disease and phenotype mapping in the domestic dog. Nat Commun 7:10460
121. Tang R, Noh HJ, Wang D et al (2014) Candidate genes and functional noncoding variants identified in a canine model of obsessive-compulsive disorder. Genome Biol 15:R25

122. Flint J, Eskin E (2012) Genome-wide association studies in mice. Nat Rev Genet 13:807–817
123. Li H, Peng Z, Yang X et al (2013) Genome-wide association study dissects the genetic architecture of oil biosynthesis in maize kernels. Nat Genet 45:43–50
124. Lin T, Zhu G, Zhang J et al (2014) Genomic analyses provide insights into the history of tomato breeding. Nat Genet 46:1220–1226
125. Nicolas SD, Péros J-P, Lacombe T et al (2016) Genetic diversity, linkage disequilibrium and power of a large grapevine (Vitis vinifera L) diversity panel newly designed for association studies. BMC Plant Biol 16:74
126. Huang X, Han B (2014) Natural variations and genome-wide association studies in crop plants. Annu Rev Plant Biol 65:531–551
127. Sharma A, Lee JS, Dang CG et al (2015) Stories and challenges ofgenome wide association studies in livestock — a review. Asian Australas J Anim Sci 28: 1371–1379
128. Terada A, Okada-Hatakeyama M, Tsuda K et al (2013) Statistical significance of combinatorial regulations. Proc Natl Acad Sci 110:12996–13001
129. Minato S, Uno T, Tsuda K et al (2014) A fast method of statistical assessment for combinatorial hypotheses based on frequent itemset enumeration. In: Machine learning and knowledge discovery in databases. Springer, Berlin, Heidelberg, pp 422–436
130. Llinares-López F, Sugiyama M, Papaxanthos L et al (2015) Fast and memory-efficient significant pattern mining via permutation testing. In: Proceedings of the 21th ACM SIGKDD International Conference on Knowledge Discovery and Data Mining. ACM, Sydney, NSW, pp 725–734

Part III

Network-Based Modeling of Cellular Phenotypes

Chapter 6

Identifying Differentially Expressed Genes Using Fluorescence-Activated Cell Sorting (FACS) and RNA Sequencing from Low Input Samples

Natalie M. Clark, Adam P. Fisher, and Rosangela Sozzani

Abstract

Cell type-specific gene expression profiles are useful for understanding genes that are important for the development of different tissues and organs. Here, we describe how to perform fluorescence-activated cell sorting (FACS) on Arabidopsis root protoplasts to isolate specific cell types in the root. We then detail how to extract and process RNA from a very low number of cells ($\geq$40 cells) for RNA sequencing (RNA seq). Finally, we describe how to process RNA seq data using TopHat and how to identify differentially expressed genes using PoissonSeq.

Key words Protoplasting, Fluorescence activated cell sorting, RNA sequencing, Differential expression, Low input samples

1 Introduction

Stem cells are crucial to the development of both plants and animals, as they divide and differentiate to form all of the tissues and organs of the organism. An important part of this differentiation is the ability of stem cells to produce daughter cells that take on different identities. For example, in the Arabidopsis root, there are four stem cell (also called Initial cells) populations that divide asymmetrically to generate all of the different tissue layers of the root [1]. Thus, obtaining a cell type-specific profile of gene expression is useful in order to identify genes that are important for the development of each of these cell types. However, depending on the cell type of interest, it can be difficult to obtain a high enough number of cells to yield sufficient RNA for gene expression analysis. Here, we provide instructions on how we can obtain

Natalie M. Clark and Adam P. Fisher contributed equally to this work.

Louise von Stechow and Alberto Santos Delgado (eds.), *Computational Cell Biology: Methods and Protocols*, Methods in Molecular Biology, vol. 1819, https://doi.org/10.1007/978-1-4939-8618-7_6,

a suitable quantity of high quality RNA from a low number of cells ($\geq$40 cells) for RNA sequencing (RNA seq). First, we report how Arabidopsis root protoplasts expressing a fluorophore can be separated using fluorescence-activated cell sorting (FACS) [2]. Separating populations of cells using FACS allows us to obtain a cell type-specific gene expression profile. Next, we provide instructions on how we process the low number of cells such that we obtain enough high quality RNA for sequencing. We use Bowtie and TopHat to map the raw read counts that come from RNA seq data to the Arabidopsis genome and obtain the number of read counts per gene. Then, we use PoissonSeq on the mapped reads to determine genes that are differentially expressed. Unlike methods such as CuffDiff [3] or edgeR [4], PoissonSeq makes no assumptions about the distribution of gene expression data, so it can be used on a wide range of datasets [5]. In addition, PoissonSeq quickly performs comparisons between more than two samples using a log-linear model rather than multiple pairwise comparisons [5]. Identifying differentially expressed genes between different cell types can reveal genes that are important for the formation of all of the tissues in the Arabidopsis root.

1.1 Materials

1.1.1 Growing Plants for Protoplasting

1. 1% sucrose, 1× Murashige and Skoog (MS) media plates (*see* **Note 1**).
2. Sucrose.
3. 2-(*N*-morpholino)-ethanesulfonic acid (MES).
4. MS salts.
5. Agar.
6. Square petri dishes.
7. Nitex mesh cut into squares that will fit into the MS media plates. Pack mesh in aluminum foil packets of 30 squares. Autoclave for 20 min on dry cycle and dry mesh at 65 °C overnight (*see* **Note 2**).
8. 50 mL conical tubes.
9. Sterile transfer or Pasteur pipettes.
10. Sterile micropipette tips.
11. 50% bleach.
12. 10% Tween 20.
13. 100% ethanol.
14. Sterile water.

1.2 Protoplasting and FACS

1. 0.2 M magnesium chloride.
2. 0.2 M calcium chloride.
3. 1 M MES.

4. 1 M potassium chloride.
5. 1 M Tris–HCl.
6. Mannitol.
7. Bovine serum albumin (BSA).
8. Cellulase.
9. Pectolyase.
10. Mini petri dishes.
11. Surgical blades.
12. 40 and 70 μm cell strainers.
13. 5 mL polypropylene and 5 mL polystyrene tubes.
14. 15 and 50 mL conical tubes.
15. Beta-mercaptoethanol.
16. RLT buffer. This is a proprietary buffer found in RNA extraction kits produced by Qiagen.
17. Dry ice.
18. Cell sorter (such as Dako Cytomation MoFlo).
19. Solution A: Add 50 mL of deionized water to a beaker with a stir bar and add the following reagents: 5.465 g of mannitol, 0.05 g of 0.01% BSA, and 500 μL of the following stock solutions: 0.2 M Magnesium chloride, 0.2 M calcium chloride, 1 M MES, 1 M potassium chloride. Prepare these solutions immediately before use.
20. Solution B: Aliquot 30 mL of solution A into another beaker with stir bar. Add 0.45 g cellulase and 0.03 g pectolyase to the 30 mL aliquot of solution A.
21. The material is ice.

1.3 Library Preparation

1. Qiagen RNeasy Micro Kit.
2. Glycoblue.
3. 7.5 M ammonium acetate.
4. Nuclease-free water.
5. SMART-Seq v4 Ultra Low Input RNA Kit for Sequencing (Clontech Laboratories).
6. Low Input Library Preparation Kit (Clontech Laboratories).

1.4 Computational Analysis of RNASeq Data

1. 64 bit machine with at least 4 GB of RAM running either Linux or Mac OS X (Tiger 10.4 or later) (*see* **Note 3**).
2. ea-utils FASTQ processing utilities (http://expressionanalysis.github.io/ea-utils/) [6].
3. Bowtie software (http://bowtie-bio.sourceforge.net/index.shtml) [3].

4. SAM tools (http://samtools.sourceforge.net/) [3].
5. TopHat software (http://ccb.jhu.edu/software/tophat/index.shtml) [3].
6. R programming language (https://www.r-project.org/) (*see* **Note 4**).
7. Rsubread package (https://bioconductor.org/packages/release/bioc/html/Rsubread.html) [7].
8. PoissonSeq package (https://cran.r-project.org/web/packages/PoissonSeq/index.html) [5].

2 Methods

2.1 Growing Plants for Protoplasting

1. Aliquot 200 mg of seeds into a 50 mL conical tube (*see* **Note 5**).
2. Add 45 mL of 50% bleach and 15 μL of 10% Tween to each tube to sterilize the seeds (*see* **Note 6**).
3. Shake the tubes for 5 min at 110 rpm on the orbital shaker to ensure all seeds come into contact with the bleach/Tween solution.
4. Briefly centrifuge the tubes at 2000 × *g* for 10–20 s or until seeds are pelleted at the bottom of the tube. This reduces seed loss during washing.
5. In a laminar flow hood, remove the bleach–Tween mixture and add 45 mL of 100% ethanol to further sterilize the seeds (*see* **Note 7**).
6. Invert the tubes for 1 min to mix and briefly centrifuge the tubes at 2000 × *g* to reduce seed loss in the next step.
7. In a laminar flow hood, remove the ethanol and add 45 mL of sterile water. Removal of any remaining bleach and ethanol is crucial for the seeds to germinate normally (*see* **Note 8**).
8. Invert the tubes until the seeds are evenly dispersed in the water and briefly centrifuge at 2000 × *g*.
9. In a laminar flow hood, remove the water and repeat **steps 7** and **8** seven times (*see* **Note 9**).
10. After the last rinse, add 45 mL of sterile water to the tubes. Place seeds at 4 °C for 48 h in order to vernalize the seeds (*see* **Note 10**).
11. Put the following items into the laminar flow hood; 1% sucrose, 1× MS media plates, transfer/Pasteur pipettes, micropipette tips, Parafilm, matches, burner, 50 mL conical tube of ethanol, forceps, and sterile Nitex mesh.
12. Sterilize forceps and carefully move one piece of sterile Nitex mesh to a 1% sucrose, 1× MS media plate. Do not use the forceps to "smooth" out the mesh (*see* **Note 11**).

13. Repeat **step 12** until between 15 and 30 1% sucrose, 1× MS media plates have mesh on them (*see* **Notes 12** and **13**).
14. Use a sterile pipette to take up the sterilized seeds.
15. Eject the seeds slowly onto the meshed plates at a density of 2–4 seeds thick. Plating the seeds at this density will result in around 15 mg of seed being dispensed on each plate.
16. Straighten the seeds with a micropipette tip. This will ensure that each root will be cut in the same general area in Subheading 3.2.7.
17. Close the plates with Parafilm and place the plates in a vertical position so the roots grow down along the mesh. Grow plants at desired growth conditions.

2.2 Protoplasting

1. Prepare Solutions A and B as described in the materials (*see* **Note 14**).
2. Place one 70 μm strainer per small petri dish and label the strainer. Aliquot 7 mL of Solution B into each small petri dish. Avoid creating bubbles, when pipetting Solution B, as bubbles can cause the cells to lyse.
3. For each sample, place a 70 μm strainer in a labeled 50 mL conical tube. The 70 μm filter will remove the large sized cellular debris from the sample.
4. For each sample, place a 40 μm strainer in a labeled 50 mL conical tube. The 40 μm filter will remove most of the remaining cellular debris from the sample.
5. Prepare a RLT buffer and beta mercaptoethanol solution that contains 1 mL RLT buffer and 10 μL beta mercaptoethanol. Aliquot 300 μL of this solution into one polystyrene tube for each of your samples and store on ice.
6. Take the plates out of the growth chamber, unwrap them, and check for contamination. Discard any contaminated plates.
7. With a surgical blade, cut the roots at the appropriate length (2–4 mm if interested in the root stem cell niche) and put those roots into the strainer in solution B (*see* **Note 15**). Change the surgical blades every ten plates to keep the blade sharp.
8. Once all of the roots are cut, place the small petri dish onto the shaker at 85 rpm. All of the samples should go onto the shaker around the same time.
9. After 20 min, stir the samples with a transfer pipette. After stirring the samples, slowly lift the filter out of the solution and use the pipette to spread the roots over the filter. Slowly place the filter back into the solution, making sure to not create any bubbles.

10. After 30 min (50 min from when the sample started shaking), stir again as described in Subheading 3.2.9.
11. After 10 additional minutes (60 min from when the sample started shaking), stir again as described in Subheading 3.2.9. After stirring, use the transfer pipette to get any liquid from the bottom filter. Transfer the liquid to the labeled 15 mL conical tube. Also, transfer a small amount of root tissue to the 15 mL conical tube. The root tissue helps the protoplasts form a pellet. Pipette the liquid in the petri dish up and down a few times, without creating bubbles, to mix the protoplasts. Then, transfer the liquid in the small petri dish to the 15 mL conical tube.
12. Spin the 15 mL tubes in a centrifuge for 6 min at 200 × *g*.
13. Starting with the sample that finished shaking the earliest, remove the supernatant using a transfer pipette without disrupting the root pellet.
14. Resuspend the pellet with 325 μL of Solution A (*see* **Note 16**).
15. Transfer the resuspended solution to the 50 mL tube that has the 70 μm filter.
16. Rinse the 15 mL tube with another 325 μL of Solution A and transfer that solution to the 50 mL tube that has 70 μm filter.
17. Transfer all of the liquid that filtered into the conical tube to the 50 mL tube that has the 40 μm filter using a micropipette.
18. Transfer all of the liquid that filtered into the conical tube to the polypropylene tube for that sample. Repeat Subheadings 3.2, **steps 14–19** for all samples.

2.3 FACS and RNA Isolation

1. Keep samples at room temperature until FACS analysis (*see* **Note 17**). Keep polystyrene collection tubes on ice and prepare a container of dry ice.
2. Run protoplasts through a high speed cell sorter, such as a Dako Cytomation MoFlo, using a 100 μm flow tip at a pressure of 20 PSI (*see* **Note 18**).
3. Use wild type Arabidopsis protoplasts obtained from the tissue of interest to set a gate for fluorophore-negative cells. Then, set a gate for protoplasts that express the fluorophore (*see* **Note 19**) (Fig. 1).
4. Collect fluorophore-positive cells in the cold polystyrene tube (*see* **Notes 20** and **21**). Collected cells should be stored on dry ice until freezing. Cells can be stored at −80 °C for up to 1 year.
5. Extract RNA from cells using the Qiagen RNeasy Micro Kit. RNA should be stored at −80 °C. RNA quality should be checked using an Agilent Bioanalyzer. RNA concentration must be >10 pg with a RIN of >7 to proceed (*see* **Notes 22** and **23**).

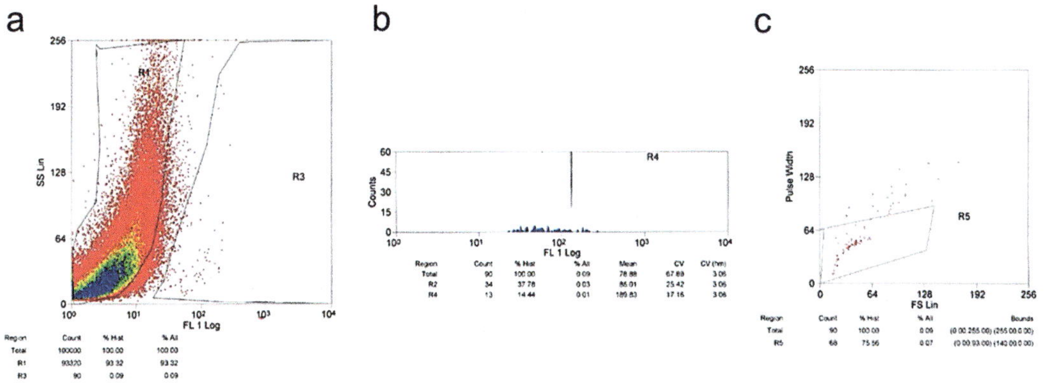

Fig. 1 Fluorescence gates used for collecting positive cells. (**a**) Side scatter (SS) versus fluorescence (FL). Cells with a greater FL value are brighter and likely to be GFP-positive cells. The R3 gate is set to take only the brightest population of cells. (**b**) Count histogram of the R3 gate. R4 is set to take only the brightest cells within R3. (**c**) Pulse width vs. forward scatter (FS) plot for the cells in R3. R5 is set so that only single cells are sorted. The final population of positive cells collected from this sort is R3 + R4 + R5

6. Prepare cDNA libraries from extracted RNA for sequencing using the SMART-Seq v4 Ultra Low Input RNA Kit for Sequencing and Low Input Library Preparation Kit (*see* **Note 24**).

2.4 Computational Analysis of RNA seq Data

1. The code used in this example represents an experiment with three different samples (1, 2, and 3), where each sample has three biological replicates (A, B, and C).
2. Download FASTQ files from RNA sequencing results to computer (*see* **Notes 25** and **26**).
3. Install ea-utils, SAM tools, Bowtie, and TopHat on computer, and copy the files to the directory that contains the fastq files.
4. Use fastq-mcf to filter samples before mapping (*see* **Notes 27** and **28**). In this example, we would run the line of code below for each of the samples (mysample1A, mysample1B, and so on).

```
$ fastq-mcf IlluminaAdaptorSeq.fasta
mysample1A.fastq.gz -q 30 -l 50 -w 5 -o
mysample1A_filtered.fastq
```

5. Map the reads for each sample to the reference genome using TopHat (*see* **Note 29**). As in Subheading 3.4.4, we would run the code below for each of the samples.

```
$ tophat2 -p 12 -G genes.gtf
--library-type=fr-
```

```
unstranded -o  mysample1A_thout genome
mysample1A_filtered.fastq
```

6. Open R and install the Rsubread and PoissonSeq packages:

```
> install.packages("Rsubread")
> library(Rsubread)
> install.packages("PoissonSeq")
> library(PoissonSeq)
```

7. Obtain gene counts for each of your samples using RSubread (*see* **Notes 30** and **31**):

```
> mycounts <-
featureCounts(files=c("mysample1A_thout/
accepted_hits.bam","mysample1B_thout/
accepted_hits.bam",
"mysample1C_thout/accepted_hits.bam",
"mysample2A_thout/accepted_hits.bam",
"mysample2B_thout/
accepted_hits.bam","mysample2C_thout/
accepted_hits.bam",
"mysample3A_thout/accepted_hits.bam",
"mysample3B_thout/
accepted_hits.bam","mysample3C_thout/
accepted_hits.bam"),
annot.ext="genes.gtf",isGTF
AnnotationFile=TRUE,
GTF.featureType="exon",GTF.
attrType="gene_id",
ignoreDup=TRUE)
> mycounttable <- mycounts$counts
> writecsv(mycounttable,"myfile.csv")
```

8. Before performing PoissonSeq to test for differentially expressed genes, create a vector that indicates how the biological replicates are grouped (*see* **Note 32**):

```
> bioreps <- c(rep(1,3),rep(2,3),rep(3,3))
```

9. Perform PoissonSeq to determine differentially expressed genes across all samples (*see* **Note 33**):

```
> pseq<-
PS.Main(dat=list(n=mycounttable,
y=bioreps,type="multiclass",
pair=FALSE,gname=row.names(mycounttable)))
> write.csv(pseq,"myresults.csv")
```

10. Obtain results for only the differentially expressed genes ($q < 0.05$) (*see* **Note 34**):

```
 > siggenes <- pseq[pseq$fdr<0.05,]
> write.csv(siggenes,"
myresults_diffgenes.csv")
```

3 Summary/Outlook

Obtaining cell type-specific gene expression profiles allows us to better understand how different cell populations develop their specialized functions. Here, we described the process of collecting protoplasts and subsequently isolating specific cell types via fluorescence-activated cell sorting (Fig. 2a). Importantly, we optimized the protocol to extract high quality RNA for sequencing from a very low amount of starting material (≥40 cells) (Fig. 2b). We then described how to identify differentially expressed genes from RNA seq data using PoissonSeq (Fig. 2c). As the cost of sequencing decreases and the number of different fluorescent marker lines that mark cell or tissue types of interest increases, we can gain further insight into the specific functions of different cells and tissues across the Arabidopsis plant.

4 Notes

1. 1× MS 1% sucrose plates are made by combining 10 g sucrose, 0.5 g MES, 4.33 g MS salts, and 5 g of Agar per 1 L of deionized water. pH solution to 5.7 and autoclave media for 20 min. Swirl media and pour into square petri dishes. Allow plates to cool and store at 4 °C.
2. We recommend packaging meshes in groups of 30 to reduce chances for contamination.
3. For deep sequencing data, we recommend using a machine with at least 16 GB of RAM [3].
4. For first time R users, we recommend the open source version of RStudio Desktop (https://www.rstudio.com/products/rstudio/).
5. The amount of seeds used in this experiment depends on the fluorescence levels of the marker of interest.
6. For sensitive seeds that have a thin seed coat or do not germinate well, we recommend reducing the concentration of bleach to 30%.
7. When pouring off the bleach from the seeds, the seeds may fall out of the tube. In this case, it is acceptable if there is some bleach remaining in the tube. This bleach will wash out during the water washes.
8. We recommend checking the tubes after 24 h to see if there is bleach still on the seeds. If the water is yellowish, wash the seeds two to three more times with sterile water to remove residual bleach.

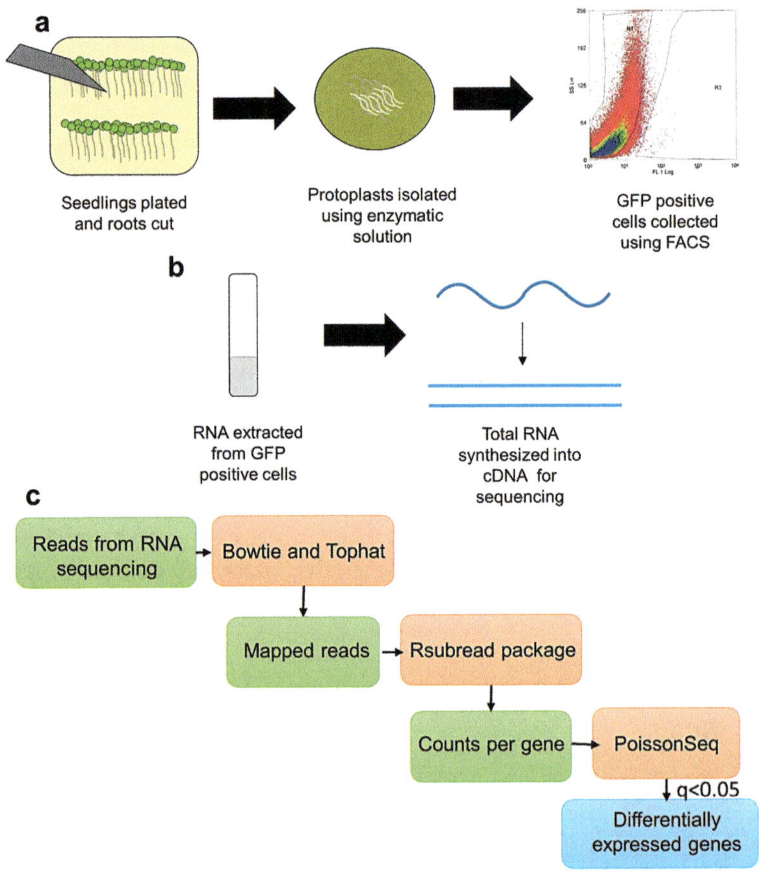

Fig. 2 RNA sequencing on cells collected using fluorescence-activated cell sorting (FACS). (**a**) First, seedlings are plated on 1× MS, 1% sucrose plates. Roots are cut using a razor blade and soaked in an enzyme solution to isolate protoplasts. Then, GFP-positive protoplasts are collected using FACS. (**b**) RNA is extracted from GFP-positive cells and then synthesized into cDNA for sequencing. (**c**) Reads from RNA sequencing are first mapped to the reference genome using Bowtie and TopHat. The Rsubread package is then used to determine the read counts per gene. Finally, PoissonSeq is used to identify differentially expressed genes between samples

9. More seeds require more water washes.
10. Vernalization is necessary to reset the internal clock of the seeds, so that germination is roughly uniform across samples. Alternatively, seeds may be plated and then stored at 4 °C for 48 h.
11. The Nitex mesh should smooth out over the plate evenly after a few seconds. Smoothing the Nitex mesh out may cause additional bubbles to form or depressions in the media, resulting in roots growing nonuniformly on the plate.

12. We do not recommend meshing and plating seeds on more than 30 plates at once, as this increases the chance of bacterial contamination.
13. To roughly calculate the number of plates you need to mesh, divide the total mass of seeds by 10. The total mass of seeds spread on each plate should be roughly between 10 and 15 mg depending on the density.
14. Solution A is the buffer used to collect the protoplasts, while solution B contains the enzymes used to isolate the protoplasts. Solution A may be frozen and thawed before sorting. However, Solution B should be prepared fresh before every sort.
15. The area of the root that is cut depends on the expression profile of the fluorescent marker.
16. The cells should be washed with 300–350 μL of Solution A. More Solution A should be used for a higher concentration of cells or to yield a longer sort time.
17. If transport is necessary, keep the protoplasts in a Styrofoam container.
18. The event rate varies depending on the sample concentration. For highly concentrated samples, the pressure should not exceed 21.5 PSI.
19. Multiple gates should be combined to set the area of positive cells. In the example we present here, we first gate using R3 (Fig. 1a). Then, we use R4 to focus on a smaller part of the R3 area (Fig. 1b). This ensures low fluorescence contamination of autofluorescent cells in the sample. Finally, we use R5 to collect only single cells from the R3 + R4 area, eliminating any cells that may be clumped together (Fig. 1c).
20. The holder for the collection tube must be refrigerated to keep the collection buffer cold.
21. When sorting, all sorts should be the same time length, and should not exceed 15 min. Using sorts of the same time length and sorting in a small amount of time will minimizes changes in gene expression, which might be caused by the fact that protoplasts are still developing. In addition, the total sort time should not exceed 1 h. After 1 h, protoplasts will begin to die, resulting in a dramatic decrease in yield.
22. When using the RNeasy Micro Kit, we recommend using 12 μL of water and eluting twice. If >100,000 cells were collected, we recommend using the RNeasy Mini Kit using 30 μL water and eluting twice.
23. If RNA is of low quality, a rescue protocol can be performed. Add 1 μL of glycoblue, 12 μL of 7.5 M ammonium acetate, and 110 μL of cold 100% ethanol to RNA. Store overnight at −20 °C. The next day, spin samples at 14,000 × *g* for

45 min at 4 °C. Pipette off supernatant and add 800 μL cold 100% ethanol. Spin samples at 14,000 × *g* for 10 min at 4 °C. Pipette off supernatant and repeat the previous spin, but this time for 2 min. Pipette off any remaining supernatant and dry pellet for 5–10 min, or until no longer glossy. Add 10 μL of nuclease-free water and resuspend pellet. Reanalyze RNA using the bioanalyzer to see if the quality is now suitable for sequencing.

24. We recommend using 100 pg–10 ng starting amount of total RNA when using these kits.

25. Code starting with $ is run from the command line, while code starting with > is run from an R environment.

26. We recommend that all files for RNA sequencing analysis are stored in the same directory. To create a new directory, use the Linux command `$ mkdir $HOME/mydir` where `mydir` is the name of the new directory. To add this new directory to the path, use the command `$ export PATH = $HOME/mydir:$PATH`. When downloading and installing eautils, SAM tools, Bowtie, and TopHat, copy their files into this directory.

27. `-q` is the quality threshold for base removal (in this example, 30). `-l` is the minimum remaining sequence length (in this example, 50). `-w` is the window size for trimming (in this example, 5). These values are the default recommended values for Bowtie [3]. We recommend manually adjusting these parameters to obtain the best mapping.

28. Filtering must be applied before mapping to remove sequencing adapters and low quality reads.

29. `genes.gtf` is the genome annotation file for the samples. Many annotation files can be found on the TopHat website (http://ccb.jhu.edu/software/tophat/igenomes.shtml). `library-type` can be adjusted depending on sequencing conditions. The default is `fr-unstranded`. `mysample1_thout` is the name of the results folder.

30. Gene counts are used for statistical analysis, however, gene expression values are generally reported using Reads/Fragments Per Kilobase of transcript per Million mapped reads (RPKM/FPKM). Multiple packages in R can convert counts to RPKM/FPKM, such as edgeR (https://bioconductor.org/packages/release/bioc/html/edgeR.html).

31. We recommend saving the count table to a file, as determining the gene counts can take hours for many samples.

32. This vector (`bioreps`) is an indicator variable that specifies which sample each biological replicate belongs to. In this example, the first three columns of the count table are the three

replicates of sample 1, so they each get "1" as their indicator variable. The next three columns belong to sample 2, so their indicator variable is "2." Finally, the final three columns are from sample 3, so their indicator variable is "3."

33. Use `type="twoclass"` when comparing two samples and `type="multiclass"` when comparing more than two samples. When comparing two samples, if the data are paired, then `pair=TRUE`.
34. When comparing more than two samples, $q < 0.05$ means that the gene is differentially expressed between at least two of the samples.

Acknowledgments

N.M.C. and A.P.F. are supported by an NSF GRF (DGE-1252376). This work was funded by an NSF CAREER grant (MCB-1453130) and by the Bilateral BBSRC NSF/BIO (MCB-1517058) to R.S.

References

1. Fisher, A.P. & Sozzani, R (2016) Uncovering the networks involved in stem cell maintenance and asymmetric cell division in the Arabidopsis root. Curr Opin Plant Biol 29:38–43.
2. Birnbaum K, Jung Jee W, Wang Jean Y et al (2005) Cell type-specific expression profiling in plants via cell sorting of protoplasts from fluorescent reporter lines. Nat Methods 2:615–619. https://doi.org/10.1038/nmeth0805-615
3. Trapnell C, Roberts A, Goff L et al (2012) Differential gene and transcript expression analysis of RNA-seq experiments with TopHat and Cufflinks. Nat Protoc 7:562–578. https://doi.org/10.1038/nprot.2012.016
4. Robinson MD, McCarthy DJ, Smyth GK (2009) edgeR: a bioconductor package for differential expression analysis of digital gene expression data. Bioinformatics 26:139–140. https://doi.org/10.1093/bioinformatics/btp616
5. Li J, Witten DM, Johnstone IM, Tibshirani R (2012) Normalization, testing, and false discovery rate estimation for RNA-sequencing data. Biostatistics 13:523–538. https://doi.org/10.1093/biostatistics/kxr031
6. Aronesty E (2013) Comparison of Sequencing Utility Programs. Open Bioinforma J 7:1–8. https://doi.org/10.2174/1875036201307010001
7. Liao Y, Smyth GK, Shi W (2013) The Subread aligner: fast, accurate and scalable read mapping by seed-and-vote. Nucleic Acids Res 41:e108. https://doi.org/10.1093/nar/gkt214

Chapter 7

Computational and Experimental Approaches to Predict Host–Parasite Protein–Protein Interactions

Yesid Cuesta-Astroz and Guilherme Oliveira

Abstract

In host–parasite systems, protein–protein interactions are key to allow the pathogen to enter the host and persist within the host. The study of host–parasite molecular communication improves the understanding the mechanisms of infection, evasion of the host immune system and tropism across different tissues. Current trends in parasitology focus on unraveling host–parasite protein–protein interactions to aid the development of new strategies to combat pathogenic parasites with better treatments and prevention mechanisms. Due to the complexity of capturing experimentally these interactions, computational approaches integrating data from different sources (mainly "omics" data) become key to complement or support experimental approaches. Here, we focus on the application of experimental and computational methods in the prediction of host–parasite interactions and highlight the potential of each of these methods in specific contexts.

Key words Host–parasite interactions, Proteomics, Computational biology, Secretome, Parasitology, Protein–protein interactions, Systems biology

1 Introduction

The word parasite originates from Greek *parasitos* meaning "a person who eats at the table of another" [1]. Parasitism is defined as the dependency of a parasite on another organism, the host, feeding at its expense during part or its entire life [2]. This relationship involves mainly a degree of metabolic dependence of the parasite upon its host and can damage the host. Parasites affect a wide range of hosts, including plants, invertebrates, and vertebrates [2] and present in some cases complex life cycles that can require intermediate hosts such as gastropods or insects during the asexual stage [3, 4]. Some parasites spend at least a portion of their life cycle within the intracellular environment of a host cell, establishing complex molecular interactions within the host cells [5]. These molecular interactions are used by the parasite for instance to migrate through various tissues, evade the host immune

Louise von Stechow and Alberto Santos Delgado (eds.), *Computational Cell Biology: Methods and Protocols*, Methods in Molecular Biology, vol. 1819, https://doi.org/10.1007/978-1-4939-8618-7_7,

system and undergo intracellular replication, all of them essential for the completion of the parasite's life cycle [5]. Some examples of intracellular parasites include some genus such as *Leishmania*, *Trypanosoma*, *Plasmodium*, and *Toxoplasma*, among others. On the other hand extracellular parasites replicate and live in the extracellular spaces, and are generally too large to be phagocytosed. An example of the extracellular parasitic lifestyle are helminths, which are almost all extracellular parasites, with the exception of *Trichinella spiralis* larvae, which live in large striated muscle cells [2].

Parasites have great socioeconomic impact and are responsible for infections in humans, animals and plants. According to the World Health Organization (WHO) more than two billion people are infected by parasites, especially in developing countries [6]. As defined by the DALYs concept (disease burden as defined by Disability-Adjusted Life-Years), schistosomiasis together with diseases caused by hookworms and leishmaniasis, are among the tropical neglected diseases with the highest epidemiological weight quantified in DALYs [7, 8]. Another example of this burden is *Plasmodium falciparum*, which is responsible for most of the malaria deaths, with an estimated 214 million cases and 438,000 deaths worldwide in 2015 [6]. An estimated seven million people are infected with *Trypanosoma cruzi* worldwide, mostly in 21 continental Latin American countries. Chagas disease causes more than 7,000 deaths per year [6]. In addition to the devastating effects mentioned above, parasites cause enormous economic loss in agriculture and livestock [9–11]. Therefore, it is essential to develop new approaches to control parasitic disease and to find new chemotherapeutic targets [12].

Genomes from different parasite species relevant in multiple fields such as medicine, veterinary and agriculture are being currently sequenced. Genome, transcriptome, proteome, and metabolome data have provided valuable information for the understanding of host–parasite interactions, immune system evasion mechanisms, molecular evolution and the metabolism of the species sequenced [13–19]. The analysis of these "omics" data has opened new frontiers in the study and control of parasites and has contributed to the development of in silico approaches that use all these data to decipher parasite's biology [20].

The pathologies caused by parasites are the end result of a complex interaction of host and parasite factors (e.g., host genetic constitution and parasite strains), coinfections, and the environment [21, 22]. Parasites affect their hosts partly by interacting with proteins of the host cells, which defines a molecular interplay between the parasite's survival mechanisms and the host defense system [8]. The prediction of molecular host–parasite interactions is essential to understand parasitic infection mechanisms and pathogenesis, which describe the cellular processes that cause the

Table 1
Summary of human–parasite PPIs detection methods reviewed in this chapter

Approach	Technique	Parasites	Reference
Experimental	Y2H^{G}	*T. gondii*	[40]
	Phage display^{G}	*C. parvum*	[48]
	Coimmunoprecipitation^{B}	*F. hepatica*	[52]
	Affinity purification^{B}	*T. cruzi*	[60]
	Cross-linking^{B}	*P. falciparum*	[63]
	Protein arrays^{B}	*S. mansoni*	[65]
Computational	Homology	*P. falciparum*	[74]
	Domains and motifs	*P. falciparum*	[85]
	Structure-based	*L. major*, *T. brucei*, *T. cruzi*, *C. hominis*, *C. parvum*, *P. falciparum*, *P. vivax*, *T. gondii*	[88]
	Machine learning	*P. falciparum*	[97]
	Coexpression	*P. falciparum*	[101]

G genetic, *B* biochemical method

diseased state. Deciphering the host–parasite interactomes can help to elucidate the mechanisms that allow parasite species to adapt to their hosts and to different microenvironments, and to shed light on different pathogenic states [23, 24].

Advances in computational and experimental methods have opened new and exciting possibilities for the study of such molecular interactions. Both approaches are complementary to each other. Computational methods can efficiently serve as a theoretical base for experiments aiming to identify relevant interactions across different parasites and their host. This chapter addresses both approaches, describing some of the most frequently used methods and their applications focusing on several human parasites (Table 1).

2 The Host–Parasite Interface: Secreted and Transmembrane Proteins

Secreted proteins play essential roles in a myriad of cellular processes in species from bacteria to mammals [25]. Such proteins represent between 8% and 20% of the organism's total proteome [26]. The interactions between parasites and their hosts are mediated in large part by secreted proteins, collectively known as the "secretome." Secreted proteins belong to several functional classes: cytokines, hormones, digestive enzymes, antibodies, proteases, toxins, antimicrobial peptides, and proteins associated with oxidative stress [25]. Some of these proteins are involved in vital biological processes, such as cell adhesion, cell migration, cell–cell communication, differentiation, proliferation, morphogenesis, and regulation of the immune response [27]. Secreted parasite proteins are of particular interest for the understanding of host–parasite interactions [28], and have been shown to play crucial roles such as regulating the host's immune response and pathological states [27]. Parasite secretomes contain proteases such as aspartate, cysteine, serine proteases and metalloproteases, which are involved in blood coagulation processes, fibrinolysis, protein metabolism, immune reaction, and tissue remodeling [21, 28]. Modulation of the host immune system during infection depends on the life span of the parasite in the host, and this process depends on proteins and other molecules secreted by the parasite that interact with the host [21].

The secretion of proteins through the endoplasmic reticulum is associated with a signal peptide in the N-terminal region and represents the classic secretion pathway [25]. Some proteins are secreted by the nonclassical pathway via extracellular vesicles and do not contain a signal peptide. Examples of proteins secreted by the nonclassical pathway include glycolytic enzymes, chaperones, and translation factors, which suggests that these proteins, whose function is normally restricted to intracellular activities, could be multifunctional (moonlighting proteins) [29]. Microscopy studies have provided experimental evidence of extracellular vesicles in Helminthes, specifically in the trematodes *Fasciola hepatica*, *Echinostoma caproni* [30], *Schistosoma japonicum* [31], and *Schistosoma mansoni* [32]. These vesicles are actively released by the parasite and are captured by the host cells playing an important role in host–parasite interaction [31].

Transmembrane proteins constitute another key subproteome, which is relevant for the host–parasite interface. Localization of these proteins is determined by hydrophobic transmembrane (TM) helices that anchor proteins to the cell membrane. Transmembrane proteins are classified as type I (N-terminus extracellular), type II (C-terminus extracellular), multipass, lipid-anchored or GPI-anchored TM proteins [33]. Such proteins are known to be involved in different processes such as nutrition, excretion, signal transduction, osmoregulation, and immune evasion and modulation, and also play a key role in host–parasite interactions, as described for the human parasite *S. mansoni* [34]. The tegument of *S. mansoni* is important for the interaction with its environment and mainly with the host, for example during the transition from free-living cercariae to the intramammalian schistosomula stages [34]. Using a combination of tegumental labeling (tegument of *S. mansoni* suffered biotinylation to label host-exposed tegumental proteins) and high-throughput quantitative proteomics techniques, Sotillo et al. [34] identified several proteins that are highly expressed on the tegument of *S. mansoni* schistosomula at different stages of development. For example, tetraspanins (TSPs) are a family of transmembrane proteins that have been shown to be essential for tegument formation, and have been shown to be potential drug and vaccine targets in animal models of schistosomiasis [35]. These proteins were detected in biotinylated and unbound tegument tissues. Such proteomic approaches are useful for the design of novel therapeutics against schistosomiasis.

3 Experimental Methods to Explore Host–Parasite Protein–Protein Interactions

Several in vivo and in vitro experimental approaches have made significant contributions to screening a large number of protein interaction partners, especially high-throughput experimental methods [36]. However, our understanding of interactions between proteins of parasites and those of human hosts is limited, as no systematic experimental studies have been performed to date. Some experimental data and large interaction datasets exist for examining intraspecies PPIs [37], but few data are available on interspecies PPIs and even less on interactions between proteins of parasites and their hosts. The experimental techniques employed in the investigation of PPIs are divided into genetic methods and biochemical methods. We will briefly describe some of the most relevant methods used in the study of host–parasite interactions.

3.1 Y2H (Yeast Two-Hybrid)

Genetic methods were first employed for studying PPIs with the development of the first genetic screen in yeast [38]. Y2H (yeast two-hybrid) is a genetic method and it is built upon the transcriptional activation of a reporter gene as a result of the interaction of protein X (fused to a DNA-binding domain and also called bait protein) with protein Y (also called prey protein and fused to a transcription activation domain) [39]. Basically, in this technique a given protein is assayed against a mixture of full-length proteins, protein domains and/or protein fragments expressed from a cDNA library, followed by isolation of the protein's interacting partners [40].

The main advantage of this method is the speed and simplicity, making it broadly applicable for the exploration of PPIs by large-scale interaction mapping. Y2H screens have led to a considerable increase in the number of protein interactions reported in the scientific literature. This technique diversified to support the analysis of PPIs in the cell membrane in the membrane Y2H (MYTH) assay [41, 42].

A study in the system host–*Toxoplasma gondii* showed the applications of Y2H in the identification of PPIs. *T. gondii* is an obligate intracellular protozoan parasite and has the ability to co-opt host cell machinery to maintain its intracellular survival. *T. gondii* can modulate signaling pathways of its host through the secretion of effector proteins localized in the rhoptry and dense granule organelles. These effectors are also essential for parasite invasion and colonization [40]. TgGRA15 is a *T. gondii* effector protein localized in the dense granule organelles. This protein is involved in the modulation of host signaling pathways. To further understand functions of TgGRA15, Liu and collaborators identified the host cellular proteins that interact with this protein by screening of a yeast two-hybrid mouse cDNA library, using TgGRA15 as the bait. The results indicated that two proteins Luzp1 (leucine zipper protein 1) and AW209491 interacted with TgGRA15. Luzp1 is a nucleus-targeted protein and is involved in regulating a subset of host noncoding RNA genes and AW209491 is of unknown function [40]. Understanding the specific mechanisms, which underlie the interactions of parasite effector proteins with host proteins may lead to the identification of novel therapeutic strategies for prevention and treatment.

3.2 Phage Display

The phage display technique is another genetic method to study protein–protein interactions. This method has been extensively employed in the identification of novel in vitro and in vivo ligands in different areas, mainly in cancer studies, vaccine development, and epitope mapping [43]. One of the most common approaches for phage display utilizes the M13 filamentous phage [43]. The DNA encoding the protein of interest is inserted into a phage coat protein gene, causing the phage to "display" the protein on its outside while containing the gene for the protein inside [44]. These displaying phages can be screened against other proteins or peptides to detect interactions with the displayed protein. Phage display has been implemented for the investigation of host–pathogen interactions and has been used for several parasites with encouraging results [43, 45, 46].

Phage display was implemented to unravel host–parasite PPIs in *Cryptosporidium* species, which are enteric coccidian parasites causing cryptosporidiosis in mammalian hosts. This disease affects the distal small intestine resulting in, among other symptoms, diarrhea. In immunocompromised individuals, the symptoms are particularly severe and can be fatal. The parasite spreads via the fecal–oral route, often through contaminated water [47]. The species *C. parvum* infects humans, especially immunocompromised individuals and the disease can become chronic and cause gastroenteritis with high mortality [43]. Infection occurs when ingested oocysts release sporozoites in the intestines of the host. The sporozoites then attach to epithelial cells of the gastrointestinal tract. The interactions between host and *C. parvum* have been investigated by panning a cDNA library of the sporozoite and oocyst stages expressed on the surface of T7 phage against intestinal epithelial cells (IECs). This study identified the CP2, a known surface protein of sporozoites involved in the invasion process, when the *C. parvum* T7 phage display library was screened by using Caco-2 (human epithelial colorectal adenocarcinoma) cells [48].

3.3 Coimmunoprecipitation and Affinity Purification

One of the most commonly applied biochemical methods to detect PPIs is coimmunoprecipitation (Co-IP). The principle is considered a "fishing expedition" to identify protein complexes in cell lysates by using an antibody directed against one of the interacting proteins and subsequent isolation of the immune complex using immobilized protein A or protein G [39]. Coimmunoprecipitation can be combined with mass spectrometry (MS) for the identification of the interacting proteins. Affinity purification (AP) is similar to co-IP. The approach combines tagging the bait protein with an affinity tag (His, glutathione *S*-transferase GST, maltose-binding protein MBP, calmodulin-binding peptide, small epitope tag such as myc) and purifying the complex by affinity or immunoaffinity and identifying interacting proteins by MS (AP-MS) [39, 49]. An innovative form of AP is tandem affinity purification (TAP) [50], in this method the tag has an immunoglobulin G binding fragment and a CBP (calmodulin-binding peptide).

Co-immunoprecipitation method has been applied to the study of the interaction of parasitic secreted proteins and host cells. Recently, Liu and collaborators investigated the key molecules in FhESPs (*Fasciola hepatica* excretory and secretory products), which were involved in suppressing and evading the host's immune responses. These secreted proteins have shown that the helminth parasite FhESPs play critical roles in modulating the host immune response [51]. To demonstrate this role, they explored the interactions between FhESPs and goat peripheral blood mononuclear cells (PBMCs) and cytokines including IL2, IL17, and IFN-γ by coimmunoprecipitation (Co-IP). This study identified proteins in FhESPs that could bind to IL2, IL17, IFN-γ, and PBMCs. These proteins can be potential targets as immune-regulators, and will be helpful to elucidate the molecular basis of host–parasite interactions and might serve as potential vaccines and drug target candidates [52].

Trypanosoma cruzi is an intracellular protozoan parasite that causes Chagas disease, which represents an important health issue in Latin America [53]. Cruzipain (Cz) the major cysteine proteinase (CP) of *T. cruzi*, contains a catalytic domain and a carboxy-terminal extension (C-T) that is responsible for the immunodominant antigenic character of the protein in natural and experimental infections [54, 55]. In previous studies it was shown that in presence of sialic acid and sulfated oligosaccharides in Cz, these sulfated structures played an important role in the immune response of the host [56–58]. In 2015, Ferrero and collaborators presented the first report of sulfates as parasite ligands enhancing Siglec-E recognition and also Siglec-E binding to different virulent strains and isolates from patients. Siglecs are a family of sialic acid-binding lectins, mainly expressed on cells of the host immune system, such as macrophages, monocytes, B cells, neutrophils, eosinophils, and basophils, among others [59]. Ferrero and collaborators performed ELISA (enzyme-linked immunosorbent assay) assays to determine the binding capacity of Siglec-E to *T. cruzi* strains and to evaluate the ability of different *T. cruzi* biomolecules to interact with Siglec-E [60]. Affinity purification of Cz by a Siglec-E column confirmed the interaction between both molecules [60]. *T. cruzi* sulfation has a determining role in the immunomodulation of the host response upon infection, and denotes the huge importance of Cz sulfation in the evolution of Chagas disease [60].

3.4 Cross-Linking

Cross-linking is another biochemical method to identify protein–protein interactions. This technique uses cross-linking chemical reagents with two or more reactive groups that are connected by a spacer or linker region to identify a protein interaction network [61]. This method can be combined with other techniques such as affinity tag and MS [61]. For cross-linking of proteins, mainly amine-, sulfhydryl-, and photo-reactive groups are used. Protein interactions are often too weak or transient to be easily detected. Cross-linking can stabilize the interactions. This method is unique in its ability to capture a "snap-shot" of an interaction [39].

Cross-linking was implemented in the study of malaria. Malaria is caused by apicomplexan parasites of the genus *Plasmodium*, which are intracellular parasites that develop inside a vacuole that separates them from the host cell cytosol [62]. The parasite–host cell interface (vacuole) is a key compartment but little is known about its molecular composition and architecture. In a 2006 study, Spielmann and collaborators analyzed the parasite–host cell interface of the most virulent human malaria parasite *Plasmodium falciparum*. The investigators employed in vivo cross-linking to analyze the molecular architecture of the malaria PVM (parasitophorous vacuole membrane). The membrane of this vacuole is the only mode of contact between the host cell and the parasite. In this study it was shown that the interface between *P. falciparum* parasites and their host cell contains ETRAMPs and EXP-1 proteins organized into oligomeric arrays and interact with either themselves or other proteins of a similar molecular weight [63].

3.5 Protein Arrays

Protein arrays are used as a high-throughput method to track the interactions and activities of proteins, and to determine their function [64]. The protein array, or protein chip, has a large number of spots of either proteins or their ligands arranged in a predefined pattern onto coated glass slides, microplates, or membranes [64]. The array may consist of antibodies that bind the proteins of interest or enzymes that will react with substrates or ligands [64]. In the study of host–parasite interactions, this technique is useful because the arrays can be probed with small volumes of host serum or plasma. Therefore the high-throughput nature of the protein array approach to antigen discovery is the key to determining the repertoire of antigens that can elicit defined immune responses in different parasitic diseases [65].

Recent advances in high order multiplexing, like protein microarrays, provide a high-throughput technology that can capture the "immunome," which is the repertoire of antibodies against the antigens or epitopes that interface with the host immune system [66]. This strategy was implemented in schistosomiasis which is among the most important of the neglected tropical diseases (NTDs), and may cause as many as 28 million Disability Adjusted Life Years (DALYs) lost and 280,000 deaths annually [67]. Five species of the genus *Schistosomiasis* (Trematoda) are involved in human infection, including *Schistosoma mansoni* and *Schistosoma japonicum* the main etiologic agents of human schistosomiasis [20]. In a 2016 study, Assis and collaborators described the production of a *S. mansoni* protein microarray, with over 900 *S. mansoni* antigens [65].

Many proteome arrays have been used to screen antigens of virus, bacteria and protozoan pathogens [68, 69], only few protein arrays have been established for multicellular pathogens. In this study, the investigators used sera from individuals either acutely or chronically infected with *S. mansoni* from endemic areas in Brazil, and sera from individuals resident outside the endemic area (the USA) to test if the array was functional and informative [65]. Of the 92 proteins printed on the pilot array, over 50 proteins were recognized by sera from individuals with either chronic or acute schistosomiasis. IgG profiles were most contrasting between nonendemic controls and *S. mansoni* chronically infected individuals [65]. Protein arrays provide an ideal platform to reveal new candidate antigens for schistosomiasis.

4 Computational Methods to Predict Host–Parasite Protein–Protein Interactions

The number of available datasets providing host–parasite PPIs is limited and challenged by the intrinsic difficulties of simultaneously analyzing the host and pathogen systems in high-throughput experiments [24]. Computational methods can be beneficial to close the gap left by experimental description of PPIs, to support the interactions that have been detected by experimental approaches or to predict new interactions. Fortunately, there are many computational resources that can be combined to predict interactions and support experiments (Table 2).

Table 2
Human and parasite databases/tools used to support or context experimental and computational predictions of host–parasite PPIs

Organism (human/parasite)	Database/tool	Summary	Source link
HP	iPfam	A database of protein family and domain interactions calculated from known structures.	http://ipfam.org/
HP	3did	A database of three-dimensional interacting domains	http://3did.irbbarcelona.org/
HP	ELM	This resource focuses on annotation and detection of eukaryotic linear motifs	http://elm.eu.org/
HP	Pfam	Protein families database	http://pfam.xfam.org/
HP	Reactome	Pathways database	http://reactome.org/
HP	KEGG	Pathways database	http://www.genome.jp/kegg/
HP	Gene Ontology	Gene annotations.	http://geneontology.org/
HP	SignalP	Prediction of signal peptides from amino acid sequences. Proteins with signal peptides are targeted to the secretory pathway	http://www.cbs.dtu.dk/services/SignalP/
H	COMPARTMENTS	Subcellular localization database	https://compartments.jensenlab.org/
H	TISSUES	Tissue expression database	https://tissues.jensenlab.org/
H	TissueNet v.2	A database of protein–protein interactions across human tissues	http://netbio.bgu.ac.il/tissuenet/
P	GeneDB	Sequence data and annotation/curation for some parasites species	http://genedb.org/
P	EupathDB	Integrative database of eukaryotic pathogens include: sequencing, expression, proteomics, metabolic and phenotype data	http://eupathdb.org/

HP human/parasite, *H* human, *P* parasites

Table 3
Intraspecies PPIs databases

Database name	Total number of interactions	Source link	Number of organisms
IntAct 4.2.7	506,367	http://www.ebi.ac.uk/intact/	8
STRING	1,380,838,440	https://string-db.org/	2031
APID	678,441	http://apid.dep.usal.es	25
DIP	81,766	http://dip.doe-mbi.ucla.edu/dip/	834
HitPredict	547,879	http://hintdb.hgc.jp/htp/	115
MINT	125,464	http://mint.bio.uniroma2.it/	611
BioGrid 3.4	1,495,320	https://thebiogrid.org/	61

Computational methods for PPI prediction are based on sequence or structural features, chromosome proximity, gene fusion, phylogenetic profile, and gene expression-based approaches [70]. Prediction methods are well investigated for the specific case of intraspecies PPIs (Table 3). Yet it remains challenging to adapt these approaches for interspecies cases such as host–parasite interactions [70]. The advances in technologies linked to an increase in computational power have led to a surge in the volume of genomic, transcriptomic and proteomic data of different parasite species. These data provide a source of information that can be exploited by computational methods to predict host–parasite interactions. Additionally, host–parasite interactions are often mediated by proteins domains and motifs, which can be also identified in the host–parasite system and used in the prediction. Parasites use interactions with the host proteins with a specific purpose, thus, these interactions are not random but target specific proteins of the host [70]. This information can as well be exploited by computational approaches to predict these interactions. In this section we will show the most commonly used computational methods applied to the prediction of host–parasite interaction.

4.1 Homology-Based Approaches

The most commonly used computational approach to predict PPIs is to map known interactions onto homologous pairs of sequences [24]. The hypothesis is that the interaction between a pair of proteins in one species is expected to be conserved homolog proteins in another species, which is called interologs [71]. This approach is implemented for intraspecies and interspecies PPI predictions. In the interspecies scenario this method starts from a known PPI (template PPI) in a species identifying the interacting proteins X and Y. Subsequently, homolog proteins in the host and pathogen

(*X'*, *Y'*) are predicted to interact as well [72]. Despite being widely used, the homology approach is not sufficient for evaluating the biological evidence of host–parasite interactions. Different filtering techniques such as expression profiles (subcellular localization, tissue), functional annotations, etc. should be considered to assess the feasibility of the interactions and to consequently decrease the identification of falsely predicted interactions [24, 73].

Based on the HomoloGene database, *P. falciparum* has the least similar genome in comparison to other 17 eukaryotic species (*H. sapiens*, *M. musculus*, *R. norvegicus*, *P. troglodytes*, *C. familiaris*, *G. gallus*, *A. thaliana*, *O. sativa*, *A. gambiae*, *D. melanogaster*, *N. crassa*, *M. grisea*, *C. elegans*, *S. cerevisiae*, *K. lactis*, *E. gossypii*, and *S. pombe*) [74]. This suggests that many cellular processes vital to other eukaryotes may be missing or replaced in *P. falciparum*. In a 2008 study, Lee et al. used experimental PPIs and interologs to infer *H. sapiens–P. falciparum* interactions. Experimental PPIs were collected from the POINT database, which has available public PPI data for a range of organisms. Most of these interactions were obtained from high-throughput techniques such as yeast two-hybrid screening [74].

Based on this PPIs source, 3,090 interspecies interactions between *P. falciparum* and *H. sapiens* were obtained. The interactions were grouped by biological process in which the interacting proteins function. Metabolic processes and cellular processes of *P. falciparum* were the most abundant in the host–parasite interaction network, likely due to the nutrient requirements of the host erythrocyte intracellular parasite. Although more than 3,000 interactions were inferred, not all of these interactions are likely to take place under physiological conditions due to spatiotemporal constraints. Filtering by specific biological processes important in the parasite's life cycle reduced the interactions to 918. The filtering helped revealing that *P. falciparum* might use calcium-modulating proteins in the host cell to maintain Ca^{2+} levels.

4.2 Domain and Motif Interaction-Based Approaches

The protein binding domain and linear motif interaction-based approach is another sequence-based approach to predict host–parasite interactions. A protein domain is a distinct, compact, and stable protein structural unit that folds independently of other such units and contributes to determine the structure and function of proteins [75]. Often, domains and motifs mediate interactions of parasite proteins with host proteins.

A resource for domain–domain interactions based on domains definition given by Pfam models is the Domain Interaction Map (DIMA) database [76]. This database integrates several methods for predicting domain–domain interactions. For instance, the Domain Pair Exclusion Method (DPEA) [77] studies the frequency of known interactions of co-occurring domain pairs. The DIP database [78] is based on machine learning techniques to determine domain interaction pairs that are predictive for protein interactions. Other methods are based on correlated mutations and search for domain pairs that contain co-evolving residues [79]. iPfam [80] and 3DiD [81] analyze PDB structures and retrieve domain pairs that are in close contact in these structures. A wide range of protein domains recognize their substrates by short linear motifs, playing an important role in eukaryotic specific signal-transduction and regulation processes, making them an interesting target for pathogens [70]. A web resource for these short linear motifs is the ELM database [82].

P. falciparum, is known for its ability to remodel host cells, particularly erythrocytes, to persist in the host environment [83, 84]. The invasion of red blood cells (RBCs) by malarial parasites is an essential step in the life cycle of *P. falciparum*. However, this mechanism remains poorly understood because numerous human–parasite interactions have not yet been identified and high-throughput screening experiments are not feasible for malarial parasites due to difficulty in expressing the parasite proteins [85]. Liu et al., performed a computational prediction of the PPIs involved in malaria parasite invasion to elucidate the mechanism by which invasion occurs based on membrane protein interactions between human and *P. falciparum* [85].

In this work Liu et al., an algorithm was used to estimate the probabilities of domain–domain interactions (DDIs) between *H. sapiens* and *P. falciparum* [85]. These probabilities were then used to infer PPI probabilities. Gene expression data was integrated to improve prediction accuracy and to reduce false positives. A network consisting of 205 PPIs, was obtained. The results of the network analysis suggested that SNARE (soluble n-ethylmaleimide-sensitive factor attachment protein receptor) proteins of parasites and APP (amyloid precursor protein) of humans are involved in the invasion of RBCs by parasites [85]. The functions of APP and SNARE in parasite invasion were not previously known.

4.3 Structure-Based Approaches

Structure-based approaches use three-dimensional structure similarity to predict protein–protein interactions. If two proteins A and B interact, then proteins A′ and B′ with similar structures to those of proteins A and B are predicted to interact as well [44]. However, the drawback of this method is that most protein structures are not known. In this case it can be helpful to predict the structure of the protein based on its sequence. This process is called homology modeling and requires an experimentally obtained three-dimensional structure of a related homolog protein (the "template"). Sequences falling below a 20% sequence identity can have very different structures [86] hence a good template must have more than 30% in terms of sequence identity, to increase the accuracy of the model [87].

Davis et al., used sequence and structural similarities to predict interactions between human proteins and proteins of ten different pathogens (eight parasites: *L. major*, *T. brucei*, *T. cruzi*, *C. hominis*, *C. parvum*, *P. falciparum*, *P. vivax*, *T. gondii* and two bacteria: *M. leprae*, *M. tuberculosis*) [88]. The investigators first scanned the host and pathogen genomes for proteins with similarities to known protein complexes. They then assessed putative interactions, using available structure, filtered the interactions using biological context, such as stage-specific expression of pathogen proteins and tissue expression of host proteins [88]. The protein structure of the host and parasite proteins were modeled using MODPIPE [89] and pairs of human–pathogen proteins with similarity to known interactions from PIBASE [90] were then identified. Using the described method, the authors identified interactions that had been previously observed experimentally such as interactions between proteases and protease inhibitors. The highest scoring interaction was between *P. falciparum* falcipain-2 protease and the human cystatin-A inhibitor. This study provides a means to mine whole-genome data and is complementary to experimental efforts in elucidating host–pathogen protein interactions [88].

4.4 Machine Learning Approaches

Applying machine learning techniques in bioinformatics has been widely used, this includes efforts to predict PPIs [91, 92]. These methods involve the use of available PPI data as features for training and classifying interacting and noninteracting protein pairs [73]. Both supervised [93, 94] and semisupervised [95] methods are used for the identification of host–pathogen PPIs. A considerable amount of interacting and noninteracting protein pairs are usually needed for training machine learning algorithms to produce good classifiers [72]. Thus, the availability of training data might constitute a limiting factor for the application of machine learning approaches to host–parasite interactions.

Random forest is one of the methods used in the prediction of PPIs. This method builds a collection of weighted decision trees and ensembles the outputs from these individual trees to obtain the final decision [96]. Wuchty and colleagues applied a random forest classifier to assess the quality of the interactions obtained by homology-based approaches. Subsequently, the predicted interactions were filtered accounting for parasite specific characteristics to support the interactions in a biological context. A random forest classifier was implemented to consider interacting candidate pairs based on their sequence composition [97]. In this study the investigators identified different chaperones (mainly HSP70) interacting with human proteins that play important roles in cell signaling such as members of the TNF pathway. This parasite chaperones might remodel protein structures in the host cell, helping the parasite to take control of the cell.

4.5 Co-expression Approach

The information about transcriptional regulation can be utilized to predict and validate PPIs by calculating the correlation coefficient of transcriptome data. Different types of transcriptome data, such as for example RNA sequencing, DNA microarrays, expressed sequence tag (EST), can be used [98]. In addition, clustering algorithms or analyses of topological structures of co-expression networks can help to reveal functional relationships and predict PPIs [99]. In a study that used both yeast expression data and proteome data as input, the authors found that proteins encoded by genes belonging to a common expression profiling cluster were more likely to interact with each other than proteins from the genes belonging to different clusters [100].

The use of gene expression data from parasite and host is useful in order to find pairs of host–parasite genes with correlating expression profiles. This approach was called interspecies interactions using gene expression measurements (ISIGEM) in a study that found detectable signatures of molecular interactions between host and parasite based on transcriptome data [101]. The ISIGEM approach appears to be accurate in identifying pairs of genes from functional modules that interact between species [101]. Predictions made by this approach found that the identified genes were enriched for functional terms related to host–parasite interaction. For example, several host genes involved in various aspects of malaria infection. The approach can be applied to study interactions between any species, for example, between host and parasite, but it requires simultaneous expression datasets to improve its reliability.

5 Conclusions and Perspectives

In this chapter we describe different computational and experimental methods to predict host–parasite protein–protein interactions. These methods may complement each other, and the complexity of identifying these PPIs may require the combination of diverse approaches (computational and experimental) and integrating different layers of data. Computational methods definitely improve the coverage and simplify the identification of host–parasite interactions in combination with existing experimental data for the host–parasite system. Furthermore, computational predictions can serve as a starting point to have a theoretical model to begin experimental efforts, which can save time and resources.

Progress in host–parasite studies will be achieved, mainly due to availability of experimental data, improved computational methods and the increase in curated PPIs databases. Those databases contribute significantly to host–parasite studies by collecting and integrating valuable and heterogeneous data, thus providing powerful tools to researchers. Another progress in this field is that genome and proteome data from different parasites species are becoming available. This will facilitate the development of accurate computational PPI predictions that integrate functional information including GO annotation, gene expression and pathway data.

Acknowledgments

We would like to thank the editors for the opportunity to contribute to this book. This work was supported by the National Institutes of Health-NIH/Fogarty International Center, USA (TW007012 and 1P50AI098507-01) to G.O., Coordenação de Aperfeiçoamento de Pessoal de Nível Superior-CAPES, Brazil (REDE 21/2015 and 070/13) to G.O., FAPEMIG (RED-00014-14 and PPM-00189-13) to G.O., and Conselho Nacional de Desenvolvimento Científico e Tecnológico-CNPq, Brazil (304138/2014-2) to G.O. G.O. is a CNPq fellow (307479/2016-1), and Y.C.A. a CAPES fellow. An EMBO short-term fellowship (400-2015) to Y.C.A is acknowledged.

References

1. Xu F, Jerlstrom-Hultqvist J, Kolisko M et al (2016) On the reversibility of parasitism: adaptation to a free-living lifestyle via gene acquisitions in the diplomonad Trepomonas sp. PC1. BMC Biol 14:62. https://doi.org/10.1186/s12915-016-0284-z
2. Gunn A, Jane Pitt S (2012) Parasitology: an integrated approach. Wiley, London, pp 86–136. https://doi.org/10.1017/S0031182012001412
3. RAUCH G, KALBE M, TBH REUSCH (2005) How a complex life cycle can improve a parasite's sex life. J Evol Biol 18:1069–1075. https://doi.org/10.1111/j.1420-9101.2005.00895.x
4. Antonovics J, Wilson AJ, Forbes MR et al (2017) The evolution of transmission mode. Philos Trans R Soc Lond Ser B Biol Sci. https://doi.org/ 10.1098/rstb.2016.0083
5. Walker DM, Oghumu S, Gupta G et al (2014) Mechanisms of cellular invasion by intracellular parasites. Cell Mol Life Sci 71:1245–1263. https://doi.org/10.1007/s00018-013-1491-1
6. WHO (2015) Investing to overcome the global impact of neglected tropical diseases. Third WHO report on neglected tropical diseases. WHO, Geneva
7. Hotez PJ, Alvarado M, Basáñez M-G et al (2014) The global burden of disease study 2010: interpretation and implications for the neglected tropical diseases. PLoS Negl Trop Dis 8:e2865. https://doi.org/10.1371/journal.pntd.0002865
8. Merrifield M, Hotez PJ, Beaumier CM et al (2016) Advancing a vaccine to prevent human schistosomiasis. Vaccine 34:2988–2991. https://doi.org/10.1016/j.vaccine.2016.03.079
9. Mehmood K, Zhang H, Sabir AJ et al (2017) A review on epidemiology, global prevalence and economical losses of fasciolosis in ruminants. Microb Pathog 109:253–262. https://doi.org/10.1016/j.micpath.2017.06.006
10. Blok VC, Pylypenko L, Phillips MS (2006) Molecular variation in the potato cyst nematode, Globodera pallida, in relation to virulence. Commun Agric Appl Biol Sci 71:637–638
11. Mantelin S, Bellafiore S, Kyndt T (2017) *Meloidogyne graminicola*: a major threat to rice agriculture. Mol Plant Pathol 18:3–15. https://doi.org/ 10.1111/mpp.12394
12. Andrews KT, Fisher G, Skinner-Adams TS (2014) Drug repurposing and human parasitic protozoan diseases. Int J Parasitol Drugs Drug Resist 4:95–111. https://doi.org/10.1016/j.ijpddr.2014.02.002
13. Brindley PJ, Mitreva M, Ghedin E, Lustigman S (2009) Helminth genomics: the implications for human health. PLoS Negl Trop Dis. https://doi.org/ 10.1371/journal.pntd.0000538

14. Lustigman S, Prichard RK, Gazzinelli A et al (2012) A research agenda for helminth diseases of humans: the problem of helminthiases. PLoS Negl Trop Dis. https://doi.org/10.1371/journal.pntd.0001582
15. Tsai IJ, Zarowiecki M, Holroyd N et al (2013) The genomes of four tapeworm species reveal adaptations to parasitism. Nature 496:57–63. https://doi.org/10.1038/nature12031
16. Zarowiecki M, Berriman M (2015) What helminth genomes have taught us about parasite evolution. Parasitology 142(Suppl):S85–S97. https://doi.org/10.1017/S0031182014001449
17. Swapna LS, Parkinson J (2017) Genomics of apicomplexan parasites. Crit Rev Biochem Mol Biol 52:254–273. https://doi.org/10.1080/10409238.2017.1290043
18. Veras P, Bezerra de Menezes J (2016) Using proteomics to understand how Leishmania parasites survive inside the host and establish infection. Int J Mol Sci 17:1270. https://doi.org/10.3390/ijms17081270
19. Greenwood JM, Ezquerra AL, Behrens S et al (2016) Current analysis of host–parasite interactions with a focus on next generation sequencing data. Zoology 119:298–306. https://doi.org/10.1016/j.zool.2016.06.010
20. Cuesta-Astroz Y, Scholte LLS, Pais FSM et al (2014) Evolutionary analysis of the cystatin family in three Schistosoma species. Front Genet. https://doi.org/10.3389/fgene.2014.00206
21. Wakelin D (1996) Helminths: pathogenesis and defenses. University of Texas Medical Branch at Galveston, Galveston
22. McCall L-I, Zhang W-W, Matlashewski G (2013) Determinants for the development of visceral leishmaniasis disease. PLoS Pathog 9:e1003053. https://doi.org/10.1371/journal.ppat.1003053
23. Salzet M, Capron A, Stefano GB (2000) Molecular crosstalk in host-parasite relationships: schistosome- and leech-host interactions. Parasitol Today 16:536–540
24. Cuesta-Astroz Y, Santos A, Oliveira G, Jensen LJ (2017) An integrative method to unravel the host-parasite interactome: an orthology-based approach. bioRxiv. https://doi.org/10.1101/147868
25. Tjalsma H, Bolhuis A, Jongbloed JD et al (2000) Signal peptide-dependent protein transport in Bacillus subtilis: a genome-based survey of the secretome. Microbiol Mol Biol Rev 64:515–547. https://doi.org/10.1128/MMBR.64.3.515-547.2000
26. Greenbaum D, Luscombe NM, Jansen R et al (2001) Interrelating different types of genomic data, from proteome to secretome:'oming in on function. Genome Res 11:1463–1468. https://doi.org/10.1101/gr.207401
27. Maizels RM, Yazdanbakhsh M (2003) Immune regulation by helminth parasites: cellular and molecular mechanisms. Nat Rev Immunol 3:733–744. https://doi.org/10.1038/nri1183
28. Cuesta-Astroz Y, Oliveira FS de, Nahum LA, Oliveira G (2017) Helminth secretomes reflect different lifestyles and parasitized hosts. Int J Parasitol doi: https://doi.org/10.1016/j.ijpara.2017.01.007
29. Nombela C, Gil C, Chaffin WL (2006) Non-conventional protein secretion in yeast. Trends Microbiol 14:15–21. https://doi.org/10.1016/j.tim.2005.11.009
30. Marcilla A, Trelis M, Cortés A et al (2012) Extracellular vesicles from parasitic helminths contain specific excretory/secretory proteins and are internalized in intestinal host cells. PLoS One 7:e45974. https://doi.org/10.1371/journal.pone.0045974
31. Zhu L, Liu J, Dao J et al (2016) Molecular characterization of S. japonicum exosome-like vesicles reveals their regulatory roles in parasite-host interactions. Sci Rep 6:25885. https://doi.org/10.1038/srep25885
32. Sotillo J, Pearson M, Potriquet J et al (2016) Extracellular vesicles secreted by Schistosoma mansoni contain protein vaccine candidates. Int J Parasitol 46:1–5. https://doi.org/10.1016/j.ijpara.2015.09.002
33. Anantharaman V, Iyer LM, Balaji S, Aravind L (2007) Adhesion molecules and other secreted host-interaction determinants in Apicomplexa: insights from comparative genomics. Int Rev Cytol 264:1–74
34. Sotillo J, Pearson M, Becker L et al (2015) A quantitative proteomic analysis of the tegumental proteins from Schistosoma mansoni schistosomula reveals novel potential therapeutic targets. Int J Parasitol 45:505–516. https://doi.org/10.1016/j.ijpara.2015.03.004
35. Loukas A, Tran M, Pearson MS (2007) Schistosome membrane proteins as vaccines. Int J Parasitol 37:257–263. https://doi.org/

10.1016/j.ijpara.2006.12.001
36. Chang J-W, Zhou Y-Q, Ul Qamar M et al (2016) Prediction of protein–protein interactions by evidence combining methods. Int J Mol Sci 17:1946. https://doi.org/10.3390/ijms17111946
37. Szklarczyk D, Morris JH, Cook H et al (2017) The STRING database in 2017: quality-controlled protein–protein association networks, made broadly accessible. Nucleic Acids Res 45:D362–D368. https://doi.org/10.1093/nar/gkw937
38. Fields S, Uetz P, Giot L et al (2000) A comprehensive analysis of protein-protein interactions in Saccharomyces cerevisiae. Nature 403:623–627. https://doi.org/10.1038/35001009
39. Ngounou Wetie AG, Sokolowska I, Woods AG et al (2014) Protein–protein interactions: switch from classical methods to proteomics and bioinformatics-based approaches. Cell Mol Life Sci 71:205–228. https://doi.org/10.1007/s00018-013-1333-1
40. Liu Q, Li F-C, Elsheikha HM et al (2017) Identification of host proteins interacting with Toxoplasma gondii GRA15 (TgGRA15) by yeast two-hybrid system. Parasit Vectors 10(1). https://doi.org/10.1186/s13071-016-1943-1
41. Gisler SM, Kittanakom S, Fuster D et al (2008) Monitoring protein-protein interactions between the mammalian integral membrane transporters and PDZ-interacting partners using a modified split-ubiquitin membrane yeast two-hybrid system. Mol Cell Proteomics 7:1362–1377. https://doi.org/10.1074/mcp.M800079-MCP200
42. Snider J, Kittanakom S, Damjanovic D et al (2010) Detecting interactions with membrane proteins using a membrane two-hybrid assay in yeast. Nat Protoc 5:1281–1293. https://doi.org/ 10.1038/nprot.2010.83
43. Tonelli RR, Colli W, Alves MJM (2012) Selection of binding targets in parasites using phage-display and aptamer libraries in vivo and in vitro. Front Immunol 3:419. https://doi.org/10.3389/fimmu.2012.00419
44. Rao VS, Srinivas K, Sujini GN, Kumar GNS (2014) Protein-protein interaction detection: methods and analysis. Int J Proteomics 2014:1–12. https://doi.org/10.1155/2014/147648
45. Ruiz A, Pérez D, Muñoz MC et al (2015) Targeting essential Eimeria ninakohlyakimovae sporozoite ligands for caprine host endothelial cell invasion with a phage display peptide library. Parasitol Res 114:4327–4331. https://doi.org/10.1007/s00436-015-4666-x
46. Carmona-Vicente N, Vila-Vicent S, Allen D et al (2016) Characterization of a novel conformational GII.4 norovirus epitope: implications for norovirus-host interactions. J Virol 90:7703–7714. https://doi.org/10.1128/JVI.01023-16
47. Clark DP (1999) New insights into human cryptosporidiosis. Clin Microbiol Rev 12:554–563
48. Guo A, Yin J, Xiang M et al (2009) Screening for relevant proteins involved in adhesion of Cryptosporidium parvum sporozoites to host cells. Zhongguo Ji Sheng Chong Xue Yu Ji Sheng Chong Bing Za Zhi 27:87–88
49. Miernyk JA, Thelen JJ (2008) Biochemical approaches for discovering protein-protein interactions. Plant J 53:597–609. https://doi.org/10.1111/j.1365-313X.2007.03316.x
50. Rigaut G, Shevchenko A, Rutz B et al (1999) A generic protein purification method for protein complex characterization and proteome exploration. Nat Biotechnol 17:1030–1032. https://doi.org/ 10.1038/13732
51. Zhang W, Moreau E, Peigné F et al (2005) Comparison of modulation of sheep, mouse and buffalo lymphocyte responses by Fasciola hepatica and Fasciola gigantica excretory-secretory products. Parasitol Res 95:333–338. https://doi.org/10.1007/s00436-005-1306-x
52. Liu Q, Huang S-Y, Yue D-M et al (2017) Proteomic analysis of Fasciola hepatica excretory and secretory products (FhESPs) involved in interacting with host PBMCs and cytokines by shotgun LC-MS/MS. Parasitol Res 116:627–635. https://doi.org/10.1007/s00436-016-5327-4
53. Manque PA, Probst CM, Probst C et al (2011) Trypanosoma cruzi infection induces a global host cell response in cardiomyocytes. Infect Immun 79:1855–1862. https://doi.org/10.1128/IAI.00643-10
54. Martinez J, Campetella O, Frasch AC, Cazzulo JJ (1991) The major cysteine proteinase (cruzipain) from Trypanosoma cruzi is anti-

genic in human infections. Infect Immun 59:4275–4277

55. Martínez J, Campetella O, Frasch AC, Cazzulo JJ (1993) The reactivity of sera from chagasic patients against different fragments of cruzipain, the major cysteine proteinase from Trypanosoma cruzi, suggests the presence of defined antigenic and catalytic domains. Immunol Lett 35:191–196
56. Barboza M, Duschak VG, Cazzulo JJ et al (2003) Presence of sialic acid in N-linked oligosaccharide chains and O-linked N-acetylglucosamine in cruzipain, the major cysteine proteinase of Trypanosoma cruzi. Mol Biochem Parasitol 127:69–72
57. Barboza M, Duschak VG, Fukuyama Y et al (2005) Structural analysis of the N-glycans of the major cysteine proteinase of Trypanosoma cruzi. FEBS J 272:3803–3815. https://doi.org/10.1111/j.1742-4658.2005.04787.x
58. Acosta DM, Arnaiz MR, Esteva MI et al (2008) Sulfates are main targets of immune responses to cruzipain and are involved in heart damage in BALB/c immunized mice. Int Immunol 20:461–470. https://doi.org/10.1093/intimm/dxm149
59. Macauley MS, Crocker PR, Paulson JC (2014) Siglec-mediated regulation of immune cell function in disease. Nat Rev Immunol 14:653–666. https://doi.org/10.1038/nri3737
60. Ferrero MR, Heins AM, Soprano LL et al (2016) Involvement of sulfates from cruzipain, a major antigen of Trypanosoma cruzi, in the interaction with immunomodulatory molecule Siglec-E. Med Microbiol Immunol 205:21–35. https://doi.org/10.1007/s00430-015-0421-2
61. Gingras A-C, Gstaiger M, Raught B, Aebersold R (2007) Analysis of protein complexes using mass spectrometry. Nat Rev Mol Cell Biol 8:645–654. https://doi.org/10.1038/nrm2208
62. Garcia-del Portillo F, Finlay BB (1995) The varied lifestyles of intracellular pathogens within eukaryotic vacuolar compartments. Trends Microbiol 3:373–380
63. Spielmann T, Gardiner DL, Beck H-P et al (2006) Organization of ETRAMPs and EXP-1 at the parasite-host cell interface of malaria parasites. Mol Microbiol 59:779–794. https://doi.org/10.1111/j.1365-2958.2005.04983.x
64. Melton L (2004) Protein arrays: proteomics in multiplex. Nature 429:101–107. https://doi.org/ 10.1038/429101a
65. de Assis RR, Ludolf F, Nakajima R et al (2016) A next-generation proteome array for Schistosoma mansoni. Int J Parasitol 46:411–415. https://doi.org/10.1016/j.ijpara.2016.04.001
66. Gaze S, Driguez P, Pearson MS et al (2014) An immunomics approach to schistosome antigen discovery: antibody signatures of naturally resistant and chronically infected individuals from endemic areas. PLoS Pathog 10:e1004033. https://doi.org/10.1371/journal.ppat.1004033
67. King CH (2010) Parasites and poverty: the case of schistosomiasis. Acta Trop 113:95–104. https://doi.org/10.1016/j.actatropica.2009.11.012
68. Cannella AP, Arlehamn CSL, Sidney J et al (2014) Brucella melitensis T cell epitope recognition in humans with brucellosis in Peru. Infect Immun 82:124–131. https://doi.org/ 10.1128/IAI.00796-13
69. Uplekar S, Rao PN, Ramanathapuram L et al (2017) Characterizing antibody responses to Plasmodium vivax and Plasmodium falciparum antigens in india using genome-scale protein microarrays. PLoS Negl Trop Dis 11:e0005323. https://doi.org/10.1371/journal.pntd.0005323
70. Arnold R, Boonen K, Sun MGF, Kim PM (2012) Computational analysis of interactomes: current and future perspectives for bioinformatics approaches to model the host–pathogen interaction space. Methods 57:508–518. https://doi.org/10.1016/j.ymeth.2012.06.011
71. Matthews LR, Vaglio P, Reboul J et al (2001) Identification of potential interaction networks using sequence-based searches for conserved protein-protein interactions or "Interologs". Genome Res 11:2120–2126. https://doi.org/ 10.1101/gr.205301
72. ZHOU H, JIN J, WONG L (2013) Progress in computational studies of host–pathogen interactions. J Bioinforma Comput Biol 11:1230001. https://doi.org/10.1142/S0219720012300018
73. Nourani E, Khunjush F, DurmuÅŸ S (2015) Computational approaches for prediction of pathogen-host protein-protein interactions. Front Microbiol 6:94. https://doi.org/10.3389/fmicb.2015.00094

74. Lee S-A, Chan C, Tsai C-H et al (2008) Ortholog-based protein-protein interaction prediction and its application to inter-species interactions. BMC Bioinformatics 9(Suppl 12):S11. https://doi.org/10.1186/1471-2105-9-S12-S11
75. Mulder NJ, Akinola RO, Mazandu GK, Rapanoel H (2014) Using biological networks to improve our understanding of infectious diseases. Comput Struct Biotechnol J 11:1–10. https://doi.org/10.1016/j.csbj.2014.08.006
76. Luo Q, Pagel P, Vilne B, Frishman D (2011) DIMA 3.0: domain interaction map. Nucleic Acids Res 39:D724–D729. https://doi.org/10.1093/nar/gkq1200
77. Riley R, Lee C, Sabatti C, Eisenberg D (2005) Inferring protein domain interactions from databases of interacting proteins. Genome Biol 6:R89. https://doi.org/10.1186/gb-2005-6-10-r89
78. Xenarios I, Salwínski L, Duan XJ et al (2002) DIP, the database of interacting proteins: a research tool for studying cellular networks of protein interactions. Nucleic Acids Res 30:303–305
79. Kass I, Horovitz A (2002) Mapping pathways of allosteric communication in GroEL by analysis of correlated mutations. Proteins Struct Funct Genet 48:611–617. https://doi.org/10.1002/prot.10180
80. Finn RD, Miller BL, Clements J, Bateman A (2014) iPfam: a database of protein family and domain interactions found in the Protein Data Bank. Nucleic Acids Res 42:D364–D373. https://doi.org/ 10.1093/nar/gkt1210
81. Mosca R, Céol A, Stein A et al (2014) 3did: a catalog of domain-based interactions of known three-dimensional structure. Nucleic Acids Res 42:D374–D379. https://doi.org/10.1093/nar/gkt887
82. Dinkel H, Van Roey K, Michael S et al (2016) ELM 2016—data update and new functionality of the eukaryotic linear motif resource. Nucleic Acids Res 44:D294–D300. https://doi.org/ 10.1093/nar/gkv1291
83. Maier AG, Cooke BM, Cowman AF, Tilley L (2009) Malaria parasite proteins that remodel the host erythrocyte. Nat Rev Microbiol 7:341–354. https://doi.org/ 10.1038/nrmicro2110
84. Mbengue A, Yam XY, Braun-Breton C (2012) Human erythrocyte remodelling during Plasmodium falciparum malaria parasite growth and egress. Br J Haematol 157:171–179. https://doi.org/10.1111/j.1365-2141.2012.09044.x
85. Liu X, Huang Y, Liang J et al (2014) Computational prediction of protein interactions related to the invasion of erythrocytes by malarial parasites. BMC Bioinformatics 15:393. https://doi.org/10.1186/s12859-014-0393-z
86. Chothia C, Lesk AM (1986) The relation between the divergence of sequence and structure in proteins. EMBO J 5:823–826
87. Fiser A (2010) Template-based protein structure modeling. Methods Mol Biol 673:73–94. https://doi.org/10.1007/978-1-60761-842-3_6
88. Davis FP, Barkan DT, Eswar N et al (2007) Host-pathogen protein interactions predicted by comparative modeling. Protein Sci 16:2585–2596. https://doi.org/10.1110/ps.073228407
89. Eswar N, John B, Mirkovic N et al (2003) Tools for comparative protein structure modeling and analysis. Nucleic Acids Res 31:3375–3380
90. Davis FP, Sali A (2005) PIBASE: a comprehensive database of structurally defined protein interfaces. Bioinformatics 21:1901–1907. https://doi.org/ 10.1093/bioinformatics/bti277
91. Jianlin Cheng J, Tegge AN, Baldi P (2008) Machine learning methods for protein structure prediction. IEEE Rev Biomed Eng 1:41–49. https://doi.org/10.1109/RBME.2008.2008239
92. Baldi P, Brunak S, Chauvin Y et al (2000) Assessing the accuracy of prediction algorithms for classification: an overview. Bioinformatics 16:412–424
93. Tastan O, Qi Y, Carbonell JG, Klein-Seetharaman J (2009) Prediction of interactions between HIV-1 and human proteins by information integration. Pac Symp Biocomput 2009:516–527
94. Dyer MD, Murali TM, Sobral BW (2011) Supervised learning and prediction of physical interactions between human and HIV proteins. Infect Genet Evol 11:917–923. https://doi.org/10.1016/j.meegid.2011.02.022
95. Qi Y, Tastan O, Carbonell JG et al (2010) Semi-supervised multi-task learning for predicting interactions between HIV-1 and human proteins. Bioinformatics 26:i645–

i652. https://doi.org/ 10.1093/bioinformatics/btq394

96. Kazan H (2016) Modeling gene regulation in liver hepatocellular carcinoma with random forests. Biomed Res Int 2016:1035945. https://doi.org/ 10.1155/2016/1035945
97. Wuchty S (2011) Computational prediction of host-parasite protein interactions between P. falciparum and H. sapiens. PLoS One 6:e26960. https://doi.org/ 10.1371/journal.pone.0026960
98. Kotlyar M, Pastrello C, Pivetta F et al (2015) In silico prediction of physical protein interactions and characterization of interactome orphans. Nat Methods 12:79–84. https://doi.org/ 10.1038/nmeth.3178
99. Pang K, Cheng C, Xuan Z et al (2010) Understanding protein evolutionary rate by integrating gene co-expression with protein interactions. BMC Syst Biol 4:179. https://doi.org/ 10.1186/1752-0509-4-179
100. Ge H, Liu Z, Church GM, Vidal M (2001) Correlation between transcriptome and interactome mapping data from Saccharomyces cerevisiae. Nat Genet 29:482–486. https://doi.org/ 10.1038/ng776
101. Reid AJ, Berriman M (2013) Genes involved in host-parasite interactions can be revealed by their correlated expression. Nucleic Acids Res 41:1508–1518. https://doi.org/ 10.1093/nar/gks1340

Chapter 8

An Integrative Approach to Virus–Host Protein–Protein Interactions

Helen V. Cook and Lars Juhl Jensen

Abstract

Since cell regulation and protein expression can be dramatically altered upon infection by viruses, studying the mechanisms by which viruses infect cells and the regulatory networks they disrupt is essential to understanding viral pathogenicity. This line of study can also lead to discoveries about the workings of host cells themselves. Computational methods are rapidly being developed to investigate viral-host interactions, and here we highlight recent methods and the insights that they have revealed so far, with a particular focus on methods that integrate different types of data. We also review the challenges of working with viruses compared with traditional cellular biology, and the limitations of current experimental and informatics methods.

Key words Viruses, Virus–host interactions, Databases, Bioinformatics, Machine learning, Orthology, Host prediction, Viral evolution, Coevolution, PPI networks, Network rewiring

1 Introduction

Viral bioinformatics and computational viral biology are still young fields in contrast with their counterparts in cellular biology [1]. As of 2013 and still today, there are only few bioinformatics tools available that enable simple analysis of virus–host interactions [2]. This provides an opportunity for those who develop such tools, especially since viruses have several characteristics that make their study attractive to computational biologists and bioinformaticians. Viruses have small genomes, are abundant, are major human pathogens, and are actively used for a variety of medical and biotechnological applications, from treating cancer [3] to gene therapy [4, 5]. Further, viruses have evolved an extremely varied set of approaches to commandeer cellular functions [6]. They act as metabolic engineers of the cells they infect and thereby present many opportunities to study perturbations of host interaction networks [7].

Louise von Stechow and Alberto Santos Delgado (eds.), *Computational Cell Biology: Methods and Protocols*, Methods in Molecular Biology, vol. 1819, https://doi.org/10.1007/978-1-4939-8618-7_8,

Integrative analyses of virus–host protein–protein interactions combine data from a variety of sources, including genomics, expression levels of the transcriptome and proteome, and the topology of protein–protein interaction networks. These analyses then use this information to provide a more complete understanding of the changes that viral infections cause in their host cells. Such approaches can be used to predict host and tissue tropism, to interpret the results of the cellular network rewiring due to viral disruption, and to predict antiviral drug targets.

1.1 Viruses Have Immense Diversity and Impact

Viruses are the most abundant organisms in the ocean and are present everywhere life is found [8]. The estimated number of virus particles on earth is 10^{31} [9], a number that is 1000 times larger than the estimated number of bacteria. This massive number of viruses consists of representatives from an extremely high diversity of species. There are an estimated 320,000 viral species that infect mammals [10], the vast majority of which we have not observed or characterized. Indeed, a recent study of the human virome revealed that nearly 95% of viral DNA reads did not match any known viral protein [11].

Viruses play key roles in the environment, including facilitating cycling carbon in the ocean away from the surface toward deeper waters [8]. Polydnaviruses are essential for the survival of their wasp host's eggs, and some plant viruses confer cold and drought tolerance [12] to their hosts. In humans, endogenous retroviral sequences are expressed in the placenta and likely drove the evolution of placental mammals, and bacteriophages that live on the skin control bacterial populations to help prevent bacterial infections [13]. Given the vast viral landscape that is as yet undiscovered, it is likely that there are many more ecological relationships to be discovered between viruses and their hosts.

However important in the formation of ecosystems, viruses are best known as major human and agricultural pathogens. 40% of the world's population is at risk for infection by *Dengue fever virus* (DENV), with a worldwide annual incidence rate of between 50 and 100 million people [14]. Other viral diseases such as *Influenza A virus* (IAV), *Hepatitis C virus* (HCV), and cervical cancer caused by *Human papillomavirus* (HPV) each cause more than a quarter of a million deaths worldwide per year [15]. Viral infection also carries a huge economic burden. Influenza costs US$16.3 billion in lost wages and US$10.4 billion in medical expenses annually in the United States [16]. The 2014 Ebola virus outbreak in West Africa killed 11,000 people, and also caused a loss of US$ 2.2 billion, 16% of the combined 2013 GDP of the affected countries.

Vaccines, where available, drastically reduce the morbidity and mortality of viruses. Vaccines for smallpox and rinderpest are responsible for the successful eradication of these viruses [17]. In all, vaccines prevent 3 million deaths per year, and the return on every dollar spent on vaccinations for measles alone and measles, mumps, and rubella (MMR) is 20 times saved in health care costs [18]. Vaccines against influenza also have been shown to reduce severity of disease in older adults [19].

With the development of antiviral drugs, *Human immunodeficiency virus* (HIV) infection has become a manageable chronic condition, providing an improvement in the lives of millions of people [20]. However, the effectiveness of antiviral drugs can decrease over time due to the development of resistance, especially for rapidly mutating RNA viruses [21]. Having a detailed understanding of host–virus protein–protein interactions will help to identify targets for new antiviral drugs [22, 23].

The human and economic impacts of viruses are likely to increase as the ranges of tropical viruses are expanding under climate change. As temperatures rise and rainfall becomes more variable, new habitats are available to vectors such as mosquitoes that carry viral disease. This introduces the virus to a new population of hosts who do not have immunity to it, allowing the virus to spread readily through the population. The rate of incidence of zoonosis, in which an animal virus "jumps species" to a human host and causes disease, is also increasing as land that was previously wildlife habitat is developed for human use [24]. These vector-borne and zoonotic viruses are those for which there are no vaccines and few antiviral drugs, providing few treatment options for infected patients. The introduction of new viruses to the western hemisphere has been repeated 4 times over the last 30 years, starting with *Dengue virus* in 1990, *West Nile virus* (WNV) in 1999, *Chikungunya virus* (CHIKV) in 2013, and most recently *Zika virus* (ZIKV) in 2014 [25]. It is likely that this trend will continue.

1.2 Classification of Viruses

Viruses are classified under the Baltimore system [26] by the type of genetic material that encodes their sequence, namely RNA or DNA. Each nucleic acid can be either single or double-stranded, and single-stranded RNA may be present in the positive or negative sense. Further, retro-transcribing viruses encode reverse transcriptase, which is able to transcribe RNA into DNA. In the case of *Human immunodeficiency virus*, the reverse-transcribed DNA is integrated into the genome as part of the viral life cycle. Not all retroviruses require integration into the genome, for example *Hepatitis B virus* (HBV), which does not need to do so to replicate. This gives seven categories of viruses, illustrated in Fig. 1.

Viruses are further classified as enveloped or non-enveloped, according to whether the protein capsid that encloses their genetic material is surrounded by a lipid bilayer or not. Non-enveloped viruses exit the host cell by lysis, but enveloped viruses exit through an intact membrane. As these new virions bud from the host cell, part of the host cell membrane coats the viral capsid forming

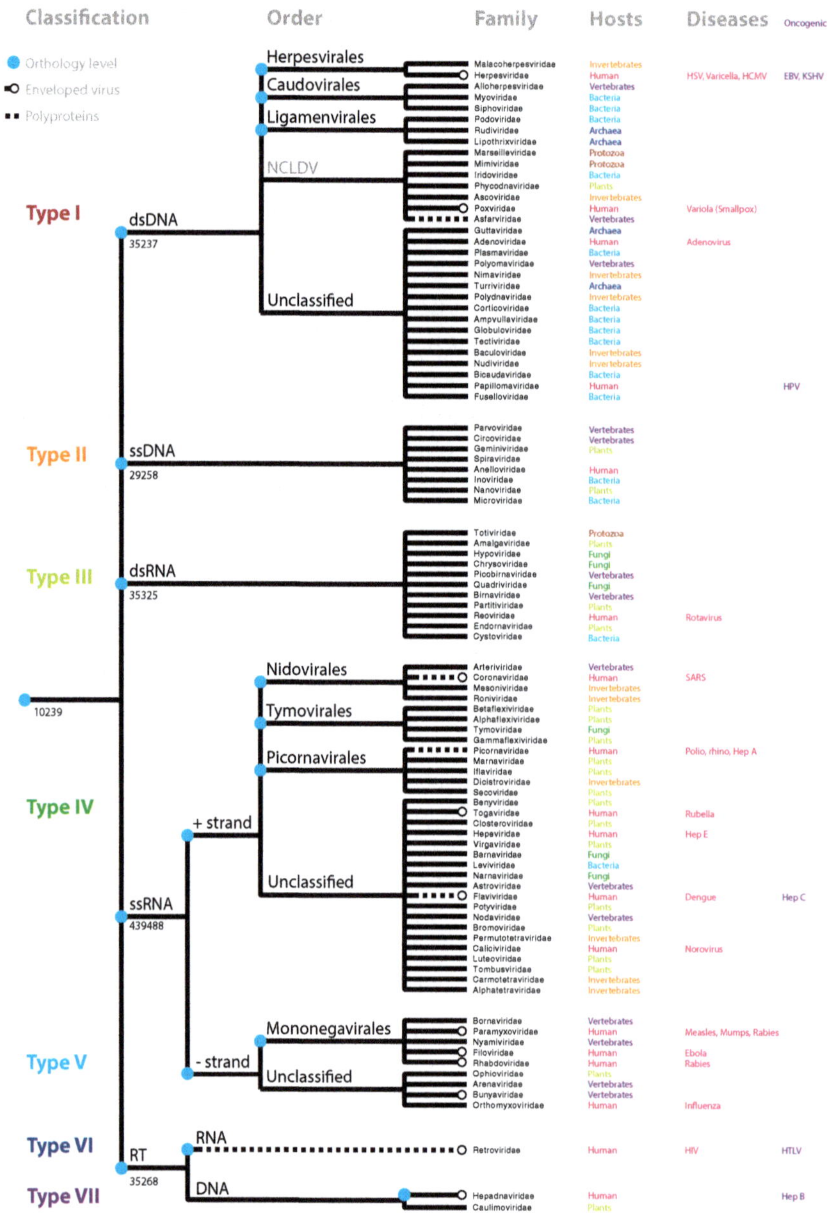

Fig. 1 Viral taxonomic tree showing Baltimore classification, whether the virus is enveloped (white circles), the levels for which orthologs are available in EggNOG (blue circles), a representative host for the virus family, whether the virus is known to cause cancer in humans (oncogenic column), and other human diseases associated with the virus

the viral envelope. The envelope is often embedded with viral glycoproteins. The host cell receptors that the glycoproteins dock to are known and well studied for some viruses such as *Influenza A virus* (IAV), but the receptor for *Vaccinia virus* (VACV) was not known until 2012 [27]. The IAV glycoproteins, hemagglutinin and neuraminidase that give their initials to influenza strain names, are responsible for interaction with the host cell, a main factor in determining host and tissue tropism.

1.3 Virus Biology

Viruses are obligate intercellular parasites that have inert and replicative phases to their life cycle. The virus particle, or virion, itself is inert, but after entering a cell the virus will begin the living phase of its life cycle and replicate. Cells must contain both the correct receptors to permit transport of the virus into the cell (such a cell is said to be susceptible) and contain all necessary machinery and be in the correct phase of the cell cycle to permit replication of the virus (such a cell is called permissive). Viruses infect all domains of life, and most viruses have a specific host range. Some phages target such a specific host range that they are used for bacterial strain typing [28]. However, the arthropod-borne arboviruses are able to replicate in both invertebrates and mammals. The range of hosts and tissues in which a virus can replicate are referred to as the host tropism or tissue tropism of the virus.

The sizes of viral particles and genomes both span a large range of sizes. Among the smallest viruses, the *Hepatitis B virus* encodes 4 genes within 3000 nucleotides. Paroviruses are about 18nm in diameter and their genomes are 4–6 kb long. On the other end of the scale, the proposed genus Megavirales [29] contains the largest viruses. *Megavirus chilensis* encodes 1120 predicted proteins in 1.3M base pairs, *Pandoravirus*' genome contains 2.5 Mbp, and the capsid of *Pithovirus sibericum* measures 1.5 μm by 0.5 μm, the size of a small bacterium.

Many viruses encode their proteins as a polyprotein, which is then auto-cleaved or cleaved by cellular or viral encoded proteases. In the case of *Hepatitis C virus* (HCV), all three mechanisms are used to cleave all proteins from the polyprotein [30].

The largest viruses are dsDNA viruses, whereas in general the viruses with the smallest genomes are RNA viruses. The larger amount of genetic material present in dsDNA viruses enables them to become more finely attuned to their hosts [31]. Many DNA viruses can enter a lysogenic phase, in which they are latent and can integrate into the genome, whereas RNA viruses, with the exception of retroviruses, do not exhibit this behavior. DNA viruses mutate at a lower rate than RNA viruses since they are transcribed with cellular polymerase, thereby incorporating error correction and proofreading into their replication strategies [32]. RNA viruses have mutation rates approximately equal to their genome lengths, so that each viral progeny contains a mutation from the original. The replication rate of RNA viruses is approximately the error rate of sequencing DNA, so new techniques such as rolling circle transcription [33] have been established to properly sequence these viruses.

1.4 Quasispecies

RNA viruses adopt a distinctly different strategy to evade detection by the host immune system than DNA viruses. With their larger genomes, DNA viruses can encode genes that enable the virus to perform immune evasion, for example, MHC receptor homologs encoded by *Human cytomegalovirus* (HCMV) that prevent NK-mediated cytotoxicity [34]. RNA viruses, on the other hand, rely on having a much higher degree of genetic diversity to avoid detection.

The mutation rate of RNA viruses is high enough, and their genomes are short enough that nearly every viral progeny resulting from a single parent sequence will contain a mutation. This population of viruses is referred to as a quasispecies. Evolution does not select the fittest individual viral sequence within the quasispecies, but rather selects for sequences that produce the most fit progeny on average, accounting for random mutations [35]. This process has been dubbed "survival of the flattest" [36].

The high mutation rate of RNA viruses enables them to adapt quickly to new hosts; however, if the mutation rate is too high, then no offspring will be viable [37]. This idea of mutagenic catastrophe is being investigated as a novel antiviral therapy [38].

In 2014, two studies synthesized the full mutation space of *Influenza virus* hemagglutinin, and sequenced the mutants that were able to replicate [39, 40]. They found that mutations in the regions that interact with host proteins were not tolerated, whereas mutations in regions that are recognized by the immune system were well tolerated. These high-throughput studies can help guide the development of universal influenza vaccines that can target immutable regions of hemagglutinin.

2 Virus–Host Protein–Protein Interactions

2.1 Studies of Viruses Elucidate Host Cellular Function

Viruses affect a variety of fundamental processes in their host cells including the cell cycle, signal transduction, and apoptosis. They antagonize or stimulate host proteins which block or overexpress essential pathways, causing a disruption of phenotype. Early discoveries about the workings of the cell cycle were enabled by the comparison of HPV, a dsDNA onocogenic virus to *Adenovirus*, a dsDNA non-oncogenic virus [41]. Since then, viral processes have been shown to be involved in many more cellular mechanisms as reviewed in [42]. For example, the study of *Rous sarcoma virus* revealed the kinase activity of one of the viral proteins, and identified the first known member of this broad protein class

[43]. Later work on VACV has elucidated the role of tyrosine kinase signaling in formation of the viral actin-tail since the virus induces actin-tail assembly from outside the cell through the plasma membrane via the kinase [44]. More recently, the study of influenza infection revealed that T-cells are guided to the site of infection by neutrophil trails [45]. This is a promising area of research that is likely to yield further discoveries.

2.2 Challenges for Protein-Centered Analysis

Viral proteins can exhibit large degrees of multi-functionality. For example, the seven *Ebola virus* (EBOV) proteins are each responsible for multiple functions. This is especially true for vp40, which regulates transcription and coordinates virion assembly and budding [46]. IAV NS1 also has a wide range of functions including limiting interferon production, regulating viral replication, and modulating cellular signaling pathways [47]. Although a challenge to analyze, these proteins may be particularly good targets for antiviral drugs due to the large number of roles they play. The cellular proteins targeted by viral proteins also tend to contain regions of higher disorder than average, perhaps hinting at their multi-functionality [48].

Intrinsically disordered proteins are widespread in viral proteomes [49], and exist at similar levels of disorder to eukaryotic cellular proteins [50]. They are extremely flexible and do not adopt a single conformation, meaning that it is very difficult to analyze their three-dimensional shape. Structured proteins generally contain domains that perform a specific function and that fold and evolve independently of each other [51]. Disordered regions, however, make up a very small percentage of known domains, even though they may exhibit a specific function such as kinase activity [52]. Due to the fact that intrinsically disordered proteins contain fewer domains for which the function is known than structured proteins, they are correspondingly more difficult to computationally analyze.

Protein–protein interactions provide one lens through which to view the interface between virus and host, but they are not the only important interactions that are occurring within the cell. Interactions of RNA and lipids are also crucial to viral replication but are by definition disregarded by investigating proteins only. MicroRNAs (miRNAs) are short noncoding RNAs that posttranscriptionally regulate gene expression. They are expressed in Metazoa and plants, and also in certain dsDNA viruses and *retroviriade* where they act to regulate viral transcripts and host networks [1, 53]. Viruses manipulate cellular lipid metabolism to facilitate replication and virion assembly [54], for example, HCV increases lipid production, likely in an attempt to provide sufficient lipids for the host under the additional demand of virus budding [54]. Interactions involving host and viral noncoding

RNA are available through the ViRBase database [55], but no corresponding database exists yet for lipid interactions.

Many of the informatics methods mentioned below rely on using data found in databases of protein–protein interactions or in the literature. Both of these sources are naturally biased toward well-studied genes [56]. Caution should be taken when making claims of network properties such as degree for these proteins, since highly studied genes will have larger numbers of connections simply because they have been well studied and more interactions have been elucidated for them.

2.3 Overview of Experimental Methods

The yeast two-hybrid method (Y2H) is the classical high-throughput approach for determining whether proteins interact [57]. This method relies on cutting apart the activation and binding domains of a transcription factor, and then fusing each domain to proteins of interest to create two libraries. These gene fusions are transformed into yeast plasmids. Yeasts carrying one plasmid each are mated to create a diploid mutant for each gene pair in the cross product of the two libraries. If the proteins interact, the binding domain will bind to a promoter for a reporter gene and the activation domain will recruit polymerase which will express the reporter transcript. While Y2H is an in vivo method, yeast may not be the native environment for the proteins that are being investigated, so chaperones or posttranslational modifications may be lacking, thereby disrupting folding and binding. Since the screen does not take place under truly physiological conditions, it can result in a high false positive rate. Further, it will detect only pairs of interacting proteins, and so will miss complexes, giving it a high false negative rate also [58].

Affinity purification followed by mass spectrometry (AP/MS) [59] is another high-throughput method, and it provides a partial solution to the false negative rate of Y2H. First, proteins of interest are fused to a protein, for example, GFP, that can be captured with an antibody. The use of a fluorescent tag also allows for imaging of the live cell, which may also be desired. The antibody is fused to a column, and cell lysate is run through the column. The tagged proteins plus their interaction partners will bind to the antibody and will remain in the column after non-stringent washing. Affinity purification may also pull down secondary or tertiary interaction partners of the tagged protein; however, if the column is washed too stringently these interactions will be disrupted.

A similar method, tandem affinity purification, fuses a second tag to the tagged protein, and follows the first purification with a second that targets the second tag, which reduces false positives [59]. Following purification, the remaining contents of the column are subjected to mass spectrometry, and the resulting interactors can be identified. AP/MS has a high false negative rate since transient and weak interactions are easily disrupted, especially if a second round of purification is performed. The tagged proteins are often overexpressed, which may lead to false positives, and the method strongly relies on the quality of the antibodies used.

To overcome the issue of not being able to stringently wash the AP column, biotin-based proximity methods such as APEX [60] and BioID [61] have been used to identify proteins in the vicinity of a protein of interest. However, membrane proteins are difficult to target with these methods so new, smaller versions of both APEX [62] and BioID [63] have been developed. These approaches are promising since they only identify proteins that are in a very small (20 nm) neighborhood of the labeled protein. It is not guaranteed that all proteins within this range do interact with the target. One example of this is that since the outer mitochondrial membrane is porous, proteins that do not cross it can be biotinylated from an APEX target on the other side of the membrane [64]. With these caveats in mind, data derived from biotin-based assays provides another axis along which to filter physical interaction data.

Many viruses have been investigated with Y2H [65] and AP/MS, including studies that focus on the interactions of viral and host proteins during viral entry [66] and studies that aim to elucidate the dynamic interactions that take place throughout the course of viral infection [59]. A range of variations on these methods and other large-scale methods are available, as reviewed in [67]. Due to their high false positive and false negative rates, overlap between the results of these high-throughput studies is often quite low [68]. AP/MS data for HCV has been validated by integrating data from RNAi screens [69] and the interaction partners of herpesvirus membrane glycoproteins have been identified by integrating data obtained from biotin-based methods [70].

2.4 Databases and Related Resources

Many databases have been developed to catalog both experimentally determined interactions and functional protein–protein interactions. VirHostNet [65], VirusMentha [71], HPIDB [72], and ViPR [73] store data specifically about experimentally derived virus–virus and virus–host protein–protein interactions. Due to the high cost and effort involved in maintaining such resources [74], these are not always updated in a timely manner, and there is a risk of those databases being deprecated as the BIND database [75] has been. The burden of database development and the risk of data loss due to deprecation of databases is reduced by the adoption of open formats such as the Proteomics Standards Initiative's MITAB format, and by open data sharing between databases [76].

The STRING database combines experimental protein–protein interaction data from many sources, and also integrates data from text mining, KEGG pathways, homology, and other predictions. It includes functional interactions between proteins that appear in the same pathway, act as transcription factors, or that otherwise have related functions but that may not physically interact. All these channels of evidence are benchmarked and integrated to a combined confidence score for each interaction. Despite the high false positive and negative rates of high-throughput experiments, these data become very useful if they are taken in aggregate and integrated together. In total, STRING version 10.5 includes information on over 9 million proteins from more than 2000 organisms, and in the future, STRING will also incorporate virus–host interactions [77].

3 Sequence-Based Methods

3.1 Challenges of Sequence-Based Methods for Viruses

Viral genomes, though small, are compact and often contain overlapping genes that are translated to proteins with multiple functions. Many RNA viruses code for a single mRNA containing one open reading frame that is translated into a long protein product. This polyprotein is then posttranslationally cleaved into the functional viral proteins [78]. Databases such as UniProt do not consider these cleavage products to be individual proteins with their own identifiers even though these are the functional units of the virus [79]. This means that the proteins must be cleaved in silico prior to analysis.

Viruses exhibit vast genetic diversity and RNA viruses specifically exhibit a much higher rate of evolution than even simple cellular organisms. To date, most bioinformatics analyses have ignored the diversity present in the quasispecies. A single sequence does not necessarily reflect the biological reality of the full diversity of the quasispecies and the interactions that take place in nature. However, much of the virus sequencing effort has come from patient samples which often are presented as a single DNA sequence, typically the consensus sequence derived from Sanger sequencing.

3.2 Codon Bias

The genetic code is redundant, but not all synonymous codons are used with the same frequency in the genome [80]. The variation in this distribution is called codon bias, and there are large differences in codon biases between species. Proteins that are highly abundant tend to contain more codons that are translated efficiently than proteins of lower abundance do. Arboviruses, which replicate in both arthropods and mammals, are adapted to both of their hosts' distinct codon biases. A recent study modified the codons used in *Dengue virus* to produce a mutant that was attenuated in humans but was still able to replicate in insects [81]. Such an approach has the potential to be used as a vaccine, since it would not cause disease in humans, but has the same antigenic profile as an infectious virus does.

Altering codon bias can also alter RNA secondary structure, which may also play a role in determining translation efficiency

[82]. Changing the RNA secondary structure can further alter the compactness of the genome, which can have consequences for packaging and delivery [83].

3.3 Bacteriophage Host Prediction

Metagenomic methods provide a wealth of information about the sequences present in environmental samples but provide little direct insight into the ecology of the discovered organisms. In silico predictions of host/phage pairs are valuable since it can be challenging to create the correct conditions to experimentally grow phages and their bacterial hosts in the lab. A variety of current methods that can be used to identify phage host pairs are reviewed in [84]. Here the authors conclude that there is a lot of room for improvement in bacteriophage host predictions. All of the reviewed methods have a low precision, and predict the incorrect host for more than half of the samples. These methods include, in the decreasing order of accuracy, finding exact sequence matches between the bacteria and the phage, finding homology between the host and the phage with BLASTN or BLASTX, finding matches specifically on CRISPR spacers that have integrated into the host, matching k-mer profiles or codon bias between the host and phage genome, and matching co-abundance profiles with the assumption that in a given environment, the most prevalent bacteria species is likely the host to the most prevalent phages.

HostPhinder [85] is a more recent host prediction method for phages that achieves very good results with 74% accuracy predicting the host species and 81% accuracy predicting the host genus. HostPhinder uses a k-mer approach due to the mosaic nature of phage genomes, in which genetic components can appear in any order. It uses a k-mer length of 16, twice as long as the longest k-mer length of 8 used in any of the methods evaluated by the previous review, after finding that shorter lengths were too unspecific. Further, HostPhinder uses a carefully selected measure of resemblance selected from four possible methods based on the results of a cross-validated assessment, and so achieves much better accuracy with k-mers than would be suggested by the results in [84]. The evaluation of phage host prediction suffers from the fact that some phages do infect multiple hosts, which can result in false positives that may instead be accurate predictions of new host ranges.

4 Evolution-Based Prediction Methods

As more genomes are being sequenced, it is possible to create very large multiple sequence alignments to compare closely neighboring species. Since experimental techniques to determine binding interfaces are time- and labor-intensive, computational techniques provide an opportunity to rapidly investigate candidates.

4.1 Orthologs

The protein CDK1 found in both the human and mouse performs the same function in both organisms to transition through the cell cycle [86]. These proteins are said to be orthologous, not because they perform the same function, but because these genes have both descended via speciation from the same gene in the last common ancestor of the human and mouse. Paralogs, however, are proteins in one species that have undergone gene duplication, resulting in two copies of the gene. Multiple outcomes for the duplicated gene are possible, from loss of function of one copy to divergence of function of the two genes, following a high rate of evolution in one of the copies [87].

Orthology groups are groups of proteins that have descended from a common ancestor, and so are strictly speaking misnamed, since they will include paralogs as well. EggNOG [88] and other databases such as OrthoDB [89] and the many reviewed in [90] use sequence homology (via all vs. all BLAST) and further refinement based on similarity to identify distantly related orthologous proteins and group them into orthology groups. Only EggNOG, as of version 4.5, provides orthology for viral proteins. The levels for which orthology is available are indicated in Fig. 1.

4.2 Challenges for Virus Orthologs

RNA viruses evolve extremely quickly, but DNA viruses also evolve faster than bacteria do [32]. This fast evolution can mask orthology, since relationships become less evident over evolutionary time. Further, viruses contain a large amount of horizontally transferred genetic material, which will give the illusion that orthology groups containing these genes are much older than they actually are. Many RNA viruses including the *corona-*, *picorna-*, *flavi-*, and *retroviridae* (and some DNA viruses such as *African swine fever virus*) translate their proteins as polyproteins that need to be cleaved before becoming functional proteins. This makes analysis difficult since sequences are generally stored as the full-length protein, and so the cleavage sites must be identified and the proteins cleaved in silico prior to aligning and analyzing the sequences.

4.3 Virus Orthologs and Paralogs

Since it is not known whether all viruses arise from a universal common ancestor, and it has been proposed that RNA viruses and DNA viruses have separate origins [91], it is perhaps disingenuous to discuss orthology groups over all viruses. However, when discussing viral orthology groups, we consider them to track the ancestry of the gene, not necessarily the ancestry of the organism that carries it. There are certain examples of clear orthology within the viruses, for example, the 30+ proteins that are common to all human herpesviruses and that exhibit conservation across the *Herpesviridae* [92]. Further, all of the dsDNA viruses share many genes. These common genes have been represented as a bipartite network illustrating the interconnectedness of these viruses [93].

Evidence that a protein interacts with an interaction partner can be transferred via homology—it is likely that orthologs of this protein will interact with orthologs of its partner in another species. The STRING protein–protein interaction database uses the EggNOG orthology data to transfer links between intra-species protein interactions via EggNOG orthology groups [77, 88].

4.4 Coevolution

Residues of proteins that are in direct contact will evolve jointly, as a mutation in one protein will be compensated for by a reciprocal mutation in the other protein to preserve the interaction interface. Methods to measure coevolution have been used to predict intra-protein interactions [94], and protein–RNA interactions [95] within species, and also across species [96]. The general approach of these methods is to generate a multiple sequence alignment of sequences that are representative of variation in the proteins of interest, and then to look for co-varying residues [97]. Since these methods rely on assembling a large variety of sequences, in this case the fast evolution of viruses is an advantage that can help amplify signals even though the divergence of host proteins may be low.

Coevolution methods include direct coupling, mutual information, and phylogeny-based methods, which are presented in detail in [97]. All methods require significant alignment depth to give statistically significant results, with alignments at least as deep as they are long being preferable [96, 97]. There are several reasons why coevolution-based binding prediction has proven difficult, in addition to insufficient alignment depth [96]. Indirect interactions that are not directly responsible for binding can lead to false positive correlated residue pairs. False positives can also result from correlations that are due to the closely related phylogeny of the chosen sequences. Finally, determining a gold standard set against which results can be evaluated is an additional challenge.

The first virus–host coevolution study was on the *Human immunodeficiency virus 1* protein Vif and APOBEC3G using 13 different methods to determine correlated residues [96]. Sites known to be responsible for binding from experimental mutational studies were found to interact by at least one coevolution method, but many false positives were also reported due to lack of alignment depth. Specific molecular examples of viral adaptation to the host that have been determined through experiments are reviewed in [6]. The specificity of host and cell tropism is one such example of the coevolution between a virus and its host. Below, we highlight two machine learning methods that predict which cells a virus is able to infect.

4.4.1 Influenza A

To predict the host tropism of *Influenza A virus*, Eng et al. [98] created a random forest predictor that aimed to distinguish avian from human strains. Previous studies have used neural networks and support vector machines to generate predictors, but this work compares a variety of machine learning methods and finds the best results with a random forest approach. They built 11 independent, parameter-optimized, tenfold cross-validated models, one for each of the 11 influenza proteins. Although only two of the 11 proteins, HA and PB2, were previously known to be species specific with clearly elucidated mechanisms of action, each of the 11 models provided good predictions of the host, showing that all of the influenza proteins have species specificity. A combined model gave very good performance when evaluated on an independent test set using the area under the receiver operating characteristic curve (AUC). This performance metric measures the trade-off between sensitivity and specificity as for a binary classifier as its discrimination threshold is varied. A high AUC means that choosing a cutoff threshold for the classifier that gives a low false positive rate will also give a high true positive rate. Solvent accessibility and charge, specifically in HA, which binds to negatively charged glycan receptors, were among the most important properties identified. The model made errors misclassifying proteins from human strains as avian which had a recent avian source, suggesting they were not yet completely adapted to their new host. Proteins from avian strains that were classified as human likely represent strains that are at risk of infecting human populations. This method could potentially be used as a monitoring system for avian strains to predict the next zoonosis in humans.

4.4.2 HIV

Machine learning has also been used to predict the cell tropism of *Human immunodeficiency virus 1* (HIV-1). HIV-1 gp120 binds to CD4+ T cells, but this binding requires that gp120 also binds to a co-receptor, either CXCR4 or CCR5. Usage of the antiviral drug Maraviroc is indicated only for patients with viruses whose gp120 binds only to CCR5. Determining the virus tropism is therefore essential to determining the treatment for the patient, and monitoring is required since the virus is able to change tropism for these receptors throughout the course of infection. Several different predictors are reviewed in [99], which use support vector machines and position-specific scoring matrices to distinguish co-receptor usage. A later study [100] shows that previous models that have been trained on HIV-1 B subtypes do not generalize to subtype A, and that new models must be built to correctly predict the tropism of subtype A. These studies highlight the success of bioinformatics methods, but also the importance of taking care when choosing training data to ensure that it is representative of the strains of interest.

5 Network-Based Prediction Methods

Protein–protein interactions (PPIs) in the cell, sometimes called the interactome [101], can be represented as a network in which the proteins are nodes and their interactions are edges. Two nodes are linked by an edge in the network if there is an interaction between the two proteins that the connected nodes represent. The interactions may be physical interactions such as binding or phosphorylation but may also be functional interactions such as genetic interactions or the presence of the proteins in the same pathway, depending on the network. The number of edges that connect to a node is called the degree of the node, and nodes that have high degree are called hubs. The shortest path between any two nodes is defined as the smallest number of edges that must be traversed to travel from one node to the other.

The structure of such PPI networks roughly follows a power law distribution across all domains of life [102]. Networks with this property, having very many nodes of low degree and few nodes of high degree, are called scale-free networks, and appear in many more contexts than just biological networks. Social, transportation, and communication networks are a few examples that also exhibit this property [103]. It has been shown that the scale-free property emerges from the iterative process of building the network [103] since as new nodes are added, they preferentially attach to the nodes that are most visible in the network—those that are already highly connected.

Scale-free networks are robust to random attacks, since hubs are rare compared to the number of nodes of small degree, and hubs will help maintain connectivity of the network as nodes of smaller degree are removed. For a PPI network, this means that if a gene is mutated at random it will most likely cause a small, local effect, and is not likely to be fatal to the cell. On the other hand, scale-free networks are vulnerable to attack against the hub nodes, since removing a few hubs will impact many other nodes and will quickly cause the network to fracture [104]. While this means that hub proteins are a liability in that mutations in them can cause diseases such as cancer, scale-free networks also present opportunities for biomedical applications such as antibiotics targeting bacterial networks [105], or treating cancer by disrupting the network in cancerous cells but not in normal tissue [106].

5.1 Network Rewiring

Viruses act as metabolic engineers of the cell [7], taking advantage of cellular proteins and commandeering gene expression to enable reproduction of viral progeny. As viral proteins are upregulated during viral infection, the host protein network will be disrupted since many viruses monopolize the transcription and translation machinery, for example by cap snatching from cellular mRNAs [107]. Further, viruses have evolved sophisticated techniques to avoid the defense mechanisms of their hosts [34, 108] and to disrupt the cell cycle [41].

Diseases can be interpreted as permutations of the cell's protein–protein interaction network, where the affected genes are clustered into modules [109]. The resulting rewiring and permutation of the native network can lead to phenotypic changes, such as oncogenesis in the case of HPV, and cell death in the case of HIV-1 infected T-cells. The *Hepatitis C virus* (HCV) protein network was overlaid with expression data from several time points during HCV infection, and distinct rewiring was observed between precancerous and cancerous infection states [110]. Viruses (and other pathogens) preferentially target cellular hubs and bottlenecks (proteins that lie on many shortest paths, and show high betweenness centrality) [111]. These proteins also tend to be those that are also associated with cancer, but care should be taken when interpreting this fact. The more a protein is studied, the more interaction partners that will be known for it, leading to the bias that proteins of high degree are well studied [112]. An analysis focusing on HIV-1 [113] found that viruses target pathways containing HIV-1 dependency factor proteins which had been previously identified as essential for HIV-1 infection from RNAi screens, although they do not interact with these proteins directly.

Drugs have historically been targeted to viral proteins, but with advances in understanding of the human–virus interactome, it is possible to also target host proteins that are central to viral infection, and to repurpose drugs to do so [68]. Since viral proteins mutate at a very high rate, drugs that target them often quickly become ineffective [22]. Therefore, with a better understanding of the intracellular and viral interaction networks, better strategies to target host proteins instead may emerge [114].

5.2 Centrality and Clustering Methods

Different centrality measures have been defined to capture different aspects of which nodes are the most "important" in the network. Here we will cover three measures. A variety of other measures exist but are correlated with those described here [115]. Degree centrality assigns the highest value to the node with the highest degree. The betweenness centrality value of a node captures whether this node is a bottleneck that joins clusters in the network together. In other words, a node has high betweenness centrality if most of the shortest paths between all the other nodes in the network pass through this node. Closeness centrality gives the node with the lowest sum of shortest paths between this node and all other nodes the highest centrality score.

Markov clustering (MCL) is a fast and effective algorithm to identify clusters in a network. Clusters are defined by starting a path at a given node, and randomly choosing to follow one of its edges. Such a random walk is much more likely to remain within a cluster than to leave it. MCODE [75] was designed specifically for finding clusters in protein interaction networks. It creates clusters by assigning a score to each node, which is a function of the number of connections between this node's immediate neighbors. MCODE then starts from a seed node and expands the cluster to its neighbors that have scores above an adjustable threshold. MCL, MCODE, and two other clustering algorithms are compared in [116], which finds that MCL is the best method to extract protein complexes from protein–protein interaction networks.

6 Conclusions

Just as having a broader view of protein–protein interactions has provided a deeper understanding of cellular function [117], having a similar understanding of the interactions between pathogens and their hosts will provide new information to treat clinically relevant infections and diseases. Despite the challenges to viral bioinformatics, integrative approaches are being used to predict host and tissue tropism and to visualize the virus–host protein–protein interaction network.

Acknowledgements

Thanks to Alberto Santos and Louise von Stechow for their thorough feedback and constructive comments.

References

1. Marz M, Beerenwinkel N, Drosten C, Fricke M, Frishman D, Hofacker IL, Hoffmann D, Middendorf M, Rattei T, Stadler PF, Töpfer A (2014) Challenges in RNA virus bioinformatics. Bioinformatics 30(13):1793–1799
2. Friedel CC (2013) Computational analysis of virus-host interactomes. In: Bailer SM, Lieber D, Walker JM (eds) Virus-host interactions methods and protocols, chap 8. Springer, Berlin, pp 115–130
3. Seymour LW, Fisher KD (2016) Oncolytic viruses: finally delivering. Br J Cancer 114(4):357–361
4. Waehler R, Russell SJ, Curiel DT (2007) Engineering targeted viral vectors for gene therapy. Nat Rev Genet 8(8):573–587
5. Thomas CE, Ehrhardt A, Kay MA (2003) Progress and problems with the use of viral vectors for gene therapy. Nat Rev Genet 4(5):346–358
6. Daugherty MD, Malik HS (2012) Rules of engagement: molecular insights from host-virus arms races. Annu. Rev. Genet. 46(1):677–700
7. Maynard ND, Gutschow MV, Birch EW, Covert MW (2010) The virus as metabolic engineer. Biotechnol J 5(7):686–694
8. Suttle CA (2005) Viruses in the sea. Nature 437(7057):356–361
9. Microbiology by numbers (2011) Nat Rev Microbiol 9(9):628–628
10. Anthony SJ, Epstein JH, Murray KA, Navarrete-Macias I, Zambrana-Torrelio C

Soloyvov A, Ojeda-Flores R, Arrigio NC, Islam A, Kahn SA, Hosseini P, Bogich TL, Mazet JK, Daszak P, Ian Lipkin W (2013) A strategy to estimate unknown viral diversity in mammals. mBio 4(5):1–15
11. Hannigan GD, Meisel JS, Tyldsley AS, Zheng Q, Hodkinson BP, Sanmiguel AJ, Minot S, Bushman FD Grice A (2015) The human skin double-stranded DNA virome: topographical and temporal diversity, genetic enrichment, and dynamic associations with the host microbiome. mBio 6(5):1–13
12. Roossinck MJ. The good viruses: viral mutualistic symbioses. Nature Reviews Microbiology, 9(2):99–108, 2011.
13. Villarreal LP (2007) Virus-host symbiosis mediated by persistence. Symbiosis 44(1–3):1–9
14. Murray NEA, Quam MB,Wilder-Smith A (2013) Epidemiology of dengue: past, present and future prospects. Clin Epidemiol 5:299–309
15. WHO (2014) WHO fact sheets: influenza, HCV, HPV
16. Molinari NAM, Ortega-Sanchez IR, Messonnier ML, Thompson WW, Wortley PM, Weintraub E, Bridges CB (2007) The annual impact of seasonal influenza in the US: measuring disease burden and costs. Vaccine 25(27):5086–5096
17. Klepac P, Metcalf CJE, McLean AR, Hampson K (2013) Towards the endgame and beyond: complexities and challenges for the elimination of infectious diseases. Philos. Trans R Soc Ser B Biol Sci 368(1623):20120137
18. Ehreth J (2003) The global value of vaccination. Vaccine 21(7–8):596–600
19. Havers F, Sokolow L, Shay DK, Farley MM, Monroe M, Meek J, Kirley PD, Bennett NM, Morin C, Aragon D, Thomas A, Schaffner W, Zansky SM, Baumbach J, Ferdinands J, Fry AM (2016) Case-control study of vaccine effectiveness in preventing laboratory-confirmed influenza hospitalizations in older adults, United States, 2010–2011. Clin Infect Dis 63(10):1304–1311
20. Blair W, Cox C (2016) Current Landscape of Antiviral Drug Discovery. F1000Research 5(5):202
21. Razonable RR (2011) Antiviral drugs for viruses other than human immunodeficiency virus. Mayo Clin Proc 86(10):1009–1026
22. De Clercq E (2002) Strategies in the design of antiviral drugs. Nat Rev Drug Discov 1(1):13–25
23. Noble CG, Chen YL, Dong H, Gu F, Lim SP, Schul W, Wang Q-Y, Shi P-Y (2010) Strategies for development of Dengue virus inhibitors. Antivir Res 85(3):450–462
24. Mills JN, Gage KL, Khan AS (2010) Potential influence of climate change on vector-borne and zoonotic diseases: a review and proposed research plan. Environ Health Perspect 118(11):1507–1514
25. Fauci AS, Morens DM (2016) Zika virus in the Americas — yet another arbovirus threat. N Engl J Med 374:601–604
26. Baltimore D (1971) Expression of animal virus genomes. Bacteriol Rev 35(3):235–241
27. Izmailyan R, Hsao J-C, Chung C-S, Chen C-H, Hsu PW-C, Liao C-L, Chang W (2012) Integrin $\beta 1$ mediates vaccinia virus entry through activation of PI3K/Akt signaling. J Virol 86(12):6677–6687
28. Clark JR, March JB (2006) Bacteriophages and biotechnology: vaccines, gene therapy and antibacterials. Trends Biotechnol 24(5):212–218
29. Colson P, De Lamballerie X, Yutin N, Asgari S, Bigot Y, Bideshi DK, Cheng X-W, Federici BA, Van Etten JA, Koonin EV, La Scola B, Raoult D (2013) "Megavirales", a proposed new order for eukaryotic nucleocytoplasmic large DNA viruses. Arch Virol 158(12):2517–2521
30. Scheel TKH, Rice CM (2013) Understanding the hepatitis C virus life cycle paves the way for highly effective therapies. Nat Med 19(7):837–849
31. Howley PM, Knipe DM (eds) (2013) Field's virology. Lippincott Williams & Williams, Philadelphia
32. Sanjuán R, Nebot MR, Chirico N, Mansky LM, Belshaw R (2010) Viral mutation rates. J Virol 84(19):9733–9748
33. Acevedo A, Andino R (2014) Library preparation for highly accurate population sequencing of RNA viruses. Nat Protoc 9(7):1760–1769
34. Vossen MTM, Westerhout EM, Söderberg-Nauclér C, Wiertz EJHJ (2002) Viral immune evasion: a masterpiece of evolution. Immunogenetics 54(8):527–542
35. Lauring AS, Andino R (2010) Quasispecies theory and the behavior of RNA viruses. PLoS Pathog 6(7):e1001005

36. Roossinck MJ (2008) Plant virus evolution. Springer, Berlin
37. Wylie CS, Shakhnovich EI. Mutation induced extinction in finite populations: lethal mutagenesis and lethal isolation. PLoS Comput Biol 8(8):2012
38. Pauly MD, Lauring AS (2015) Effective lethal mutagenesis of influenza virus by three nucleoside analogs. J Virol 89(7):3584–3597
39. Wu NC, Young AP, Al-Mawsawi LQ, Olson CA, Feng J, Qi H, Chen S-H, Lu I-H, Lin C-Y, Chin RG, Luan HH, Nguyen N, Nelson SF, Li X, Wu T-T, Sun R (2014) High-throughput profiling of influenza A virus hemagglutinin gene at single-nucleotide resolution. Sci Rep 4:1–8
40. Thyagarajan B, Bloom JD (2014) The inherent mutational tolerance and antigenic evolvability of influenza hemagglutinin. eLife 2014(3):1–26
41. Sumedha B, Bouchard MJ (2014) Cell cycle regulation during viral infection. In: Sivakumar S, Daum JR, Gorbsky GJ (eds) Cell cycle control, vol 1170. Springer, Berlin
42. Welch MD (2015) Why should cell biologists study microbial pathogens? Mol Biol Cell 26(24):4295–4301
43. Hunter T, Sefton BM (1980) Transforming gene product of Rous sarcoma virus phosphorylates tyrosine. Proc Natl Acad Sci USA 77(3):1311–1315
44. Frischknecht F, Cudmore S, Moreau V, Reckmann I, Röttger S, Michael W (1999) Tyrosine phosphorylation is required for actin-based motility of vaccinia but not Listeria or Shigella. Curr Biology 9(2):89–92
45. Lim K, Hyun Y-M, Lambert-Emo K, Capece T, Bae S, Miller R, Topham DJ, Kim M (2015) Neutrophil trails guide influenza-specific CD8(+) T cells in the airways. Science 349(6252):aaa4352
46. Madara JJ, Han Z, Ruthel G, Freedman BD, Harty RN (2015) The multifunctional Ebola virus VP40 matrix protein is a promising therapeutic target. Future Virol 10(5):537–546
47. Hale BG, Randall RE, Ortin J, Jackson D (2008) The multifunctional NS1 protein of influenza A viruses. J Gen Virol 89(10):2359–2376
48. Meyniel-Schicklin L, de Chassey B, Andre P, Lotteau V (2012) Viruses and interactomes in translation. Mol Cell Proteomics 11(7):M111.014738–M111.014738
49. Charon J, Theil S, Nicaise V, Michon T (2016) Protein intrinsic disorder within the Potyvirus genus: from proteome-wide analysis to functional annotation. Mol BioSyst 12(2):634–652
50. Peng Z, Yan J, Fan X, Mizianty MJ, Xue B, Wang K, Hu G, Uversky VN, Kurgan L (2014) Exceptionally abundant exceptions: comprehensive characterization of intrinsic disorder in all domains of life. Cell Mol Life Sci 72(1):137–151
51. Punta M, Coggill PC, Eberhardt RY, Mistry J, Tate J, Boursnell C, Pang N, Forslund K, Ceric G, Clements J, Heger A, Holm L, Sonnhammer ELL, Eddy SR, Bateman A, Finn RD (2011) The Pfam protein families database. Nucleic Acids Res 40(D1):D290–D301
52. Van Der Lee R, Buljan M, Lang B, Weatheritt RJ, Daughdrill GW, Dunker AK, Fuxreiter M, Gough J, Gsponer J, Jones DT, Kim PM, Kriwacki RW, Oldfield CJ, Pappu RV, Tompa P, Uversky VN, Wright PE, Babu MM (2014) Classification of intrinsically disordered regions and proteins. Chem Rev 114(13):6589–6631
53. Skalsky RL, Cullen BR (2010) Viruses, microRNAs, and host interactions. Annu Rev Microbiol 64:123–141
54. Mazzon M, Mercer J (2014) Lipid interactions during virus entry and infection. Cell Microbiol 16(10):1493–1502
55. Li X, Li Y, Wang C, Miao Z, Bi X, Wu D, Jin N, Wang L, Wu H, Qian K, Li C, Zhang T, Zhang C, Yi Y, Lai H, Hu Y, Cheng L, Leung KS, Li X, Zhang F, Li K, Wang D (2015) ViRBase: a resource for virus-host ncRNA-associated interactions. Nucleic Acids Res 43(D1):D578–D582
56. Schaefer MH, Serrano L, Andrade-Navarro MA (2015) Correcting for the study bias associated with protein-protein interaction measurements reveals differences between protein degree distributions from different cancer types. Front Genet 6(Aug):1–8
57. Ito T, Chiba T, Ozawa R, Yoshida M, Hattori M, Sakaki Y (2001) A comprehensive two-hybrid analysis to explore the yeast protein interactome. Proc Natl Acad Sci USA 98(8):4569–4574
58. Brückner A, Polge C, Lentze N, Auerbach D, Schlattner U (2009) Yeast two-hybrid, a powerful tool for systems biology. Int J Mol Sci 10(6):2763–2788

59. Lum KK, Cristea IM (2016) Proteomic approaches to uncovering virus-host protein interactions during the progression of viral infection. Expert Rev Proteomics 13(3): 325–340
60. Martell JD, Deerinck TJ, Sancak Y, Poulos TL, Mootha VK, Sosinsky GE, Ellisman MH, Ting AY (2012) Engineered ascorbate peroxidase as a genetically encoded reporter for electron microscopy. Nat Biotechnol 30(11):1143–1148
61. Roux KJ, Kim DI, Burke B (2013) BioID: a screen for protein-protein interactions. Curr Protoc Protein Sci 19:74
62. Lam SS, Martell JD, Kamer KJ, Deerinck TJ, Ellisman MH, Mootha VK, Ting AY (2014) Directed evolution of APEX2 for electron microscopy and proximity labeling_Supplementary. Nat Methods 12(1):51–54
63. Kim DI, Jensen SC, Noble KA, Birendra KC, Roux KH, Motamedchaboki K, Roux KJ (2016) An improved smaller biotin ligase for BioID proximity labeling. Mol Biol Cell 27(8):1188–1196
64. Hung V, Udeshi ND, Lam SS, Loh KH, Cox KJ, Pedram K, Carr SA, Ting AY (2016) Spatially resolved proteomic mapping in living cells with the engineered peroxidase APEX2. Nat Protoc 11(3):456–475
65. Guirimand T, Delmotte S, Navratil V (2015) VirHostNet 2.0: surfing on the web of virus/host molecular interactions data. Nucleic Acids Res 43(Database issue):D583–D587
66. Gerold G, Bruening J, Pietschmann T (2015) Decoding protein networks during virus entry by quantitative proteomics. Virus Res 218:25–39
67. Mehta V, Trinkle-Mulcahy L (2016) Recent advances in large-scale protein interactome mapping. F1000Research 5(0):782
68. de Chassey B, Meyniel-Schicklin L, Vonderscher J, André P, Lotteau V (2014) Virus-host interactomics: new insights and opportunities for antiviral drug discovery. Genome Med 6(11):115
69. Germain M-A, Chatel-Chaix L, Gagné B, Bonneil E, Thibault P, Pradezynski F, de Chassey B, Meyniel-Schicklin L, Lotteau V, Baril M, Lamarre D (2014) Elucidating novel Hepatitis C virus-host interactions using combined mass spectrometry and functional genomics approaches. Mol Cell Proteomics 13(1):184–203
70. Lajko M, Haddad AF, Robinson CA, Connolly SA (2015) Using proximity biotinylation to detect herpesvirus entry glycoprotein interactions: limitations for integral membrane proteins. J Virol Methods 221:395–401
71. Calderone A, Licata L, Cesareni G (2014) VirusMentha: a new resource for virus-host protein interactions. Nucleic Acids Res 43(D1):1–5
72. Ammari MG, Gresham CR, McCarthy FM, Nanduri B (2016) HPIDB 2.0: a curated database for host-pathogen interactions. Database 2016:baw103
73. Pickett BE, Sadat EL, Zhang Y, Noronha JM, Squires RB, Hunt V, Liu M, Kumar S, Zaremba S, Gu Z, Zhou L, Larson CN, Dietrich J, Klem B, Scheuermann RH (2012) ViPR: An open bioinformatics database and analysis resource for virology research. Nucleic Acids Res 40(D1):593–598
74. Attwood T, Agit B, Ellis L 2015 Longevity of biological databases. EMBnet.journal 21(0):e803
75. Bader GD, Hogue CWV (2003) An automated method for finding molecular complexes in large protein interaction networks. BMC Bioinf 4:2
76. HUPO Proteomics Standards Initiative. Molecular Interactions - HUPO Proteomics Standards Initiative.
77. Szklarczyk D, Morris JH, Cook H, Kuhn M, Wyder S, Simonovic M, Santos A, Doncheva NT, Roth A, Bork P, Jensen LJ, von Mering C (2016) The STRING database in 2017: quality-controlled protein-protein association networks, made broadly accessible. Nucleic Acids Res 45(D1):D362–D368
78. Yost SA, Marcotrigiano J (2013) Viral precursor polyproteins: keys of regulation from replication to maturation. Curr Opin Virol 3(2):137–142
79. The UniProt Consortium (2014) UniProt: a hub for protein information. Nucleic Acids Res 43(D1):D204–D212
80. Plotkin JB, Kudla G (2011) Synonymous but not the same: the causes and consequences of codon bias. Nat Rev Genet 12(1):32–42
81. Shen SH, Stauft CB, Gorbatsevych O, Song Y, Ward CB, Yurovsky A, Mueller S, Futcher B, Wimmer E (2015) Large-scale recoding of an arbovirus genome to rebalance its insect

versus mammalian preference. Proc Natl Acad Sci USA 112(15):4749–4754
82. Tuller T, Waldman YY, Kupiec M, Ruppin E (2010) Translation efficiency is determined by both codon bias and folding energy. Proc Natl Acad Sci USA 107(8):3645–3650
83. Reddy T, Sansom MSP (2016) Computational virology: from the inside out. Biochim Biophys Acta Biomembr 1858(7):1610–1618
84. Edwards RA, McNair K, Faust K, Raes J, Dutilh BE (2016) Computational approaches to predict bacteriophage-host relationships. FEMS Microbiol Rev 40(2):258–272
85. Villarroel J, Kleinheinz K, Jurtz V, Zschach H, Lund O, Nielsen M, Larsen M (2016) HostPhinder: a phage host prediction tool. Viruses 8(5):116
86. Harper JW, Adams PD (2001) Cyclin-dependent kinases. Chem Rev 101(8):2511–2526
87. Conrad B, Antonarakis SE (2007) Gene duplication: a drive for phenotypic diversity and cause of human disease. Annu Rev Genomics Hum Genet 8:17–35
88. Huerta-Cepas J, Szklarczyk D, Forslund K, Cook H, Heller D, Walter MC, Rattei T, Mende DR, Sunagawa S, Kuhn M, Jensen LJ, von Mering C, Bork P (2015) eggNOG 4.5: a hierarchical orthology framework with improved functional annotations for eukaryotic, prokaryotic and viral sequences. Nucleic Acids Res 44(Database issue):286–293
89. Kriventseva EV, Tegenfeldt F, Petty TJ, Waterhouse RM, Simão FA, Pozdnyakov IA, Ioannidis P, Zdobnov EM (2015) OrthoDB v8: update of the hierarchical catalog of orthologs and the underlying free software. Nucleic Acids Res 43(D1):D250–D256
90. Altenhoff AM, Boeckmann B, Capella-Gutierrez S, Dalquen DA, DeLuca T, Forslund K, Huerta-Cepas J, Linard B, Pereira C, Pryszcz LP, Schreiber F, da Silva AS, Szklarczyk D, Train C-M, Bork P, Lecompte O, von Mering C, Xenarios I, Sjölander K, Jensen LJ, Martin MJ, Muffato M, Gabaldón T, Lewis SE, Thomas PD, Sonnhammer E, Dessimoz C (2016) Standardized benchmarking in the quest for orthologs. Nat Methods 13(5):425–430
91. Forterre P (2006) The origin of viruses and their possible roles in major evolutionary transitions. Virus Res 117(1):5–16
92. Mocarski ES (2007) Comparative analysis of herpesvirus-common proteins. In: Arvin A, Campadelli-Fiume G, Mocarski E, Moore PS, Roizman B, Whitley R, Yamanishi K (eds) Human herpesviruses: biology, therapy, and immunoprophylaxis, chap 4. Cambridge University Press, Cambridge
93. Iranzo J, Krupovic M, Koonin EV (2016) The double-stranded DNA virosphere as a modular hierarchical network of gene sharing. mBio 7(4):e00978–16
94. Marks DS, Hopf TA, Sander C (2012) Protein structure prediction from sequence variation. Nat Biotechnol 30(11):1072
95. Brandman R, Brandman Y, Pande VS (2012) Sequence coevolution between RNA and protein characterized by mutual information between residue triplets. PLoS ONE 7(1):e30022
96. Avila-Herrera A, Pollard KS (2015) Coevolutionary analyses require phylogenetically deep alignments and better null models to accurately detect inter-protein contacts within and between species. BMC Bioinf 16(1):268
97. de Juan D, Pazos F, Valencia A (2013) Emerging methods in protein co-evolution. Nat Rev Genet 14(4):249–261
98. Eng CLP, Tong JC, Tan TW (2014) Predicting host tropism of influenza A virus proteins using random forest. BMC Med Genomics 7(Suppl 3):S1
99. Lengauer T, Sander O, Sierra S, Thielen A, Kaiser R (2007) Bioinformatics prediction of HIV coreceptor usage. Nature Biotechnol 25(12):1407–1410
100. Riemenschneider M, Cashin KY, Budeus B, Sierra S, Shirvani E, Bayanolhagh S, Kaiser R, Gorry PR, Heider D (2016) Genotypic prediction of co-receptor tropism of HIV-1 subtypes A and C. Sci Rep 6:1–9
101. Sanchez C, Lachaize C, Janody F, Bellon B, Röder L, Euzenat J, Rechenmann F, Jacq B (1999) Grasping at molecular interactions and genetic networks in Drosophila melanogaster using FlyNets, an internet database. Nucleic Acids Res 27(1):89–94
102. Jeong H, Tombor B, Albert R, Oltvai ZN, Barabasi A-L (2000) The large-scale organization of metabolic networks. Nature 407(6804):651–654
103. Barabási A-L (2009) Scale-free networks: a decade and beyond. Science 325(5939):412–413
104. Albert R, Jeong H, Barabási A-L (2000) Error and attack tolerance of complex networks. Nature 406:387–382

105. Kohanski MA, Dwyer DJ, Collins JJ (2010) How antibiotics kill bacteria: from targets to networks. Nat Rev Microbiol 8(6): 423–435
106. Ivanov AA, Khuri FR, Fu H (2013) Targeting protein-protein interactions as an anticancer strategy. Trends Pharmacol Sci 34(7): 393–400
107. Dias A, Bouvier D, Crepin T, McCarthy AA, Hart DJ, Baudin F, Cusack S, Ruigrok RW (2009) The cap-snatching endonuclease of influenza virus polymerase resides in the PA subunit. Nature 458(7240):914–918
108. Finlay BB, McFadden G (2006) Anti-immunology: evasion of the host immune system by bacterial and viral pathogens. Cell 124(4):767–782
109. Menche J, Sharma A, Kitsak M, Ghiassian SD, Vidal M, Loscalzo J, Barabási A-L (2015) Uncovering disease-disease relationships through the incomplete interactome. Science 347(6224)
110. Zheng S, Tansey WP, Hiebert SW, Zhao Z (2011) Integrative network analysis identifies key genes and pathways in the progression of Hepatitis C virus induced hepatocellular carcinoma. BMC Med Genet 4(1):62
111. Dyer MD, Murali TM, Sobral BW (2008) The landscape of human proteins interacting with viruses and other pathogens. PLoS Pathog 4(2):e32
112. IDG Knowledge Management Center (2016) Unexplored opportunities in the druggable human genome. Nat Rev Drug Discov
113. Wuchty S, Siwo G, Ferdig MT (2010) Viral organization of human proteins. PLoS ONE 5(8):e11796
114. Murali TM, Dyer MD, Badger D, Tyler BM, Katze MG (2011) Network-based prediction and analysis of HIV dependency factors. PLoS Comput Biol 7(9):e1002164
115. Valente TW, Coronges K, Lakon C, Costenbader E (2008) How correlated are network centrality measures? Connections 28(1): 16–26
116. Brohée S, van Helden J (2006) Evaluation of clustering algorithms for protein-protein interaction networks. BMC Bioinf 7:488
117. Gaballa A, Newton GL, Antelmann H, Parsonage D, Upton H, Rawat M, Claiborne A, Fahey RC, Helmann JD (2010) Biosynthesis and functions of bacillithiol, a major low-molecular-weight thiol in Bacilli. Proc Natl Acad Sci USA 107(14):6482–6486

Chapter 9

The SQUAD Method for the Qualitative Modeling of Regulatory Networks

Akram Méndez, Carlos Ramírez, Mauricio Pérez Martínez, and Luis Mendoza

Abstract

The wealth of molecular information provided by high-throughput technologies has enhanced the efforts dedicated to the reconstruction of regulatory networks in diverse biological systems. This information, however, has proven to be insufficient for the construction of quantitative models due to the absence of sufficiently accurate measurements of kinetic constants. As a result, there have been efforts to develop methodologies that permit the use of qualitative information about patterns of expression to infer the regulatory networks that generate such patterns. One of these approaches is the SQUAD method, which approximates a Boolean network with the use of a set of ordinary differential equations. The main benefit of the SQUAD method over purely Boolean approaches is the possibility of evaluating the effect of continuous external signals, which are pervasive in biological phenomena. A brief description and code on how to implement this method can be found at the following link: https://github.com/caramirezal/SQUADBookChapter.

Key words Regulatory networks, Network modeling, Cell fate, Expression pattern

1 Introduction

One of the challenges in biology is to understand the relationship between genotype and phenotype. The phenotype of a cell emerges from the complex interactions between molecules, the genome, and environmental cues that act in a nonlinear manner. Understanding how these interactions act in concert to regulate cellular phenotypes requires an integrative view of the processes that control cell fate decisions. The advent of the "omics" era has contributed to the study of the genotype–phenotype relationship, as new molecular biology methods and high-throughput technologies allow the identification and mapping of an increasing number of molecules and their interactions, thus offering valuable information regarding the regulatory networks involved in biological processes.

Louise von Stechow and Alberto Santos Delgado (eds.), *Computational Cell Biology: Methods and Protocols*, Methods in Molecular Biology, vol. 1819, https://doi.org/10.1007/978-1-4939-8618-7_9,

Knowledge of the connectivity of regulatory networks, while valuable, only provides a static view of the complex spatiotemporal behavior observed in any biological phenomenon [1]. Therefore, it is necessary to develop and employ mathematical models to understand the functionality and dynamical properties of regulatory networks. This allows us to understand questions related to the number, nature, and stability of the possible patterns of activation, the role of specific molecules and interactions in the establishment of such patterns, as well as the effects of external stimuli or permanent perturbations such as gain- and loss-of-function mutations. This information could eventually be used to devise control mechanisms with the aim of driving the system to a desired state [2, 3].

Network-based modeling approaches have been shown to be valuable tools to integrate biological information to understand the dynamical behavior observed in multiple cellular processes [4–10]. For example, the establishment of molecular patterns in floral morphogenesis of both wild-type and mutants in *Arabidopsis thaliana* [11, 12], the phenotype stability and plasticity of T helper lymphocytes and other blood cell subpopulations in mammals [6, 13, 14], the cell cycle in yeast or mammals [7, 15, 16], and apoptosis [17].

There are several approaches for modeling regulatory networks (for reviews on qualitative modeling approaches, see [18–21]). In particular, logical models are used to describe systems with a finite number of possible states. In contrast, ODE-based models are able to describe systems with a possibly infinite number of states [22, 23]. As a result, logical approaches are well suited for qualitative descriptions, while ODE-based models are more appropriate for quantitative descriptions.

When dealing with biological problems, it is often convenient to start with a simple model describing a few basic characteristics, and then gradually transform the model into a more refined and complex one. In the case of a regulatory network, one basic characteristic is the number and nature of the steady states of activation or expression generated by the system. In some cases, steady states have been shown to be largely insensitive to the precise values of parameters [24], or even the formalism chosen [5] to find them. In such cases, qualitative modeling can be implemented as a first approximation to model regulatory networks [19].

One approach to make a qualitative model more realistic is to incorporate continuous variables, so as to provide the model with the capacity of describing graded variations. There are several methods to transform Boolean networks into continuous dynamical systems. All of them rely on the general principle of interpolating or replacing the discrete functions by continuous mappings defined in the [0, 1] closed interval [25]. Interpolation functions can be linear [26], continuous logical extensions [27], polynomial approximations [28], or polynomial composed with Hill or exponential functions [1, 3] (see [25] for a comparative study). In this chapter, we describe the Standardized Qualitative Dynamical Systems (SQUAD) method, which is used to formalize regulatory networks as a continuous dynamical system focusing on their qualitative behavior rather than on their detailed kinetic parameters [1, 12].

The SQUAD method can be used to either automatically transform a Boolean network model to a continuous system of ordinary differential equations (ODEs), or to directly define a continuous model from the available information regarding the interactions among the components of a regulatory network. In this way, the modeler can simulate a regulatory network based mostly on the network architecture by making a few basic assumptions regarding the response of the nodes to their regulators, or, alternatively, to construct more elaborated models based on known regulatory mechanisms obtained from experimental data.

In the next section, we briefly describe how to define a Boolean model, and then we show how to transform it into a continuous model using the SQUAD method. Finally, we show how to analyze the dynamical behavior of regulatory networks to identify its steady states, as well as how to study the effect of diverse perturbations.

2 Methods

2.1 Reconstruction of the Regulatory Network

To construct a regulatory network model, it is necessary to identify the key molecules involved in the control of a biological process of interest (genes, proteins, molecular complexes, etc.), as well as the regulatory relationships among them (activating or inhibitory). This is done by integrating available experimental data regarding the function of the components of the network under different experimental conditions, such as gain- or loss-of-function mutants, epistasis analysis, and known expression patterns, for instance [18, 21]. While it is very important to have the most comprehensive data possible on the list of molecular markers that are present or absent under a given condition, it is also absolutely necessary to have a notion about the flux of information among the nodes to be included into the network model. That is, besides knowing that markers A and B are co-expressed, it is necessary to know if A regulates B, B regulates A, another node regulates both A and B, or even if there is no regulatory relationship between A and B.

While a detailed discussion on the different network inference methods is beyond the scope of this work, we refer the interested reader to some relevant literature on the topic [29–31]. Most of these methods rely on the estimation of the probability of dependency measures between gene expression values (Pearson correlation, mutual information, and Bayesian) [29]. In general, these methods are not well suited to infer the specific combinations of regulators that turn a particular gene *ON* or *OFF.* More importantly, such methodologies are very inefficient at inferring the presence of regulatory circuits [29]. As a result, most of the published Boolean models were inferred manually, using carefully selected experiments reported in the literature.

The proposed molecules and interactions forming a regulatory network can be incorporated into a table of interactions summarizing its architecture. Once we have a static representation of the regulatory network, it is necessary to postulate a set of logical rules or functions controlling the activation of the nodes of the network and simulate its dynamical behavior to qualitatively analyze the temporal expression profiles that can be attained under multiple conditions.

2.2 Definition of the Discrete Model

A deterministic Boolean model is the simplest formalism that can be used to study regulatory networks as dynamical systems. Despite their conceptual simplicity, Boolean networks show a nonlinear behavior resulting in interesting and nonintuitive dynamical properties such as multistability, cyclic trajectories, and robustness against perturbations [13, 32–34]. In Boolean models, the nodes of a regulatory network are represented by binary variables, i.e., they can attain one of two possible levels of activity (*ON/OFF*) at any given time (Fig. 1a). Also, regulatory interactions are formalized by means of Boolean functions that can be described by logic rules or truth tables (Fig. 1b). The Boolean function assigned to a node determines its value of activity at the next time step depending on the values of their regulators at a given time. Formally,

$$x_k(t+1) = f_k(x_{k,1}(t),\ x_{k,2}(t),\ x_{k,3}(t),\ \ldots,\ x_{k,r}(t)) \tag{1}$$

Where k represents the index of a node with r input regulators. The variables $x_{k,1}(t), \ldots, x_{k,r}(t)$ represent the set of values of the regulators of the node x_k at time t, and f_k is the Boolean function defining the activation of the node at time $t + 1$. For example, the node B in Fig. 1a has two regulators, given that nodes A and C activate and inhibit B, respectively. These interactions can be translated into a logical form to represent the regulatory mechanisms that control the activation levels of node B. Thus, node B becomes active only when its activator (node A) is present and if its inhibitor (node C) is absent (Fig. 1b). For each node, representing molecules or molecular complexes in the network, it is possible to formulate a logical rule summarizing the regulatory mechanisms behind each interaction, which can then be used to construct a dynamical model of the regulatory network. In the network shown in Fig. 1, the Boolean function for the node B is given by

$$B(t+1) = f_B(A(t),\ C(t)) = A(t) \wedge \neg\, C(t) \tag{2}$$

Where the symbols $\wedge$ and $\neg$ represent the logic functions AND and NOT, respectively. It is important to note that any logical rule can be expressed in terms of the Boolean operators AND, OR, and NOT, see Table 1.

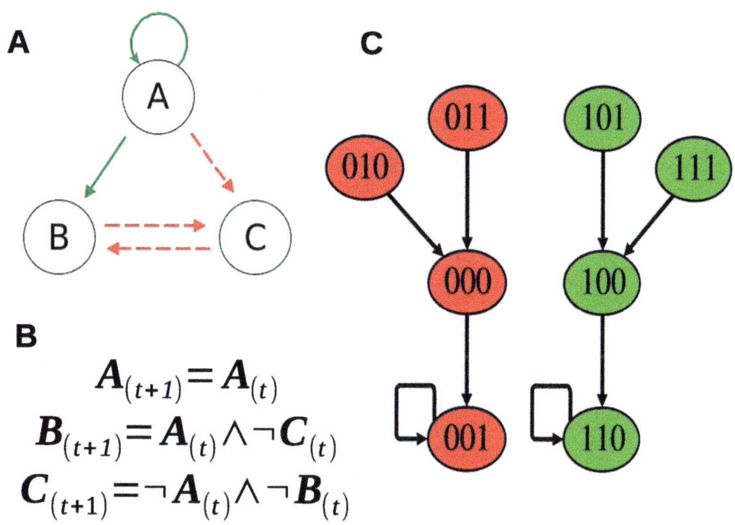

Fig. 1 Boolean networks. (**a**) A regulatory network comprising three factors A, B, and C, depicted as nodes in the graph. Activations and inhibitions are represented as green and red arrows, respectively. Every node has an associated state, which can be either 0/OFF/false or 1/ON/true. (**b**) Logic rules can be used to represent regulatory interactions in the regulatory network. (**c**) Dynamical behavior of the synchronous Boolean network model. In this case, nodes represent states of the network which are labeled by the binary values of the nodes A, B, and C, in that order. States are colored according to the attractor they converge to

Table 1
The logic rules describing the response of a node to its regulators expressed with the logic operators $\wedge$ (AND), $\vee$ (OR), and $\neg$ (NOT) can be transformed into a continuous form with the use of their equivalent fuzzy logic functions

Boolean operator	Fuzzy logic function
$A \wedge B$	$\min(A, B)$
$A \vee B$	$\max(A, B)$
$\neg A$	$1 - A$

If n is the number of nodes of the regulatory network, we call any binary vector $\hat{x} = (x_1, x_2, x_3, \ldots, x_n)$ a state of the network. In a Boolean system, there exist 2^n possible states, this set of configurations is called the phase space [18]. Equation (1) determines the transition of the network through the phase space, where successive states of the network are determined by the present state of the system. Therefore, we can rewrite the system of Boolean equations, if $F = (f_1, f_2, f_3, \ldots, f_n)$, then Eq. (1) becomes

$$\hat{x}_{t+1} = F(\hat{x}_t). \tag{3}$$

where $\hat{x}_t$ and $\hat{x}_{t+1}$ are the predecessor and successor states in the dynamics of the network.

The updating of the values of the nodes can be done in different ways, the simplest of which is the *synchronous* updating scheme, in which all nodes are updated at the same time. Otherwise, if nodes of the network are updated in a different order, the Boolean network is said to be *asynchronous* [35]. We will focus in the rest of the chapter on the synchronous updating [35, 36]. The dynamics of the Boolean network can be represented as a transition diagram (Fig. 1c), in which the phase space is drawn along with all the possible transitions between network states. The dynamical behavior of a Boolean network starting from an initial state x_0 can be easily seen by just following the arrows until the system converges to a closed cycle, i.e., a set of states that repeats consecutively for more than one time step during a trajectory.

Cycles are called *attractors* (i.e., solutions of the network) and capture the long-term behavior of the system [18, 21, 37]. If an attractor contains only one network state, it is called a *fixed point* attractor, otherwise it is referred as a *cyclic* attractor [37].

The network in Fig. 1 possesses two fixed point attractors. There are two sets of network states, shown as red and green nodes, each set of states converges to one of the fixed point attractors (Fig. 1c). The set of all network states that converge to a specific attractor is known as its basin of attraction. In the example of Fig. 1, the set of states $S = \{010, 011, 000, 001\}$, shown in red, correspond to the basin of the attractor 001. Because the system once in an attractor will remain in it, attractors are interpreted as patterns of activation associated with specific stable phenotypes [32, 38].

Boolean networks have the advantage that the entire phase space can be explored, so as to find all the possible attractors of the system. This is very useful because it permits understanding all possible phenotypes allowed by a given network model. Although the computation time of search scales very fast with the number of nodes of the network, there are algorithms that search for attractors in large networks comprising hundreds to several thousands of nodes [1, 39].

2.3 Construction of a Continuous Model Using SQUAD

For most biological systems, relatively few quantitative data is available regarding the kinetics and stoichiometry of biochemical reactions. Nonetheless, there is a wealth of qualitative data regarding the molecular interactions, and the effect of genetic mutations on the establishment of particular cell phenotypes. Qualitative data also offer valuable functional information about the regulation of some biological processes. The available information has limited the development of dynamical models of regulatory networks to only a small number of well-characterized systems [19].

To overcome these restrictions, and thus facilitate the systematic construction of regulatory network models, the SQUAD method was developed to simulate a dynamical system without the need for detailed kinetic parameters [40]. Instead, the method relies mostly on the connectivity of the regulatory network, i.e, the flow of information among the nodes, as well as on the rules describing the regulatory mechanisms among the molecules they represent. The method approximates a Boolean network with the use of a set of ordinary differential equations. Importantly, a continuous approximation of a Boolean system allows the construction of complex dynamical models even in the absence of quantitative information regarding the precise molecular regulatory mechanisms of the biological system.

The continuous system describes the rate of change of activation of a node x_k with the following ODE:

$$\frac{dx_k}{dt} = \frac{-e^{0.5h_k} + e^{-h_k(\omega_k - 0.5)}}{(1 - e^{0.5h_k})(1 + e^{-h_k(\omega_k - 0.5)})} - \gamma_k x_k \tag{4}$$

Where $0 < h_k, \gamma_k$ and $0 \leq x_k, \omega_k \leq 1$. The nonlinear function on the right-hand side of the differential equation defines a sigmoid curve constrained in the interval $[0, 1]$. The parameter h_k determines the steepness of the sigmoid, for high h_k values the sigmoid function approaches the step function characteristic of Boolean models (Fig. 2).

The parameter γ_k is the decay rate for the node x_k. Although its specific numerical value may vary according to the necessities of the modeler and depends on the biological system under study, it has been found that the attractors of a model are rather robust to variations in the variation of this parameter [4, 41].

The total regulatory input of a node is represented by the variable ω_k. In the first implementation of the SQUAD method, ω_k was defined as a weighted sum of positive regulations multiplied by a weighted sum of negative regulators [40, 42]. This approach permitted the automatic creation of a dynamical system based exclusively on the topology of the network. While the instant transformation of a diagrammatic representation into a dynamical system was a very convenient feature, the chosen definition of ω_k had two assumptions: First, that any positive regulator is strong enough to activate its target in the absence of inhibitors;

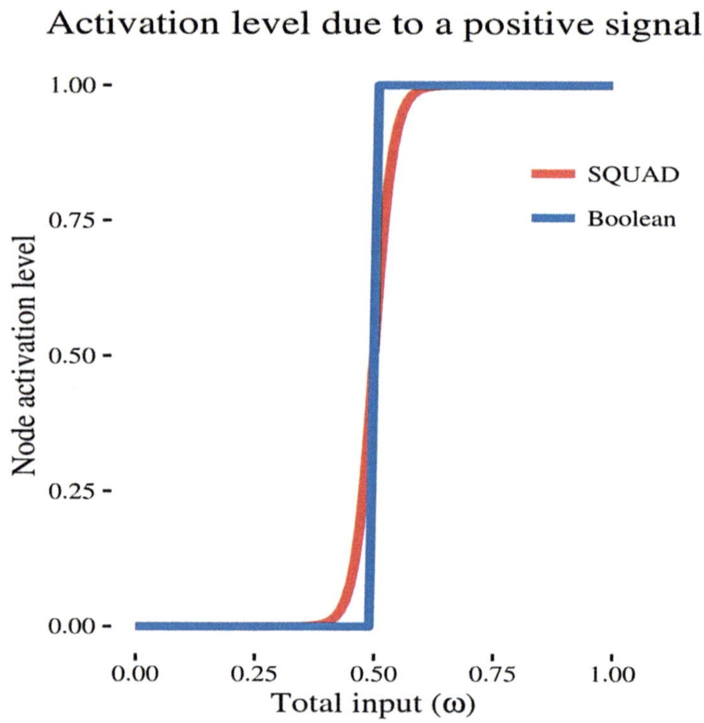

Fig. 2 Discrete versus continuous models. Response to a positive input in a discrete Boolean context (blue), and the continuous SQUAD version (red). The parameters h in SQUAD can be modified resulting in a steeper or shallower sigmoid

and second, that any negative regulator is stronger than any combination of positive regulators. These assumptions, however, are very strong and they are not necessarily true for most biological systems.

A new SQUAD version was proposed in [12] with the aim to relax the assumptions mentioned above. In this second version, ω_k is defined in terms of an interpolated Boolean function with the use of fuzzy logic operators [27] (Table 1). The first thing to do is to define the regulatory input of a node x_k as a Boolean function f_k of its regulators, just as in the case of the Boolean networks described in the previous section. It can be shown that any f_k can be rewritten in the following form:

$$f_k(x_1, \ldots, x_r) = \bigvee_{i=1}^{n} (x_1 \wedge x_2 \ldots \wedge x_l) \wedge (\neg x_{l+1} \wedge \neg x_{l+2} \wedge \ldots \wedge \neg x_r) \quad (5)$$

where r denotes the total number of regulators of the node x_k and l represents the positive regulators of the node x_k, whereas the remaining $r - l$ nodes indicate negative regulators of such node. This equation states that any Boolean function can be represented as a disjunction of n Boolean input configurations (2^r in total), conforming the set of states of the regulators that result in a value of f_k of $1/ON$. Then, the Boolean function can be interpolated by using fuzzy logic operators [1, 40], see Table 1, thus defining ω_k as

$$\omega_k(x_1, \ldots, x_r) = \max_{i=1}^{n}(\min(x_1, \ldots, x_l, 1 - x_{l+1}, \ldots, 1 - x_r)) \quad (6)$$

where, as in Eq. (5), the fuzzy logic rule for each x_k node is constructed according to its x_r regulators, of which l input nodes are positive regulators and the remaining $r - l$ nodes correspond to negative regulators. An example of how to translate a Boolean model into its continuous fuzzy logic form is shown in Fig. 3. This methodology thus provides a straightforward manner to translate a Boolean model into a continuous model in the form of a set of ODEs.

Given that the SQUAD method approximates a Boolean model, there is a close correspondence between the steady states

A

$\omega_1 = \mathbf{A}(t)$

$\omega_2 = min(\mathbf{A}(t), [1 - \mathbf{C}(t)])$

$\omega_3 = min(1 - \mathbf{A}(t), 1 - \mathbf{B}(t))$

B

$$\frac{dx_1}{dt} = \frac{-e^{0.5h} + e^{-h(w_1-0.5)}}{(1-e^{0.5h})(1+e^{-h(w_1-0.5)})} - \gamma x_1$$

$$\frac{dx_2}{dt} = \frac{-e^{0.5h} + e^{-h(w_2-0.5)}}{(1-e^{0.5h})(1+e^{-h(w_2-0.5)})} - \gamma x_2$$

$$\frac{dx_3}{dt} = \frac{-e^{0.5h} + e^{-h(w_3-0.5)}}{(1-e^{0.5h})(1+e^{-h(w_3-0.5)})} - \gamma x_3$$

C

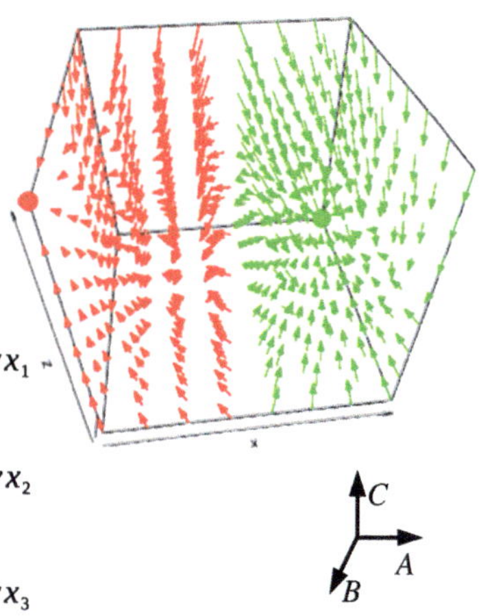

Fig. 3 SQUAD method. Example of the SQUAD methodology applied to the regulatory network given in Fig. 1. (**a**) Determination of the parameter ω_k summarizing the regulation for each node in terms of fuzzy logic operators. (**b**) Definition of the continuous dynamical system in the form of a set of ODEs. The parameters h and γ are set to 50 and 1, respectively. (**c**) Dynamical behavior of the regulatory network as simulated by the continuous model. Attractors are shown as red and green dots

found in Boolean discrete models and the steady states found in the continuous model. Notice, however, that SQUAD may find (usually unstable) steady states, or cyclic behaviors not recovered in the Boolean system [12, 43]. As in the discrete Boolean model the phase space is split into attractor basins. This can be seen in Fig. 3c, where the dynamics of the continuous model analogous to the regulatory network given in Fig. 1 are shown. The phase space is split into two basins, observed as green and red partitions, according to the convergence to the corresponding attractors $(A, B, C) = (0, 0, 1)$ and $(1, 1, 0)$, that coincides with that of the discrete model.

The ability of SQUAD to incorporate graded signals makes it a suitable tool to study the effect of extracellular signals on the establishment of expression patterns during differentiation processes [12, 44]. Figure 4 shows the difference in effect between small and large perturbations on the steady state configurations

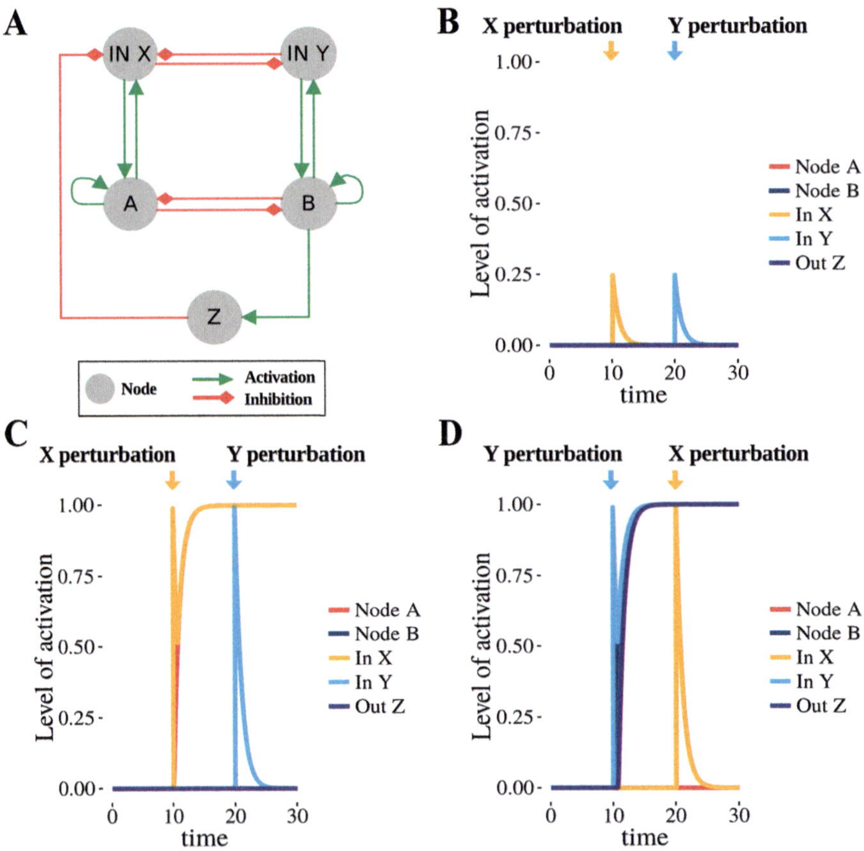

Fig. 4 Effects of node perturbations on the dynamics of a regulatory network. (**a**) Regulatory network module. (**b**) Small perturbations are given to node X and Y, they are absorbed by the system and vanish. (**c**)–(**d**). Larger perturbations cause permanent network state transitions. (**c**) If X is perturbed first, the network transits to a state in which node Z is inactive. (**d**) If the order of stimuli is inverted, the node Z becomes active

attained by a regulatory network. Also, it is possible to observe the effect of the order of such external signals. In this example, when the regulatory network represented in Fig. 4 is perturbed by a small increase of inputs X and Y, the perturbations are absorbed by the system (Fig. 4b). However, with stronger perturbations there is a transition to another network state, and thus the effect on the system is permanent (Fig. 4c). Moreover, the long-term behavior of the network depends on the order in which the perturbations are given (Fig. 4c, d). The interested reader may find in https://github.com/caramirezal/SQUADBookChapter the implementation of this example in the programming language R.

3 Modeling Biological Networks Using SQUAD

In this section, we describe the construction of a model of a particular biological process, so as to give a clearer view of the steps needed to make use of SQUAD. Specifically, we will focus on the regulatory network controlling terminal B cell differentiation, see [45]. A brief description of the model is provided in R language at https://github.com/caramirezal/SQUADBookChapter/. The model is also available in a standard SBMLqual format in The Cell Collective platform, see "B cell differentiation" model at https://cellcollective.org/.

B cells are the main effectors of humoral immune response in vertebrates, which is responsible for the recognition of foreign agents by the production of highly specific antibodies. Terminal B cell differentiation is achieved by the concerted action of several transcription factors in response to antigen recognition and extracellular signals provided by other blood cell types, like T-helper cells, dendritic cells, etc. This process of cell differentiation is characterized by the transition from a progenitor cell type (Naive B cell) into specialized cell types such as germinal center B cells (GC) and memory B cells (Mem), responsible for the editing and improvement of antibodies and its subsequent differentiation into antibody-producing plasma cells (PC) in response to antigen recognition [46].

The wealth of published experimental data regarding the molecules and signals involved in the control of B cell differentiation allowed us to reconstruct a regulatory network that incorporates several molecules known to be necessary for the control of the differentiation process, namely, Bach2, Bcl6, Blimp1, Irf4, and Pax5, see Fig. 5.

The information regarding the regulatory mechanisms controlling the activation of each node can be expressed in the form of logical rules. For example, it is known that Pax5 is regulated

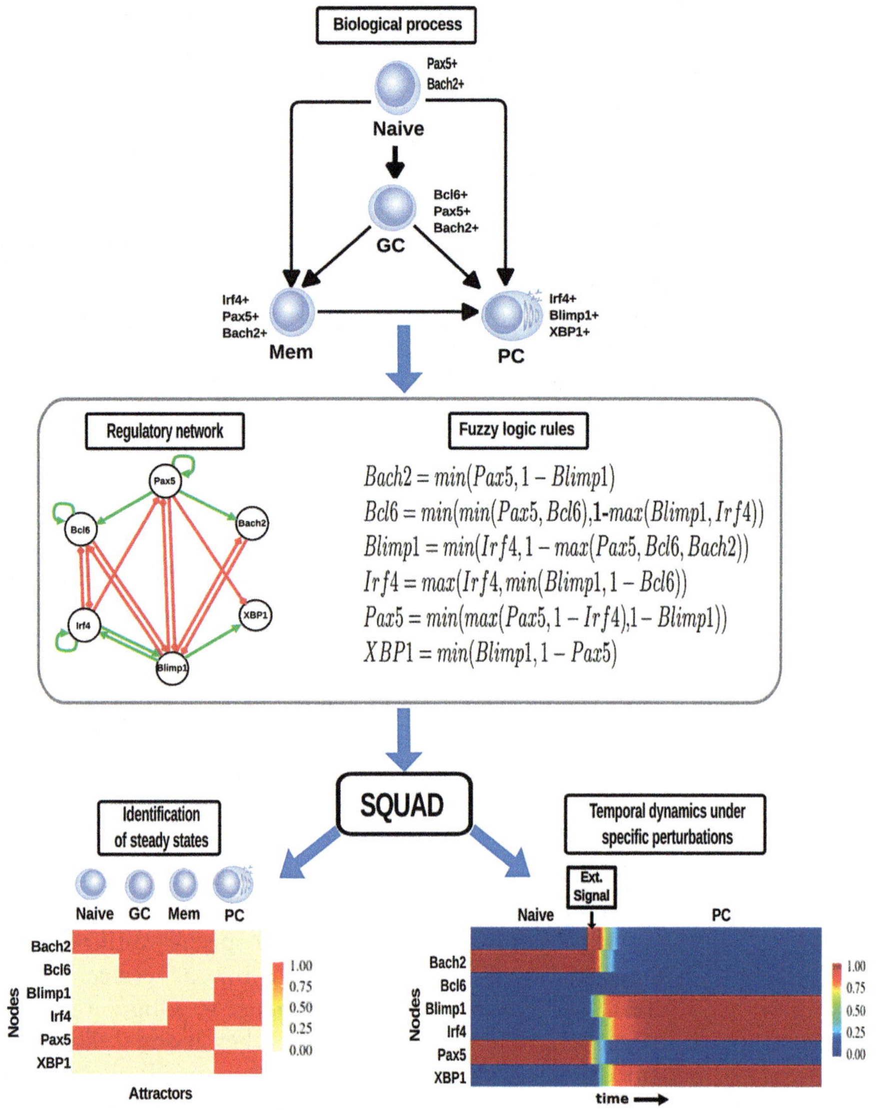

Fig. 5 Modeling B cell differentiation using SQUAD. Based on available information from a biological process, it is possible to infer a regulatory network taking into account the information flow among molecules (nodes), by activations (green arrows) and inhibitions (red arrows) between them. The regulatory mechanisms are incorporated to the network as logical rules. The system is translated into a set of ODEs, and the dynamical behavior is analyzed. The basic analysis consists in the identification of steady states, under wild-type and simulated mutations, as well as the analysis of the response of the system to certain signals

by itself and by low levels of Irf4 [47, 48], and also that Pax5 is negatively regulated by Blimp1 [49, 50]. This information can be expressed in the form of logical rules as follows:

$$Pax5(t+1) = (Pax5(t) \vee \neg\, Irf4(t)) \wedge \neg\, Blimp1(t)) \tag{7}$$

This expression, in turn, can be expressed in fuzzy logic terms as

Table 2
The logic rules summarizing the regulatory interactions in the B cell differentiation network expressed by the use of the logic operators and their translation into fuzzy logic functions

Node	Description	Logical rule	Fuzzy logic function
Bach2	Bach2 is activated by Pax5 if the supressor Blimp1 is absent	Pax5 ***AND NOT*** Blimp1	min(Pax5, 1−Blimp1)
Bcl6	Bcl6 regulates its own expression and is activated by Pax5 only if their repressors Blimp1 and Irf4 are inactive	Pax5 ***AND*** Bcl6 ***AND NOT*** (Blimp1 ***OR*** Irf4)	min(Pax5, Bcl6, 1−max(Blimp1, Irf4))
Blimp1	Blimp1 is activated by Irf4 if all its inhibitors, Pax5, Bcl6, and Bach2, are absent	Irf4 ***AND NOT*** (Pax5 ***OR*** Bcl6 ***OR*** Bach2)	min(Irf4, 1−max(Pax5, Bcl6, Bach2))
Irf4	Irf4 regulates its own activation and is positively regulated by Blimp1 if its inhibitor Bcl6 is inactive	Irf4 ***OR*** (Blimp1 ***AND NOT*** Bcl6)	max(Irf4, min(Blimp1, 1−Bcl6))
Pax5	Pax5 positively regulates its own expression and is maintained by low levels of Irf4. Pax5 is repressed by Blimp1	(Pax5 ***OR NOT*** Irf4) ***AND NOT*** Blimp1	min(max(Pax5, 1−Irf4), 1−Blimp1)
XBP1	XBP1 is activated by Blimp1 and repressed by Pax5	Blimp1 ***AND NOT*** Pax5	min(Blimp1, 1−Pax5)

$$Pax5(t+1) = \min(\max(Pax5(t), 1 - Irf4(t)), 1 - Blimp1(t)) \tag{8}$$

The complete set of logical rules summarizing the regulatory mechanisms controlling the activation of each node is presented in Table 2.

Once the complete set of fuzzy logic rules is established, they can be inserted as the corresponding ωs in the skeleton Eq. (4). That is, for the equation $dPax5/dt$ the corresponding ω_{Pax5} is $\min(\max(Pax5(t), 1 - Irf4(t)), 1 - Blimp1(t))$. Then, once the complete set of ODEs is determined, the resulting system is numerically integrated by sampling a large number of random initial states. The asymptotic solutions of the system of ODEs can then be compared with the known activation patterns characterizing each of the aforementioned B cell phenotypes.

Specifically, the method allowed the identification of attractors corresponding to the reported activation patterns of Naive, GC, Mem, and PC B cell phenotypes. Additionally, the SQUAD method was used to study the temporal response of the regulatory network to external stimuli with varying levels of intensity to simulate the effect of relevant biological signals and to evaluate the effect of multiple perturbations such as gain- and loss-of-function mutations.

The model was able to recapitulate the dynamical behavior of the set of key molecules involved in the differentiation of B cells [45]. Moreover, the model allowed the prediction of regulatory interactions that are necessary for the correct specification of multiple cell types during terminal B cell differentiation. Furthermore, the model gave theoretical support for the instructive role of cytokines involved in this differentiation process. And finally, the model explained the mechanism underlying the dynamical robustness of the PC attractor, a property that closely resembles the stability of the terminally differentiated plasma cells.

4 Summary and Outlook

There is a large variety of network modeling approaches; the selection of a particular methodology mostly depends on the available information and the kind of question one is seeking to answer. Specifically, the SQUAD method allows the construction of qualitative dynamical models of regulatory networks by making use of available experimental information where there is a lack of kinetic parameters. This method was developed to analyze the nature and number of the stationary states found in a regulatory network under different scenarios, such as gain- and loss-of-function mutants, or the presence of external stimuli.

Despite the usefulness of the SQUAD method, there is ample room for improvement. For example, some future refinements might incorporate stochastic effects on the concentration of molecules by adding noise to the nodes. Also, it would be desirable to include modifications that permit the transformation of multivalued discrete systems into ODE systems. These and other modifications might be re-implemented in the form of a user-friendly software package to facilitate the construction and exchange of models. Moreover, a long-term desirable goal would be to seek integration of SQUAD with other methods and platforms that complement it, such as GINsim [51], BoolNet [52], MaBoSS [53], JIMENA [54], and The Cell Collective [55].

5 Conclusions

SQUAD was developed to provide a flexible methodology to develop continuous models of regulatory networks, whenever only qualitative information is available. It can also be used as an initial tool for understanding the qualitative behavior of a network before embarking in the use of more refined, complex, and time-consuming modeling techniques. The successful use of SQUAD to understand the dynamical behavior of different biological systems shows its value as a tool of first choice for modelers.

In agreement with this, a comparative study carried out in [56] found SQUAD to be computationally more efficient than Odefy and CellNetAnalyzer tools [28, 57]. Recently, the group of Thomas Dandekar and colleagues implemented an optimization algorithm using RBDD Boolean representation methods and developed a tool called JIMENA to simulate regulatory networks. They found that with their optimization procedure the method performs as efficiently as SQUAD (old version) [54]. Although comparisons with the new version of SQUAD has not been carried out, in our hands we have been able to analyze regulatory networks comprising more than 80 nodes efficiently [14].

SQUAD was developed for qualitative modeling, thus it would not be an ideal tool for use in those systems with extensive presence of cooperative binding or allosteric regulation, where the kinetic details are of importance to determine the behavior of the system [58, 59]. Only very general molecular mechanisms are able to be included in the SQUAD equations; details of such mechanisms cannot be incorporated into the equations. The SQUAD method focuses on the modeling on system characteristics derived by the flux of information of the regulatory circuits, rather than in the details of molecular mechanisms.

SQUAD was developed to better understand the role of regulatory networks in the process of cellular differentiation. The flexibility and performance of the methodology, however, has allowed it to be used as a general tool in the analysis of the dynamical properties of regulatory networks. Its main strength is its capacity to analyze graded signals whenever the available information allows only for the construction of Boolean regulatory network models. Due to the large diversity of biological systems, modeling methods must be continuously created and modified to make the best possible use of all the available experimental information.

Acknowledgements

Akram Méndez thanks Programa de Doctorado en Ciencias Bioquímicas, UNAM. Mauricio Pérez Martínez thanks Programa de Doctorado en Ciencias Biológicas, UNAM. Carlos Ramírez thanks

Programa de Doctorado en Ciencias Biomédicas, UNAM. Luis Mendoza acknowledges the sabbatical scholarships from PASPA-DGAPA-UNAM and CONACYT 251420.

References

1. Garg A, Xenarios I, Mendoza L, DeMicheli G (2007) An efficient method for dynamic analysis of gene regulatory networks and in silico gene perturbation experiments. In: Speed T., Huang H. (eds) Research in computational molecular biology. RECOMB 2007. Lecture Notes in Computer Science, vol 4453. Springer, Berlin, pp 62–76
2. Zañudo JG, Albert R (2015) Cell fate reprogramming by control of intracellular network dynamics. PLoS Comput Biol 11(4): e1004193
3. Karl S, Dandekar T (2015) Convergence behaviour and control in non-linear biological networks. Scientific Reports 5
4. Martínez-Sosa P, Mendoza L (2013) The regulatory network that controls the differentiation of t lymphocytes. Biosystems 113(2):96–103
5. Álvarez-Buylla ER, Chaos Á, Aldana M, Benítez M, Cortes-Poza Y, Espinosa-Soto C, Hartasánchez DA, Lotto RB, Malkin D, Santos GJE et al (2008) Floral morphogenesis: stochastic explorations of a gene network epigenetic landscape. Plos One 3(11):e3626
6. Martinez-Sanchez ME, Mendoza L, Villarreal C, Alvarez-Buylla ER (2015) A minimal regulatory network of extrinsic and intrinsic factors recovers observed patterns of cd4+ t cell differentiation and plasticity. PLoS Comput Biol 11(6):e1004324
7. Fauré A, Naldi A, Chaouiya C, Thieffry D (2006) Dynamical analysis of a generic boolean model for the control of the mammalian cell cycle. Bioinformatics 22(14):e124–e131
8. Remy E, Rebouissou S, Chaouiya C, Zinovyev A, Radvanyi F, Calzone L (2015) A modeling approach to explain mutually exclusive and co-occurring genetic alterations in bladder tumorigenesis. Cancer Res 75(19):4042–4052
9. Naldi A, Carneiro J, Chaouiya C, Thieffry D (2010) Diversity and plasticity of th cell types predicted from regulatory network modelling. PLoS Comput Biol 6(9):e1000912
10. Calzone L, Tournier L, Fourquet S, Thieffry D, Zhivotovsky B, Barillot E, Zinovyev A (2010) Mathematical modelling of cell-fate decision in response to death receptor engagement. PLoS Comput Biol 6(3):e1000702
11. Mendoza L, Alvarez-Buylla ER (1998) Dynamics of the genetic regulatory network for arabidopsis thaliana flower morphogenesis. J Theor Biol 193(2):307–319
12. Sanchez-Corrales YE, Alvarez-Buylla ER, Mendoza L (2010) The arabidopsis thaliana flower organ specification gene regulatory network determines a robust differentiation process. J Theor Biol 264(3):971–983
13. Mendoza L (2006) A network model for the control of the differentiation process in Th cells. Biosystems 84(2):101–114
14. Mendoza L, Méndez A (2015) A dynamical model of the regulatory network controlling lymphopoiesis. Biosystems 137:26–33
15. Li F, Long T, Lu Y, Ouyang Q, Tang C (2004) The yeast cell-cycle network is robustly designed. Proc Natl Acad Sci USA 101(14):4781–4786
16. Davidich MI, Bornholdt S (2008) Boolean network model predicts cell cycle sequence of fission yeast. PloS One 3(2):e1672
17. Schlatter R, Schmich K, Vizcarra IA, Scheurich P, Sauter T, Borner C, Ederer M, Merfort I, Sawodny O (2009) On/off and beyond-a boolean model of apoptosis. PLoS Comput Biol 5(12):e1000595
18. Karlebach G, Shamir R (2008) Modelling and analysis of gene regulatory networks. Nat Rev Mol Cell Biol 9(10):770–780
19. Le Novère N (2015) Quantitative and logic modelling of molecular and gene networks. Nat Rev Genet 16(3):146–158
20. Abou-Jaoudé W, Traynard P, Monteiro PT, Saez Rodriguez J, Helikar T, Thieffry D, Chaouiya C (2016) Logical modeling and dynamical analysis of cellular networks. Front Genet 7:94
21. Albert R, Thakar J (2014) Boolean modeling: a logic-based dynamic approach for understanding signaling and regulatory networks and for making useful predictions. Wiley Interdiscip Rev Syst Biol Med 6(5):353–369
22. De Jong H (2002) Modeling and simulation of genetic regulatory systems: a literature review. J Comput Biol 9(1):67–103

23. Vijesh N, Chakrabarti SK, Sreekumar J et al (2013) Modeling of gene regulatory networks: a review. J Biomed Sci Eng 6(02):223
24. Gutenkunst RN, Waterfall JJ, Casey FP, Brown KS, Myers CR, Sethna JP (2007) Universally sloppy parameter sensitivities in systems biology models. PLoS Comput Biol 3(10):e189
25. Wittmann DM, Krumsiek J, Saez-Rodriguez J, Lauffenburger DA, Klamt S, Theis FJ (2009) Transforming boolean models to continuous models: methodology and application to t-cell receptor signaling. BMC Syst Biol 3(1):1
26. Glass L, Kauffman SA (1973) The logical analysis of continuous, non-linear biochemical control networks. J Theor Biol 39(1): 103–129
27. Zadeh L (1965) Fuzzy sets. Inf Control 8(3):338–353
28. Krumsiek J, Pölsterl S, Wittmann DM, Theis FJ (2010) Odefy-from discrete to continuous models. BMC Bioinf 11(1):1
29. Villaverde AF, Banga JR (2014) Reverse engineering and identification in systems biology: strategies, perspectives and challenges. J R Soc Interface 11(91):20130505
30. Basso K, Margolin AA, Stolovitzky G, Klein U, Dalla-Favera R, Califano A (2005) Reverse engineering of regulatory networks in human B cells. Nat Genet 37(4):382–390
31. Hecker M, Lambeck S, Toepfer S, Van Someren E, Guthke R (2009) Gene regulatory network inference: data integration in dynamic models—a review. Biosystems 96(1):86–103
32. Kauffman SA (1969) Metabolic stability and epigenesis in randomly constructed genetic nets. J Theor Biol 22(3):437–467
33. Derrida B, Pomeau Y (1986) Random networks of automata: a simple annealed approximation. Europhys Lett 1(2):45
34. Thieffry D (2007) Dynamical roles of biological regulatory circuits. Brief in Bioinf 8(4):220–225
35. Garg A, Di Cara A, Xenarios I, Mendoza L, De Micheli G (2008) Synchronous versus asynchronous modeling of gene regulatory networks. Bioinformatics 24(17):1917–1925
36. Bornholdt S (2008) Boolean network models of cellular regulation: prospects and limitations. J R Soc Interface 5(Suppl 1):S85–S94
37. Aldana M (2003) Boolean dynamics of networks with scale-free topology. Phys D 185(1):45–66
38. Huang S, Eichler G, Bar-Yam Y, Ingber DE (2005) Cell fates as high-dimensional attractor states of a complex gene regulatory network. Phys Rev Lett 94(12):128701
39. Dubrova E, Teslenko M (2011) A sat-based algorithm for finding attractors in synchronous boolean networks. IEEE/ACM Trans Comput Biol Bioinf 8(5):1393–1399
40. Mendoza L, Xenarios I (2006) A method for the generation of standardized qualitative dynamical systems of regulatory networks. Theor Biol Med Model 3(1):1
41. Sankar M, Osmont KS, Rolcik J, Gujas B, Tarkowska D, Strnad M, Xenarios I, Hardtke CS (2011) A qualitative continuous model of cellular auxin and brassinosteroid signaling and their crosstalk. Bioinformatics 27(10):1404–1412
42. Di Cara A, Garg A, De Micheli G, Xenarios I, Mendoza L (2007) Dynamic simulation of regulatory networks using squad. BMC Bioinf 8(1):1
43. Ortiz-Gutiérrez E, García-Cruz K, Azpeitia E, Castillo A, de la Paz Sánchez M, Álvarez-Buylla ER (2015) A dynamic gene regulatory network model that recovers the cyclic behavior of arabidopsis thaliana cell cycle. PLoS Comput Biol 11(9):e1004486
44. Mendoza L, Pardo F (2010) A robust model to describe the differentiation of T-helper cells. Theory Biosci 129(4):283–293
45. Méndez A, Mendoza L (2016) A network model to describe the terminal differentiation of b cells. PLoS Comput Biol 12(1):e1004696
46. Nutt SL, Hodgkin PD, Tarlinton DM, Corcoran LM (2015) The generation of antibody-secreting plasma cells. Nat Rev Immunol 15(3):160–171
47. O'Riordan M, Grosschedl R (1999) Coordinate regulation of B cell differentiation by the transcription factors EBF and E2A. Immunity 11(1):21–31
48. Decker T, di Magliano MP, McManus S, Sun Q, Bonifer C, Tagoh H, Busslinger M (2009) Stepwise activation of enhancer and promoter regions of the B cell commitment gene pax5 in early lymphopoiesis. Immunity 30(4):508–520
49. Mora-López F, Reales E, Brieva JA, Campos-Caro A (2007) Human bsap and blimp1 conform an autoregulatory feedback loop. Blood 110(9):3150–3157
50. Lin KI, Angelin-Duclos C, Kuo TC, Calame K (2002) Blimp-1-dependent repression of pax-5 is required for differentiation of b cells to immunoglobulin m-secreting plasma cells. Mol Cell Biol 22(13):4771–4780

51. Chaouiya C, Naldi A, Thieffry D (2012) Logical modelling of gene regulatory networks with GINsim. In: Bacterial molecular networks: methods and protocols. Springer, New York, pp 463–479
52. Müssel C, Hopfensitz M, Kestler HA (2010) Boolnet—an R package for generation, reconstruction and analysis of boolean networks. Bioinformatics 26(10):1378–1380
53. Stoll G, Viara E, Barillot E, Calzone L (2012) Continuous time boolean modeling for biological signaling: application of gillespie algorithm. BMC Syst Biol 6(1):116
54. Karl S, Dandekar T (2013) Jimena: efficient computing and system state identification for genetic regulatory networks. BMC Bioinf 14(1):1
55. Helikar T, Kowal B, McClenathan S, Bruckner M, Rowley T, Madrahimov A, Wicks B, Shrestha M, Limbu K, Rogers JA (2012) The cell collective: toward an open and collaborative approach to systems biology. BMC Syst Biol 6(1):96
56. Schlatter R, Philippi N, Wangorsch G, Pick R, Sawodny O, Borner C, Timmer J, Ederer M, Dandekar T (2012) Integration of boolean models exemplified on hepatocyte signal transduction. Brief Bioinf 13(3):365–376
57. Klamt S, Saez-Rodriguez J, Gilles ED (2007) Structural and functional analysis of cellular networks with cellnetanalyzer. BMC Syst Biol 1(1):1
58. Kahlem P, DiCara A, Durot M, Hancock JM, Klipp E, Schächter V, Segal E, Xenarios I, Birney E, Mendoza L (2011) Strengths and weaknesses of selected modeling methods used in systems biology. In: Systems and computational biology-bioinformatics and computational modeling. InTech, Rijeka
59. Kutejova E, Briscoe J, Kicheva A (2009) Temporal dynamics of patterning by morphogen gradients. Curr Opin Genet Dev 19(4): 315–322

Chapter 10

miRNet—Functional Analysis and Visual Exploration of miRNA–Target Interactions in a Network Context

Yannan Fan and Jianguo Xia

Abstract

To gain functional insights into microRNAs (miRNAs), researchers usually look for pathways or biological processes that are overrepresented in their target genes. The interpretation is often complicated by the fact that a single miRNA can target many genes and multiple miRNAs can regulate a single gene. Here we introduce miRNet (www.mirnet.ca), an easy-to-use web-based tool designed for creation, customization, visual exploration and functional interpretation of miRNA–target interaction networks. By integrating multiple high-quality miRNA-target data sources and advanced statistical methods into a powerful network visualization system, miRNet allows researchers to easily navigate the complex landscape of miRNA–target interactions to obtain deep biological insights. This tutorial provides a step-by-step protocol on how to use miRNet to create miRNA–target networks for visual exploration and functional analysis from different types of data inputs.

Key words miRNA–target interaction, miRNA functional analysis, Network analysis, Visual analytics, Empirical sampling, Microarray, RNA-seq, RT-qPCR, Differential expression analysis

1 Introduction

MicroRNAs (miRNAs) are endogenous small noncoding RNAs that can bind to their target mRNAs to suppress mRNA translation or induce mRNA degradation [1]. MicroRNAs modulate many fundamental biological processes such as cell-cycle control, development, and apoptosis. Their deregulations have been implicated in cancer, cardiovascular disease, neurological disorders, etc. [2]. Recent studies have shown that miRNAs may also participate in cross-species communications between host, parasites, and gut microbiota [3, 4].

Identification of miRNA targets is the first step toward understanding miRNA functions. Both computational and experimental approaches are commonly employed to perform the task. The computational algorithms are based primarily on estimating the binding affinities of miRNA–target interactions [5–7]. Though

Louise von Stechow and Alberto Santos Delgado (eds.), *Computational Cell Biology: Methods and Protocols*, Methods in Molecular Biology, vol. 1819, https://doi.org/10.1007/978-1-4939-8618-7_10,

very useful, the predicted interactions suffer from high false positive rates [8]. In responding to this issue, recent years have witnessed a growing number of large-scale studies on miRNA–target interactions using high-throughput experimental methods such as cross-linking and immunoprecipitation coupled to deep sequencing (CLIP-seq) or high-throughput sequencing coupled with cross-linking and immunoprecipitation (HITS-CLIP) [9, 10]. The resulting high-quality miRNA–gene interactions datasets are available in several databases [11–14]. More specialized databases have also been developed in recent years, connecting miRNAs to diseases, small compounds, long noncoding RNAs (lncRNAs), and epigenetic modifiers [15–20].

After target genes have been identified, researchers need to annotate these genes for their participation in pathways and biological processes, followed by testing whether some of those predefined functions are overrepresented in the target genes [21]. Recently, Bleazard et al. showed that strong biases could be introduced when directly applying the enrichment analysis on target genes, and proposed to use an empirical sampling based approach to identify enriched functions within miRNA target genes [22].

It is well established that a single miRNA can target multiple genes and one gene is often regulated by many miRNAs. It is through these complex interactions that miRNAs exert their collective influences to fine-tune the expression profiles of their target genes. An efficient way to navigate such complex many-to-many relationship is to use a network-based visualization approach to obtain a global understanding of the system [23–25].

We have recently developed miRNet [26], a user-friendly web-based tool designed to help researchers to intuitively perform the three common tasks during miRNA functional analysis: (1) miRNA target identification and refinement, (2) network-based visual exploration, and (3) functional enrichment analysis. miRNet currently supports miRNA-target analysis and visualization for nine different organisms using a comprehensive miRNA-target database (Table 1). In addition, miRNet also provides support for differential expression analysis for data generated from RT-qPCR, microarray, or RNA-seq experiments. In this protocol, we show how to create miRNA–target interaction networks from a list of miRNAs, followed by functional analysis on target genes; Secondly, we show how to perform miRNA–target interaction network analysis from a RT-qPCR data; Finally, we show how to create a functional interaction network from a list of bioactive compounds and their perturbed miRNAs.

2 Methods

2.1 Network Creation and Visual Analysis from a List of miRNAs

Section I: Data upload and network creation

1. Go to the miRNet home page (http://www.mirnet.ca). At the center of the page, there are nine round buttons corresponding to nine different data inputs supported by miRNet (Fig. 1). In this protocol, we show how to create miRNA–target gene interaction networks from a list of miRNA IDs.
2. Click on the "miRNA" button to start. In the data upload page, users need to first specify organism, miRNA ID type and target type. miRNet currently supports nine organisms (*H. sapiens*, *M. musculus*, *R. norvegicus*, *B. taurus*, *G. gallus*, *D. melanogaster*, *C. elegans*, *D. rerio*, and *S. mansoni*), two types of miRNA IDs (miRBase ID and accession number), and five types of miRNA targets (genes, diseases, lncRNAs, small molecules, and epigenetic modifiers). Note that some miRNA targets are only available for certain organisms (Table 1).
3. Copy and paste a list of miRNA IDs (miRBase IDs or accession numbers) to the data input area, with one ID per row. In this case, we will use the example data provided by miRNet. Click the "Try Example" button on the bottom-left of the page. On the pop-up dialog, click the "OK" button to upload the example list, and miRNet will set the parameters for this

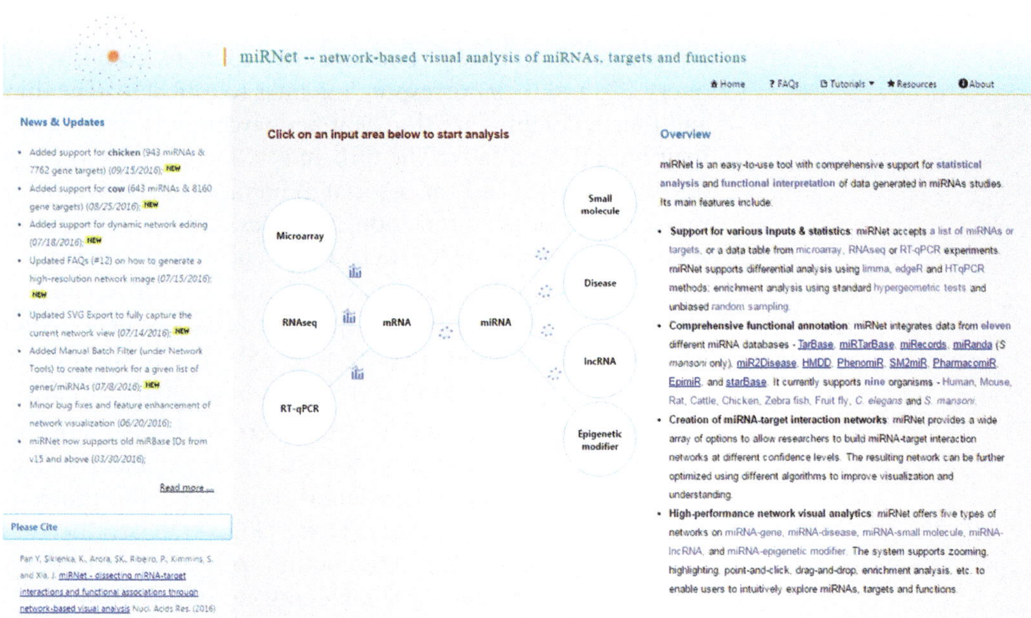

Fig. 1 The miRNet home page. The center is the navigation chart with nine buttons corresponding to nine different types of data inputs. Users need to click on a button based on their data to start analysis

Table 1
The summary statistics of miRNA targets/associations available in miRNet

	Genes	Compounds	Diseases	LncRNAs	Epigenetic Modifiers
Human	327,209	3110	19,341	10,212	1932
Mouse	57,921	1062	-	-	23
Rat	610	581	-	-	-
Cattle	48,878	-	-	-	-
Chicken	82,164	-	-	-	-
Zebrafish	175	148	-	-	-
Fruit fly	283	-	-	-	-
C. elegans	3215	34	-	-	-
S. mansoni	1274	-	-	-	-
Total	521,729	4935	19,341	10,212	1955

example data (Fig. 2). Click the "Submit" button to upload the miRNA list. An "OK" message will show up on the top-right corner of the page after a few seconds.

4. Click the "Proceed" button on the bottom-right of the page to search gene targets for those uploaded miRNAs.
5. The miRNA–target interaction table will be displayed after a few seconds (Fig. 3). The first two columns comprise the query IDs and their corresponding links to miRBase. The third and fourth columns are the identified target gene symbols and their links to GenBank. The fifth and sixth columns show supporting evidence and the associated literature publications for the miRNA–target interactions. The last column ("Action") allows users to manually delete a particular interaction. Users can click the column headers of the first or third column to sort the table or to search for specific miRNA or gene target. The table can be downloaded as a CSV file by clicking the "Download" button on the top-right of the table.
6. (Optional) Users can apply data filters to further improve the quality of the default miRNA–target interaction data. To do this, click the "Data Filter" button on the top-right of the table to bring up the dialog (Fig. 4). Users need to specify three parameters—the column to be filtered ("Target Column"); a keyword and matching criteria—*"(Character) Matching," "(Character) Containing,"* or *"(Numeric) At least"*; and whether to *"remove"* or *"keep"* rows that meet the specified criteria. Note the numeric value filter is mainly

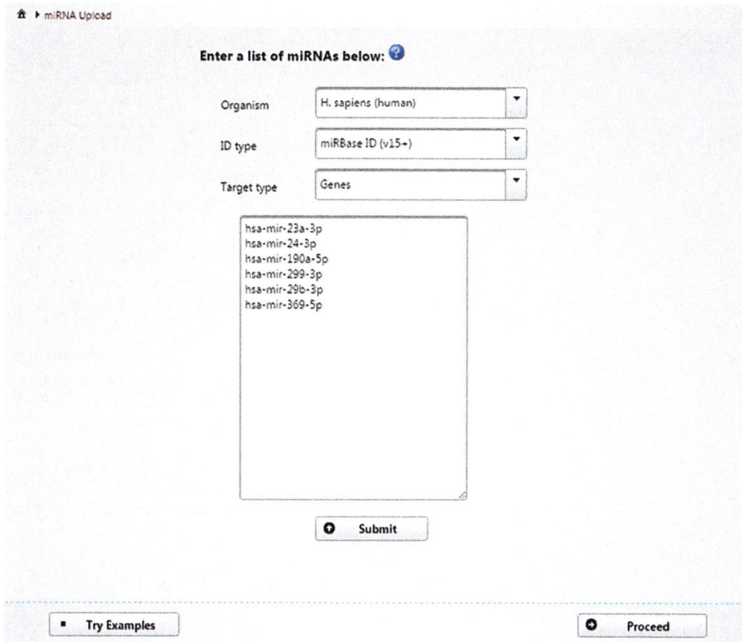

Fig. 2 A screenshot of the miRNA upload page using the example miRNA list

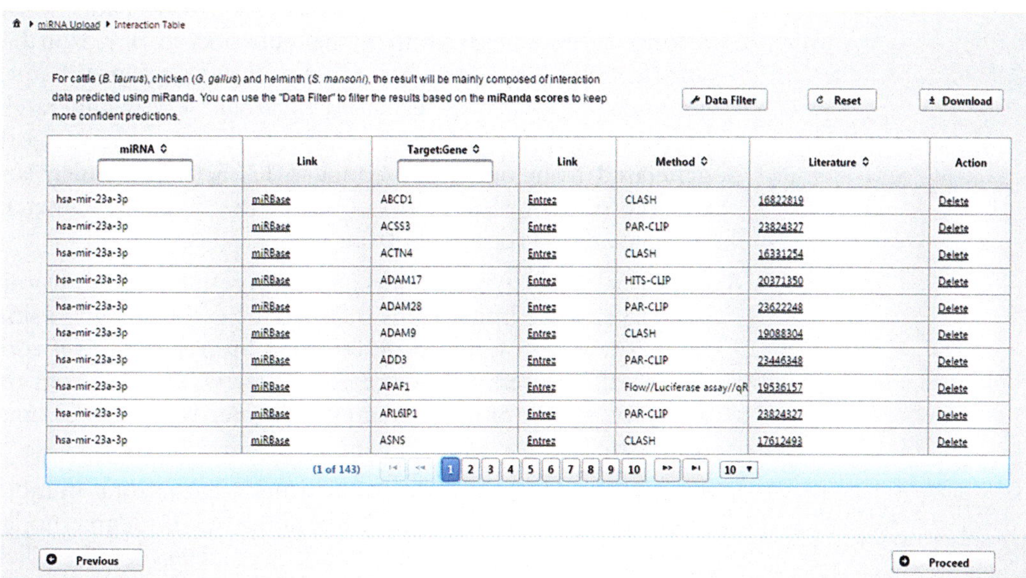

Fig. 3 The miRNA–target interaction table. It has seven columns that can be used for searching and data filtering. The "Data Filter" button is on the top-right with "Reset" and "Download"

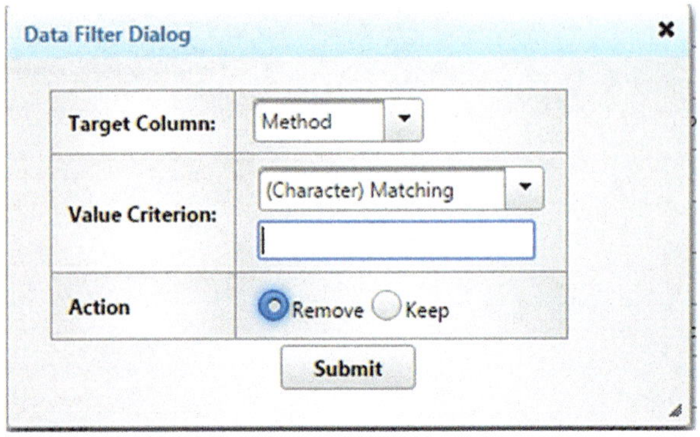

Fig. 4 A screenshot showing the Data Filter Dialog

designed to filter the miRNA gene targets for cattle, chicken and *S. mansoni*, predicted using the well-established mirSVR algorithm [5].

7. Click the "Proceed" button to the "Network Builder" page to create the miRNA–gene interaction networks (Fig. 5). This page summarizes the overall statistics of the miRNA–target interaction networks, the individual network(s) and the number of nodes and edges. It also offers a set of functions to further manipulate the networks ("Network Tools"). We recommend users to control the networks at a reasonable size (<2000 nodes) for better visualization experience. Larger networks typically will lead to a "hairball"-effect and rarely produce any informative outcome. In this case, the network generated from our query contains 1355 nodes, which is a suitable size for visualization. Accept the default parameters and click the "Proceed" button.

8. (Optional) For large networks, users can return to the previous page to exclude interactions that may be of low quality using the "Data Filter" function described in **Step 6**. The "Network Tools" allow users to further reduce network size based on the network topology measures - degree, betweenness, and shortest path.

   (a) *Degree Filter*. The degree of a node is the total number of connections it has to other nodes. High-degree nodes are considered important "hubs" in a network [27]. Click the "Degree Filter" button to bring up the dialog (Fig. 6). Users can apply this filter to different node types (miRNA or targets) to remove the nodes with degrees lower than a specified threshold (default 1.0).

   (b) *Betweenness Filter*. The betweenness measures the number of shortest paths going through a node. Nodes with higher betweenness are important "bottlenecks" in a network [27]. Click the "Betweenness Filter" button to bring up

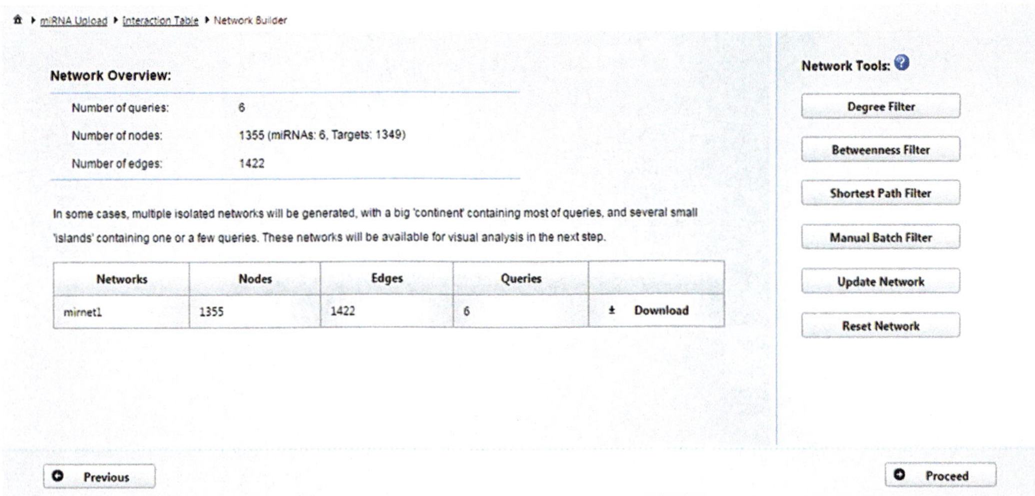

Fig. 5 A screenshot of the Network Builder page. Network Overview shows the summary of the network. The right section is Network Tools for network refinement

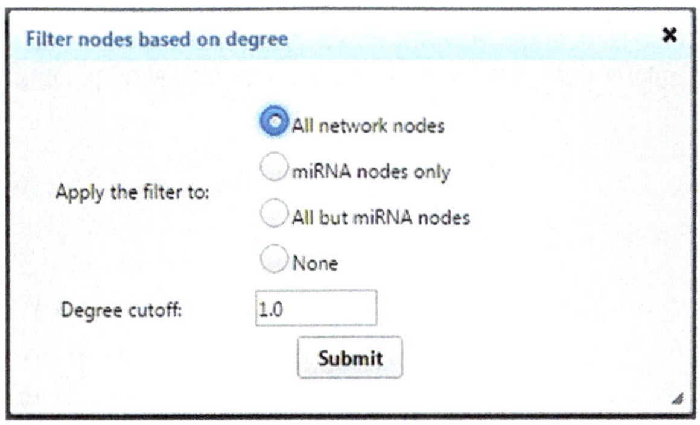

Fig. 6 A screenshot showing various options in Degree Filter

the dialog and use the similar steps as described in the **Step 8a** to filter the networks.

(c) *Shortest Path Filter*. For very dense networks, the number of edges within the network can be reduced significantly by keeping only one shortest path between hub nodes. That is, if there are multiple paths connecting two nodes, only one shortest path will remain to keep only essential connection patterns [27].

After testing different network tools, click the "Reset Network" button to return to the default network. Click the "Proceed" button to visualize the network.

Section II: Network visual exploration

9. The "Network Viewer" page is composed of four main components—the top tool menu, the "Node Explorer" panel

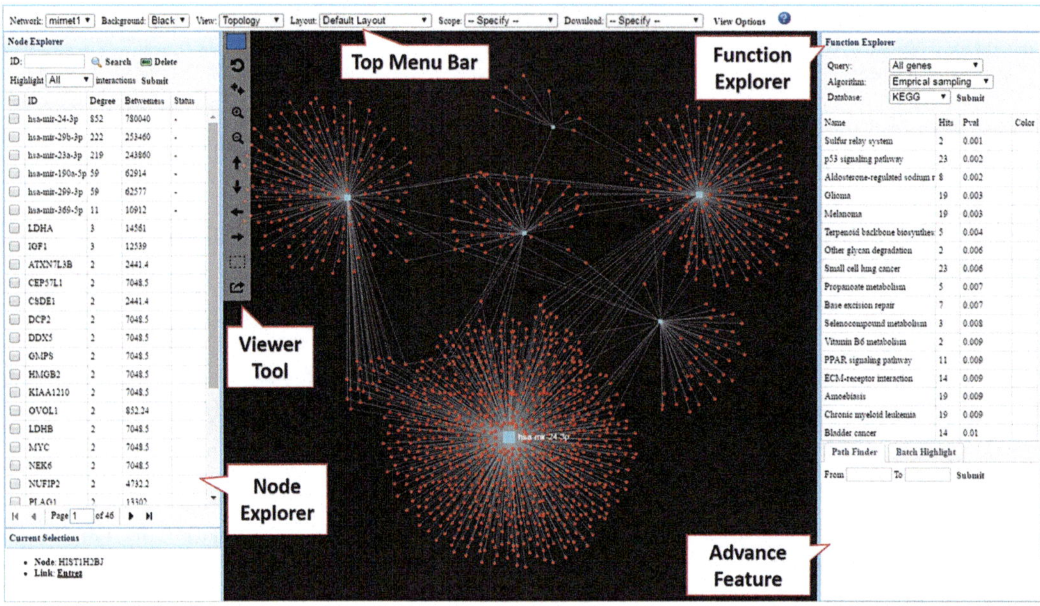

Fig. 7 A screenshot of the Network Viewer page. There are four main components—the main menu bar (*top*), the Node Explorer (*left*), the Network Explorer (*center*), and the Function Explorer (*right*)

on the left, the central "Network Explorer," and the "Function Explorer" panel on the right (Fig. 7). The top tool menu allows users to specify which network to display and to control the overall style of the network.

(a) *Network.* This option provides a dropdown menu listing all networks that have been detected. Users can specify the currently displayed network. The largest network ("mirnet1") is displayed by default.

(b) *Background.* Users can switch between a black (default) and a white background. The black background provides better contrast for online interactive visual exploration, while the white background is preferred for publication.

(c) *View.* The default network uses blue squares to represent miRNAs and red circles for their targets, with bigger sizes for nodes with more connections (degrees). When the expression levels (i.e., log fold changes) are available, users can also choose the "Expression" view with red color for upregulated nodes, green for down-regulated nodes and grey for nodes without expression information.

(d) *Layout.* This option allows users to control the default node positions of the network. Five different algorithms have been implemented: *Default Layout*, *Fruchterman–Reingold Layout*, *Graphopt Layout*, *Large Graph Layout*, and *Random Layout* [28]. The default approach will choose an

Table 2
Mouse actions for network visual exploration

***Scope* option**	**Purpose**	**Mouse event**
Single node	Display node name	Mouse over the node
	Show more details	Click the node
	Change color	Set the highlight color, and then double-click the node
	Increase size	Repeatedly click the node
	Change position	Mouse over the node and then drag
Node-neighbors	Change positions	Mouse over the central node and then drag
	Change colors	Set the highlight color, and then double-click the central node
	Change sizes	Repeatedly click the central node
Current highlights	Change positions	Mouse over any current highlighted nodes (with yellow circle around), and then drag
	Change colors	Set the highlight color and then reperform highlighting
	Change sizes	Open the *View options* dialog, change the node size for "Selected nodes" and then click "Submit"
All highlights	Change positions	Mouse over any highlight nodes and then drag
	Change colors	Set the highlight color and then reperform highlighting
	Change sizes	Open the *View options* dialog, change the node size for "Selected nodes" and then click "Submit"

appropriate approach based on the size of the network. Users can explore different algorithms to view the effect.

(e) *Scope*. This option allows users to control the nodes being affected when they manually drag or highlight a single or a group of nodes. Its usage is defined together with mouse events in Table 2.

(f) *Styles for nodes, edges and highlighting*. Click the "View Options" on the right end of the tool bar to bring up the dialog (Fig. 8). The "Node" option allows users to change the size of all nodes or only those highlighted nodes. Node labels will appear automatically when node sizes reach a certain threshold. The "Edge" option allows users to adjust both the shape and the thickness of edges. When a group of nodes is highlighted, users can use the "Highlight" option to control whether the other remaining

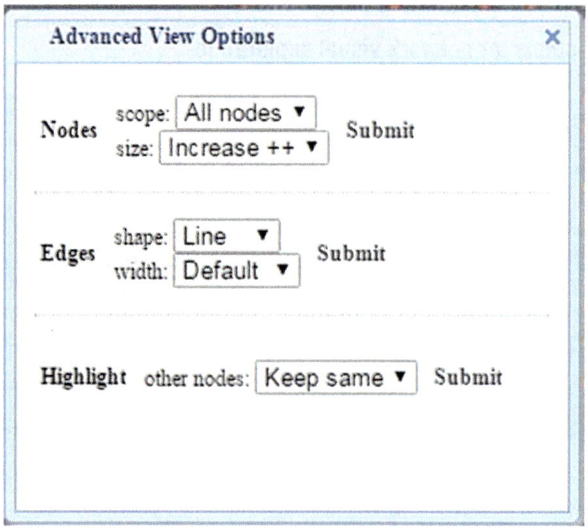

Fig. 8 A screenshot showing the Advance View Options dialog. Users can easily adjust the styles of nodes, edges and highlight effects

nodes should "Keep same," "Dim down," or "Hide" for better visualization.

10. *Visual exploration of the network.* The current network is displayed on the central area with a vertical tool bar on the left. The color palette located at the top left allows users to set the current color for node selection or highlighting. Users can perform zooming, highlighting, drag-and-drop, or extracting the highlighted nodes using a mouse in combination with functions in the tool bar. These mouse events are defined in Table 2. The button with a dotted rectangle icon allows users to manually select a group of nodes. Click on the icon. When the cursor becomes a cross mark, press down on the left mouse button and select a group of nodes. Release the mouse and the nodes will be highlighted (Fig. 9). Users can now drag these nodes to a new position.
11. *Node visualization and manipulation.* The "Node Explorer" on the left panel shows all the nodes in the current network (Fig. 7). Nodes are identified by their IDs or names, together with degree and betweenness values, as well as expression levels (when available). The table supports several common tasks. For instance, users can click a column header to sort the node table, or click a node name (i.e., *LDHA*) to view it within the network. To search for a particular node, enter the node name in the "ID" input box and click "Search" to highlight it in both the table and the network. Users can easily highlight hub nodes by first sorting the nodes according to their degrees and then selecting the checkboxes of the top ranked nodes in the table. Another useful function is to highlight genes targeted by

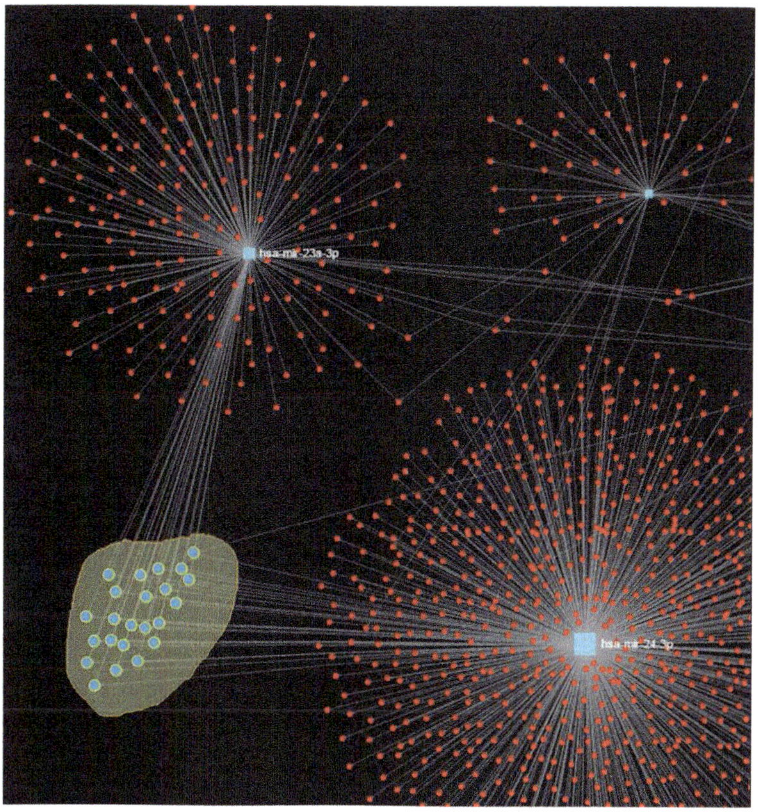

Fig. 9 A screenshot showing how to manually select-and-drag the shared nodes between "hsa-mir-23a-3p" and "hsa-mir-24-3p"

multiple miRNAs. To do this, use the checkboxes to select the miRNAs of interest, specify to highlight "Shared" interactions and then click "Submit." An example is shown in the Fig. 10, in which 17 genes targeted by both "hsa-mir-24-3p" and "hsa-mir-29b-3p" are highlighted in light blue. Users can also remove certain nodes from the network by first selecting them, and then clicking the "Delete" button.

Section III: Functional enrichment analysis

12. *Exploring different options for enrichment analysis.* The "Function Explorer" in the right panel allows users to identify enriched functions as defined by gene ontology, KEGG/Reactome pathways (for target genes); or to identify enriched miRNA families (for miRNAs). Two algorithms have been implemented—the conventional *hypergeometric tests* and the *empirical sampling* as recently proposed by Bleazard et al. [22]. Note that the empirical sampling is only defined for the scenario when all genes are used. For demonstration, choose "All genes," "Empirical sampling," and "KEGG" database, and then click "Submit" button. The results will

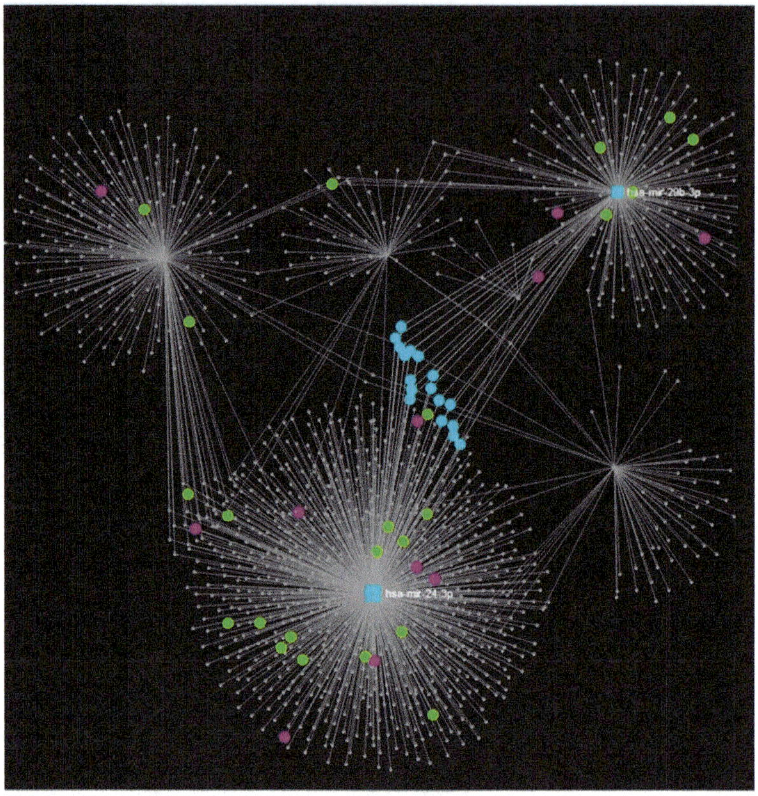

Fig. 10 A screenshot illustrating how to highlight nodes. The nodes highlighted in light blue are shared by "hsa-mir-24-3p" and "hsa-mir-29b-3p". The nodes highlighted in purple belong to "PPAR signaling pathway," and the nodes in light green are from "p53 signaling pathway"

return after 30 s. Due to the nature of random sampling, users may get slightly different p-values and rankings each time. The empirical sampling method has been developed to estimate the null distribution of target genes selected based on a list of miRNAs against a given miRNA-gene target database. The procedures can be divided into three steps: (1) A list of miRNAs (the same size as the original queries) are randomly selected from all miRNAs in the current database; (2) Their gene targets are identified and functional enrichment analysis is then performed for these genes; (3) Repeat the above two steps 1000 times; (4) Compare the hits in each GO category or KEGG pathway to compute an empirical p value (Emp. p). A *p* value <0.001 will be reported if no results based on the random miRNAs are better than the one using the original miRNAs. The empirical sampling algorithm is a computationally intensive procedure. Users can improve the accuracy of the empirically estimated *p*-value by clicking "Submit" multiple times. For instance, clicking five times will generate an empirical p values calculated based on the accumulated results from 5000 random samplings.

13. *Visualizing the pathway genes within the network.* Use the color palette within the "Network Explorer" to set a new highlight color (i.e., purple). Click the "PPAR signaling pathway." The corresponding 11 genes from the pathway will be highlighted in purple within the current network (Fig. 10). Follow the same procedure to highlight the 23 genes involved in "p53 signaling pathway" in light green color. Note that selecting the same pathway again will reset the genes to their original states. If a particular gene is involved in multiple highlighted pathways, the most recently highlighted color prevails.
14. *Module extraction and display.* When a network is too complex, users can extract a module (subnetwork) containing only nodes of interest. Here we show how to extract a module containing only those nodes from the two pathways ("p53 signaling pathway" and "PPAR signaling pathway"). Make sure all nodes involved in the two pathways are currently highlighted, and click the "Extract" button on the Network Explorer tool bar (the bottom one). This process may take a few seconds to finish. The extracted module is now listed as "module1" in the "Network Options" on the top tool bar, and is displayed in the center of "Network Viewer." We can perform further customization for this module. For instance, set the background as "White" and set layout using "Fruchterman–Reingold" algorithm. Open the "View Options" dialog and set the edge shape as "Curve" and width as "Thick," and click the "Submit" button. The result is shown in Fig. 11.
15. *Save the results.* Use the "Download" option in the top menu bar to save the current network as a portable network graphics (PNG) image or as scalable vector graphics (SVG) format. The SVG file can be used to generate a high-resolution image for publication purpose. Refer to miRNet FAQs #12 for detailed instructions (http://www.mirnet.ca/faces/docs/FaqView.xhtml#hires).

2.2 MicroRNA–Target Interaction Analysis from a RT-qPCR Expression Dataset

1. Click the home icon to return to miRNet home page (http://www.mirnet.ca). Three types of expression data (microarray, RNAseq, and RT-qPCR) are currently supported in miRNet. The data analysis procedures for all three data types share a similar workflow. In this section, we will use an example RT-qPCR data to demonstrate the available data analysis procedures. Click the "RT-qPCR" button to start.
2. The *qPCR Profiling* page shows the six sequential steps for RT-qPCR data analysis. At each step, users need to specify the parameters and then click the "Submit" button to proceed. A notification message will pop up on the top-right corner to

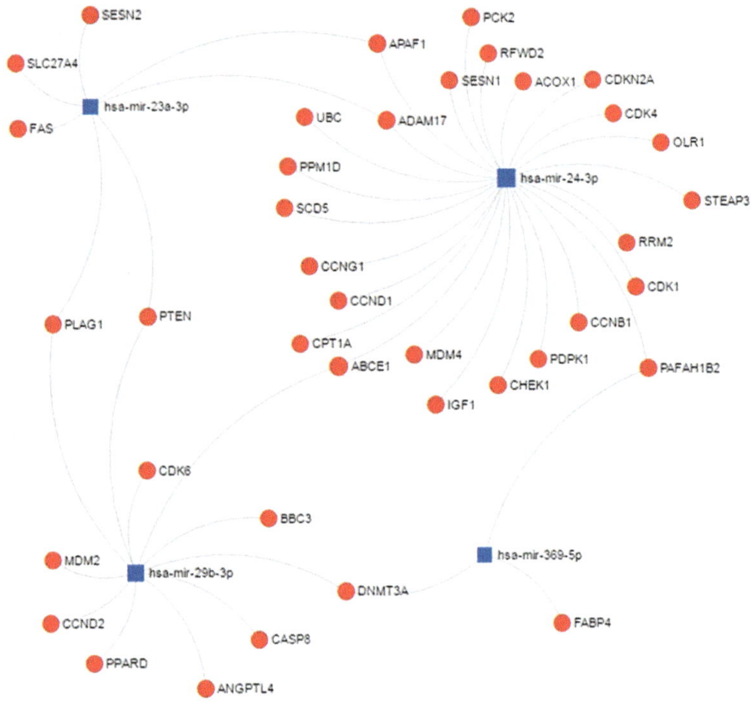

Fig. 11 A screenshot of an extracted module. The module was created by extracting nodes in "PPAR signaling pathway" and "p53 signaling pathway," with further style customizations of nodes, edges, layout, and background

indicate the status. If successful, a green check mark will appear beside the step name. Figure 12 shows a screenshot when all six steps were completed.

(a) *Data Upload*. The expression data should be uploaded as a tab-delimited file (.txt) with features (genes, probes, or miRNAs) in rows and samples in columns. The first row contains sample names beginning with "#NAME". The second row contains class labels beginning with "#CLASS". Refer to miRNet FAQs #1–3 for more detailed instructions (http://www.mirnet.ca/faces/docs/FaqView.xhtml). Click "Choose File" to select a data file from your computer and then click "Submit." To use our example data, click the "Try example" button on the bottom-left to bring up the Example Dataset dialog. The data contains six miRNA expression samples from 384-well RT-qPCR analysis. The miRNAs are identified by miRNA IDs. There are three groups—*Ctrl*, *GrpA*, and *GrpB*. Click "Yes" to allow the system to upload the data and set up the default parameters.

(b) *Annotation*. miRNet currently supports nine organisms (*H. sapiens*, *M. musculus*, *R. norvegicus*, *B. taurus*, *G. gallus*, *D. melanogaster*, *C. elegans*, *D. rerio*, and *S. mansoni*),

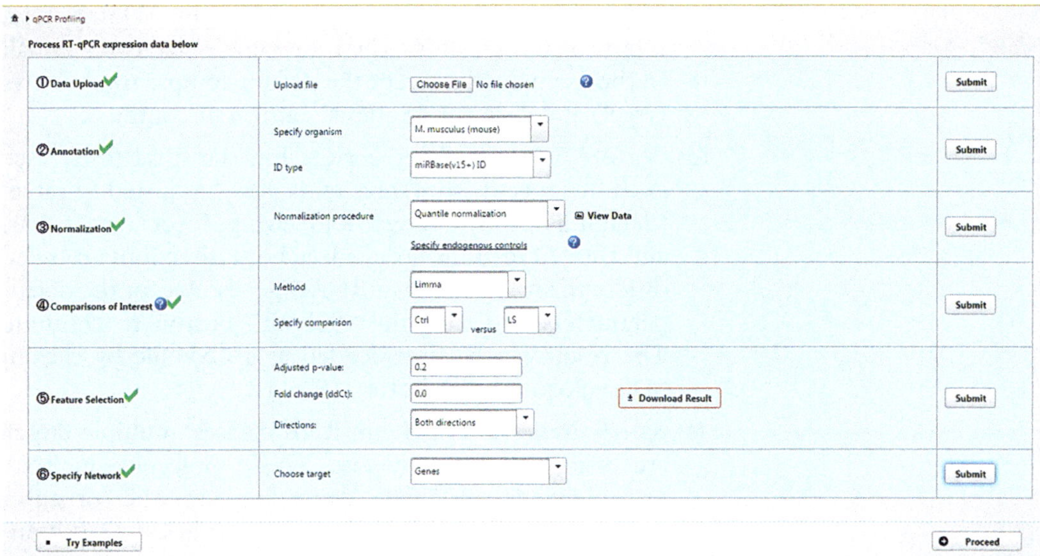

Fig. 12 A screenshot of the RT-qPCR Profiling page when the six data analysis steps have been successfully completed

two types of miRNA IDs (miRBase ID and accession number), and three generic gene IDs (Ensembl gene ID, Entrez ID, and official gene symbol). For microarray data from human and mouse, miRNet also supports probe IDs from 45 common microarray platforms. Users need to choose parameters suitable for their datasets. As we are using the example data, these parameters have already been set by the system.

(c) *Normalization*. Data normalization is a critical step in differential expression analysis, as raw expression values are usually not suitable for statistical analysis. For this purpose, miRNet provides several well-established normalization algorithms—log-transformation in combination with quantile normalization for microarray data analysis using limma [29], trimmed mean of M values (TMM) in combination with a dispersion measure for RNAseq data analysis using edgeR [30], and five common normalization methods for RT-qPCR data using HTqPCR [31]. In this case, we will use the "Quantile" normalization to enforce the same distributions across all samples. If users are not sure whether the data is already normalized or not, click "View Data" to visualize a boxplot summary of the data.

(d) *Comparisons of interest*. When data contains multiple groups, users need to specify which two groups are to be compared. Three statistical methods are available for identification of significant features in RT-qPCR data, including limma, standard t-tests and Mann–Whitney tests. Limma is more robust when the sample size is very small. In the example, we accept the default comparison (*Ctrl* vs. *GrpA*) and click the “Submit” button to continue.

(e) *Feature Selection*. After performing statistical tests, users need to specify the thresholds for “Adjusted *p*-value” (default is 0.05), “Log2 fold change” (default is 1.0), and the “Directions” (default is both directions) to select differentially regulated miRNAs/genes. Accept the default parameters and click the “Submit” button to continue. The result can be downloaded as a CSV file by clicking “Download Result” button (Fig. 12).

(f) *Specify Network*. For the miRNA dataset, multiple targets are available—gene, disease, small molecule, lncRNA, or epigenetic modifier—depending on the organism selected. Choose “Gene” as target and click the “Submit” button.

(g) Now the RT-qPCR data has been analyzed and the significant miRNAs have been identified for downstream analysis. A screenshot is shown in Fig. 12. Click on the “Proceed” button to enter the Interaction Table page.

3. Follow **Steps 5** and **6** in Subheading 2.1 to view and refine the miRNA–gene interaction table.
4. Follow **Steps** 7 and **8** in Subheading 2.1 to create and customize the miRNA–gene network in the Network Builder.
5. Follow **Steps 9–15** in Subheading 2.1 to perform visual exploration and functional enrichment analysis of the miRNA–gene target interaction network.

2.3 Creation and Analysis of Other Types of miRNA–Target Interaction Networks

1. In addition to identifying gene targets from a given list of miRNAs and vice versa, miRNet also allows users to easily create networks from a list of disease names, small compounds, lncRNAs or epigenetic modifiers. As there are no widely accepted IDs for these data and using free text tends to be error-prone, miRNet allows users to select from the available data to specify their items of interest. In this protocol, we will use “Small molecule” as an example to show the procedures. Go to the miRNet home page (http://www.mirnet.ca), and click the “Small molecule” button to start (Fig. 1).
2. The *Item Selection* page is displayed (Fig. 13). The miRNA and small compound interaction data is available for “Human” (default), “Mouse,” and “Rat.” The data is obtained from the SM2miR database [16]. Users need to manually select their compounds of interest from the “Available” entries shown on the left box to the “Selected” box on the right,

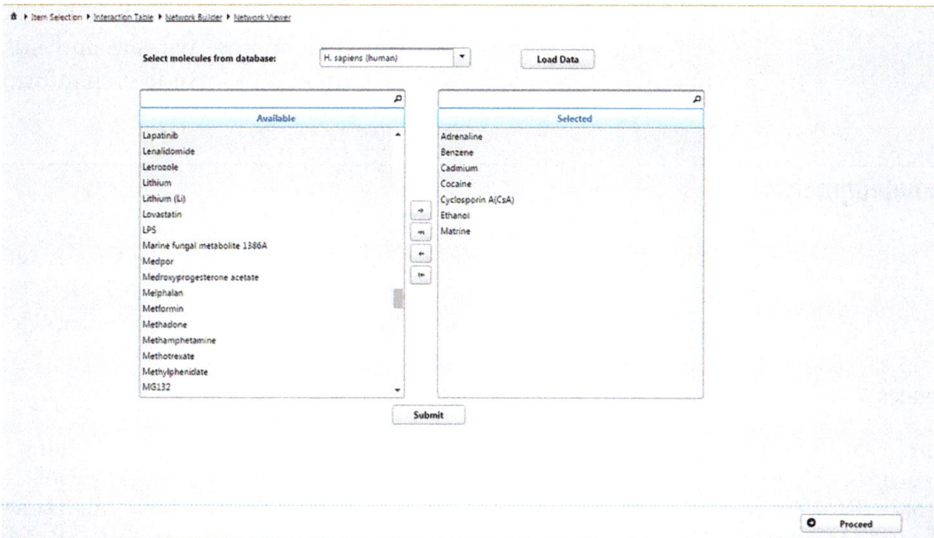

Fig. 13 A screenshot of the Item Selection page for small molecules. Users can directly drag and drop items of interest from the left "Available" box to the "Selected" box on the right

using the "Add" button located at the center. Here we select *"Adrenaline," "Benzene," "Cadmium," "Cocaine," "Cyclosporin A (CsA)," "Ethanol,"* and *"Matrine."* After the selection, click "Submit" and then click the "Proceed" button to view the miRNA–small compound interaction table.

3. Follow **Steps 5** and **6** in Subheading 2.1 to view and refine the miRNA–small compound interaction table.
4. Follow **Steps** 7 and **8** in Subheading 2.1 to create and customize the miRNA–small compound network.
5. Follow **Steps 9–15** in Subheading 2.1 for visual exploration of the miRNA–small compound interaction network. Note that users can only perform miRNA family enrichment analysis in this case, as no functional categories have been defined for these small molecules.

3 Summary and Outlook

miRNet has been developed as an easy-to-use high-performance web server to support miRNA functional analysis and visual exploration. Through its intuitive user interface, researchers can perform various tasks including expression profiling, miRNA target identification and refinement, as well as network-based visual analytics and/or enrichment analysis. The underlying miRNA knowledge base will be updated regularly to keep pace with the rapid progress of the fields. Given our increasing understanding of miRNAs in cross-species communications such as host–parasite and host–gut microbiota interactions, we also intend to include such information in the future release.

Acknowledgments

The McGill Graduate Excellence Fellowship (GEF) Award (Y. Fan), the NSERC Discovery Grant, and the FRQNT New University Researcher Start-up Program grant (J. Xia) are acknowledged.

References

1. Bartel DP (2004) MicroRNAs: genomics, biogenesis, mechanism, and function. Cell 116(2):281–297
2. Mendell JT, Olson EN (2012) MicroRNAs in stress signaling and human disease. Cell 148(6):1172–1187. https://doi.org/10.1016/j.cell.2012.02.005
3. Manzano-Roman R, Siles-Lucas M (2012) MicroRNAs in parasitic diseases: potential for diagnosis and targeting. Mol Biochem Parasitol 186(2):81–86. https://doi.org/10.1016/j.molbiopara.2012.10.001
4. Liu S, da Cunha AP, Rezende RM, Cialic R, Wei Z, Bry L, Comstock LE, Gandhi R, Weiner HL (2016) The host shapes the gut microbiota via fecal MicroRNA. Cell Host Microbe 19(1):32–43. https://doi.org/10.1016/j.chom.2015.12.005
5. Betel D, Koppal A, Agius P, Sander C, Leslie C (2010) Comprehensive modeling of microRNA targets predicts functional non-conserved and non-canonical sites. Genome Biol 11(8):R90. https://doi. org/10.1186/gb-2010-11-8-r90
6. Krek A, Grun D, Poy MN, Wolf R, Rosenberg L, Epstein EJ, MacMenamin P, da Piedade I, Gunsalus KC, Stoffel M, Rajewsky N (2005) Combinatorial microRNA target predictions. Nat Genet 37(5):495–500. https://doi. org/10.1038/ng1536
7. Rehmsmeier M, Steffen P, Hochsmann M, Giegerich R (2004) Fast and effective prediction of microRNA/target duplexes. RNA 10(10):1507–1517. https://doi.org/10.1261/rna.5248604
8. Alexiou P, Maragkakis M, Papadopoulos GL, Reczko M, Hatzigeorgiou AG (2009) Lost in translation: an assessment and perspective for computational microRNA target identification. Bioinformatics 25(23):3049–3055. https://doi.org/10.1093/bioinformatics/btp565
9. Chi SW, Zang JB, Mele A, Darnell RB (2009) Argonaute HITS-CLIP decodes microRNA-mRNA interaction maps. Nature 460(7254):479–486. https://doi.org/10.1038/nature08170
10. Leung AK, Young AG, Bhutkar A, Zheng GX, Bosson AD, Nielsen CB, Sharp PA (2011) Genome-wide identification of Ago2 binding sites from mouse embryonic stem cells with and without mature microRNAs. Nat Struct Mol Biol 18(2):237–244. https://doi.org/10.1038/nsmb.1991
11. Vergoulis T, Vlachos IS, Alexiou P, Georgakilas G, Maragkakis M, Reczko M, Gerangelos S, Koziris N, Dalamagas T, Hatzigeorgiou AG (2012) TarBase 6.0: capturing the exponential growth of miRNA targets with experimental support. Nucleic Acids Res 40(Database issue):D222–D229. https://doi.org/10.1093/nar/gkr1161
12. Hsu SD, Lin FM, Wu WY, Liang C, Huang WC, Chan WL, Tsai WT, Chen GZ, Lee CJ, Chiu CM, Chien CH, Wu MC, Huang CY, Tsou AP, Huang HD (2011) miRTarBase: a database curates experimentally validated microRNA-target interactions. Nucleic Acids Res 39(Database issue):D163–D169. https://doi.org/10.1093/nar/gkq1107
13. Yang JH, Li JH, Shao P, Zhou H, Chen YQ, Qu LH (2011) starBase: a database for exploring microRNA-mRNA interaction maps from Argonaute CLIP-Seq and Degradome-Seq data. Nucleic Acids Res 39(Database issue):D202–D209. https://doi.org/10.1093/nar/gkq1056
14. Xiao F, Zuo Z, Cai G, Kang S, Gao X, Li T (2009) miRecords: an integrated resource for

microRNA-target interactions. Nucleic Acids Res 37(Database):D105–D110. https://doi.org/10.1093/nar/gkn851
15. Rukov JL, Wilentzik R, Jaffe I, Vinther J, Shomron N (2014) Pharmaco-miR: linking microRNAs and drug effects. Brief Bioinform 15(4):648–659. https://doi.org/10.1093/bib/bbs082
16. Liu X, Wang S, Meng F, Wang J, Zhang Y, Dai E, Yu X, Li X, Jiang W (2013) SM2miR: a database of the experimentally validated small molecules' effects on microRNA expression. Bioinformatics 29(3):409–411. https://doi.org/10.1093/bioinformatics/bts698
17. Jiang Q, Wang Y, Hao Y, Juan L, Teng M, Zhang X, Li M, Wang G, Liu Y (2009) miR2Disease: a manually curated database for microRNA deregulation in human disease. Nucleic Acids Res 37(Database):D98–D104. https://doi. org/10.1093/nar/gkn714
18. Ruepp A, Kowarsch A, Theis F (2012) PhenomiR: microRNAs in human diseases and biological processes. Methods Mol Biol 822:249–260. https://doi.org/10.1007/978-1-61779-427-8_17
19. Li Y, Qiu C, Tu J, Geng B, Yang J, Jiang T, Cui Q (2014) HMDD v2.0: a database for experimentally supported human microRNA and disease associations. Nucleic Acids Res 42(Database issue):D1070–D1074. https://doi. org/10.1093/nar/gkt1023
20. Dai E, Yu X, Zhang Y, Meng F, Wang S, Liu X, Liu D, Wang J, Li X, Jiang W (2014) EpimiR: a database of curated mutual regulation between miRNAs and epigenetic modifications. Database (Oxford) 2014:bau023. https://doi. org/10.1093/database/bau023
21. Gusev Y, Schmittgen TD, Lerner M, Postier R, Brackett D (2007) Computational analysis of biological functions and pathways collectively targeted by co-expressed microRNAs in cancer. BMC Bioinformatics 8(Suppl 7):S16. https://doi.org/10.1186/1471-2105-8-S7-S16
22. Bleazard T, Lamb JA, Griffiths-Jones S (2015) Bias in microRNA functional enrichment analysis. Bioinformatics 31(10):1592–1598. https://doi.org/10.1093/ bioinformatics/btv023
23. Ma'ayan A (2011) Introduction to network analysis in systems biology. Sci Signal 4(190):tr5. https://doi.org/10.1126/scisignal.2001965
24. Xia J, Benner MJ, Hancock RE (2014) NetworkAnalyst–integrative approaches for protein-protein interaction network analysis and visual exploration. Nucleic Acids Res 42 (Web Server issue):W167–W174. https://doi.org/10.1093/nar/gku443
25. Xia J, Gill EE, Hancock RE (2015) NetworkAnalyst for statistical, visual and network-based meta-analysis of gene expression data. Nat Protoc 10(6):823–844. https://doi.org/10.1038/nprot.2015.052
26. Fan Y, Siklenka K, Arora SK, Ribeiro P, Kimmins S, Xia J (2016) miRNet - dissecting miRNA-target interactions and functional associations through network-based visual analysis. Nucleic Acids Res 44(W1):W135–W141. https://doi. org/10.1093/nar/gkw288
27. Zhu X, Gerstein M, Snyder M (2007) Getting connected: analysis and principles of biological networks. Genes Dev 21(9):1010–1024. https://doi. org/10.1101/gad.1528707
28. Csardi G, Nepusz T (2006) The igraph software package for complex network research. InterJournal, Complex Systems 1695(5):1–9
29. Ritchie ME, Phipson B, Wu D, Hu Y, Law CW, Shi W, Smyth GK (2015) Limma powers differential expression analyses for RNA-sequencing and microarray studies. Nucleic Acids Res 43(7):e47. https://doi.org/10.1093/nar/gkv007
30. Robinson MD, McCarthy DJ, Smyth GK (2010) edgeR: a bioconductor package for differential expression analysis of digital gene expression data. Bioinformatics 26(1):139–140. https://doi.org/10.1093/bioinformatics/btp616
31. Dvinge H, Bertone P (2009) HTqPCR: high-throughput analysis and visualization of quantitative real-time PCR data in R. Bioinformatics 25(24):3325–3326. https://doi.org/10.1093/bioinformatics/btp578

Chapter 11

Systems Biology Analysis to Understand Regulatory miRNA Networks in Lung Cancer

Meik Kunz, Andreas Pittroff, and Thomas Dandekar

Abstract

Lung cancer has currently the highest cancer-related mortality rate worldwide. MicroRNAs (miRNAs) are small noncoding RNAs that play a fundamental role in gene expression and are linked to disease progression of different cancer types such as lung cancer. However, functional characterization is made difficult by the fact that miRNAs generally regulate several mRNA interaction partners, resulting in complex regulatory networks. Thus, analysis of the network biology of miRNAs is essential for comprehensive understanding of their regulatory effects in lung cancer. A deeper understanding of miRNA networks in cancer could finally serve as a basis for the development of new therapeutic interventions. Here, we present a systems biology approach to analyze regulatory miRNA interaction networks to get better insight into their function.

Key words Systems biology, miRNAs, Networks, Bioinformatics, Algorithm, Lung cancer

1 Introduction

Lung cancer is the main cause of cancer-associated mortality worldwide. Non-small cell lung cancer (NSCLC) is with 85% the most often diagnosed subtype, showing deregulation of gene expression such as EGFR (epidermal growth factor receptor) and the tumor protein P53 [1–3]. The 5-year survival rate for lung cancer (all types) is less than 15% as diagnosis is often too late and existing treatment strategies are not sufficient to prolong patient life [2, 3]. Thus, new clinical interventions for a better patient survival of lung cancer patients are required.

MicroRNAs (miRNAs) have emerged as new regulators of the human genome with high potential as diagnostic markers and therapeutic targets. This small noncoding RNAs (22 nucleotides) are found in coding and noncoding regions of the genome and

Meik Kunz and Andreas Pittroff contributed equally to this work.

Louise von Stechow and Alberto Santos Delgado (eds.), *Computational Cell Biology: Methods and Protocols*, Methods in Molecular Biology, vol. 1819, https://doi.org/10.1007/978-1-4939-8618-7_11,

are highly conserved among different species [3, 4]. miRNAs regulate gene expression by binding to complementary messenger RNA (mRNA), resulting in translational inhibition or mRNA degradation [3, 4]. Notably, one miRNA can regulate several mRNAs and can thus be associated with several signaling pathways and biological functions, leading to complex regulatory effects [3, 4]. Interestingly, miRNAs are deregulated in lung cancer and easily accessible in body fluids, which makes them attractive biomarker candidates [3, 5]. Consequently, several miRNAs have been investigated for their role in lung cancer including noninvasive diagnostic blood tests based on deregulated miRNA expression signatures [3, 6]. For instance, the oncogenic miRNA-21 is upregulated in lung cancer and associated with metastasis and bad prognosis [3, 7, 8]. An extensive review of miRNAs in lung cancer can be found in Kunz et al. [3].

To date thousands of miRNAs have been identified (*see* for example miRBase [9]). Nevertheless, detailed experimental characterization is challenging due to the high number of mRNA interaction partners one miRNA can have, and the fact that validating predicted interactions biochemically can be laborious [10]. Most miRNAs are therefore not well characterized [3, 4]. For this reason, the systematic analysis of the miRNA regulatory network context is essential to understand their function. We recently introduced an analysis approach to unravel immune responses by combining transcriptome and interactome data [11, 12]. Here, we focus on array data annotation and statistics with network analysis for efficient analysis of regulatory miRNA networks in lung cancer. For this, we developed a systems biology methodology combining large-scale data and integrated network biology analysis approaches (*see* Fig. 1). However, integration of different transcriptome and interactome data not only capture previously unknown regulatory effects of miRNAs but also can preselect and rank for example best potential target candidates for further experimental testing (*see* Table 1; Figs. 2 and 3). Thus, our in silico approaches save work and reduce the need for experimentation and animal experiments.

2 Materials

2.1 Databases

1. Gene Expression Omnibus (GEO, [13]): database containing vast amount of data on large-scale gene expression measurement experiments

2.2 Software

1. R version 3.2.2 [14]: Program language. It is particularly strong to analyze statistical parameters but has also strong further calculation capabilities.
2. Bioconductor [15]: Compilation of tools for the analysis of genomic data with R.

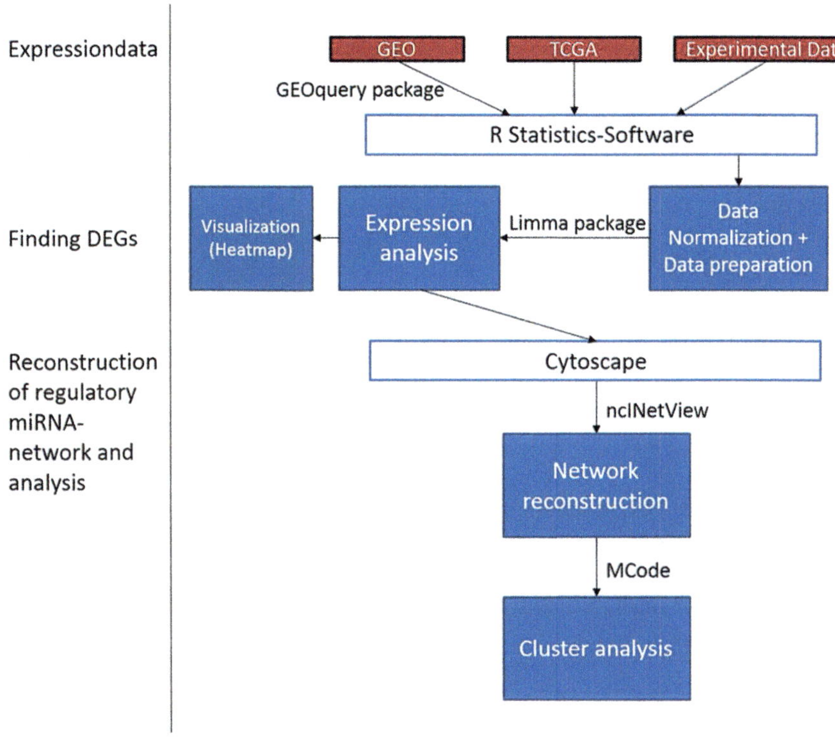

Fig. 1 Workflow of our systems biology methodology. Our method combines expression data (e.g., from GEO) analysis with systems biology approaches to understand regulatory miRNA networks. The figure summarizes important steps with software and programs (external data in red rectangles, own analysis in blue rectangles)

Table 1
Overview of selected miRNAs showing different regulation between nontumor and lung tumor samples. The data shown are miRNA expression profiling array data of lung tumor and nontumor tissues from the GEO database (GSE63805). The table shows five deregulated miRNAs with their corresponding expression (log2 values; *p*-value<0.05). The *p*-values given are together with their effect size for comparison. The *p*-value helps to identify significant differences and the effect size suggests the importance of these differences, in particular if considered in comparison to the mRNA effects and *p*-values. This indicates striking regulatory differences which then should be analyzed further including interaction partners and regulatory network effects (*see* text and Fig. 3 for details)

	hsa-miR-196a	hsa-miR-21	hsa-miR-451	hsa-miR-520e	hsa-miR-9
p-value	2.859963e-03	2.215532e-16	6.621423e-16	1.247681e-11	9.700031e-07
Control	2.523104	13.166980	13.691610	8.536191	3.712503
Tumor	4.642499	15.250346	10.766707	6.205290	5.884301

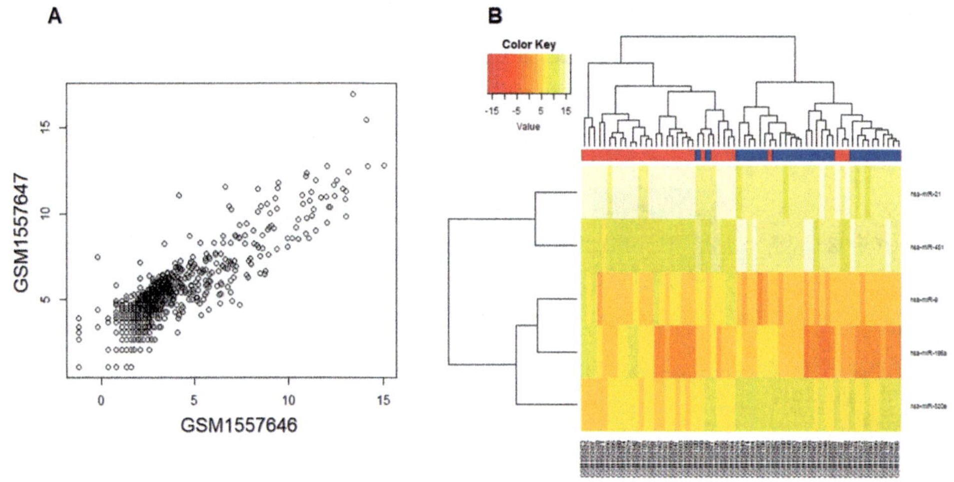

Fig. 2 Gene expression data analysis. (**a**) Pairs plot of the expression data of nontumor (e.g., GSM1557646; x-axis) and lung tumor (e.g., GSM1557647; y-axis) tissue samples from GEO (GSE63805; 734 genes, each gene is plotted as single dot). (**b**) Heat map of the five DEGs (p-value<0.05; logFC of 2). The heat map indicates a difference in gene expression (miRNAs at y-axis) between the tumor and nontumor tissue samples (x-axis; 30 control in blue, 32 lung tumor in red)

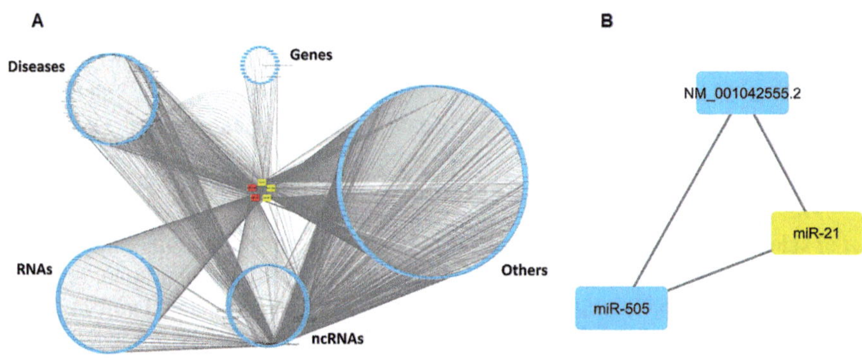

Fig. 3 Overview of the regulatory miRNA network. (**a**) We set up a miRNA network (1339 nodes in blue, 3438 edges in grey; yellow: upregulated miRNAs; red: downregulated miRNAs) based on five DEGs (miRNA-196a, miRNA-21, miRNA-451, miRNA-520e, and miRNA-9) and their interaction partners using the Cytoscape plugin ncINetView. Here, connectivity to other miRNAs (e.g., miRNA-1276 and miRNA-93 showing a deregulation in the array) become visible, but also to non-array ncRNAs and genes such as FRS2 (Fibroblast Growth Factor Receptor Substrate 2, NM_001042555.2), CKAP4 (Cytoskeleton-Associated Protein 4), and CSBP (Csaids Binding Protein, known as MAPK14) that play a role in tumorigenesis and metastasis (not shown in detail). Moreover, the network also shows associated diseases (lung cancer and other tumors, associations for other diseases) for the five miRNAs and their interactors. Numerous interactions are indicated by the lines. The detailed data are available by zooming in (not shown). (**b**) We found a functional cluster around the miRNA-21 (in yellow), miRNA-505, and FRS2 (NM_001042555.2) which indicates a strong regulatory effect in the reconstructed miRNA network

- Packages: GEOquery, limma, and gplots.

Packages are precompiled or preassembled programs. Specific references explain GEOquery [16], limma [17], and gplots [18].

3. Cytoscape 3.4.0 [19]: software and framework program for visualization and system biology analysis of biological networks.

- Plugins:

Plugins are subroutines which can easily communicate with popular frameworks such as Cytoscape, a framework program (originally a commenting editor) for cellular interaction networks.

- MCODE [20]: plugin to find highly connected regions in a network which "can talk" with Cytoscape, i.e., directly pipe its output into the Cytoscape visualization of the network.
- ncINetView [21]: plugin for ncRNA network reconstruction.

3 Methods

We provide the R code at http://www.bioinfo.biozentrum.uni-wuerzburg.de/computing/scripts/.

3.1 Software and Package Installation (see Note 1)

1. Download and install the R software package, version 3.2.2.
2. Install the Bioconductor packages: Open R. This needs to be run in a terminal window.
3. Type in the command line source ("http://bioconductor.org/biocLite.R").
4. Type in the next line biocLite()

   (biocLite() with empty brackets just calls the biocLite() package.).
5. Type in the next line biocLite("GEOquery").
6. Type in the next line biocLite("limma").
7. Type in the next line biocLite("gplots").

3.2 Load the R Libraries (see Note 1)

1. Type in the command line library("GEOquery").
2. Type in the next line library("limma").
3. Type in the next line library("gplots").

3.3 Downloading the miRNA Expression Dataset from GEO (see Note 1)

1. Type in the command line GSE63805<-getGEO("GSE63805").
2. Type in the next line eset1=GSE63805[[1]].
3. Type in the next line eset1b=eset1.

3.4 Data Preparation (see Note 2)

1. Type in the command line colnames(pData(eset1b))[8]="Tumor".
2. Type in the next line levels(eset1b$Tumor)=c("Control","Lung Cancer").
3. Type in the next line dim(eset1b).
4. Type in the next line length(eset1b$Tumor[eset1b$Tumor =="Control"]).
5. Type in the next line length(eset1b$Tumor[eset1b$Tumor=="Lung Cancer"]).

3.5 Check Normalization (see Note 1)

1. Type in the command line boxplot(exprs(eset1b)).
2. Type in the next line ctrl<-which(eset1b$Tumor=="Control")[1].
3. Type in the next line test<-which(eset1b$Tumor=="Lung Cancer")[1].
4. Type in the next line exp<-exprs(eset1b)[,c(ctrl,test)].
5. Type in the next line plot(exp).
6. Type in the next line exprs(eset1b)=log2(exprs(eset1b)).
7. Visualize the log2 transformed expression values: repeat Subheading 3.5, **steps 1–5** to generate the plots.

3.6 Selecting Differentially Expressed Genes (see Note 3)

1. Type in the command line labels=pData(eset1b)[,"Tumor"].
2. Type in the next line design=model.matrix(labels).
3. Type in the next line fit=eBayes(lmFit(eset1b, design)).
4. Type in the next line t<-topTable(fit,coef=2,number=Inf, *p*.value=0.05,lfc=2).
5. Type in the next line l.deg=list().
6. Type in the next line l.deg<-intersect(rownames(exprs(eset1b)), rownames(t)).
7. Type in the next line write.table(l.deg, file="topDEGs.txt", quote=FALSE, sep= "\t", row.names=FALSE, col.names=FALSE).

3.7 Data Visualization of Differentially Expressed Genes (see Note 4)

1. Type in the command line color.map<-function(Tumor) {if(Tumor=="Control") "#0000FF" else "#FF0000"}

   (colors are now blue for the control, code "#0000FF", and red for the tumor, code "#FF0000").
2. Type in the next line patientcolors<-unlist(lapply(eset1b$Tumor, color.map)).
3. Type in the next line

   heatmap.2(exprs(eset1b[l.deg,]),ColSideColors= patientcolors, key=TRUE, symkey=FALSE, density.info= "none", trace="none", cexRow=0.8).

3.8 Network Reconstruction and Regulatory Process and Signaling Pathway Analysis (see Note 5)

1. Open the Cytoscape 3.4.0 software ➔ click "With Empty Network" ➔ name "miRNA_regulatory_network" ➔ click "OK".
2. Click "Apps" ➔ Install all plugins.
3. Add two columns to the node table of the empty network ➔ click on the "+" button above the node table ➔ click "New Single Column" ➔ click "String" and name it "Type Column" ➔ click on the "+" button ➔ click "New List Column" ➔ click "String" and name it "Alias Column".
4. Reconstruct a regulatory network of the Differentially Expressed Genes (DEGs) and their direct interaction partners: Click "Apps" ➔ select "*ncINetView*" ➔ click "Add nodes" ➔ open "topDEGs.txt" and select the "hsa-miR-196a, hsa-miR-21, hsa-miR-451, hsa-miR-520e, hsa-miR-9" ➔ copy them in the text field (separated by space) ➔ click "Search" ➔ select in the Network column "miRNA_regulatory_network", in the Type column "Type Column", in the Alias column "Alias Column" ➔ select "Include neighbours", *then* "ncRNA", "RNA", "*Gene*", "*Disease*", "*Others*" ➔ click "Import".
5. Group the nodes by their type: click "Layout" ➔ click "Group Attribute Layout" and select "Type Column".
6. Edit node names: select one node ➔ right click ➔ click edit ➔ click "rename node".
7. Change color code of the miRNAs: select the "Style" option in the Control Panel ➔ click "Fill color" ➔ select the red and yellow color manually.
8. Analyze the regulatory network effect using MCODE: Click "Apps" ➔ select "MCODE" ➔ click "Open MCODE" ➔ click "Analyze Current Network" (*see* **Note 5**).

3.9 Interpretation of the Results

Using our approach the connectivity and regulatory network effect of the array data become visible. This can be further analyzed in more detail (e.g., RNAseq, miRNA and mRNA profiling data, experimental modulation). Our example dataset contains 62 samples (30 control, 32 lung tumor) and 734 genes (features). Using the limma package we found the five DEGs (log Fold Change (logFC) of 2, p-value less than 0.05), hsa-miRNA-196a, hsa-miRNA-21, hsa-miRNA-451, hsa-miRNA-520e, and hsa-miRNA-9 (*see* Table 1; similar deregulation of these miRNAs in lung tumors is reported in the literature; for more information about the miRNAs *see* [3, 7]).

About 0.7% of the total genes from the array show a significant (p-value<0.05) differential expression of log fold 2 between the nontumor and tumor tissues, which can be used for further analysis. Our graphical plotting of the DEGs as a heat map shows a clear difference in gene expression between both tissue types (*see* Fig. 2b; each sample at x-axis, expression of each gene at y-axis. Top and left: the dendrogram shows similar sample and gene expression clusters; healthy tissues in blue, tumor tissues in red; color key plots logFC ranging from −15 to 15). Next, based on the DEGs we reconstructed a miRNA regulatory network, which contains 1339 nodes (e.g., ncRNAs) and 3438 edges (interactions) (*see* Fig. 3a). This provides an overview regarding the complete interaction context of the five selected miRNAs. Moreover, the connectivity to other regulated miRNAs from the array becomes obvious, for example the interaction with miRNA-1276 (p-value = 0.003; logFC = −0.85) as well as with miRNA-93 (p-value = 0.0057; logFC = −0.51). Moreover, the network shows non-array interacting ncRNAs and genes, for instance the interaction to FRS2 (Fibroblast Growth Factor Receptor Substrate 2, NM_001042555.2) [22], CKAP4 (Cytoskeleton-Associated Protein 4) [23], and CSBP (Csaids Binding Protein, known as MAPK14) [24]. The proteins play a role in the MAP kinase/EGFR and p53 signaling and are linked to tumorigenesis and metastasis (not shown in detail). Furthermore, using ncINetView [21] we found in our network disease terms associated with lung cancer and other tumors (*see* Fig. 3a). Moreover, using the MCODE plugin we found a cluster around the miRNA-21 (in yellow), miRNA-505 and FRS2 (NM_001042555.2), highlighting an important regulatory role in the reconstructed lung tumor miRNA network (*see* Fig. 3b).

As an alternative to our network analysis approach, interactome data from the NPInter [25], miRTarBase [10], KEGG [26], or WikiPathway [27] databases can be downloaded, in which a network can be reconstructed by manually mapping the DEGs to the interactome, for instance using a MySQL (https://www.mysql.de/) database (here not shown). However, it is worth mentioning, when users select other databases reported interactions should be preferentially experimentally validated. In addition to our approach the reconstructed network can be further analyzed for biological processes and signaling pathways using the Cytoscape plugin ClueGO [28] or the R packages FGNet, topGO, and KEGGprofile (here not shown). For more information and manually mapping of nodes using a MySQL database see our methods paper [11, 12].

4 Notes

1. As a first step, it is important to consider that the correct R version is used and that the installed and loaded bioconductor libraries work without any errors. To keep it simple and reduce the complexity of the code, we use here miRNA expression profiling data of stage I NSCLC patients and nontumor tissues from the GEO database (GSE63805; 30 nontumor, 32 tumor

samples). This database contains already normalized expression set objects (normalization followed the processing protocol of the manufacturer for the Nanostring nCounter Human miRNA Expression Assay Kit version 1.6 [29]). After downloading (getGEO function) each dataset has to be carefully checked for data format and normalization before processing it with our methodology (e.g., if the expression values are on a log2 scale; here not shown). As log transformed data are symmetric for reciprocals and better represent the expression changes [30], we apply a log2 transformation using the log2 function and plot the values using a pairs plot which shows the expression of each gene of the tumor and the control sample (*see* Fig. 2a; example nontumor sample GSM1557646 on x-axis; tumor sample GSM1557647 on y-axis). In the pairs plot we can see a dot cloud, this indicates normalized and transformed data format. If the dataset is not normalized and needs further processing, for example, the Robust Multichip Average (RMA) function can be applied (*see* [31] for usage). Furthermore, as an alternative to GEO [13], there are datasets from "The Cancer Genome Atlas" (https://cancergenome.nih.gov/) using the RTCGA data packages [32] or raw data (.CEL-files) from (own) experimental data using the read.table function (including set up of an expression set object and normalization) that can be uploaded and further be processed using our presented methodology.

2. Next, as the pData (phenotype annotation) is inconsistent for different datasets available at the GEO database, manual sample editing and preparation for example for columns of interest is needed. For this, we renamed the Tumor column manually using the colnames function so that the column name and its values are similar over all samples to avoid errors. Moreover, we labeled the gene samples according to lung cancer and nontumor (control) using the levels function. If the order of levels is unknown it can also be checked via the levels function. To check if the labeling was done correctly the dim and length function can be used. Notably, it is important to check that the sum of the control and lung cancer samples (length) is identical with the total number of samples (dim). In our example the datasets contains 62 samples (30 control, 32 lung tumor) and 734 genes (features). It has to be mentioned that we only used one dataset, however, in order to find deregulated genes over several datasets a list of genes, which are common between the individual datasets is required. For this, the pData and the corresponding expression values of the datasets have to be merged using for example the inSilicoDb and inSilicoMerging package [33], in which genes that are not present in all datasets were automatically excluded (not shown here in detail).

3. To identify DEGs in the dataset the limma package is applied. This is an important step as DEGs indicate differences in gene expression between the tumor and nontumor samples, which can be used for further analysis regarding the regulatory effects and experimental testing. The limma package applies an empirical linear Bayes model to the standard errors and fits data using t-test statistics of differential expression. However, limma identifies top DEGs based on a p-value cutoff using the topTable function (standard parameter is infinite number of obtained *p*-values, but also allows selection based on an individually chosen number of interested top-ranked DEGs). The topTable function uses the adjustment method (adjust.method) to adjust the *p*-values for multiple testing, in which the default method is the Bonferroni correction (bonferroni; also NULL value) (*see* https://www.rdocumentation.org/packages/limma/versions/3.28.14/topics/toptable and help("p.adjust") for more details and other available correction methods). We used a *p*-value of less than 0.05 and a logFold change (logFC) of 2 as standard cutoff parameter, but both can be changed manually depending on the analysis (e.g., lfc cutoff of 1 results in 27 DEGs, whereas no lfc cutoff (replace lfc) shows 85 DEGs; here not shown). However, as our example dataset already contains the gene names we did not show platform identifier mapping. For different platforms with unknown gene names the corresponding identifiers have to be mapped to gene symbols, which can be done using the R packages AnnotationDBi, annotate and affy (here not shown). Finally, the DEGs can be saved as a text file using the write.table function for further analysis.

4. As a next step, the identified DEGs can be graphically plotted. For this, a heat map is optimal as a first visualization as it plots for each sample (x-axis) the expression of each gene (y-axis). This provides an overview of samples and gene expression and indicates which genes and samples lie in clusters of similar expression pattern (shown by a dendrogram). Moreover, to better represent the tumor and nontumor tissue samples different colors can be assigned to each sample based on their value in the Tumor column of their pData using the color.map function. We used for the nontumor samples blue and tumors red. Subsequently, the heat map based on this color key can be generated using the heatmap.2 function (see Fig. 2b).

5. miRNAs can have multiple interaction partners and are associated with several biological processes and signaling pathways. For this, the reconstruction of a regulatory interaction network is an essential step to understand the whole regulatory effect of miRNAs. We applied ncINetView [21] for the reconstruction of a miRNA regulatory network using Cytoscape, a well-known systems biology tool for network analysis based on several plugins [19]. Notably, ncINetView [21] also maps diseases to the

nodes of the network. The ncINetView plugin is based on the ncRNA-DB database [21] which integrates ncRNA interactions from several sources such as NPInter [25] and miRTarBase [10] including information on diseases [21]. We used our five DEGs to reconstruct a miRNA regulatory network (*see* Fig. 3a). To label the different regulation (upregulated or downregulated) of the DEGs we used the Style option in the Control Panel. Moreover, the reconstructed network can further be analyzed using the Cytoscape plugins MCODE to find highly connected functional clusters with a strong regulatory network effect [20].

5 Conclusion and Outlook

miRNAs have a fundamental biological importance in health and disease. Here we investigate how they are associated with lung cancer, the leading cause of cancer-related deaths. The complex functional role of miRNAs is often not or only partly understood due to their high number of interaction partners and difficulties in experimental validation of miRNA interactions. Bioinformatics protocols for comprehensive functional analysis are still limited. Our presented systems biology approach efficiently combines high-throughput data with regulatory network analysis. We therefore hope that our approach helps in functional understanding of the complex regulatory effects of miRNAs with the genome and the identification of target candidates from large-scale experimentally screenings. Future efforts for our approach should address the tissue-specific context and the combination with drug-target interaction modeling. This could eventually lead to new clinical strategies for lung cancer therapy. In addition, as shown for lung cancer, our methodology reduces experimental work and animal experiments and has the potential to be applied to other tumors and diseases.

Acknowledgment

We thank BMBF for funding (FKZ: 031L0129B).

References

1. Herbst RS, Heymach JV, Lippman SM (2008) Lung cancer. N Engl J Med 359(13):1367–1380. https://doi.org/10.1056/NEJMra0802714
2. Yao Y, Fan Y, Wu J, Wan H, Wang J, Lam S, Lam WL, Girard L, Gazdar AF, Wu Z, Zhou Q (2012) Potential application of non-small cell lung cancer-associated autoantibodies to early cancer diagnosis. Biochem Biophys Res Commun 423(3):613–619. https://doi.org/10.1016/j.bbrc.2012.06.050
3. Kunz M, Wolf B, Schulze H, Atlan D, Walles T, Walles H, Dandekar T (2016) Non-coding RNAs in lung Cancer: contribution of bioinformatics analysis to the development of non-invasive diagnostic tools. Genes 8(1).

https://doi. org/10.3390/genes8010008
4. Kunz M, Xiao K, Liang C, Viereck J, Pachel C, Frantz S, Thum T, Dandekar T (2015) Bioinformatics of cardiovascular miRNA biology. J Mol Cell Cardiol 89(Pt A):3–10. https://doi.org/10.1016/j.yjmcc.2014.11.027
5. Kasinski AL, Kelnar K, Stahlhut C, Orellana E, Zhao J, Shimer E, Dysart S, Chen X, Bader AG, Slack FJ (2015) A combinatorial microRNA therapeutics approach to suppressing non-small cell lung cancer. Oncogene 34(27):3547–3555. https://doi. org/10.1038/onc.2014.282
6. Montani F, Marzi MJ, Dezi F, Dama E, Carletti RM, Bonizzi G, Bertolotti R, Bellomi M, Rampinelli C, Maisonneuve P, Spaggiari L, Veronesi G, Nicassio F, Di Fiore PP, Bianchi F (2015) miR-Test: a blood test for lung cancer early detection. J Nat Cancer Inst 107(6):djv063. https://doi.org/10.1093/jnci/djv063
7. Leidinger P, Keller A, Meese E (2011) MicroRNAs – important molecules in lung Cancer research. Front Genet 2:104. https://doi.org/10.3389/fgene.2011.00104
8. Vosa U, Vooder T, Kolde R, Vilo J, Metspalu A, Annilo T (2013) Meta-analysis of microRNA expression in lung cancer. Int J Cancer 132(12):2884–2893. https://doi.org/10.1002/ijc.27981
9. Kozomara A, Griffiths-Jones S (2014) miRBase: annotating high confidence microRNAs using deep sequencing data. Nucleic Acids Res 42(D1):D68–D73. https://doi.org/10.1093/nar/gkt1181
10. Chou CH, Chang NW, Shrestha S, Hsu SD, Lin YL, Lee WH, Yang CD, Hong HC, Wei TY, Tu SJ, Tsai TR, Ho SY, Jian TY, Wu HY, Chen PR, Lin NC, Huang HT, Yang TL, Pai CY, Tai CS, Chen WL, Huang CY, Liu CC, Weng SL, Liao KW, Hsu WL, Huang HD (2016) miRTarBase 2016: updates to the experimentally validated miRNA-target interactions database. Nucleic Acids Res 44(D1):D239–D247. https://doi.org/10.1093/nar/gkv1258
11. Kunz M, Dandekar T, Naseem M (2017) A systems biology methodology combining transcriptome and interactome datasets to assess the implications of cytokinin signaling for plant immune networks. Methods in Mol Biol 1569:165–173. https://doi.org/10.1007/978-1-4939-6831-2_14
12. Naseem M, Kunz M, Dandekar T (2014) Probing the unknowns in cytokinin-mediated immune defense in Arabidopsis with systems biology approaches. Bioinform Biol Insights 8:35–44. https://doi.org/10.4137/BBI.S13462
13. Barrett T, Wilhite SE, Ledoux P, Evangelista C, Kim IF, Tomashevsky M, Marshall KA, Phillippy KH, Sherman PM, Holko M, Yefanov A, Lee H, Zhang N, Robertson CL, Serova N, Davis S, Soboleva A (2013) NCBI GEO: archive for functional genomics data sets—update. Nucleic Acids Res 41(D1):D991–D995. https://doi.org/10.1093/nar/gks1193
14. Team RDC (2008) R: a language and environment for statistical computing. R Foundation for Statistical Computing, Vienna
15. Gentleman RC, Carey VJ, Bates DM, Bolstad B, Dettling M, Dudoit S, Ellis B, Gautier L, Ge Y, Gentry J, Hornik K, Hothorn T, Huber W, Iacus S, Irizarry R, Leisch F, Li C, Maechler M, Rossini AJ, Sawitzki G, Smith C, Smyth G, Tierney L, Yang JY, Zhang J (2004) Bioconductor: open software development for computational biology and bioinformatics. Genome Biol 5(10):R80. https://doi.org/10.1186/gb-2004-5-10-r80
16. Davis S, Meltzer PS (2007) GEOquery: a bridge between the gene expression omnibus (GEO) and BioConductor. Bioinformatics (Oxford, England) 23(14):1846–1847. https://doi.org/10.1093/bioinformatics/btm254
17. Ritchie ME, Phipson B, Wu D, Hu Y, Law CW, Shi W, Smyth GK (2015) Limma powers differential expression analyses for RNA-sequencing and microarray studies. Nucleic Acids Res 43(7):e47. https://doi.org/10.1093/nar/gkv007
18. Warnes GR, Bolker B, Bonebakker L, Gentleman R, Huber W, Liaw A, Lumley T, Maechler M, Magnusson A, Moeller S (2009) gplots: various R programming tools for plotting data. R package version 2 (4)
19. Shannon P, Markiel A, Ozier O, Baliga NS, Wang JT, Ramage D, Amin N, Schwikowski B, Ideker T (2003) Cytoscape: a software environment for integrated models of biomolecular interaction networks. Genome Res 13(11):2498–2504. https://doi.org/10.1101/gr.1239303
20. Bader GD, Hogue CW (2003) An automated method for finding molecular complexes in large protein interaction networks. BMC

Bioinformatics. Jan 13;4:2. Epub 2003 Jan 13. (https://www.ncbi.nlm.nih.gov/pubmed/12525261)
21. Bonnici V, Russo F, Bombieri N, Pulvirenti A, Giugno R (2014) Comprehensive reconstruction and visualization of non-coding regulatory networks in human. Front Bioeng Biotechnol 2:69. https://doi.org/10.3389/fbioe.2014.00069
22. Kitai H, Ebi H (2016) Key roles of EMT for adaptive resistance to MEK inhibitor in KRAS mutant lung cancer. Small GTPases 8:1–5. https://doi.org/10.1080/21541248.2016.1210369
23. Kimura H, Fumoto K, Shojima K, Nojima S, Osugi Y, Tomihara H, Eguchi H, Shintani Y, Endo H, Inoue M, Doki Y, Okumura M, Morii E, Kikuchi A (2016) CKAP4 is a Dickkopf1 receptor and is involved in tumor progression. J Clin Invest 126(7):2689–2705. https://doi.org/10.1172/jci84658
24. Park SH, Seong MA, Lee HY (2016) p38 MAPK-induced MDM2 degradation confers paclitaxel resistance through p53-mediated regulation of EGFR in human lung cancer cells. Oncotarget 7(7):8184–8199. https://doi.org/10.18632/oncotarget.6945
25. Hao Y, Wu W, Li H, Yuan J, Luo J, Zhao Y, Chen R (2016) NPInter v3.0: an upgraded database of noncoding RNA-associated interactions. Database 2016:baw057-baw057. doi:https://doi.org/10.1093/database/baw057
26. Kanehisa M, Sato Y, Kawashima M, Furumichi M, Tanabe M (2016) KEGG as a reference resource for gene and protein annotation. Nucleic Acids Res 44(D1):D457–D462. https://doi. org/10.1093/nar/gkv1070
27. Kutmon M, Riutta A, Nunes N, Hanspers K, Willighagen Egon L, Bohler A, Mélius J, Waagmeester A, Sinha Sravanthi R, Miller R, Coort SL, Cirillo E, Smeets B, Evelo Chris T, Pico AR (2016) WikiPathways: capturing the full diversity of pathway knowledge. Nucleic Acids Res 44(D1):D488–D494. https://doi.org/10.1093/nar/gkv1024
28. Bindea G, Mlecnik B, Hackl H, Charoentong P, Tosolini M, Kirilovsky A, Fridman WH, Pages F, Trajanoski Z, Galon J (2009) ClueGO: a Cytoscape plug-in to decipher functionally grouped gene ontology and pathway annotation networks. Bioinformatics 25(8):1091–1093. https://doi.org/10.1093/bioinformatics/btp101
29. Robles AI, Arai E, Mathé EA, Okayama H, Schetter AJ, Brown D, Petersen D, Bowman ED, Noro R, Welsh JA, Edelman DC, Stevenson HS, Wang Y, Tsuchiya N, Kohno T, Skaug V, Mollerup S, Haugen A, Meltzer PS, Yokota J, Kanai Y, Harris CC (2015) An integrated prognostic classifier for stage I lung adenocarcinoma based on mRNA, microRNA and DNA methylation biomarkers. J Thoracic Oncol 10(7):1037–1048. https://doi.org/10.1097/JTO.0000000000000560
30. Quackenbush J Microarray data normalization and transformation. Nature Genetics 32:496–501. https://www.nature.com/articles/ng1032
31. Irizarry RA, Hobbs B, Collin F, Beazer-Barclay YD, Antonellis KJ, Scherf U, Speed TP (2003) Exploration, normalization, and summaries of high density oligonucleotide array probe level data. Biostatistics (Oxford, England) 4(2):249–264. https://doi.org/10.1093/biostatistics/4.2.249
32. Biecek MKaP (2016) RTCGA: the cancer genome atlas data integration. R package version 1.6.0.
33. Taminau J, Meganck S, Lazar C, Steenhoff D, Coletta A, Molter C, Duque R, de Schaetzen V, Weiss Solis DY, Bersini H, Nowe A (2012) Unlocking the potential of publicly available microarray data using inSilicoDb and inSilicoMerging R/Bioconductor packages. BMC bioinformatics 13:335. https://doi.org/10.1186/1471-2105-13-335

Chapter 12

Spatial Analysis of Functional Enrichment (SAFE) in Large Biological Networks

Anastasia Baryshnikova

Abstract

Spatial analysis of functional enrichment (SAFE) is a systematic quantitative approach for annotating large biological networks. SAFE detects network regions that are statistically overrepresented for functional groups or quantitative phenotypes of interest, and provides an intuitive visual representation of their relative positioning within the network. In doing so, SAFE determines which functions cocluster in a network, which parts of the network they are associated with and how they are potentially related to one another.

Here, I provide a detailed stepwise description of how to perform a SAFE analysis. As an example, I use SAFE to annotate the genome-scale genetic interaction similarity network from *Saccharomyces cerevisiae* with Gene Ontology (GO) biological process terms. In addition, I show how integrating GO with chemical genomic data in SAFE can recapitulate known modes of action of chemical compounds and potentially identify novel drug mechanisms.

Key words Yeast, Networks, Functional annotation, Genetic interactions, Chemical genomics, Drug target, Bortezomib, Cell map

1 Introduction

Network diagrams have long been used to represent molecular structures, signaling cascades, biochemical reactions and many other types of connections between components of biological systems. In all of their variations, ball-and-stick models had unparalleled success in illustrating complex relationships and revealing underlying organizational structures. The rise of genomic technologies empowered the discovery of biological relationships on a large scale and led to an exponential increase in size and density of known biological networks. Yet our understanding of global network organization did not improve at the same rate. One likely reason is entailed in the difficulty to read and interpret large networks. While in small networks it is possible to manually locate familiar nodes and infer their relationship to unfamiliar ones

Louise von Stechow and Alberto Santos Delgado (eds.), *Computational Cell Biology: Methods and Protocols*, Methods in Molecular Biology, vol. 1819, https://doi.org/10.1007/978-1-4939-8618-7_12,

based on distance and connectivity, using the same approach with genome-scale networks is highly impractical. This is due to the mere number of nodes, the intricacy of their connections and the deficiency of our knowledge about biological functions, which prevents us from discerning familiar and unfamiliar nodes on a network-wide scale.

To address this problem and expand our understanding of complex biological networks, I have recently developed a strategy named spatial analysis of functional enrichment (SAFE) [1]. This method is designed to provide a bird's-eye view of the functional organization of a network by mapping the distribution of biological knowledge throughout all of its local neighborhoods. For doing so, SAFE relies on three basic concepts:

1. That nodes in a large network can be spatially organized in a way that reflects their connectivity.
2. That the resulting node positions can be used to identify network neighborhoods associated with prior knowledge.
3. That these associations can be visually represented within the network to drive its biological interpretation.

In practice, SAFE detects network regions that are statistically overrepresented for a functional group or a quantitative phenotype of interest, and provides an intuitive visual representation of their relative positioning within the network.

In a proof-of-concept experiment, I used SAFE to systematically examine the genome-wide genetic interaction similarity (GIS) network of *Saccharomyces cerevisiae* [2]. By annotating the network with 4400 Gene Ontology (GO) biological process terms (i.e., all those associated with at least one yeast gene), SAFE uncovered 20 large-scale domains associated with major cellular processes [1]. SAFE also showed that different GO terms have different patterns of distribution throughout the GIS network and that these patterns can be potentially used to define term specificity [1]. Furthermore, we employed SAFE to combine GO annotations with gene expression, protein abundance, chemical genomic and genetic interaction data to explore unfolded protein response [3], compare protein expression levels across biological processes [4] and identify a potentially novel cause of resistance to chemotherapy [1].

Here, I provide a practical demonstration of how to use SAFE for annotating a biological network. As an example, I use the most recent yeast GIS network [5] that includes 4000 nodes and 28,200 edges and is therefore 40% larger and 50% denser than the network examined previously [2]. In the first example (Subheading 3.2), I describe the steps required to reproduce the GO annotation of the new GIS network [5]. In the second example (Subheading 3.3), I illustrate how SAFE can be used to integrate

GO annotations with chemical genomic data to potentially uncover drug mode-of-action.

2 Materials

2.1 Software Requirements

1. SAFE: available as MATLAB code at https://bitbucket.org/abarysh/safe.
2. MATLAB R2014b or later.
3. Bioinformatics Toolbox for MATLAB.
4. Statistics and Machine Learning Toolbox for MATLAB.
5. Optional. Cytoscape 3.3 or later.

2.2 Input Data Files

1. Cytoscape session file containing the genetic interaction similarity network [5]: included as part of the SAFE package (`/data/`) and available separately at https://github.com/baryshnikova-lab/safe-data/raw/master/networks/Costanzo_Science_2016.cys.
2. Gene Ontology (GO) biological process annotations for yeast: included as part of the SAFE package (`/data/`) and available separately at https://github.com/baryshnikova-lab/safe-data/raw/master/attributes/go_bp_140819.mat.
3. The functional annotation file containing the chemical genomic screen for camptothecin [6]: available at https://github.com / baryshnikova-lab / safe-data / raw / master / attributes/hoepfner_movva_2014_camptothecin.txt.

3 Methods

3.1 SAFE Installation, Setup, and Basic Workflow

1. Download the latest version of SAFE from Bitbucket and save it (e.g., in `/AAA/BBB/safe/`).
2. Open MATLAB and add the SAFE folder to the path: `addpath(genpath('/AAA/BBB/safe/'));`
3. Create a folder that will store the results of the SAFE analysis: `mkdir('/CCC/DDD/');`
4. Modify `/AAA/BBB/safe.m` (the main SAFE script) to record the absolute path to the results folder: `path_to_results_folder = '/CCC/DDD/'`.
5. Make a copy of the SAFE settings file (`/AAA/BBB/safe/safe.ini`) and save it in the results folder (`/CCC/DDD/safe.ini`). This file defines all tunable parameters and serves as a log record for every session of the SAFE analysis. Do not change the name of the file. Also, do not remove the original file as it can be used as a reference for default parameters.

6. Modify `/CCC/DDD/safe.ini` in any text editor to adjust input/output parameters as needed (*see* Subheadings 3.2 and 3.3 for the most common settings).
7. Run `safe.m` (*see* also **Note 1** and the examples below).
8. The results will be visually displayed in MATLAB and stored in `/CCC/DDD/`.

3.2 Annotating the GIS Network with GO Biological Process Terms

In this example, SAFE is used to annotate the yeast genetic interaction similarity (GIS) network with GO biological process terms. The steps of the analysis are run manually one by one to observe and explain the progression of the algorithm.

1. Perform Subheading 3.1, **steps 1–5**.
2. Make sure that the input network file (`Costanzo_Science_2016.cys`) is located at `/AAA/BBB/data/`. If not, *see* Subheading 2.2, **item 1** to download the file.
   (a) In this example, we are using an existing Cytoscape session file in which the genetic interaction similarity network has been loaded and spatially organized using the edge-weighted spring embedded layout [7], an implementation of the Kamada–Kawai algorithm [8]. This Cytoscape session file was originally used for SAFE analysis in Costanzo et al. [5]. As an alternative to importing the network from this file, the user can choose one of several options.
      - Create a new Cytoscape session by loading the network from a different file format and applying the same or a different layout algorithm (see Cytoscape documentation for a list of available options [9]).
      - Load the network directly from a tab-delimited text file. SAFE accepts networks in list and adjacency matrix formats. In all cases, all edges with weights greater than 0 will be used.
         – Three-column list format: <Node 1 systematic label><Node 2 systematic label><Edge weight> (see example at https://github.com/baryshnikova-lab/safe-data/raw/master/networks/Costanzo_Science_2010_3col.txt).
         – Five-column list format: <Node 1 label><Node 1 systematic label><Node 2 label><Node 2 systematic label><Edge weight> (see example at https://github.com/baryshnikova-lab/safe-data/raw/master/networks/Costanzo_Science_2010_5col.txt).
         – Adjacency matrix: One row header (<Node 1 systematic label>), one column header (<Node 2 systematic label>), edges weights (see example at

https://github.com/baryshnikova-lab/safe-data/raw/master/networks/Costanzo_Science_2010_adjacencyMatrix.txt).

3. Modify `/CCC/DDD/safe.ini` to include the following changes (the parameters are explained in detail in the sections below):
   (a) `networkfile = /AAA/BBB/data/Costanzo_Science_2016.cys`
   (b) `neighborhoodRadius = 7.5`
   (c) `neighborhoodRadiusType = diameter`
   (d) `unimodalityType = hartigans`
   (e) `groupMinimize = 2`
4. Run part 1 of `safe.m` ("Load settings") (*see* also **Note 1**).
   (a) A subfolder with a unique SAFE session ID is automatically created inside the results folder (e.g., `/CCC/DDD/safe-2016-11-02-16-08-08/`).
   (b) All parameters from `safe.ini` are loaded into the main structure array (`layout`). This array will contain all intermediate and final SAFE results, and will be automatically stored in `/CCC/DDD/<safe-session-id>/` together with other output files.
5. Run part 2 of `safe.m` ("Load network").
   (a) The network is extracted from `networkfile` (specified in `safe.ini`) and loaded into `layout`.
   (b) In this example, the genetic interaction similarity network containing 3971 nodes and 28,202 undirected edges is loaded from the Cytoscape session file `Costanzo_Science_2016.cys`. A CYS file is effectively an archive of numerous XML and XGMML files. SAFE uncompresses the archive into the `/CCC/DDD/safe-session-id/` subfolder and parses the XGMML files containing information about nodes (labels and coordinates) and edges (*see* also **Note 2**).
   (c) The following key fields in `layout` are populated:
      - `label`: a list of 3971 node labels.
      - `label_orf`: a list of 3971 systematic node labels that will be used to match nodes to functional annotations; multiple nodes can have the same systematic node label.
      - `x` and `y`: a list of 3971 node coordinates.
      - `edges`: a binary 3971 × 3971 network adjacency matrix.

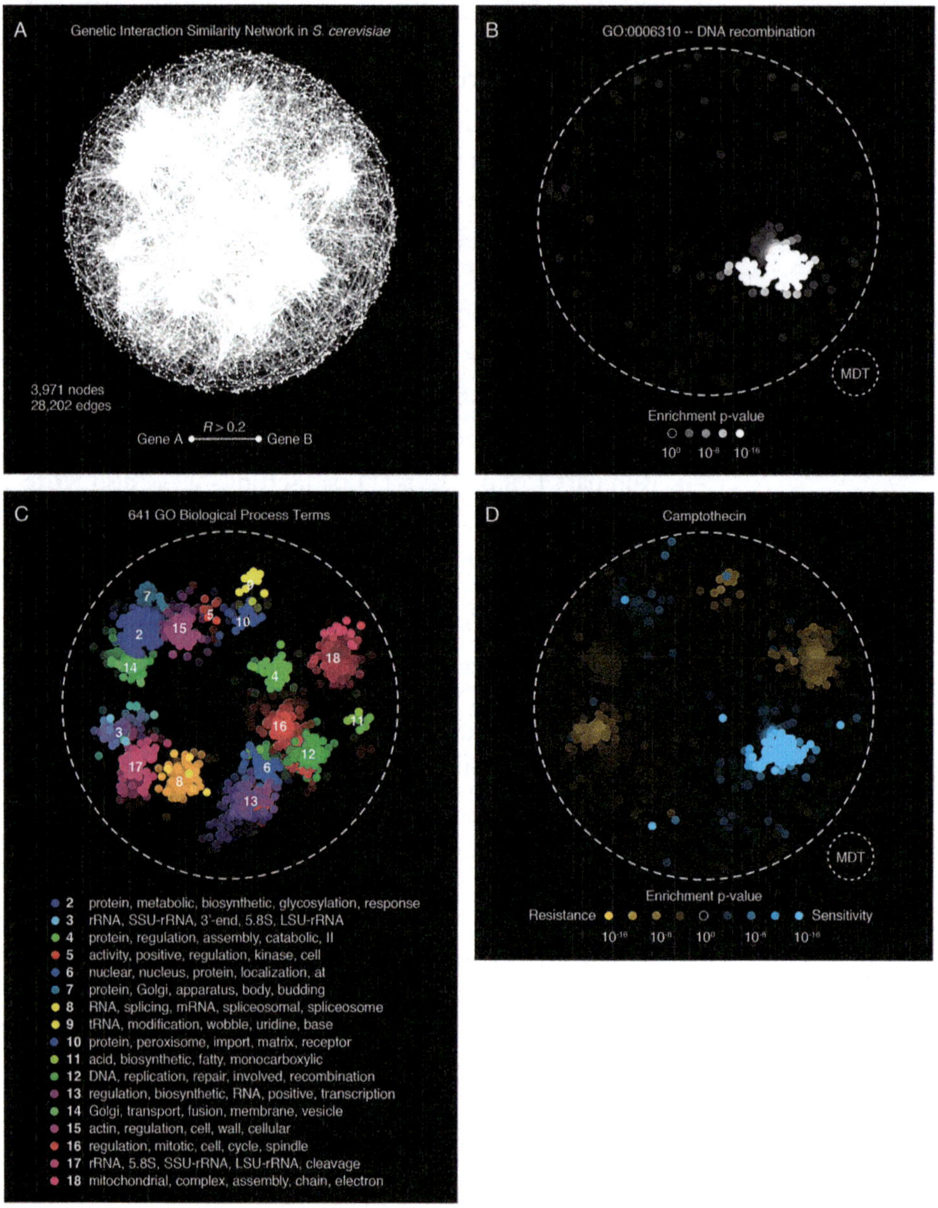

Fig. 1 (**a**) The genetic interaction similarity (GIS) network, containing 3971 nodes and 28,202 edges, was originally constructed in [5]. Nodes represent genes. Two genes are connected if their genetic interaction profiles showed a Pearson correlation coefficient (R) greater than 0.2. The positions of the nodes in two-dimensional space were determined by the spring-embedded layout algorithm, implemented in Cytoscape [9]. (**b**) An example of a network enrichment landscape produced by SAFE. The local neighborhood of every node in the network is tested for enrichment for the GO biological process term "DNA recombination" (GO:0006310). The resulting enrichment p-value is Bonferroni-corrected for multiple testing across all GO terms, log-transformed into an enrichment score and assigned back to the node. The GO:0006310 enrichment scores of all nodes constitute the term's enrichment landscape. In this example, the landscape is region-specific. All edges are hidden. All nodes are colored based on their enrichment scores. The small circle in the bottom right corner illustrates the approximate size of a neighborhood, defined by a set of nodes that are within a maximum distance threshold (MDT) from each other. (**c**) The final annotated network.

6. Skip part 3 of `safe.m` ("Apply network layout, if necessary").
   (a) In this example, the edge-weighted spring embedded layout has already been applied (*see* also **Note 3**).
7. Run part 4 of `safe.m` ("Plot network").
   (a) A figure displays a black-and-white representation of the network (Fig. 1a).
8. Skip part 5 of `safe.m` ("Translate node labels, if necessary").
   (a) In this example, the list of systematic node labels (`label_orf`) has already been obtained from the Cytoscape session (*see* also **Note 4**).
9. Run part 6 of `safe.m` ("Load functional annotations").
   (a) The functional annotations are extracted from `annotationfile` (specified in `safe.ini`) and loaded into `layout`.
   (b) In this example, `annotationfile` is empty, indicating that the default functional annotation standard should be loaded. The default standard corresponds to 4373 GO biological process terms that were associated with at least one yeast gene as of August 19, 2014. When a gene is annotated to a GO term, it is automatically annotated to all its parent terms (*see* also **Note 5**).
   (c) The following key fields are populated in `layout`:
      - `group_ids`: a list of 4373 GO term IDs.
      - `group_names`: a list of 4373 GO term names.
      - `label2group`: a binary 3971 x 4373 matrix of node-to-GO term annotations.

Fig. 1 (continued) SAFE combined 641 region-specific GO terms into 17 functional domains (labeled 2–18) based on the similarity of their enrichment landscapes. Different colors represent different domains. Each domain is labeled with a tag list, composed of the five words that occur most frequently within the names of the associated GO terms. The complete list of GO terms in each domain is provided in one of the output files. (**d**) The enrichment landscape of camptothecin, a chemical compound that inhibits DNA topoisomerase I. A chemical genomic screen measured the relative growth of 5000 yeast knockout mutants in the presence of camptothecin [6]. SAFE used the quantitative mutant growth data to annotate the GIS network. All edges are hidden. All nodes are colored based on the enrichment of their neighborhoods for genes that confer sensitivity (blue) or resistance (yellow) to camptothecin. The brightness of the color is proportional to the node's enrichment score

10. Run part 7 of `safe.m` ("Calculate distances between node pairs").
    (a) Distance between all pairs of nodes is calculated based on the `nodeDistanceType` metric specified in `safe.ini`.
    (b) In this example, `nodeDistanceType = shortpath_weighted_layout`. According to this metric, edges are weighted by their physical lengths resulting from the network layout. The distance between two nodes is defined by the shortest physical path connecting them (*see* also **Note 6**).
    (c) The following key field is populated in `layout`:
    - `nodeDistance`: a quantitative 3971 × 3971 matrix of node distances.
11. Run part 8 of `safe.m` ("Calculate enrichments").
    (a) Every node in the network is associated with a local neighborhood, defined as the set of nodes located within a certain distance from it. The distance threshold is set by `neighborhoodRadiusType` and `neighborhoodRadius` in `safe.ini`. Each neighborhood is tested for enrichment of each attribute using a one-tailed Fisher's exact test (*see* also **Note 7**), using the complete network as background (*see* also **Note 8**). The corresponding *p*-values are Bonferroni-corrected for multiple testing and log-transformed into enrichment scores. The set of enrichment scores for a given GO term across all neighborhoods constitutes the term's enrichment landscape.
    (b) In this example, `neighborhoodRadiusType = diameter` and `neighborhoodRadius = 7.5`. These settings indicate that two nodes are part of the same neighborhood if their distance is equal to or shorter than 7.5% of the network diameter. Network diameter is the distance between the two farthest nodes in the network. The threshold of 7.5% was defined arbitrarily; however, systematic evaluations showed that enrichment scores are robust to twofold variations of the distance threshold in either direction [1] (*see* also **Note 9**).
    (c) The following key fields are populated in `layout`:
    - `R`: the distance threshold used to define neighborhoods.
    - `neighborhoods`: a binary 3971 × 3971 matrix that defines neighborhood membership.
    - `pval`: a quantitative 3971 × 4373 matrix of enrichment *p*-values.

- `opacity`: a quantitative 3971 × 4373 matrix of enrichment scores.
- `thresholdOpacity`: the lowest enrichment score to be considered significant (corresponds to *p*-value = 0.05).
- `opacity_01`: a binary 3971 × 4373 matrix of significant enrichments.

12. Run part 9 of `safe.m` ("Plot sample enrichment landscapes").
    (a) A series of figures display the enrichment landscapes of several GO terms. These terms can be selected at random and/or specified by the user. The figures are exported to PDF and saved in the `/CCC/DDD/<safe-session-id>/` folder. The enrichment scores for all GO terms are exported into a tab-delimited text file and saved in the `/CCC/DDD/<safe-session-id>/` folder as well.
    (b) For example, the enrichment landscape of the GO term "DNA recombination" (GO:0006310, Fig. 1b) can be plotted by executing the following command:
    - `plot_sample_groups(layout, 'Range', [find (layout.group_ids == 6310]);`
    (c) Note that this step is optional. To suppress the figures when running `safe.m` as a script, set `plotSampleGroups` in `safe.ini` to `0`.
13. Run part 10 of `safe.m` ("Combine functional attributes into domains").
    (a) First the attributes are filtered based on the size and the pattern of their enrichment landscapes.
    - SAFE retains only attributes that are enriched in a minimum number of node neighborhoods (`groupEnrichmentMinSize` in `safe.ini`). In this example, `groupEnrichmentMinSize = 10`.
    - If required, SAFE retains only attributes that are enriched in a single region of the network and excludes all attributes that are enriched in multiple regions or are broadly distributed (*see* **Note 10**). The method for measuring the enrichment distribution is set by `unimodalityType` in `safe.ini`. In this example, it is `unimodalityType = hartigans`.
    (b) Next, SAFE constructs functional domains by grouping attributes with redundant enrichment landscapes, i.e., landscapes that are specific to the same region of the network. The grouping is done by computing pairwise similarity between enrichment landscapes and performing

hierarchical clustering. The similarity metric and the cluster threshold are set by `groupDistanceType` and `groupDistanceThreshold` in `safe.ini`, respectively. In this example, `groupDistanceType` = `jaccard` and `groupDistanceThreshold` = `0.75` (*see* **Note 11**).

(c) Finally, functional domains are filtered by size and consolidated. The size of a functional domain is determined by how many node neighborhoods have their highest enrichment for an attribute assigned to that domain. Note that a node neighborhood can be enriched for multiple attributes in multiple domains; however, the attribute and the domain that show the highest enrichment will be considered predominant and officially "claim" the neighborhood. If a domain has claimed less than ten node neighborhoods, it is excluded from further evaluation. After all the exclusions, the domains are renumbered consecutively (*see* **Note 12**).

(d) The following key fields are populated in `layout`:

- `groupIsTop`: a list of 4373 values that indicate whether or not an attribute is included in the analysis (i.e., is region-specific and meets the minimum size criteria).
- `groupColor`: a list of 4373 domain IDs, one for each attribute ("1" denotes attributes that were not assigned to any domain).
- `labelColor`: a list of 3971 predominant domain IDs, one for each node ("1" denotes nodes that were not claimed by any domain).
- `labelOpacity`: a list of 3971 predominant enrichment scores, one for each node.

14. Run part 11 of `safe.m` ("Generate automatic domain labels").

(a) Each domain is provided with a summary label composed of the five most frequently occurring words within the labels of its attributes. Articles and other common words are excluded.

15. Run part 12 of `safe.m` ("Plot annotated network").

(a) A figure displays the final annotated network (Fig. 1c). Nodes are positioned at their original x and y coordinates and colored by their average RGB value. For every domain, a numeric ID is placed next to its center.

(b) To produce the figure, a random color is assigned to each domain and, by extension, to all of its associated attributes. Note that this random assignment results in a different (but equivalent) color palette at every run.

(c) For every node, attribute colors are combined proportionally to the enrichment score of each attribute in the node's neighborhood. For example, if a node's neighborhood is significantly enriched for attributes A, B, and C, assigned to domains 3, 3, and 5, respectively, the color of the node will equal to $(c_3 \times w_A + c_3 \times w_B + c_5 \times w_C)/3$, where c_3 and c_5 are the colors of domains 3 and 5 (in RGB format), and w_X is the square of the node's enrichment score for attribute X.

16. Run part 13 of `safe.m` ("Print results").
    (a) The annotated network is exported to PDF format and saved in `/CCC/DDD/<safe-session-id>/figure3.pdf`.
    (b) Three additional output files are added to `/CCC/DDD/<safe-session-id>` (*see* also **Note 13**):
    - `domain_properties_annotation-highest.txt`: lists the properties of all functional domains (domain id, domain label and domain color, as it appears on the annotated network, in RGB format).
    - `attribute_properties_annotation-highest.txt`: lists the properties of all functional attributes (attribute id, attribute label and its corresponding domain id).
    - `node_properties_annotation-highest.txt`: lists the properties of all nodes (node unique label, node systematic label, id of predominant domain, enrichment score of predominant attribute, total number of associated domains, and number of associated attributes per domain).
17. Run part 14 of `safe.m` ("Save SAFE session").
    (a) This step saves the `layout` array into the `/CCC/DDD/<safe-session-id>/safe_session.mat` file.

3.3 Annotating the GIS Network with Camptothecin Chemical Genomic Data

In this example, SAFE is used to annotate the yeast GIS network with a genome-wide chemical genomic screen for camptothecin, a DNA topoisomerase inhibitor that blocks DNA synthesis [6]. In a typical chemical genomic experiment, a pool of 5000 yeast knockout mutants is grown in the presence of a chemical compound [10]. The growth of each individual mutant, relative to other mutants and to an untreated control, is measured via microarray hybridization or en masse sequencing of mutant-specific molecular barcodes [11]. The sets of mutants that are relatively sensitive or resistant to the compound are often indicative on the compound's mode of action and have been instrumental for identifying direct drug targets [12].

Here, we annotate the GIS network with camptothecin growth data to identify network regions that are specifically enriched for sensitivity and resistance phenotypes. The steps of the analysis are run manually one by one to observe and explain the progression of the algorithm.

1. Perform Subheading 3.1, **steps 1–5** (*see* above).
2. Make sure that the input network file (`Costanzo_Science_2016.cys`) is located at `/AAA/BBB/data/`. If not, *see* Subheading 2.2, **item 1** to download the file. *See* also Subheading 3.2, **step 1** for alternative network file formats.
3. Modify `/CCC/DDD/safe.ini` to include the following changes (the parameters are explained in detail in the sections below):
   (a) `networkfile = /AAA/BBB/data/Costanzo_Science_2016.cys`
   (b) `annotationfile = /CCC/DDD/hoepfner_movva_2014_camptothecin.txt`
   (c) `annotationsign = both`
   (d) `unimodalityType = <empty>`
   (e) `groupDistanceType = <empty>`
   (f) `groupMinimize = 0`
4. Repeat Subheading 3.2, **steps 4–8** (*see* above).
5. Run part 6 of `safe.m` ("Load functional annotations").
   (a) The functional annotations are extracted from `annotationfile` (specified in `safe.ini`) and loaded into `layout`.
   (b) In this example, `annotationfile` points to a file containing one functional attribute, i.e., the chemical genomic screen for camptothecin. The file lists 4960 yeast knockout mutants and their relative growth values in the presence of 200 μM camptothecin. Negative values indicate sensitivity, while positive values indicate resistance. Considering that we are interested in detecting network regions that are enriched for both phenotypes, we set `annotationsign = both`.
   (c) The following key fields are populated in `layout`:
      - `group_ids`: 1 attribute ID.
      - `group_names`: 1 attribute name.
      - `label2group`: a quantitative 3971 × 1 matrix of node-to-attribute annotations.
6. Run part 7 of `safe.m` ("Calculate distances between node pairs").
   (a) The following key field is populated in `layout`:
      - `nodeDistance`: a quantitative 3971 × 3971 matrix of node distances.

7. Run part 8 of `safe.m` ("Calculate enrichments").

   (a) Every node in the network is associated with a local neighborhood, defined as the set of nodes located within a certain distance from it. The distance threshold is set by `neighborhoodRadiusType` and `neighborhoodRadius` in `safe.ini`. Within each neighborhood, the attribute values are summed up to produce a quantitative neighborhood score. This score is compared to the distribution of 1000 random scores obtained by randomizing node labels, and thus attribute values, across the entire network. The resulting empirical p-values are Bonferroni-corrected for multiple testing and log-transformed into neighborhood enrichment scores. The set of enrichment scores for a given attribute across all neighborhoods constitutes the attribute's enrichment landscape.

   (b) In this example the local neighborhoods are determined by `neighborhoodRadiusType = percentile` and `neighborhoodRadius = 0.5`. These settings indicate that two nodes are part of the same neighborhood if their distance is within the lowest 0.5th percentile of all pairwise node distances. Also, given that `annotationsign = both` (*see* Subheading 3.3, **steps 3** and **5** above), SAFE tests both tails of the random score distribution and produces 2 *p*-values for each neighborhood:

      - *p*-value #1 that measures the probability of observing a neighborhood score this low or lower in a random score distribution.
      - *p*-value #2 that measures the probability of observing a neighborhood score this high or higher in a random score distribution.

   (c) The following key fields are populated in `layout`:

      - `R`: the distance threshold used to define neighborhoods.
      - `neighborhoods`: a binary 3971 × 3971 matrix that defines neighborhood membership.
      - `pval`: a quantitative 3971 × 1 × 2 matrix of enrichment *p*-values (3971 is the number of node neighborhoods, 1 is the number of attributes, 2 is the number of tails of the random distribution that were considered).
      - `opacity`: a quantitative 3971 × 1 × 2 matrix of enrichment scores.
      - `thresholdOpacity`: the lowest enrichment score to be considered significant (corresponds to *p*-value = 0.05).

- `opacity_01`: a binary 3971 × 1 × 2 matrix of significant enrichments.

8. Run part 9 of `safe.m` ("Plot sample enrichment landscapes").
   (a) A figure displays the enrichment landscape for camptothecin (Fig. 1d). Given that `annotationsign = both` and each node has two enrichment scores (one for sensitivity and one for resistance), the landscape is produced by selecting the score with the highest value.
   (b) The camptothecin enrichment landscape shows that the network region associated with DNA replication, recombination and repair (Fig. 1c, domain 12), is statistically enriched for genes that confer sensitivity to the drug, when mutated (Fig. 1d, blue). Similarly, network regions associated with ribosome biogenesis, tRNA modification, and mitochondrial functions (Fig. 1c, domains 3, 9, and 18) are enriched for resistance to the drug (Fig. 1d, yellow).
   (c) The figure is exported to PDF and saved in `/CCC/DDD/<safe-session-id>/figure2.pdf`.
   (d) The enrichment scores are exported into two tab-delimited text files (one for the highest attribute values and one for the lowest) and saved in the `/CCC/DDD/<safe-session-id>/` folder as well.
9. Given than only one attribute is considered in this analysis, the remaining parts of `safe.m` ("Combine functional attributes into domains" and others) can be skipped.

4 Conclusions and Discussion

The rapid and systematic mapping of network domains enriched for specific functional attributes is a distinctive feature of SAFE that encourages deep exploration of biological networks through iterative annotation with multiple complementary functional standards. Here, I showed that annotating the yeast genetic interaction similarity network with GO biological process and chemical genomic data can link genes and their functions to bioactive compounds and suggest possible modes-of-action for uncharacterized chemicals.

In contrast to other methods, such as graph clustering [13] or identification of active subnetworks [14], SAFE was specifically designed to annotate complete networks rather than their separate parts. To achieve this goal, SAFE takes advantage of network layout algorithms that position nodes in a two- or three-dimensional space based on an equilibrium of forces that reflects network topology. Using layouts, instead of clustering algorithms, is advantageous because they do not require hard boundaries between network domains and enable the annotation of sparse, yet functionally coherent, neighborhoods.

So far, SAFE has been used primarily to characterize genetic interaction networks in yeast [3, 4, 5, 15]. This is largely due to the fact that genetic interactions have a clear modular structure that provided early evidence of a potential global functional organization and prompted a systematic analysis. The global topological structure of other biological networks, such as protein-protein interaction or gene regulatory networks [16–21], is much less obvious. However, early results indicate that their functional structure may be just as rich: for example, SAFE suggests that the global protein-protein interaction network in yeast, containing 5700 nodes and 78,000 edges, is organized in 21 large functional domains, each enriched for a distinctive set of GO biological process terms [15].

Despite these early findings, a standard approach for exploring the functional organization of a new network is not yet fully established. One of the limiting factors is our understanding of layout algorithms. It is yet to be determined which algorithm is the most appropriate for a particular network, which parameters produce the best results and how stable the final outcome is. Another potential limitation is the topological differences between the networks, such as edge density, degree distribution, and modularity. While it is generally true that sparser networks require larger neighborhood definitions (higher `neighborhoodRadius` values), the exact parameter settings currently depend on the specific network at hand. Addressing these questions will unlock powerful comparative analyses of numerous biological networks and provide new insight into the global functional organization of the systems they represent.

5 Notes

1. The main SAFE file `/AAA/BBB/safe.m` can be executed as a regular MATLAB script (`run('/AAA/BBB/safe.m');`). However, it is often more useful to manually run the different parts of the file, one step at a time, to get a better understanding of the algorithm and examine intermediate results.
2. The `Costanzo_Science_2016.cys` file contains only one network collection (`Costanzo_Science_2016_network_collection`) and only one network within that collection (`Costanzo_Science_2016_network`). However, in general, Cytoscape sessions can contain multiple network collections with multiple networks each. By default, SAFE will load the first network of the first network collection. To change the default, the user can modify `networkname` and `viewname` in

`safe.ini` (see a more detailed explanation within `safe.ini` itself).

3. To reapply the layout and obtain a different configuration of the network, the user should set `layoutAlgorithm` in `safe.ini` to `Kamada-Kawai (Cytoscape)`. Due to an initial randomization step, the Kamada–Kawai algorithm is nondeterministic and results in a different network configuration at every run. To make layouts reproducible, the user should set `randomSeed` in `safe.ini` to any number (e.g., `randomSeed = 3`). This setting will make sure that, for every SAFE run the nodes are initialized at the same random positions and will end up in the same final configuration.
4. Systematic node labels (`label_orf`) are used to match nodes to annotations (Subheading 3.2, **step 9** and Subheading 3.3, **step 5**) and calculate enrichments (Subheading 3.2, **step 11** and Subheading 3.3, **step 7**). As a result, it is important that `label_orf` is populated with the same label type that is used by the functional annotation file. By default, SAFE assumes that all node labels are systematic labels and thus `label_orf` is populated directly from `label`. However, in three scenarios, SAFE populates `label_orf` from elsewhere:
   (a) If the user is loading a network from a Cytoscape session file and the network has a node attribute named "ORF".
   (b) If the user is loading a network from a five-column tab-delimited text file (*see* Subheading 3.2, **step 2**).
   (c) If the default node labels in the network match gene names or allele names from *S. cerevisiae* and can be translated into their corresponding systematic ORF names.
5. As an alternative to the yeast GO biological process standard, the user can import any other annotation file that is provided in a tab-delimited matrix format. In this format, rows are node systematic labels, columns are functional attribute labels and the data are binary or quantitative values indicating the association between a node and an attribute. By default, SAFE will consider only the highest values of the attributes to be of interest: e.g., the 1's in a binary standard such as GO or the top positive values of a quantitative standard such as [6] (Subheading 3.2). If only the lowest or both lowest and highest values of the attributes should be examined, the user should set `annotationsign` in `safe.ini` to `lowest` or `both`, respectively. Examples of binary and quantitative annotation files are provided as part of the SAFE package (`/data/`) and also at https://github.com/baryshnikova-lab/safe-data/tree/master/attributes.
6. Other distance metrics include `shortpath`, which measures the unweighted shortest path length between two nodes,

and the `shortpath_weighted_edge`, which weights edges according to their original values supplied by the user. Unlike the default `shortpath_weighted_layout`, neither of these distance metrics depends on the layout algorithm and the resulting node positions.

7. Fisher's exact test is used to calculate enrichments for binary functional annotations, such as GO terms. When the annotation standard is quantitative, enrichments are automatically calculated using random permutations and empirical p-values (*see* Subheading 3.2 for a more detailed explanation).

8. By default, neighborhood enrichment is assessed by comparing the frequency of an attribute within a neighborhood to the frequency of the attribute in the entire network (`background = map` in `safe.ini`). In case of functional biases in the network, enrichment can also be calculated relative to the total frequency of the attribute, which is defined by the number of labels associated with the attribute in the annotation standard (`background = standard`).

9. In addition to `diameter`, the distance threshold can be defined in two other ways. The default `neighborhoodRadiusType` setting is `percentile` (with a default `neighborhoodRadius` value of `0.5`), which indicates that two nodes are part of the same neighborhood if their distance is in the lowest 0.5th percentile of all pairwise distances in the network. Alternatively, the user can define the distance threshold in absolute terms by setting `neighborhoodRadiusType` to `absolute` and `neighborhoodRadius` to a fixed value (e.g., `200`). This option is particularly useful when node distances are measured by the unweighted shortest path length (`nodeDistanceType = shortpath`): for example, by setting `neighborhoodRadiusType` to `absolute` and `neighborhoodRadius` to `2`, the neighborhoods will contain only nodes that are within two hops from one another.

10. The filter for region-specific attributes is set by `unimodalityType` in `safe.ini` and can have one of four values:
    (a) `<empty>`: no filter is applied, all attributes are retained.
    (b) `hartigans`: the Hartigan's dip test of unimodality [22] retains an attribute only if the distribution of pairwise distances between its significantly enriched neighborhoods is unimodal.
    (c) `radius`: the radius-based method retains an attribute only if at least 65% of its significantly enriched neighborhoods are within a distance `2R`, where `R` is the distance threshold that defines the size of the neighborhood (*see* Subheading 3.2, **step 11** and Subheading 3.3, **step 7**).

(d) `subclust`: the subtractive clustering-based method [23, 24] retains an attribute only if there is at most one cluster center among its significantly enriched neighborhoods.

11. Functional domains are groups of attributes that have similar enrichment landscapes. These groups are defined by a distance metric and a distance threshold (`groupDistanceType` and `groupDistanceThreshold` in `safe.ini`, respectively). The `groupDistanceType` can be set to `<empty>`, indicating that no grouping will be performed. Alternatively, it can be set to `jaccard` or `correlation` to compute distances via Jaccard similarity or Pearson correlation, respectively. Pearson correlation considers all enrichment scores, whereas Jaccard similarity takes into account only the significant ones. The `groupDistanceThreshold` defines the number and size of functional domains. It is expressed as a fraction of the total height of the hierarchical clustering tree and can assume any value between `0` and `1` (higher values indicate fewer and bigger clusters).
12. The domain filtering and consolidation step is optional and can be skipped by setting `groupMinimize` in `safe.ini` to `0`. Alternatively, two options are available: the default option (`groupMinimize` = `1`) and a "legacy" option (`groupMinimize` = `2`). The main difference between the two options is in how the predominant domain for a node is chosen. In the "legacy" implementation, domain X is predominant for a node if X is associated with the attribute that has the highest enrichment score in the node's neighborhood. In the newer implementation, domain X is predominant for a node if X is associated with the largest number of attributes that are significantly enriched in the node's neighborhood.
13. In this example, only the highest attribute values were considered of interest and thus the output files are labeled with "highest" (e.g., `domain_properties_annotation-highest.txt`). If only the lowest values were considered, the files would be named "lowest". If both lowest and highest values were included in the analysis, two sets of files would be printed.

References

1. Baryshnikova A (2016) Systematic functional annotation and visualization of biological networks. Cell Syst 2(6):412–421. https://doi.org/10.1016/j.cels.2016.04.014
2. Costanzo M, Baryshnikova A, Bellay J, Kim Y, Spear ED, Sevier CS, Ding H, Koh JL, Toufighi K, Mostafavi S, Prinz J, St Onge RP, VanderSluis B, Makhnevych T, Vizeacoumar FJ, Alizadeh S, Bahr S, Brost RL, Chen Y, Cokol M, Deshpande R, Li Z, Lin ZY, Liang W, Marback M, Paw J, San Luis BJ, Shuteriqi E, Tong AH, van Dyk N, Wallace IM, Whitney JA, Weirauch MT, Zhong G, Zhu H, Houry WA, Brudno M, Ragibizadeh S, Papp B, Pal

C, Roth FP, Giaever G, Nislow C, Troyanskaya OG, Bussey H, Bader GD, Gingras AC, Morris QD, Kim PM, Kaiser CA, Myers CL, Andrews BJ, Boone C (2010) The genetic landscape of a cell. Science 327(5964):425–431. https://doi.org/10.1126/science.1180823
3. DeRossi C, Vacaru A, Rafiq R, Cinaroglu A, Imrie D, Nayar S, Baryshnikova A, Milev MP, Stanga D, Kadakia D, Gao N, Chu J, Freeze HH, Lehrman MA, Sacher M, Sadler KC (2016) trappc11 is required for protein glycosylation in zebrafish and humans. Mol Biol Cell 27(8):1220–1234. https://doi.org/10.1091/mbc.E15-08-0557
4. Ho B, Baryshnikova A, Brown GW (2017) Comparative analysis of protein abundance studies to quantify the Saccharomyces cerevisiae proteome. bioRxiv. https://doi.org/10.1101/104919
5. Costanzo M, VanderSluis B, Koch EN, Baryshnikova A, Pons C, Tan G, Wang W, Usaj M, Hanchard J, Lee SD, Pelechano V, Styles EB, Billmann M, van Leeuwen J, van Dyk N, Lin ZY, Kuzmin E, Nelson J, Piotrowski JS, Srikumar T, Bahr S, Chen Y, Deshpande R, Kurat CF, Li SC, Li Z, Usaj MM, Okada H, Pascoe N, San Luis BJ, Sharifpoor S, Shuteriqi E, Simpkins SW, Snider J, Suresh HG, Tan Y, Zhu H, Malod-Dognin N, Janjic V, Przulj N, Troyanskaya OG, Stagljar I, Xia T, Ohya Y, Gingras AC, Raught B, Boutros M, Steinmetz LM, Moore CL, Rosebrock AP, Caudy AA, Myers CL, Andrews B, Boone C (2016) A global genetic interaction network maps a wiring diagram of cellular function. Science 353(6306). https://doi. org/10.1126/science.aaf1420
6. Hoepfner D, Helliwell SB, Sadlish H, Schuierer S, Filipuzzi I, Brachat S, Bhullar B, Plikat U, Abraham Y, Altorfer M, Aust T, Baeriswyl L, Cerino R, Chang L, Estoppey D, Eichenberger J, Frederiksen M, Hartmann N, Hohendahl A, Knapp B, Krastel P, Melin N, Nigsch F, Oakeley EJ, Petitjean V, Petersen F, Riedl R, Schmitt EK, Staedtler F, Studer C, Tallarico JA, Wetzel S, Fishman MC, Porter JA, Movva NR (2014) High-resolution chemical dissection of a model eukaryote reveals targets, pathways and gene functions. Microbiol Res 169(2-3):107–120. https://doi.org/10.1016/j.micres.2013.11.004
7. Cytoscape.org (2016) Cytoscape user manual. http://wiki.cytoscape.org/Cytoscape_3/UserManual -Cytoscape_3.2BAC8 -UserManual.2BAC8-Navigation_Layout. Automatic_Layout_Algorithms. Accessed 6 Feb 2016
8. Kamada T, Kawai S (1989) An algorithm for drawing general undirected graphs. Inf Process Lett 31(1):7–15
9. Shannon P, Markiel A, Ozier O, Baliga NS, Wang JT, Ramage D, Amin N, Schwikowski B, Ideker T (2003) Cytoscape: a software environment for integrated models of biomolecular interaction networks. Genome Res 13(11):2498–2504. https://doi.org/10.1101/gr.1239303
10. Giaever G, Chu AM, Ni L, Connelly C, Riles L, Veronneau S, Dow S, Lucau-Danila A, Anderson K, Andre B, Arkin AP, Astromoff A, El-Bakkoury M, Bangham R, Benito R, Brachat S, Campanaro S, Curtiss M, Davis K, Deutschbauer A, Entian KD, Flaherty P, Foury F, Garfinkel DJ, Gerstein M, Gotte D, Guldener U, Hegemann JH, Hempel S, Herman Z, Jaramillo DF, Kelly DE, Kelly SL, Kotter P, LaBonte D, Lamb DC, Lan N, Liang H, Liao H, Liu L, Luo C, Lussier M, Mao R, Menard P, Ooi SL, Revuelta JL, Roberts CJ, Rose M, Ross-Macdonald P, Scherens B, Schimmack G, Shafer B, Shoemaker DD, Sookhai-Mahadeo S, Storms RK, Strathern JN, Valle G, Voet M, Volckaert G, Wang CY, Ward TR, Wilhelmy J, Winzeler EA, Yang Y, Yen G, Youngman E, Yu K, Bussey H, Boeke JD, Snyder M, Philippsen P, Davis RW, Johnston M (2002) Functional profiling of the Saccharomyces cerevisiae genome. Nature 418(6896):387–391. https://doi. org/10.1038/nature00935
11. Smith AM, Ammar R, Nislow C, Giaever G (2010) A survey of yeast genomic assays for drug and target discovery. Pharmacol Ther 127(2):156–164. https://doi.org/10.1016/j.pharmthera.2010.04.012
12. Ho CH, Piotrowski J, Dixon SJ, Baryshnikova A, Costanzo M, Boone C (2011) Combining functional genomics and chemical biology to identify targets of bioactive compounds. Curr Opin Chem Biol 15(1):66–78. https://doi.org/10.1016/j.cbpa.2010.10.023
13. Newman ME (2006) Modularity and community structure in networks. Proc Natl Acad Sci U S A 103(23):8577–8582. https://doi.org/10.1073/pnas.0601602103
14. Mitra K, Carvunis AR, Ramesh SK, Ideker T (2013) Integrative approaches for finding modular structure in biological networks.

Nat Rev Genet 14(10):719–732. https://doi.org/10.1038/nrg3552
15. Baryshnikova A (2016) Systematic functional annotation and visualization of biological networks. bioRxiv. https://doi.org/10.1101/030551
16. Gavin AC, Aloy P, Grandi P, Krause R, Boesche M, Marzioch M, Rau C, Jensen LJ, Bastuck S, Dumpelfeld B, Edelmann A, Heurtier MA, Hoffman V, Hoefert C, Klein K, Hudak M, Michon AM, Schelder M, Schirle M, Remor M, Rudi T, Hooper S, Bauer A, Bouwmeester T, Casari G, Drewes G, Neubauer G, Rick JM, Kuster B, Bork P, Russell RB, Superti-Furga G (2006) Proteome survey reveals modularity of the yeast cell machinery. Nature 440(7084):631–636. https://doi.org/10.1038/nature04532
17. Krogan NJ, Cagney G, Yu H, Zhong G, Guo X, Ignatchenko A, Li J, Pu S, Datta N, Tikuisis AP, Punna T, Peregrin-Alvarez JM, Shales M, Zhang X, Davey M, Robinson MD, Paccanaro A, Bray JE, Sheung A, Beattie B, Richards DP, Canadien V, Lalev A, Mena F, Wong P, Starostine A, Canete MM, Vlasblom J, Wu S, Orsi C, Collins SR, Chandran S, Haw R, Rilstone JJ, Gandi K, Thompson NJ, Musso G, St Onge P, Ghanny S, Lam MH, Butland G, Altaf-Ul AM, Kanaya S, Shilatifard A, O'Shea E, Weissman JS, Ingles CJ, Hughes TR, Parkinson J, Gerstein M, Wodak SJ, Emili A, Greenblatt JF (2006) Global landscape of protein complexes in the yeast Saccharomyces cerevisiae. Nature 440(7084):637–643. https://doi.org/10.1038/nature04670
18. Zhu X, Gerstein M, Snyder M (2007) Getting connected: analysis and principles of biological networks. Genes Dev 21(9):1010–1024. https://doi.org/10.1101/gad.1528707
19. Tarassov K, Messier V, Landry CR, Radinovic S, Serna Molina MM, Shames I, Malitskaya Y, Vogel J, Bussey H, Michnick SW (2008) An in vivo map of the yeast protein interactome. Science 320(5882):1465–1470. https://doi.org/10.1126/science.1153878
20. Yu H, Braun P, Yildirim MA, Lemmens I, Venkatesan K, Sahalie J, Hirozane-Kishikawa T, Gebreab F, Li N, Simonis N, Hao T, Rual JF, Dricot A, Vazquez A, Murray RR, Simon C, Tardivo L, Tam S, Svrzikapa N, Fan C, de Smet AS, Motyl A, Hudson ME, Park J, Xin X, Cusick ME, Moore T, Boone C, Snyder M, Roth FP, Barabasi AL, Tavernier J, Hill DE, Vidal M (2008) High-quality binary protein interaction map of the yeast interactome network. Science 322(5898): 104–110. https://doi.org/10.1126/science.1158684
21. Mo ML, Palsson BO, Herrgard MJ (2009) Connecting extracellular metabolomic measurements to intracellular flux states in yeast. BMC Syst Biol 3:37. https://doi.org/10.1186/1752-0509-3-37
22. Hartigan JA, Hartigan PM (1985) The dip test of unimodality. Ann Stat 13(1):70–84
23. Chiu S (1994) Fuzzy model identification based on cluster estimation. J Intell Fuzzy Syst 2(3):267–278
24. Yager R, Filev D (1994) Generation of fuzzy rules by mountain clustering. J Intell Fuzzy Syst 2(3):209–219

Part IV

Mathematical Modeling of Cellular Phenotypes

Chapter 13

Toward Large-Scale Computational Prediction of Protein Complexes

Simone Rizzetto and Attila Csikász-Nagy

Abstract

Cellular functions are often performed by multiprotein structures called protein complexes. These complexes are dynamic structures that evolve during the cell cycle or in response to external and internal stimuli, and are tightly regulated by protein expression in different tissues resulting in quantitative and qualitative variation of protein complexes. Advances in high-throughput techniques, such as mass-spectrometry and yeast two-hybrid provided a large amount of data on protein–protein interactions. This sparked the development of computational methods able to predict protein complex formation under a variety of biological and clinical conditions. However, the challenges that need to be addressed for successful computational protein complex prediction are highly complex.

The post-genomic era saw an emerging number of algorithms and software, which are able to predict protein complexes from protein–protein interaction networks and a variety of other sources. Despite the high capacity of these methods to qualitatively predict protein complexes, they could provide only limited or no quantitative information of the predicted complexes. Recently, a new large-scale simulation of protein complexes was able to achieve this task by simulating protein complex formation on the proteome scale.

In this chapter, we review representative methods that can predict multiple protein complexes at different scales and discuss how these can be combined with emerging sources of data in order to improve protein complex characterization.

Key words Protein complexes, Protein interactions, Proteome-wide simulations, Complexome, Interactome, Disease-associated protein complexes

1 Introduction

Proteins are the major actors in performing catalytic cellular functions in an organism. Most of these biomolecules execute their functions interacting with other proteins and forming larger multiprotein complexes. Characterization of protein complexes in a cell (complexome) will improve our understanding of the mechanism that drive cellular processes.

Over the recent years, many techniques have been adopted to identify protein complexes on a large scale. Tandem affinity purification (TAP) combined with mass spectrometry (MS) was one

Louise von Stechow and Alberto Santos Delgado (eds.), *Computational Cell Biology: Methods and Protocols*, Methods in Molecular Biology, vol. 1819, https://doi.org/10.1007/978-1-4939-8618-7_13,

of the first high-throughput experimental techniques that allowed researchers to generate a protein–protein interaction network in yeast identifying 257 novel protein complexes by looking at the co-occurrence of stably bound protein partners in the dataset [1]. A similar approach was adopted by Havugimana et al., who identified 13,993 physical protein–protein interactions covering 3006 soluble human proteins [2] and predicting 622 protein complexes. More recently, another proteomics based method, called QUBIC (Quantitative BAC InteraCtomics), has been used to quantify the stoichiometry of protein complex subunits in HeLa cell line [3]. This method is based on MS quantification of 1125 bait proteins along with their interactors.

The information generated in these studies can now be used to train and develop computational methods to predict protein complexes in various biological settings not investigated yet by wet-lab experiments.

Protein complex prediction is a very challenging task, because it can be affected by different processes that operate at different biological scales. Therefore, in order to successfully predict protein complexes, computational methods need to account for the high complexity of the mechanisms and constraints that drive complex formation. These mechanisms can be divided into two main categories: (1) those that are ruled by physiochemical properties of the interacting proteins and are mostly acting on a local scale; (2) global scale processes governed by the cell using transcriptional, posttranslational and epigenetic modifications in response to various stimuli, in different cellular phenotypes or phases during the cell cycle (Fig. 1). Indeed, protein complexes are dynamic structure that evolve during the cell cycle [4] or in response to external stimuli and are tightly regulated by protein expression profiles in different tissues. With the increasing amount of proteome-wide data becoming available many computational methods have been proposed to predict protein complexes from the available protein–protein interaction and transcriptomic or proteomic profile.

Starting from one of the first methods to predict protein complexes from protein–protein interaction (PPI) networks in 2002 [5], a number of different methods have been developed to improve the quality of the predictions. Methods that predict protein complexes on a large scale are often based on static PPI networks and can provide only qualitative prediction of protein complexes [6]. Over the recent years, most of these protein complex prediction algorithms have focused on codifying protein–protein interaction information (e.g., protein coexpression and shared gene ontology) as edge weights in the PPI network [6]. They use temporal and structural information to give a specific weight to each interaction, but the problem is that it is not easy to integrate both temporal and structural information on the weight of a single edge or to cluster a network with multiple edge types.

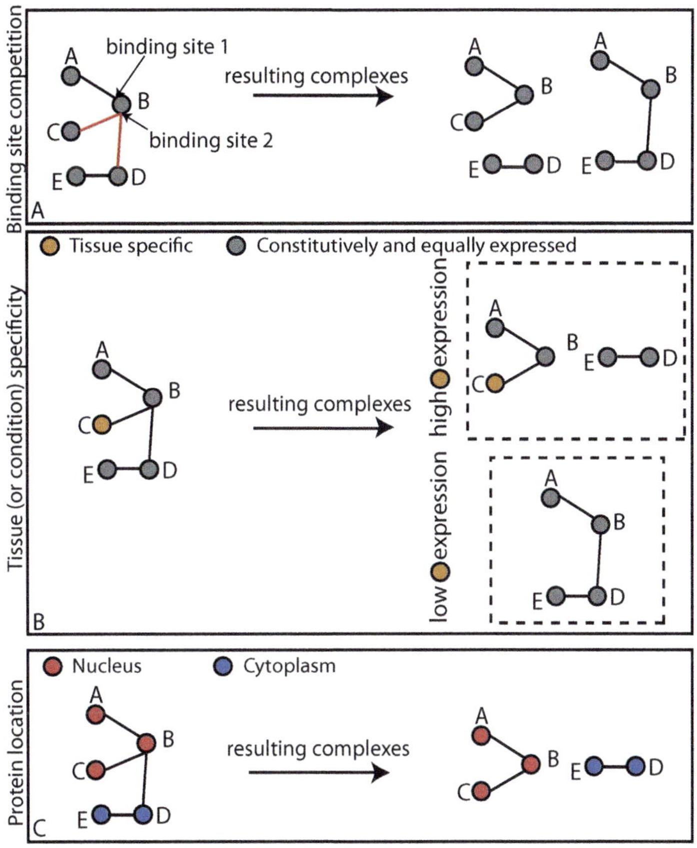

Fig. 1 Constraints that restrict protein complex formation. Example of how different constraints can generate different protein complex structures. Multiple interacting proteins often compete for the same binding sites (binding site competition), meaning that not every interaction can happen simultaneously (**a**). Moreover, the proteins involved in a PPI are not necessarily expressed at the same time (tissue or condition specificity) (**b**) and in the same location (protein location) (**c**). All these constraints can lead to different complexes

A more natural way to incorporate all this information is by performing protein–protein interaction simulations. PPI simulations have been performed at multiple scales according to the level of detail included. Molecular dynamics are now able to simulate the physical interactions on the atomic scale in the ribosome [7] as well as in other systems of similar complexity [8]. By reducing the level of detail in the simulations it is possible to increase the number of predicted complexes up to the level of an entire pathway or cellular compartment [9, 10].

The limitation of such detailed simulations is dictated by the complexity of protein complex formation. Due to the requirement of extremely high computational power it is possible to simulate only a few protein complexes in isolation. Only few studies performed proteome-wide simulations, but without predicting which proteins are involved in various protein complexes [11, 12].

Recently, a new simulation model, called SiComPre [13], has been proposed to simulate protein complexes on the whole proteome, considering a sufficiently detailed model of protein complex formation to predict more than half of the known yeast protein complexes and outperforming network-based methods. In this approach, different types of proteins diffuse, bind and unbind in a two-dimensional space based on stochastic rules. The interactions between two proteins are localized on specific interfaces defined in the model for each protein. Protein copy numbers, at a given tissue and cell cycle timepoint, can be included by defining different amounts of proteins in the simulation. Structural information, such as domain–domain interaction, is also included assigning domains as interfaces used by proteins to interact in the simulation.

In the next section, we review current methods that are able to simulate multiple protein complexes and highlight, types of information that have been generated from high-throughput techniques that can help increasing the quality of the prediction.

1.1 Protein Complex Prediction at Different Levels of Abstraction

As described in the introduction, current computational methods to predict protein complexes are ranging from highly detailed and low-scale methods to techniques that are less detailed but have the ability to consider the full proteome.

The development of computational methods followed two main streams of modeling. The first one focuses on a low-resolution proteome-wide network such as PPI and DDI (domain–domain interactions) to identify modules that resemble protein complexes. The second stream is mostly based on simulating the dynamic binding between proteins.

In Table 1, we summarize some of the key methods that have been proposed for each level of abstraction. Other computational methods that try to identify protein complexes directly from Affinity Purification—MS experiments are reviewed elsewhere [14].

1.2 Network-Based Complex Prediction

Over the recent years, a number of methods based on clustering of PPI networks have been developed to predict protein complexes. As described in a recent review, methods to predict protein complexes from PPI networks can be categorized into methods, which are based only on clustering the PPI like MCODE [15] and ClusterOne [16] and methods that integrate multiple biological information sources with the PPI network.

1.2.1 Methods Based on Clustering PPI Networks

Methods only based on clustering aim to identify regions of the PPI that form a dense subgraph that resembles protein complexes. MCODE (2003) [15] was one of the first methods utilized to cluster the PPI and identify protein complexes. In the following years a large number of clustering methods have been developed

Table 1
Representative methods for each approach used to predict protein complexes ordered by scale of prediction

Methods	Scale	Input information	Prediction
Multiscale modeling, Cα-based coarse-grained models + rigid body (RB) [29]	Multiprotein complexes with limited number of protein types	Binding rates	Dynamic formation of quaternary structures
Rule-based stochastic simulation [9, 10]	Pathway or organelle subset	Structural information network	Complexes formed during pathway cascade signaling or in a subset of a cellular compartment
Rigid-body simulation [12]	Full proteome	–	Effect of nonspecific interactions
ProtNet [27]	Full proteome	Static PPI	Complexome robustness and organization
MCODE [15]	Full proteome	Static PPI	Qualitative prediction of protein complexes
ClusterOne [16]	Full proteome	Static PPI	Qualitative prediction of protein complexes
Zhang et al. [23]	Full proteome	Dynamic PPI	Qualitative prediction of protein complexes
DPC-NADPIN [33]	Full proteome	Dynamic PPI	Qualitative prediction of protein complexes
SiComPre [13]	Full proteome	PPI, DDI, Protein Abundances, and GO	Qualitative and quantitative prediction of protein complexes

and protein complex detection from PPIs became a test case to prove the quality of clustering algorithms [16, 17]. Many of the methods in this category work both for weighted and unweighted networks. Therefore, even if these methods do not directly consider biological information, interaction confidences can be added through edge weights.

1.2.2 Clustering Methods That Include DDI and Other Structural Information

PPI networks often include hub proteins that can participate in a large number of interactions, however it is not feasible for a single protein to have so many simultaneous interactions, because interactors compete with each other for binding sites However, identification of binding sites for each pair of interacting proteins is extremely complicated. One of the easiest solutions for this problem is to decompose the PPIs into domain–domain interactions (DDIs). Domains are regions of the protein that fold and evolve independently [18]. Since proteins are often composed of multiple domains and binding sites usually correspond to one of its domain, protein domains are often used as a resource to approximate binding sites (Fig. 1a). Protein domain information was used in multiple studies to improve protein complex prediction. For example, Ozawa et al. [19] filtered out conflicting interactions and refined the prediction of clustering-based methods. Subsequently, other network-based approaches used that information to improve protein complexes. Ma et al. [20] also used the idea that a domain can participate only once in a complex, limiting the number of simultaneous PPI. Differently from Ozawa et al. [19], authors used the maximum matching algorithm to identify a set of DDI that does not violate the condition to have at maximum one interaction per domain. Another approach was implemented by Xu et al. [21] who first reconstructed a new protein–protein interaction network (PPIN) combining DDI information with GO annotation, gene coexpression, MS experiments, PPI network properties along with STRING database and then applied multiple state-of-the-art protein complex prediction algorithms to extract complexes from the new PPIN.

1.2.3 Clustering Methods That Include Spatiotemporal Information (Gene Expression)

As mentioned earlier, protein complexes are dynamic structures that change in response to external or internal stimuli, and during cell differentiation or cell-cycle progression. Indeed, proteins are usually only active at a specific temporal point and in specific positions of the cell. Therefore, not all PPIs in a PPI network take place simultaneously (Fig. 1b). A simple method to find proteins, which are active at a given time point is to assume that temporal distribution of the expression of each gene is normal and the probability of a gene to be active at a given time point is 0.68, 0.95, or 0.99 if the gene expression exceed k times the standard deviation, for $k = 1, 2, 3$ respectively [22]. Zhang et al. [23] proposed an alternative method to calculate the probability of proteins being active at a given time point from gene expression data obtained during the cell cycle and they used the resulting dynamic PPI network to predict protein complexes.

1.2.4 Methods That Integrate Multiple Types of Information

More recently, a number of groups developed methods to integrate multiple types of information in a multirelationship protein interaction network (MPIN), which contains different types of edges based on different information types [24, 25]. This is not a trivial task, because combining different edges into a single interaction is often misleading. To solve this problem, Zhang et al. [24] created a new PPI network combining high-throughput PPI data, gene ontology terms (as node attributes) with protein coexpression, genomic data, and text mining. Finally, the authors proposed a method called IMBP to identify protein complexes on this attributed network. After constructing the weighted MPIN, IMBP clusters the network starting from a set of seed cliques (a subset of nodes that are all connected to each other) and adding the closest node based on the closeness measure.

Li et al. [25] decomposed a weighted MPIN generated from gene ontology categories and PPIs into multiple single-relationship networks and predicted protein complexes with a novel clustering algorithm, called MINE which predict protein complexes by identifying dense subgraphs in each single-relationship network.

Network-based approaches can be also divided into other categories intersecting with the categories mentioned above. For instance, methods that consider edge weights [16] and those that use only binary information of PPI [15]. Furthermore, some network clustering methods allow proteins to belong to multiple complexes (overlapping clusters) such as ClusterOne [16], while others generate disjoint clusters like Markov Clustering (MCL) [26].

A more exhaustive review on network-based algorithms to predict protein complexes has been presented by Srihari and colleagues [6].

2 Simulation-Based Complex Prediction

A different approach to predict protein complexes is by simulating PPI and analyzing the complexes that can be observed during the simulation. One of the first studies adopting this approach was a dynamic simulation of protein interactions, which was able to reproduce the protein complex size distribution of complexes determined from a TAP experiment in yeast [11]. More recently, a rigid-body simulation was performed to study the impact of non-specific interactions on protein complex formation [12]. A network simulation software (ProtNet) considering the whole PPI dataset, but not actual protein abundances, enables to study properties of the complexome (complex size and reconstruction of some protein complexes) with different simulation settings [27]. Galeota et al. used the same software to simulate a PPI network showing that the organization and robustness of the protein complexes in the simulated PPI network are higher than in a random network [28].

To simulate protein complexes, it is important to consider that interactions occur at specific binding sites (Fig. 1a) of proteins.

Therefore, multiple interacting proteins are in competition for a limited number of binding sites. To model this phenomenon, it is common to decompose PPIs into domain–domain interactions. Using this approach Deeds et al. implemented a rule-based simulation of a subset of cytoplasmic proteins (167 proteins) and integrated this with protein copy numbers. In this model, the rules were controlling the association and dissociation of proteins. Stochastic simulations showed that simulations starting from single initial conditions can lead to formation of dissimilar complexes. One problem encountered in this simulation was the unexpected formation of extremely large polymers, which were deemed to not be physically possible in vivo.

To explain this problem of the simulation, consider a case in which a circular protein complex formed by the proteins A–B–C, such that A interacts with B, B with C and C interacts with A. All these interactions happen on different binding sites. The problem is that when the first two interactions occur (A with B and B with C) then the binding between A and C of the same complex is very unlikely, since there are many other copies of A and C competing with these. As a result of this, A, B, and C proteins form very long polymers with the repetition of this order. To avoid this issue the authors implemented a local rule to prevent long polymers [10]. However, this issue occurs only because spatial information is not considered as all proteins are considered in a well-mixed single compartment.

A multiscale approach utilized by Xie et al. [29] integrates a coarse-grained rigid body simulation to model the mesoscopic behavior of the system at a low scale. This method used binding rates of protein interactions as input and was employed to model the dynamic formation of heterotetramers.

These methods provide predictions of various properties of the complexome or a subset of it, but none of these generated protein complex predictions, which are comparable to proteome-wide, manually curated reference datasets of protein complexes [30–32], as it has been shown for network-based approaches [15, 16, 23, 33].

Recently, a proteome-wide spatial stochastic simulation (called SiComPre) was implemented to overcome some of these limitations [13]. SiComPre integrates protein copy numbers and protein domains with PPI networks and gene ontology information to predict protein complexes with higher performance scores than any other network-based methods. SiComPre runs a proteome-wide simulation where proteins can diffuse in a discretized two-dimensional space, performing complexation and decomplexation with other proteins (Fig. 2). Diffusion is calculated according Fick's law based on the diffusion rate of each protein, while complexation and decomplexation are simulated by Gillespie's algorithm [34]. Interactions between proteins are retrieved from

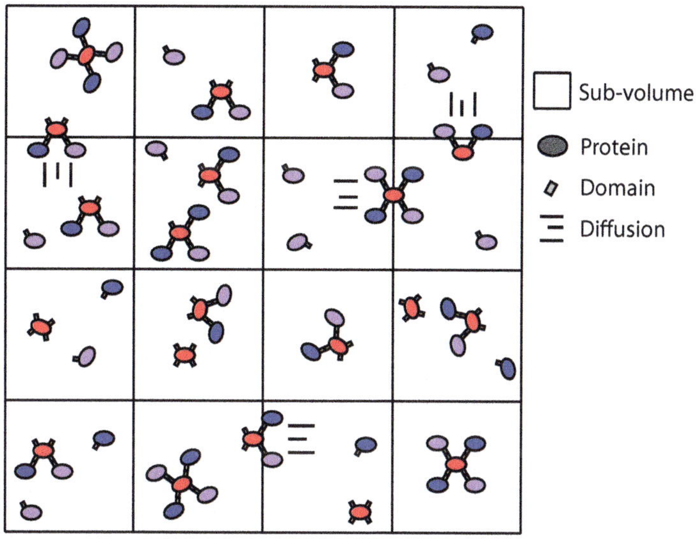

Fig. 2 SiComPre framework. In SiComPre [13] proteins are represented as abstract entities that reside in a 2D space divided into multiple subvolumes and multiple copies for each protein can be created. Proteins can interact with other proteins (represented by PPI and DDI data) in the same subvolume through specific interfaces that represent domains. The possible interactions are complexation and decomplexation forming protein complexes. Furthermore, complexes and proteins diffuse across subvolumes at discrete timepoints. Diffusion events occur according to Fick's law are calculated using the diffusion rate for each protein, while complexation and decomplexation are simulated with Gillespie algorithm [34]

an initial PPI network and decomposed into multiple DDIs. Protein domains are identified with SMART [35]. Often, pairs of interacting proteins in the PPI network do not contain any known interacting domains. In this cases, a fictitious pair of interacting domains will be added to the model, but only for those pairs that have in common at least one of the most nested term in the gene ontology tree (GO) database for Human and MIPS tree for Yeast [31].

It is also possible to add multiple copies of each protein to the simulation; therefore it is possible to observe multiple copies of protein complexes during the simulation. The number of copies of each simulated protein complexes is used as quantitative prediction of protein complexes. These predicted values can be compared to the limited information on abundances of protein complexes reported in the literature [36–41]. Based on the eight protein complexes abundances retrieved from the literature, this comparison showed a Pearson's correlation of 0.46 with the literature-derived values. In contrast to other techniques, allowing multiple copies of individual proteins gives the possibility to predict how many proteins of each type appear in a single complex. Comparing the results from different models is possible to predict the effect of perturbations on the complexome (e.g., predicting the effect of inhibiting a binding site). Furthermore, with this methodology the problem of long polymers was solved by considering spatial distribution of proteins in the simulation.

3 Data Sources That Can Be Integrated to Improve Predictions

Many of the recent algorithms are already integrating different proteome-scale resource in order to improve protein complex prediction. In this paragraph we list some of those resources and databases.

3.1 Binding Sites and Other Structural Information

Nearly all the methods presented above start from a PPIN. Some methods provide a score that quantifies the likelihood of the interactions, but most methods do not include any information related to the structure of proteins (e.g., protein domains). For instance, Hippie uses a confidence score of a PPI as weight, which is based on the amount and reliability of evidence that support each PPI [42]. Indeed, only a few of the interactions in a PPI network can occur simultaneously due to a limited number of binding sites available for each protein, and the fact that they are not necessarily present at the same time and cellular space. As mentioned earlier, a good strategy is to decompose a PPI network into DDIs. This information was used by network-based approaches and simulation-based methods [13] as binding interface of proteins during the simulation. A database storing such DDI networks is called DOMINE and contains a total of 26,219 interactions (among 5410 domains) derived from 3D structure and computational predictions [43]. Another database is IDDI and integrates more than 200,000 DDI from multiple datasets assigning also a score to the interaction based on the accuracy of each dataset, datasets similarity, and reliability of the DDI within the dataset [44]. In order to decompose a PPI network into DDIs it is also necessary to know which domains are in a protein. SMART [35] is a web-service that allows the identification of protein domains by manually integrating curated hidden Markov models. There could be also an issue if a protein has too many binding sites, it might not be able to provide enough binding surface to all its partners and steric inhibition can occur between sites. Only very detailed models consider this problem, because it requires taking into account the geometry of proteins and their biding sites.

Many complexes might not be observed because there are no interactions reported in the initial PPIs (*see* Table 2). To overcome this issue, one can include novel interactions from databases of computationally predicted protein complexes to extend the

Table 2
Types of proteome-scale datasets that can be included in computational methods to predict protein complexes

Type of information	Type of data	Method	Information
Structural	Database	DOMINE [43]	Domain–domain interactions
Structural	Database	Integrated domain-domain interactions (IDDI) [44]	Domain–domain interactions
Structural	Database/Web-service	SMART [35]	Protein domains annotation
Structural	Experimental procedure	Hein et al. [3]	Protein complex stoichiometries
Protein copy number	Database	Moped DB [59]	Tissue and organism specific protein copy number
Protein copy number	Database	Pax-DB [60]	Tissue and organism specific protein copy number
Protein copy number	Database	Human proteome map [54]	Tissue specific protein copy number in human
Protein expression and localization	Database	Protein atlas [49]	Protein expression and localization in multiple conditions
Protein localization	Database	COMPARTMENTS [61]	Protein subcellular localization
Structural	Computational prediction	PrePPI [45]	Predicted PPIs
Spatiotemporal	Experimental procedure	Single cell RNA-sequencing [66]	Genes co-expressed at specific time and condition
Posttranslational modifications	Database	dbPTM [75]	Experimentally verified post translational modifications

prediction. An example is PrePPI [45], a database containing more than 1.35 million predicted PPIs.

Anhert et al. [46] showed that the process of protein complex assembly follows three basic types of complexation: dimerization, cyclization and subunit addition. These findings can be used to build better models of protein complex formation. For instance, agent-based models [10] can define rules to match these three types of assembly.

Another important consideration is that binding and unbinding of proteins can happen at different rates. However, those rates are dependent on the tertiary structure of the proteins involved in the interaction and this information is not available on a large-scale yet. A compromised alternative for the moment is to split PPIs into two categories, stable and transient interactions. Stable interactions are almost irreversible and lead to the formation of stable protein complexes, while transient interactions disassemble quickly after a binding event. The latter are common in signaling pathways, serving to transmit the signal from the receptors to the nucleus or other intracellular signaling cascade [47]. Methods to identify transient protein interactions are summarized elsewhere [47].

Stoichiometry of protein complexes also plays an important role, not every protein appears with a 1:1 ratio in a complex. Analysis of these data can also help validation of protein complex predictions [2, 3].

3.2 Tissue Specificity and Temporal Information

Since there is competition between various proteins that bind to the same site on another protein, one of the key determinants of which protein complexes will form is the abundance of proteins. Abundance can depend on tissue type, cell cycle stage or cellular compartmentalization. In 2002, Ghaemmaghami et al. determined absolute abundance of Yeast proteins expressed during log growth phase [48] showing that protein abundances ranged from 50 to 10^6 copies per cell. An analysis performed on 32 different human tissues, which covered 90% of the putative protein-coding genes revealed that almost half of the genes are expressed in every tissue, but still more than 7000 genes are expressed only under specific conditions [49]. The constitutive, housekeeping genes are mostly associated with essential functions of the cell such as the ones performed by genes encoding ribosomal proteins or genes involved in key metabolic functions [50]. Tissue-specific gene expression will result in the inhibition or promotion of specific PPIs in a cell, therefore cells can use these regulatory mechanisms to alter the formation of protein complexes. Tissue-specificity provides valuable information to improve protein complex prediction (Fig. 1b).

Recent improvements in the sensitivity of MS-based proteomics have led to the development of many techniques for relative and absolute protein quantification [51], such as SILAC, iTRAQ [52], and iBAQ [53]. SILAC is a method to measure differences in protein abundance among samples using nonradioactive isotopic labeling, instead iTRAQ quantify relative and absolute quantitation of proteins with isobaric tags. iBAQ allows for intensity-based, absolute quantification of protein through MS without the need of labeling peptides. Over the recent years these methods have been applied in a variety of different systems generating a large number of protein abundance datasets [54–56]. Recently, iBAQ was used to quantify protein expression in 20 different breast cancer samples [57] and high-resolution Fourier-transform MS was used to measure protein abundances in 30 different human samples based on spectral count data, a common approach where the number of matched spectra to peptides is used as measurement of abundance [54]. Combining different MS-based proteome quantification methods it was possible to map protein abundances in *E. coli* in 22 different conditions. iBAQ-based proteomics was also used to measure the absolute protein quantification of cytotoxic T cells before and after mTORC1 inhibition with rapamycin treatment [58]. These types of analyses provide information that can be used to predict protein complexes in specific conditions.

Protein quantification datasets have been deposited into a variety of freely accessible databases. MOPED-DB [59] stores protein expression from four different organisms (*Homo sapiens*, *Mus musculus*, *S. cerevisiae*, and *C. elegans*) in more than 400 biological conditions. Another important repository of protein quantification datasets is Pax-db [60].

Computational methods which predict protein complexes can use this information to predict tissue-specific protein complexes. For instance, SiComPre can be initiated with tissue-specific protein abundances to predict differences in protein complex abundances.

3.3 Protein Compartmental Localization

Spatiotemporal protein expression is also an important feature that is becoming more readily available in recent years [61]. Indeed, proteins need to be expressed at the same time and in the same cellular localization in order to interact [61].

De Lichtenberg et al. [4] showed that many protein complexes assemble and disassemble dynamically during the cell cycle. Moreover, complexes are rarely synthesized entirely when needed but rather assembled "just-in-time" meaning that only some dynamic subunits of a complex are synthetized at specific cell cycle stages, allowing the reusability of the static component of the complex (Fig. 1c).

Interacting proteins not only have to be expressed in the same tissue at the same time, they also need to be located in the same cellular compartment. More protein localization datasets are being assembled, driven by high-throughput techniques such as microscopy-based screens, cDNA tagging ,and other high-throughput localization studies [61]. Another source of protein localization data comes from methods that predict their position in the cell based on the amino-acid sequence, which contain information for protein sorting [62]. The different format and nature in which protein localization data are provided makes challenging building an up-to-date database. To integrate all this information and keep it updated with the current literature, Binder et al. [61] created COMPARTMENTS, a database that collects data from high-throughput experiments, predictions, and automatic text mining and generates a comprehensive dataset of protein localization in multiple eukaryotes, including human and yeast.

3.4 Single-Cell Omics

Most of the data explained so far can be used to check if a PPI can occur in a specific condition and a specific time. However, the majority of these studies have been performed in bulk experiments, averaging the measurements of a heterogeneous cellular population. In recent years there has been an explosion in the number of different omics studies that operate at the single cell level [63–66]. Although the coverage of proteins that can be simultaneously measured at the single cell level is still low (for instance mass cytometry allows up to 45 simultaneous measurements [67]), gene expression can already be quantified at the full genomic scale. In many cases, single cell data can be directly integrated in the methods presented here to improve protein complex prediction.

3.5 Other Proteome-Wide Information

Interestingly, other types of Omics datasets show correlation with protein complexes [68]. Analysis of genetic interaction networks identified modules corresponding to protein complexes and predicted how perturbation in proteins form a single complex lead to common phenotypes [68]. Therefore, integrating genetic interactions can provide useful insight for predicting protein complexes. Furthermore, it is not always the case that proteins bind each other directly. For many interactions the binding is mediated by nonprotein molecules such as RNA, DNA, or lipids [69, 70]. Although this will severely increase the complexity of the model, integrating datasets of RNA, DNA or lipid-binding proteins such as ATtRACT [71] and NPIDB [72] can be beneficial for protein complex prediction.

PPIs can also be affected by posttranslational modifications (PTMs) which regulate protein complex formation. Yachie et al. [73] showed that phosphoproteins are more likely have a larger number of interactions and occupy a more central position in the PPIN than non-PTM proteins. Daun et al. [74] analyzed twelve types of PTMs in the context of PPI, revealing a preferential binding of specific PTMs with different regions of the PPIN. For instance, phosphorylated proteins are mostly located in the center of the PPIN, while glycolized proteins are more likely to be in the periphery. DbPTM integrates experimentally verified PTMs from several databases including also PTM sites and PPI affected by PTMs [75].

Recently, a few studies simultaneously measured various types of information, which are relevant for protein complex formation. Hein et al. [3] applied an experimental method called QUBIC to measure interaction stoichiometry, specificity and protein abundance in a HeLa cell. Another study [76] quantified protein abundances and stoichiometry in 11 human cell lines (including HeLa and HEK293 cells) and from cells sorted at five time points during the reprogramming of mouse embryonic fibroblasts into induced pluripotent stem cells.

Other tissue-specific information of protein expression in different organs and subcellular localization are collected in the Human Protein Atlas (www.proteinatlas.org) and contains information at the single cell level as well [49].

4 Prediction of Altered Complexes in Diseases

4.1 Dysfunctional Protein Complexes in Various Diseases

Protein complexes are the upper layer unit that performs biological functions. Genomic mutations lead to different protein structures, which can ultimately alter protein complex stability and functioning. When a protein complex is compromised in its function it can lead to diseases. For instance, the Crumb protein complex has been specifically associated with multiple types of neurodegenerative diseases [77]. Furthermore, protein localization is also regulated by the protein trafficking machinery and alterations in this structure lead to mislocalization of protein that can lead to a vast array of diseases including neurodegenerative diseases and cancer [78]. In a recent study, Rodina and colleagues [79] found that in some cancer cell lines, the heat shock protein 90 (HSP90) forms a large structure (called epichaperome) together with other proteins involved in protein folding. In other cancer and non-cancer cell lines HSP90 interacted with only a small set of proteins, but the prevalence of the epichaperome is not associated to cancer subtype, origin and genome in at least half of the tumors tested.

Identification of dysfunctional complexes that lead to disease requires a huge effort. Various computational tools have been deployed to help biologist identifying such complexes.

Many of these methods rely on the assumption that mutations in different subunits of a protein complex lead to the same or similar diseases. Under this assumption Lage et al. [80] implemented a Bayesian prediction to identify protein complexes associated with disease based on mutated genes in the protein complex that were previously associated with a similar phenotype. A different method to identify disease-associated protein complexes was able to rank candidate protein complexes associated with prostate cancer. Sixty-nine among the top 100 ranked complexes were known to be associated with this type of cancer [81].

Vinayagan et al. [82] combined RNA inference screens and protein complex enrichment analysis to identify that the SWI/SNF chromatin-remodeling complex participates in the insulin response.

Many diseases lead to both altered protein expression and structural mutations affecting the PPI network. SiComPre can simulate this behavior using as input the altered protein expression and PPIs predicting complexes that differ (in their composition or abundance) between healthy and diseased cells [13].

4.2 Altered Protein Complexes as Therapeutic Targets

Identification of dysfunctional protein complexes represents an opportunity to suggest novel therapeutic targets. Networks of "disease–protein complex" interactions have been analyzed to find new therapeutic targets [83, 84]. In these networks a node represents either a disease or a protein complex, and an edge means that a protein complex might be implicated in the disease. Wu et al. [83] reconstructed a network of drug-protein complexes–disease interactions, where nodes are single drugs, complexes or diseases and edges represent a connection between them [83]. Using this network as an example, the authors found a link between imatinib, a chemotherapy used for the treatment of chronic myelogenous leukemia and acute lymphocytic leukemia, and a protein complex associated with type I diabetes.

Computational prediction of protein complexes will improve our understanding of the role of dysfunctional protein complexes in at least two ways. First, most of the discussed prediction methods rely on a predefined dataset of protein complexes. Computational methods can further extend the dataset of protein complexes. Second, simulations-based prediction can also mimic the effect of a disease. For instance, by using protein abundances quantified in a pathogenic condition and simulating complexes while considering the interference with bindings caused by mutations.

Similar to a single genomic mutation which can remodel the cell [85], interference with complex formation reaction can have implications for other complexes as well. When a member of a protein complex is targeted by drugs, this can reshape the structure of the protein complex due to competition for binding partners. This effect can propagate throughout the whole complexome. Simulation-based methods, such as SiComPre, have the potential to capture this complexity and provide a prediction of protein complexes altered downstream of such perturbations [13].

5 Protein Complex Prediction with SiComPre

This section contains the instructions to run SiComPre and how to predict qualitative and quantitative protein complexes.

5.1 Installation and Requirement

1. Current SiComPre version supports only Windows and requires a graphics card supporting CUDA (https://developer.nvidia.com/cuda-gpus). To generate the model and analyze the results is also necessary to have Java (only version 1.8 was tested), Python (packages MySQLdb and SciPy are required), and MySQL installed.
2. From COSBI website (http://www.cosbi.eu/research/prototypes/sicompre) download SiComPre by filling the form and the database (DumpDB). Next, unzip both files.
3. The database contains the dump of the database that can be installed using one of the various interfaces for MySQL such as MySQL Workbench (https://dev.mysql.com/doc/workbench/en/wb-admin-export-import-management.html).

5.2 Model Generation

1. The next step is to generate the model. The java script will query the database to collect protein information such as protein abundance and PPI. To generate the model, open a terminal in the same directory where SiComPre is located and type

```
$ java -jar GenModel.jar -s 2 -i Collins -f
data/prot_collins.txt -o model_collins.xml
```

Parameter "-f" indicates the location of the file containing the list of proteins (one for each line) to include in the model. Output model is saved in the file indicated after the parameter "-o".

2. A manual explaining each parameter can be found in the README.txt file in the SiComPre directory.
3. It is possible to run SiComPre with a personalized input by manually changing the values of the following tables in the database:
   - Table "abundance" contains the protein abundance for each protein.
   - Table "collinsppi" stores the binary PPI information [86].
   - "iddi" contains DDI from the IDDI dataset [44].
   - In the table "protdomain" there are all protein-domain association collected from SmartDB [35].

5.3 Simulation

1. The simulation step is performed in the command line:

```
$ SiComPre10.exe -f model_collins.xml
```

where "model_collins.xml" is the generated model. For large models, it is possible to use instead "SiComPre_1_1b.exe" script.

2. Once the simulation has finished, the protein complexes can be converted in a readable format with the command:

```
$ java -jar Results.jar
```

The resulting complexes are stored in file "results_C1.txt".

3. To improve qualitative prediction of the obtained complexes, SiComPre allows a refinement step:

```
$ java -jar FrequencyMatrix.jar
complex_structure.out
$ IPCAw.exe -Gint_network.txt -S2 -P2 -T0.5
-Oresults_quali.txt
```

complex_structure.out is generated during the simulation, int_network.txt is generated from the FrequencyMatrix.jar script and results_quali.txt contains the refined complexes.

5.4 Results Analysis

1. The script generateCSV.py can be used to write a report of the predicted complexes together with their quantitative prediction

```
$ python generateCSV.py results_quali.txt
results_C1.txt int_network.txt fictitious_
interactions.txt funcat.txt n > prediction.csv
```

results_quali.txt is generated from the IPCAw.exe file, while results_C1.txt is the output of Results.jar. "int_network.txt" is the FrequencyMatrix output and funcat.txt is a file that stored in the "data" folder containing the protein function used for functional enrichment of protein complexes.

2. It is possible to further optimize the results by removing large complexes with low abundance using the command.

```
$ python optimize.py results_quali.txt
results_C1_1.txt 6 16 c > final_
prediction.txt
```

This part filters out complexes having an abundance > =6 or complex size < =16.

3. Finally, users can check the results by comparing predicted complexes to a reference dataset of protein complexes.

```
$ python match_standalone.py -n collins2007.txt
CYC08.txt final_prediction.txt
```

collins2007.txt is the initial PPI network, CYC08.txt is the reference dataset of protein complexes, and final_prediction.txt is the final qualitative prediction of protein complex.

After these steps, proteins of each predicted complex can be found in the "final_prediction.txt" file.

6 Discussion

Computational prediction can expand the identified number of protein complexes and complement conditions covered by experimental detection of protein complexes. In *D. melanogaster*, only a 15% overlap has been identified between the set of complexes detected from computational-based predictions and a manually curated list [82]. SiComPre was able to identify 890 new complexes that were not found in the CORUM dataset of manually curated protein complexes [13]. Therefore, computational methods have the potential to predict complexes not yet identified by experimental techniques, although some of these might be artifacts of the high-throughput data source used and might not appear in the in vivo situation.

In this chapter, we briefly explore the two main streams of methods that have been developed to predict protein complexes at the proteome scale. These two streams have always been separated in term of type of predictions.

Over the recent years, network-based approaches have mostly focused on codifying multiple resources, such as spatiotemporal dynamics and structural properties into PPI networks. However, simulation based approaches, such as SiComPre can explicitly incorporate in their models protein abundances, and domains or other binding sites, providing condition-specific prediction of protein complexes. This depth, however, comes at the cost of higher complexity. Simulation-based methods require much higher computational power than network based methods and can further increase the level abstraction used to model protein complex formation.

Differences in gene expression or therapies that inhibit specific proteins generate perturbations that are very unlikely to affect only the local structure of protein complexes. PPI networks are robust to random perturbations [87], but small variations in the complexome and in the initial conditions can lead to variations that propagate through the whole system. Namely, variation in the complexome cannot be measured just by averaging the difference of its components, but individual reactions have to be considered. For this reason, simulation based methods could be a more suitable to analyze these kinds of perturbations.

As these computational methods improve in accuracy, predicted protein complexes from context-specific information will be used to quickly test the effects of perturbations, such as gene knockouts or inhibition of specific interactions by drug treatment providing insight into possible therapeutic targets.

Unfortunately, it is not yet possible to use highly detailed models to perform proteome-wide simulations. Therefore, predictions should mostly be considered as novel hypotheses that should be validated by more accurate methods (Fig. 3).

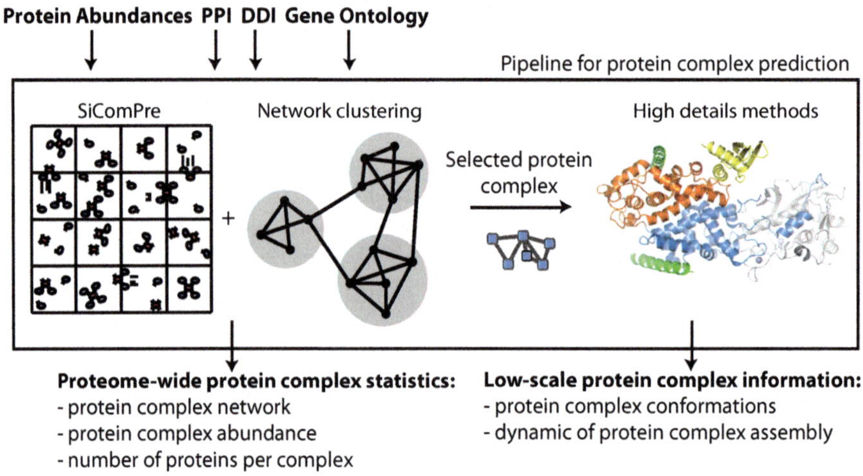

Fig. 3 Global workflow for improved protein complex predictions. Large scale models such as SiComPre [13] and ClusterOne [16] can be used to quickly generate hypothesis on possible new protein complexes or those that are altered in diseases. Then, these selected complexes can be analyzed in more detail to draw stronger predictions on protein complexes structure

References

1. Gavin AC, Aloy P, Grandi P, Krause R, Boesche M, Marzioch M, Rau C, Jensen LJ, Bastuck S, Dumpelfeld B, Edelmann A, Heurtier MA, Hoffman V, Hoefert C, Klein K, Hudak M, Michon AM, Schelder M, Schirle M, Remor M, Rudi T, Hooper S, Bauer A, Bouwmeester T, Casari G, Drewes G, Neubauer G, Rick JM, Kuster B, Bork P, Russell RB, Superti-Furga G (2006) Proteome survey reveals modularity of the yeast cell machinery. Nature 440(7084):631–636. https:// doi.org/10.1038/nature04532
2. Havugimana Pierre C, Hart GT, Nepusz T, Yang H, Turinsky Andrei L, Li Z, Wang Peggy I, Boutz Daniel R, Fong V, Phanse S, Babu M, Craig Stephanie A, Hu P, Wan C, Vlasblom J, Dar V-u-N, Bezginov A, Clark Gregory W, Wu Gabriel C, Wodak Shoshana J, Tillier Elisabeth RM, Paccanaro A, Marcotte Edward M, Emili A (2012) A census of human soluble protein complexes. Cell 150(5):1068–1081. https:// doi.org/10.1016/j.cell.2012.08.011
3. Hein Marco Y, Hubner Nina C, Poser I, Cox J, Nagaraj N, Toyoda Y, Gak Igor A, Weisswange I, Mansfeld J, Buchholz F, Hyman Anthony A, Mann M (2015) A human interactome in three quantitative dimensions organized by stoichiometries and abundances. Cell 163(3):712–723. https:// doi.org/10.1016/j.cell.2015.09.053
4. de Lichtenberg U, Jensen LJ, Brunak S, Bork P (2005) Dynamic complex formation during the yeast cell cycle. Science 307(5710):724–727. https:// doi.org/10.1126/science.1105103
5. Bader GD, Hogue CW (2002) Analyzing yeast protein–protein interaction data obtained from different sources. Nat Biotechnol 20(10):991–997
6. Srihari S, Yong CH, Patil A, Wong L (2015) Methods for protein complex prediction and their contributions towards understanding the organisation, function and dynamics of complexes. FEBS Lett. https:// doi.org/10.1016/j.febslet.2015.04.026
7. Sanbonmatsu KY (2012) Computational studies of molecular machines: the ribosome. Curr Opin Struct Biol 22(2):168–174. https:// doi.org/10.1016/j.sbi.2012.01.008
8. Perilla JR, Goh BC, Cassidy CK, Liu B, Bernardi RC, Rudack T, Yu H, Wu Z, Schulten K (2015) Molecular dynamics simulations of large macromolecular complexes. Curr Opin Struct Biol 31:64–74. https:// doi.org/10.1016/j.sbi.2015.03.007
9. Suderman R, Deeds EJ (2013) Machines vs. ensembles: effective MAPK signaling through heterogeneous sets of protein complexes. PLoS Comput Biol 9(10):e1003278. https:// doi.org/10.1371/journal.pcbi.1003278

10. Deeds EJ, Krivine J, Feret J, Danos V, Fontana W (2012) Combinatorial complexity and compositional drift in protein interaction networks. PLoS One 7(3):e32032. https://doi.org/10.1371/journal.pone.0032032
11. Beyer A, Wilhelm T (2004) Dynamic simulation of protein complex formation on a genomic scale. Bioinformatics 21(8):1610–1616. https://doi.org/10.1093/bioinformatics/bti223
12. Osmanović D, Rabin Y (2016) Effect of non-specific interactions on formation and stability of specific complexes. J Chem Phys 144(20):205104. https://doi.org/10.1063/1.4952981
13. Rizzetto S, Priami C, Csikasz-Nagy A (2015) Qualitative and quantitative protein complex prediction through proteome-wide simulations. PLoS Comput Biol 11(10):e1004424. https://doi.org/10.1371/journal.pcbi.1004424
14. Clancy T, Hovig E (2014) From proteomes to complexomes in the era of systems biology. Proteomics 14(1):24–41. https://doi.org/10.1002/pmic.201300230
15. Bader GD, Hogue CW (2003) An automated method for finding molecular complexes in large protein interaction networks. BMC Bioinformatics 4:2
16. Nepusz T, Yu H, Paccanaro A (2012) Detecting overlapping protein complexes in protein-protein interaction networks. Nat Methods 9(5):471–472
17. Palla G, Derényi I, Farkas I, Vicsek T (2005) Uncovering the overlapping community structure of complex networks in nature and society. Nature 435(7043):814–818. https://doi.org/10.1038/nature03607
18. Richardson JS (1981) The anatomy and taxonomy of protein structure. Adv Protein Chem 34:167–339. https://doi.org/10.1016/s0065-3233(08)60520-3
19. Ozawa Y, Saito R, Fujimori S, Kashima H, Ishizaka M, Yanagawa H, Miyamoto-Sato E, Tomita M (2010) Protein complex prediction via verifying and reconstructing the topology of domain-domain interactions. BMC Bioinformatics 11:350. https://doi.org/10.1186/1471-2105-11-350
20. Ma W, McAnulla C, Wang L (2012) Protein complex prediction based on maximum matching with domain-domain interaction. Biochim Biophys Acta 1824(12):1418–1424. https://doi.org/10.1016/j.bbapap.2012.06.009
21. Xu B, Lin H, Chen Y, Yang Z, Liu H (2013) Protein complex identification by integrating protein-protein interaction evidence from multiple sources. PLoS One 8(12):e83841. https://doi.org/10.1371/journal.pone.0083841
22. Wang J, Peng X, Li M, Pan Y (2013) Construction and application of dynamic protein interaction network based on time course gene expression data. Proteomics 13(2):301–312. https://doi.org/10.1002/pmic.201200277
23. Zhang Y, Lin H, Yang Z, Wang J (2016) Construction of dynamic probabilistic protein interaction networks for protein complex identification. BMC Bioinformatics 17(1). https://doi.org/10.1186/s12859-016-1054-1
24. Zhang Y, Lin H, Yang Z, Wang J, Xu B (2013) Integrating multiple biomedical resources for protein complex prediction. IEEE, Shanghai, pp 456–459. https://doi.org/10.1109/bibm.2013.6732535
25. Li X, Wang J, Zhao B, Wu F-X, Pan Y (2016) Identification of protein complexes from multi-relationship protein interaction networks. Hum Genomics 10(S2). https://doi.org/10.1186/s40246-016-0069-z
26. SV Dongen (2000) Graph clustering by flow simulation. Ph.D. thesis, University of Utrecht
27. Bernaschi M, Castiglione F, Ferranti A, Gavrila C, Tinti M, Cesareni G (2007) ProtNet: a tool for stochastic simulations of protein interaction networks dynamics. BMC Bioinformatics 8(Suppl 1):S4. https://doi.org/10.1186/1471-2105-8-s1-s4
28. Galeota E, Gravila C, Castiglione F, Bernaschi M, Cesareni G (2015) The hierarchical organization of natural protein interaction networks confers self-organization properties on pseudocells. BMC Syst Biol 9(Suppl 3):S3. https://doi.org/10.1186/1752-0509-9-s3-s3
29. Xie Z-R, Chen J, Wu Y (2016) Multiscale model for the assembly kinetics of protein complexes. J Phys Chem B 120(4):621–632. https://doi.org/10.1021/acs.jpcb.5b08962
30. Ruepp A, Waegele B, Lechner M, Brauner B, Dunger-Kaltenbach I, Fobo G, Frishman G, Montrone C, Mewes HW (2010) CORUM: the comprehensive resource of mammalian protein complexes-2009. Nucleic Acids Res 38:D497–D501. https://doi.org/10.1093/Nar/Gkp914
31. Mewes H-W, Frishman D, Mayer KF, Münsterkötter M, Noubibou O, Pagel P, Rattei T, Oesterheld M, Ruepp A, Stümpflen V (2006)

MIPS: analysis and annotation of proteins from whole genomes in 2005. Nucleic Acids Res 34(suppl 1):D169–D172
32. Pu S, Wong J, Turner B, Cho E, Wodak SJ (2009) Up-to-date catalogues of yeast protein complexes. Nucleic Acids Res 37(3):825–831. https:// doi.org/10.1093/nar/gkn1005
33. Shen X, Yi L, Jiang X, Zhao Y, Hu X, He T, Yang J (2016) Neighbor affinity based algorithm for discovering temporal protein complex from dynamic PPI network. Methods 110:90–96. https:// doi.org/10.1016/j.ymeth.2016.06.010
34. Gillespie DT (2001) Approximate accelerated stochastic simulation of chemically reacting systems. J Chem Phys 115(4):1716. https:// doi.org/10.1063/1.1378322
35. Letunic I, Doerks T, Bork P (2014) SMART: recent updates, new developments and status in 2015. Nucleic Acids Res 43(D1):D257–D260. https:// doi.org/10.1093/nar/gku949
36. von der Haar T (2008) A quantitative estimation of the global translational activity in logarithmically growing yeast cells. BMC Syst Biol 2:87. https:// doi.org/10.1186/1752-0509-2-87
37. Jackson DA, Pombo A, Iborra F (2000) The balance sheet for transcription: an analysis of nuclear RNA metabolism in mammalian cells. FASEB J 14(2):242–254. https:// doi.org/10.1096/fj.1530-6860
38. Lee TI, Young RA (2000) Transcription of eukaryotic protein-coding genes. Annu Rev Genet 34:77–137. https:// doi.org/10.1146/annurev.genet.34.1.77
39. Adam SA (2001) The nuclear pore complex. Genome Biol 2(9):REVIEWS0007. https:// doi.org/10.1186/gb-2001-2-9-reviews0007
40. Deng C, Xiong X, Krutchinsky AN (2009) Unifying fluorescence microscopy and mass spectrometry for studying protein complexes in cells. Mol Cell Proteomics 8(6):1413–1423. https:// doi.org/10.1074/mcp.M800397-MCP200
41. Mavrich TN, Ioshikhes IP, Venters BJ, Jiang C, Tomsho LP, Qi J, Schuster SC, Albert I, Pugh BF (2008) A barrier nucleosome model for statistical positioning of nucleosomes throughout the yeast genome. Genome Res 18(7):1073–1083. https:// doi.org/10.1101/gr.078261.108
42. Schaefer MH, Fontaine JF, Vinayagam A, Porras P, Wanker EE, Andrade-Navarro MA (2012) HIPPIE: integrating protein interaction networks with experiment based quality scores. PLoS One 7(2):e31826. https:// doi.org/10.1371/journal.pone.0031826
43. Yellaboina S, Tasneem A, Zaykin DV, Raghavachari B, Jothi R (2010) DOMINE: a comprehensive collection of known and predicted domain-domain interactions. Nucleic Acids Res 39(Database):D730–D735. https:// doi.org/10.1093/nar/gkq1229
44. Kim Y, Min B, Yi GS (2012) IDDI: integrated domain-domain interaction and protein interaction analysis system. Proteome Sci 10(Suppl 1):S9. https:// doi.org/10.1186/1477-5956-10-S1-S9
45. Garzón JI, Deng L, Murray D, Shapira S, Petrey D, Honig B (2016) A computational interactome and functional annotation for the human proteome. eLife 5:pii: e18715. https:// doi.org/10.7554/eLife.18715
46. Ahnert SE, Marsh JA, Hernandez H, Robinson CV, Teichmann SA (2015) Principles of assembly reveal a periodic table of protein complexes. Science 350(6266):aaa2245. https:// doi.org/10.1126/science.aaa2245
47. Acuner Ozbabacan SE, Engin HB, Gursoy A, Keskin O (2011) Transient protein-protein interactions. Protein Eng Des Sel 24(9):635–648. https:// doi.org/10.1093/protein/gzr025
48. Ghaemmaghami S, Huh W-K, Bower K, Howson RW, Belle A, Dephoure N, O'Shea EK, Weissman JS (2003) Global analysis of protein expression in yeast. Nature 425(6959):737–741
49. Uhlen M, Fagerberg L, Hallstrom BM, Lindskog C, Oksvold P, Mardinoglu A, Sivertsson A, Kampf C, Sjostedt E, Asplund A, Olsson I, Edlund K, Lundberg E, Navani S, Szigyarto CAK, Odeberg J, Djureinovic D, Takanen JO, Hober S, Alm T, Edqvist PH, Berling H, Tegel H, Mulder J, Rockberg J, Nilsson P, Schwenk JM, Hamsten M, von Feilitzen K, Forsberg M, Persson L, Johansson F, Zwahlen M, von Heijne G, Nielsen J, Ponten F (2015) Tissue-based map of the human proteome. Science 347(6220):1260419–1260419. https:// doi.org/10.1126/science.1260419
50. Eisenberg E, Levanon EY (2013) Human housekeeping genes, revisited. Trends Genet 29(10):569–574. https:// doi.org/10.1016/j.tig.2013.05.010
51. Mayne J, Ning Z, Zhang X, Starr AE, Chen R, Deeke S, Chiang C-K, Xu B, Wen M, Cheng K, Seebun D, Star A, Moore

JI, Figeys D (2016) Bottom-up proteomics (2013–2015): keeping up in the era of systems biology. Anal Chem 88(1):95–121. https://doi.org/10.1021/acs.analchem.5b04230
52. Zieske LR (2006) A perspective on the use of iTRAQTM reagent technology for protein complex and profiling studies. J Exp Bot 57(7):1501–1508. https://doi.org/10.1093/jxb/erj168
53. Schwanhäusser B, Busse D, Li N, Dittmar G, Schuchhardt J, Wolf J, Chen W, Selbach M (2011) Global quantification of mammalian gene expression control. Nature 473(7347):337–342. https://doi.org/10.1038/nature10098
54. Kim MS, Pinto SM, Getnet D, Nirujogi RS, Manda SS, Chaerkady R, Madugundu AK, Kelkar DS, Isserlin R, Jain S, Thomas JK, Muthusamy B, Leal-Rojas P, Kumar P, Sahasrabuddhe NA, Balakrishnan L, Advani J, George B, Renuse S, Selvan LD, Patil AH, Nanjappa V, Radhakrishnan A, Prasad S, Subbannayya T, Raju R, Kumar M, Sreenivasamurthy SK, Marimuthu A, Sathe GJ, Chavan S, Datta KK, Subbannayya Y, Sahu A, Yelamanchi SD, Jayaram S, Rajagopalan P, Sharma J, Murthy KR, Syed N, Goel R, Khan AA, Ahmad S, Dey G, Mudgal K, Chatterjee A, Huang TC, Zhong J, Wu X, Shaw PG, Freed D, Zahari MS, Mukherjee KK, Shankar S, Mahadevan A, Lam H, Mitchell CJ, Shankar SK, Satishchandra P, Schroeder JT, Sirdeshmukh R, Maitra A, Leach SD, Drake CG, Halushka MK, Prasad TS, Hruban RH, Kerr CL, Bader GD, Iacobuzio-Donahue CA, Gowda H, Pandey A (2014) A draft map of the human proteome. Nature 509(7502):575–581. https://doi.org/10.1038/nature13302
55. Beck M, Schmidt A, Malmstroem J, Claassen M, Ori A, Szymborska A, Herzog F, Rinner O, Ellenberg J, Aebersold R (2011) The quantitative proteome of a human cell line. Mol Syst Biol 7(1):549. https://doi.org/10.1038/msb.2011.82
56. Schmidt A, Kochanowski K, Vedelaar S, Ahrne E, Volkmer B, Callipo L, Knoops K, Bauer M, Aebersold R, Heinemann M (2016) The quantitative and condition-dependent Escherichia coli proteome. Nat Biotechnol 34(1):104–110. https:// doi.org/10.1038/nbt.3418
57. Lawrence Robert T, Perez Elizabeth M, Hernández D, Miller Chris P, Haas Kelsey M, Irie Hanna Y, Lee S-I, Blau CA, Villén J (2015) The proteomic landscape of triple-negative breast cancer. Cell Rep 11(4):630–644. https://doi.org/10.1016/j.celrep.2015.03.050
58. Hukelmann JL, Anderson KE, Sinclair LV, Grzes KM, Murillo AB, Hawkins PT, Stephens LR, Lamond AI, Cantrell DA (2016) The cytotoxic T cell proteome and its shaping by the kinase mTOR. Nat Immunol 17(1):104–112. https:// doi.org/10.1038/ni.3314
59. Kolker E, Higdon R, Haynes W, Welch D, Broomall W, Lancet D, Stanberry L, Kolker N (2012) MOPED: model organism protein expression database. Nucleic Acids Res 40(Database issue):D1093–D1099. https://doi.org/10.1093/nar/gkr1177
60. Wang M, Weiss M, Simonovic M, Haertinger G, Schrimpf SP, Hengartner MO, von Mering C (2012) PaxDb, a database of protein abundance averages across all three domains of life. Mol Cell Proteomics 11(8):492–500. https://doi.org/10.1074/mcp.O111.014704
61. Binder JX, Pletscher-Frankild S, Tsafou K, Stolte C, O'Donoghue SI, Schneider R, Jensen LJ (2014) COMPARTMENTS: unification and visualization of protein subcellular localization evidence. Database 2014:bau012. https://doi.org/10.1093/database/bau012
62. Emanuelsson O, Nielsen H, Brunak S, von Heijne G (2000) Predicting subcellular localization of proteins based on their N-terminal amino acid sequence. J Mol Biol 300(4):1005–1016. https:// doi.org/10.1006/jmbi.2000.3903
63. Darmanis S, Gallant Caroline J, Marinescu Voichita D, Niklasson M, Segerman A, Flamourakis G, Fredriksson S, Assarsson E, Lundberg M, Nelander S, Westermark B, Landegren U (2016) Simultaneous multiplexed measurement of RNA and proteins in single cells. Cell Rep 14(2):380–389. https://doi.org/10.1016/j.celrep.2015.12.021
64. Dey SS, Kester L, Spanjaard B, Bienko M, van Oudenaarden A (2015) Integrated genome and transcriptome sequencing of the same cell. Nat Biotechnol 33(3):285–289. https://doi.org/10.1038/nbt.3129
65. Hou Y, Guo H, Cao C, Li X, Hu B, Zhu P, Wu X, Wen L, Tang F, Huang Y, Peng J (2016) Single-cell triple omics sequencing reveals genetic, epigenetic, and transcriptomic heterogeneity in hepatocellular carcinomas. Cell Res 26(3):304–319. https://doi.org/10.1038/cr.2016.23
66. Satija R, Farrell JA, Gennert D, Schier AF, Regev A (2015) Spatial reconstruction

of single-cell gene expression data. Nat Biotechnol 33(5):495–502. https:// doi.org/10.1038/nbt.3192
67. Newell EW, Cheng Y (2016) Mass cytometry: blessed with the curse of dimensionality. Nat Immunol 17(8):890–895. https:// doi.org/10.1038/ni.3485
68. Costanzo M, VanderSluis B, Koch EN, Baryshnikova A, Pons C, Tan G, Wang W, Usaj M, Hanchard J, Lee SD, Pelechano V, Styles EB, Billmann M, van Leeuwen J, van Dyk N, Lin ZY, Kuzmin E, Nelson J, Piotrowski JS, Srikumar T, Bahr S, Chen Y, Deshpande R, Kurat CF, Li SC, Li Z, Usaj MM, Okada H, Pascoe N, San Luis BJ, Sharifpoor S, Shuteriqi E, Simpkins SW, Snider J, Suresh HG, Tan Y, Zhu H, Malod-Dognin N, Janjic V, Przulj N, Troyanskaya OG, Stagljar I, Xia T, Ohya Y, Gingras AC, Raught B, Boutros M, Steinmetz LM, Moore CL, Rosebrock AP, Caudy AA, Myers CL, Andrews B, Boone C (2016) A global genetic interaction network maps a wiring diagram of cellular function. Science 353(6306). https:// doi.org/10.1126/science.aaf1420
69. Chiu YL, Cao H, Rana TM (2007) Quantitative analysis of RNA-mediated protein-protein interactions in living cells by FRET. Chem Biol Drug Des 69(4):233–239. https:// doi.org/10.1111/j.1747-0285.2007.00501.x
70. Nilsson T, Lundin CR, Nordlund G, Adelroth P, von Ballmoos C, Brzezinski P (2016) Lipid-mediated protein-protein interactions modulate respiration-driven ATP synthesis. Sci Rep 6:24113. https:// doi.org/10.1038/srep24113
71. Giudice G, Sánchez-Cabo F, Torroja C, Lara-Pezzi E (2016) ATtRACT—a database of RNA-binding proteins and associated motifs. Database 2016:baw035. https:// doi.org/10.1093/database/baw035
72. Zanegina O, Kirsanov D, Baulin E, Karyagina A, Alexeevski A, Spirin S (2016) An updated version of NPIDB includes new classifications of DNA–protein complexes and their families. Nucleic Acids Res 44(D1):D144–D153. https:// doi.org/10.1093/nar/gkv1339
73. Yachie N, Saito R, Sugiyama N, Tomita M, Ishihama Y (2011) Integrative features of the yeast phosphoproteome and protein-protein interaction map. PLoS Comput Biol 7(1):e1001064. https:// doi.org/10.1371/journal.pcbi.1001064
74. Duan G, Walther D (2015) The roles of post-translational modifications in the context of protein interaction networks. PLoS Comput Biol 11(2):e1004049. https:// doi.org/10.1371/journal.pcbi.1004049
75. Huang KY, Su MG, Kao HJ, Hsieh YC, Jhong JH, Cheng KH, Huang HD, Lee TY (2016) dbPTM 2016: 10-year anniversary of a resource for post-translational modification of proteins. Nucleic Acids Res 44(D1):D435–D446. https:// doi.org/10.1093/nar/gkv1240
76. Ori A, Iskar M, Buczak K, Kastritis P, Parca L, Andrés-Pons A, Singer S, Bork P, Beck M (2016) Spatiotemporal variation of mammalian protein complex stoichiometries. Genome Biol 17(1). https:// doi.org/10.1186/s13059-016-0912-5
77. Gosens I, den Hollander AI, Cremers FP, Roepman R (2008) Composition and function of the crumbs protein complex in the mammalian retina. Exp Eye Res 86(5):713–726. https:// doi.org/10.1016/j.exer.2008.02.005
78. Hung MC, Link W (2011) Protein localization in disease and therapy. J Cell Sci 124(20):3381–3392. https:// doi.org/10.1242/jcs.089110
79. Rodina A, Wang T, Yan P, Gomes ED, Dunphy MPS, Pillarsetty N, Koren J, Gerecitano JF, Taldone T, Zong H, Caldas-Lopes E, Alpaugh M, Corben A, Riolo M, Beattie B, Pressl C, Peter RI, Xu C, Trondl R, Patel HJ, Shimizu F, Bolaender A, Yang C, Panchal P, Farooq MF, Kishinevsky S, Modi S, Lin O, Chu F, Patil S, Erdjument-Bromage H, Zanzonico P, Hudis C, Studer L, Roboz GJ, Cesarman E, Cerchietti L, Levine R, Melnick A, Larson SM, Lewis JS, Guzman ML, Chiosis G (2016) The epichaperome is an integrated chaperome network that facilitates tumour survival. Nature 538(7625):397–401. https:// doi.org/10.1038/nature19807
80. Lage K, Karlberg EO, Størling ZM, Ólason P′i, Pedersen AG, Rigina O, Hinsby AM, Tümer Z, Pociot F, Tommerup N, Moreau Y, Brunak S (2007) A human phenome-interactome network of protein complexes implicated in genetic disorders. Nat Biotechnol 25(3):309–316. https:// doi.org/10.1038/nbt1295
81. Le DH (2015) A novel method for identifying disease associated protein complexes based on functional similarity protein complex networks. Algorithms Mol Biol 10:14. https:// doi.org/10.1186/s13015-015-0044-6

82. Vinayagam A, Hu Y, Kulkarni M, Roesel C, Sopko R, Mohr SE, Perrimon N (2013) Protein complex-based analysis framework for high-throughput data sets. Sci Signal 6(264):rs5. https:// doi.org/10.1126/scisignal.2003629
83. Wu M, Yu Q, Li X, Zheng J, Huang JF, Kwoh CK (2013) Benchmarking human protein complexes to investigate drug-related systems and evaluate predicted protein complexes. PLoS One 8(2):e53197. https:// doi.org/10.1371/journal.pone.0053197
84. Nacher JC, Schwartz JM (2012) Modularity in protein complex and drug interactions reveals new polypharmacological properties. PLoS One 7(1):e30028. https:// doi.org/10.1371/journal.pone.0030028
85. Hart JR, Zhang Y, Liao L, Ueno L, Du L, Jonkers M, Yates JR, Vogt PK (2015) The butterfly effect in cancer: a single base mutation can remodel the cell. Proc Natl Acad Sci 112(4):1131–1136. https:// doi.org/10.1073/pnas.1424012112
86. Collins SR, Kemmeren P, Zhao X-C, Greenblatt JF, Spencer F, Holstege FC, Weissman JS, Krogan NJ (2007) Toward a comprehensive atlas of the physical interactome of Saccharomyces cerevisiae. Mol Cell Proteomics 6(3):439–450
87. Azevedo H, Moreira-Filho CA (2015) Topological robustness analysis of protein interaction networks reveals key targets for overcoming chemotherapy resistance in glioma. Sci Rep 5:16830. https:// doi.org/10.1038/srep16830

Chapter 14

Computational Models of Cell Cycle Transitions

Rosa Hernansaiz-Ballesteros, Kirsten Jenkins, and Attila Csikász-Nagy

Abstract

The cell cycle is one of the best understood cellular processes in biology. Many of the key interactions occurring throughout the cell cycle have already been identified. This feature makes the system ideally suited for modelers who can use all the available interaction knowledge to build a systems level model of the underlying molecular regulatory network. This model can serve to identify gaps in our knowledge and to test theoretical assumptions or constrain the space of possible solutions. The cell cycle is a repetitive chain of events that goes through several checkpoints. Thus, the cell cycle can be studied under the perspective of an oscillator with checkpoints built into it, or as a series of switch-like transitions that goes from one state to another, converging on a closed loop. We shall discuss that latter position and present a framework for building and analyzing differential equation models of switch-like behavior. We shall then apply and review diverse models for each of the cell cycle transitions and discuss how multiple switches are combined in the cell cycle to create fast and robust transitions.

Key words Cell cycle, Mathematical modeling, Biological switches, Bistability, Systems biology

1 Introduction

Growth and replication are two of the defining characteristics of life, and therefore are universal to all living organisms. Not only are these processes observed for organisms but also are present within each single cell. A cell's reproduction pathway (division), referred to as the cell cycle, is defined by two major events: the copy of the hereditary material and its partition to pass it to the next generation.

Despite the universality of the reproduction process, the three domains of life (bacteria, archaea, and eukaryotes) use different, albeit related, mechanisms especially in terms of cell cycle regulation [1]. The best studied and understood regulatory network for cell division is the eukaryotic cell cycle. In this chapter we focus

---

Rosa Hernansaiz-Ballesteros and Kirsten Jenkins contributed equally to this work.

Louise von Stechow and Alberto Santos Delgado (eds.), *Computational Cell Biology: Methods and Protocols*, Methods in Molecular Biology, vol. 1819, https://doi.org/10.1007/978-1-4939-8618-7_14,

on models of the eukaryotic system; for reviews of the prokaryotic system see [2, 3] and for the archaea system see [4–6].

The eukaryotic cell cycle is traditionally divided into four phases: G1, S, G2, and M. The two major events (1) the copying of the hereditary material and (2) its separation, take place in S phase (Synthesis phase) and M phase (Mitosis), respectively. These events are separated by two "gap" phases, G1 and G2, which allow the cell to grow and at the same time assess if the internal conditions and the external environment are suitable to proceed to the next phase [7, 8].

All phases are temporally separated, and the regulatory network, or *cell cycle control system*, coordinates the phases to ensure they occur in the correct sequence [7, 8]. The main molecules that drive the progression of the cell cycle are the cyclin-dependent kinases (Cdks), a family of heterodimeric serine/threonine kinases [9, 10]. The concentration of inactive Cdk is constant throughout the cell cycle. The first step to activate Cdks is the binding to their cyclin partners [11]. The cell uses multiple cyclins, which are regulated in a phase-dependent fashion [7]. Through the synthesis and degradation of individual cyclins in specific phases, the cell can track individual cell cycle phases and regulate its growth and reproduction accordingly. Cdk/cyclin complexes are also controlled through regulators, to ensure that full activation of the Cdk/cyclin complex only occurs when required. Figure 1 displays the concentrations of cyclins binding to their Cdk partner (Cdc28 or Cdc2) for each phase, as well as the key regulators, for two important test organisms, budding and fission yeast.

Cdk/cyclin complexes trigger phase transitions in the cell cycle. They are precisely controlled by numerous inhibitors and activators, many of which are themselves, kinases, and phosphatases. To progress to the next phase, the balance between Cdk/cyclin inhibitors and activators must be permutated. Before any transition, the level of inhibitor must outweigh that of the activator. Once a phase is completed, a signal promotes activators to inhibit inhibitors, primarily by phosphorylation, so that the transition can occur. Thus, the cell cycle control system is a series of connected biochemical switches [12, 13].

There are three major biochemical transitions that occur during the cell cycle. Each can be treated as a switch and is connected to a checkpoint. The function of a checkpoint is to prevent the transition until certain processes are complete. The G1 checkpoint, also known as "Start" in yeast or "restriction point" in mammalian cells, enables the cell to decide whether conditions are favorable for reproduction and to advance to S phase. The "G2 checkpoint" controls the early mitotic events that lead to chromosome alignment. The "spindle-assembly checkpoint" controls the metaphase-to-anaphase transition, which induces sister-chromatid separation and the completion of the cell cycle. If the cell detects

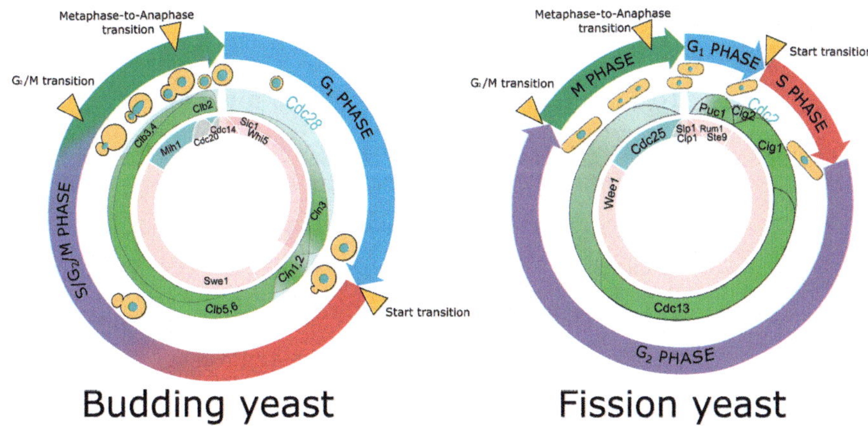

Fig. 1 Cell cycle regulation in budding (left) and fission (right) yeast cells. A typical eukaryotic cell cycle is discretized into four phases (external arrows): Gap 1 (G_1, blue), Synthesis phase (S, red), Gap 2 (G_2, purple), and Mitosis (M, green). Each organism or cell type has different lengths for each phase, and therefore different duration of the cell cycle. Budding yeast (*Saccharomyces cerevisiae*, left cycle) has a long G_1 phase, while S, G_2, and M phases are not clearly separated; fission yeast (*Saccharomyces pombe*, right cycle) has a long G_2 phase, and short G_1, S, and M phases. Each organism has a main cyclin-dependent kinase (Cdk) driver of the cell cycle, Cdc28 for budding yeast and Cdc2 for fission yeast. They need to bind tightly to their activator partners (cyclins) as a first step of activation (the concentration of cyclins throughout the cell cycle for each organism is presented in the middle circle). Cyclins are usually phase-specific, and they are synthesized and degraded in a specific phase. When Cdc28/Cdc2 is in a complex with its partner cyclin, its activity is controlled by many regulators. In each phase, there are different primary regulators (the names and concentrations are presented in the inner cycle), most of them play an inhibitory role (red part of the inner cycle) while a few promote the transitions as activators (blue part of the inner cycle). The yellow triangles denote the three critical checkpoints, in which the cell must decide whether to continue to the next phase: Start, G_2–M and metaphase-to-anaphase transitions

any problem in a phase prior to a checkpoint, the cycle will be blocked until it is resolved [7, 8, 14–16].

A lot is known about the eukaryotic cell cycle and its regulation, however simple knowledge of the biochemical behavior of the molecules, and their interaction partners, is not enough to create system-level understanding of how the cell cycle works as a unit. Mathematical models are a powerful tool to summarize our existing knowledge and test if these models can indeed match experimental results. The model can then also be used to make predictions, which are experimentally tested and used to broaden our knowledge through comparisons to observed physiology [17–19].

In this chapter we will guide the readers on how mathematical models of biological networks are built, and provide key features and concepts to understand mathematical models. We will also explore models for the different cell cycle checkpoints and indicate how their similarities appear to present a common fundamental design in the regulation of cell cycle transitions.

2 Methodology

To model a biological system, we must go through four major phases:

1. Identify the network architecture.
2. Decide on the best modeling approach.
3. Consider how we want to visualize the results.
4. Compare those with experimental data.

Figure 2 provides a step-by-step visualization of the whole modeling process, using a simple model to exemplify all the steps of the process.

Network architecture arises from interaction between the molecules that compose the biological system under study. The architecture of a biological network defines the global behavior of the system. Depending on the interactions in the network under study, different motifs can be identified [23]. The cell cycle regulatory network relies on negative and positive feedback loop motifs to control the transitions and the periodicity of the cycle. Negative feedback loops (NFBL) contain an odd number of negative interactions (inhibitions), and they usually allow maintenance of homeostasis [24–29]. They can furthermore create oscillations (if there is long enough delay in the feedback motif) [24–30]. A long negative feedback loop regulates the periodicity of the cell cycle [31]. Positive feedback loops (PFBL) contain either an even number of negative interactions or purely positive interactions (activations); they force the systems to choose between possible states, and may create switches [24–29, 32].

The example we present here is a system based on mutual inhibition, where a kinase and a phosphatase are self-activated, and mutually inhibit each other (Fig. 2a, b). This network contains three positive feedback loops, two autocatalytic loops and one loop created by an even number of negative interactions. This system should choose between two different states (activated kinase—inhibited phosphatase, or inhibited kinase—activated phosphatase). Such decision-making systems exist in the cell and they are also present in cell cycle transitions. In these cases, the cell converts graded inputs (i.e., quantity of activated Cdk/cyclin) into on/off responses (i.e., change in phase), upon the input obtaining a given threshold [24–29, 33–36]. In the mutual inhibition system (Fig. 2b), the input helps the kinase activation, while inhibiting phosphatase activation (basically representing the effect of an upstream kinase). The threshold activates the phosphatase and suppresses kinase activity (representing the effects of a constitutive phosphatase). The ratio between threshold and input will therefore determine which molecule becomes activated for a given set of initial conditions.

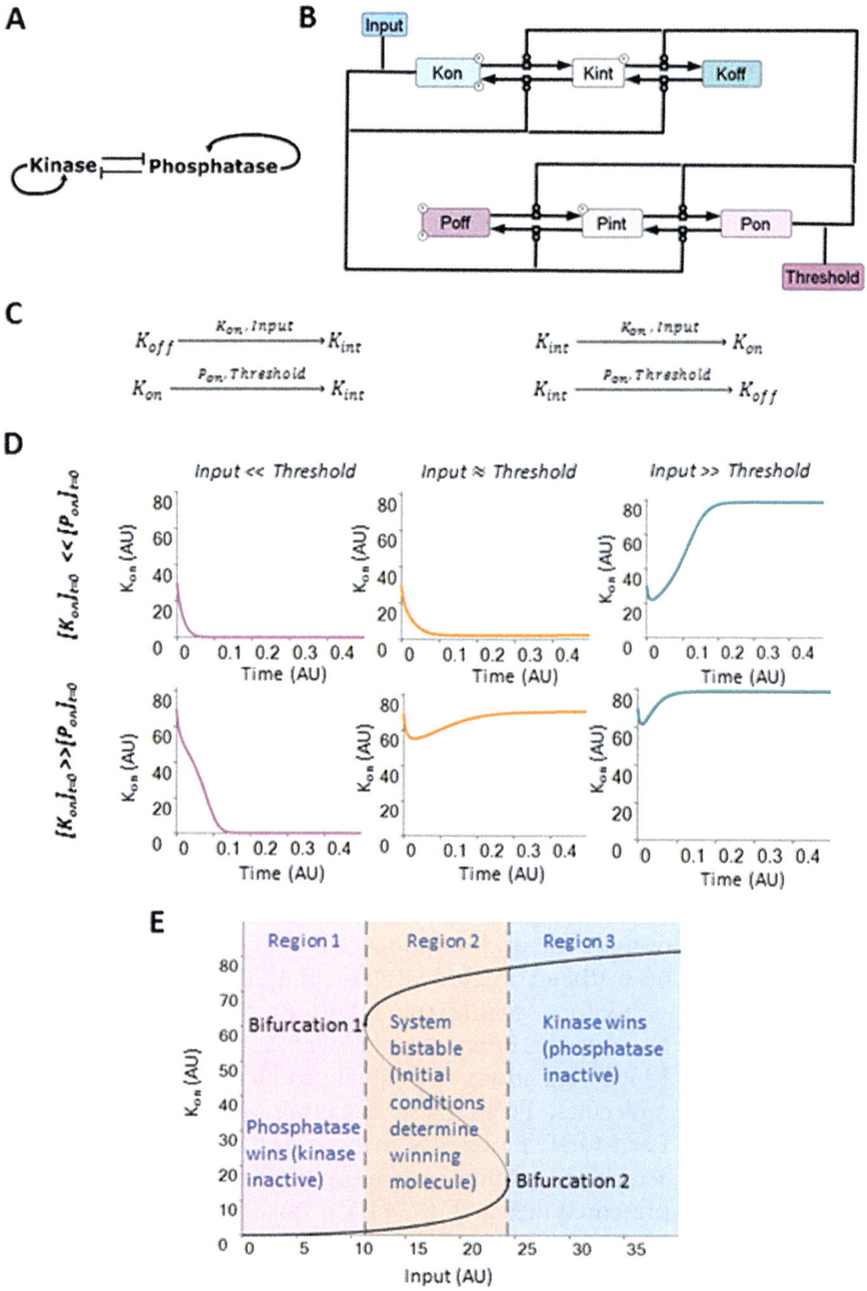

Fig. 2 Model of a simple biological switch. (**a**) A kinase and a phosphatase pair cross inhibit each other and self-activate by phosphorylating and dephosphorylating crucial residues. (**b**) Detailed model drawn in the systems biology graphical notation of Celldesigner [20–22]. (**c**) Kinetic equations for the kinase reactions of the network shown in B (the phosphate equations can be obtained by switching $K \leftrightarrow P$ and *input* $\leftrightarrow$ *threshold* due to network symmetry). (**d**) Time course diagrams for the amount of activated kinase when the parameter input has low (pink), medium (orange) and high (blue) values, for high and low initial activated kinase. (**e**) The bifurcation diagram shows the steady states of activated kinases as the input is varied. Bistability is seen in region 2 (orange), whereas regions 1 (pink) and 3 (blue) are monostable. Simulations and bifurcation diagram are calculated by Oscill8 (http://oscill8.sourceforge.net/)

The changes in state of a biological switch created by PFBLs often create hysteretic behavior and bistability. Those phenotypes occur when the system reaches a new state and subsequently resists further state changes; only major changes in the input are able to create a reversion to the previous state [37]. The feature of bistability ensures that cell cycle transitions are stable against perturbations [38]. Once a phase change is made, the cell is unable to revert back to the previous phase using the same set of conditions that caused the switch, and must wait until the system is reset in the next cell cycle.

Once the network has been created, we can move on to building the mathematical model. There are multiple approaches for modeling a biological system. The choice of model should depend not only on the properties of the system but also on the goals of the modeling effort and the availability of biological data to identify the right level of abstraction [39]. Qualitative approaches, such as Boolean models, represent discrete states of a molecule (usually in an On/Off way with a binary variable), and they are ideal for modeling gene regulatory networks [40–44]. On the other hand, quantitative approaches track the concentration of molecules over time. For example, stochastic solutions solve models by explicitly including a degree of uncertainty through random probability distributions for interactions. This creates noise within the system, enabling the simulation of individual cell behaviors and variations inside a population [45–47]. Deterministic solutions solve models by ordinary differential equations (ODEs), which represent how the concentrations of the interacting molecules evolve over time in an average cell. Since deterministic approaches have been the primary method to model cell cycle transitions, we will focus here on mathematical models based on ODEs.

Before writing the ODEs of the model, it is advisable to translate the interaction network into kinetic equations (Fig. 2c). Kinetic equations show how products are obtained from reacting molecules. From these equations, ODEs can be easily written. Each ODE represents changes over time in the concentration of a protein, or any other molecule within the cell. For the model presented in Fig. 2, the ODE equations for the kinase would be:

$$\frac{\mathrm{d}K_{\mathrm{on}}}{\mathrm{d}t} = a_1\left(K_{\mathrm{on}} + \mathrm{Input}\right)K_{\mathrm{int}} - a_2\left(P_{\mathrm{on}} + \mathrm{Threshold}\right)K_{\mathrm{on}}$$

$$\frac{\mathrm{d}K_{\mathrm{int}}}{\mathrm{d}t} = a_3\left(K_{\mathrm{on}} + \mathrm{Input}\right)K_{\mathrm{off}} - a_4\left(P_{\mathrm{on}} + \mathrm{Threshold}\right)K_{\mathrm{int}} - a_1\left(K_{\mathrm{on}} + \mathrm{Input}\right)K_{\mathrm{int}} + \dots$$

$$\dots + a_2\left(P_{\mathrm{on}} + \mathrm{Threshold}\right)K_{\mathrm{on}}$$

where terms with positive signs stand for activation, and negative signs for inhibition reactions. We assume that the total concentration of a kinase, K, is not changing throughout the time, so we can calculate the concentration of the inactive form algebraically from the total concentration (totK):

$$K_{\text{off}} = \text{totK} - K_{\text{on}} - K_{\text{int}}$$

The phosphatase equations are symmetric to the kinase equations, and therefore can be generated easily from the equations given by switching $K \leftrightarrow P$ and input $\leftrightarrow$ threshold.

As time is the independent variable that drives ODEs, localization or different conformations of molecules cannot be independent, as well. They can be independent if we move to partial differential equations (PDEs), which are not generally used for cell cycle modeling as they require detailed spatiotemporal information on all modeled proteins. To couple with this constraint in ODE based models, a common simplification is to treat the various localizations of molecules as a separate species, for example cytosolic protein X and nuclear protein X. Thus, models with multiple equations will represent the same molecule as it changes in location or conformation. This is shown in Fig. 2, where kinases and phosphatase can be in an ON, OFF or intermediary state, depending upon their phosphorylation state.

Nonlinearity is required for the positive feedback loop to induce bistability [30]. In Fig. 2, we achieve this by having 2 cooperative modification sites on each enzyme. This is then mathematically described through mass action kinetics, as seen in the above equations. Another method to describe nonlinear behaviour is Goldbeter-Koshland kinetics [48], which is used to create the cell cycle models provided in Subheading 3. In Goldbeter-Koshland kinetics, the molecule also switches between on and off states, but the transitions are driven by an enzymatic reaction, with the kinetics approximated by the Michaelis-Menten formula. Both methods enable the molecules to sharply convert between these 2 states (on and off), each state serving as a steady state of the bistable system.

The presented model in Fig. 2 has as parameters all kinetic rates (a_1, a_2 ...) and total population of molecules (Input, Threshold, total amount of K and P, etc.). Often the parameters within cell cycle models are not known. If they are known, the concentrations and ratios of a reaction in vivo may not be same as in vitro, due to the different buffering conditions [49, 50]. It is the problem of unknown parameters, which particularly complicates modeling biological systems accurately. Some parameters within a system, when altered, will result in vastly different global behavior of the system, while other parameters can be changed without any major change in model behavior. All models presented in this chapter are downloadable from https://github.com/CsikaszNagyLab or http://www.csikasznagylab.org/.

By running time course diagrams from different parameter sets (Fig. 2d), we can observe how the model reaches a steady state (constant concentration of molecules over time). Figure 2d shows that, if a parameter set is chosen and the *input* is low in comparison to the *threshold*, then regardless of the initial conditions, the kinase will not become activated and in the steady state of the system K_{on} is close to zero (Fig. 2d, input is low). Contrarily, if the *input* is much higher than the *threshold*, then the kinase will always turn on (Fig. 2d, input is high). The region that is particularly interesting is the range where *threshold* and *input* are comparable (Fig. 2d, input is medium). In this region, the steady state of the system will vary depending upon the initial conditions, as there are two fixed points in this region, which is the cause of bistability.

The effect of a parameter on a system's behavior can be investigated through bifurcation diagrams (Fig. 2e). A bifurcation diagram is created by varying a parameter and watching the evolution of fixed points within the system. A position in parameter space where the behavior of the model changes its qualitative behavior is known as a bifurcation point [29, 51]. Some of the most common bifurcations include a creation or destruction of a fixed point (saddle node bifurcations), or a change in stability of a fixed point. The switch is a transition between states at bifurcation points. The system initially rests in one of the states and parameter changes push the system through a bifurcation where this steady state disappears and the system is attracted to the other steady state. This forces the system to rapidly jump to a different steady state.

The bifurcation diagram in Fig. 2e presents two bifurcations in the system as *input* is varied. The first bifurcation destroys the steady state at low amounts of activated kinase (Bifurcation 1), and the second bifurcation creates a new fixed point at a high value of activated kinase (Bifurcation 2). At low and high values of *input*, there is only one steady state (Region 1 and 3) so that the system does not depend upon initial conditions. Region 2 is where there are 2 steady states for a given value of *input*, creating the dependence upon initial conditions seen in the time course diagrams in Fig. 2d.

The switch in Fig. 2 starts with a low value for *input*, where the system is allowed time to find the steady state. The *input* is then slowly increased such that the system is always able to find the new steady state. Eventually, the low active kinase steady state curve reaches bifurcation 1, and any further increase in the *input* then causes the whole system to switch to the highly active kinase steady state. Once on the highly active kinase steady state curve, the *input* would need to decrease to a value equal to bifurcation 1 for the kinase to turn off, thus there is bistability and hysteresis in Region 2. The bifurcation diagram also shows how the switches are robust to noise. If the values of *input* for bifurcation 1 and 2 are vastly different, it is unable to switch back with small perturbations to *input*.

In order to create realistic models, their behaviors need to match experimental results, and therefore it is essential to optimize the parameters. Optimization requires an objective function that needs to be minimized as parameters are changed. The objective function is written such that the closer the model gets to matching the experimental results, the smaller the number obtained from the function is; if the model exactly fits all the data, the function is evaluated to be 0. There are many methods for parameter optimization, and when running optimization algorithms, it is preferable to use existing packages within the chosen programming language, for example *scipy.optimize* in Python, or various optimization methods of COPASI [52].

When cell cycle transition models need to be optimized, there are two primary methods, and they are dependent upon the experimental data being used to optimize the model. Gradient based optimization is primarily utilized with time course data of concentrations of proteins during the cell cycle. This category of optimizers ensures that each subsequent set of parameters chosen by the algorithm will evaluate the minimization function to be closer to zero than the last set. The algorithm continues until it finds that, in whichever direction it moves in parameter space, the function will be larger and it has reached the minimum. Gradient based methods should also be incorporated into a larger search, as they do not account for the possibility that the minimum they locate is a local rather than global minimum. This can be achieved by running many optimization runs from random initial parameter sets. Many cell cycle models do not use time course data to optimize their parameters [53, 54]. Instead, they investigate the effect of different phenotypes upon the transition of interest. For this data, non-gradient-based optimization methods are required. It is worth noting that some non-gradient-based methods do not always move in the direction of the minimum, and thus do not have the problem of getting stuck in local minima.

3 Cell Cycle Models

The cell cycle can be viewed from two different perspectives, providing two methods for modeling. As the cell cycle is repeated, some researchers assume it is a free running oscillator that goes through different phases [31], whereas others see it as a sequence of biological switches closing on a periodic loop [19].

In this section we will be following the idea that the cell cycle oscillator works through the combination of positive and negative feedback loops, as it moves through several switches to close up on a robust oscillator [55]. We will explain the key features of the cell cycle transitions and the similarities in their regulation. For wider reviews on cell cycle modeling, we refer the reader to the bibliography [56, 57].

3.1 Models of the G1–S Transition

The G1–S transition is known as "Start" in yeast and as the "restriction point" (R) in multicellular eukaryotes. The model organism, which will be used as an example to present this transition in this section will be the budding yeast *Saccharomyces cerevisiae*. In *S. cerevisiae*, the G1–S transition ensures that the cell has obtained a critical size prior to DNA replication. In late G1, the increasing levels of the cytoplasmic cyclin Cln3 increase and bind to the cyclin-dependent kinase (Cdk) Cdc28. The Cln3/Cdc28 complex enters the nucleus activating the transcription factor complex SBF that induces the production of the cyclins Cln1 and Cln2 (Fig. 3a). Cln1 and Cln2, when bound to Cdc28, further activate SBF, closing a positive feedback loop [58]. Furthermore Cdk/Cln1,2 induces budding and initiates the later steps of the cell cycle [59]. One of the targets of Cdc28/Cln1,2 is the Cdk inhibitor Sic1, which is involved in an another biological switch, where Cdk/Clb5,6 inhibition by Sic1 is overtaken by Cdk/Clb5,6 inducing the degradation of Sic1 [60], closing a second positive (double negative or antagonistic) feedback loop [61]. In most models of the "Start" transition growth controls the amount of Cln3 in the nucleus which then activates these two positive feedback loops [62–64].

To ensure that SBF is not activated too early or by very small concentrations of nuclear Cln3, the retinoblastoma-likeWhi5 protein inhibits the activation of SBF [65–68]. Whi5 is also considered as an integrator of the widely fluctuating Cln3 activity, reducing the noise that emerges from the low molecule numbers of Cln3 [69]. Besides, Whi5 is a dilution factor to measure cell size, as it is only degraded but not produced during G1 phase [66]. Modeling has shown that the Whi5 inhibitor contributes to the irreversibility of the SBF $\longrightarrow$ Cln2 $\longrightarrow$ SBF positive feedback loop [70]. Charvin et al. found that "Start" activation through the SBF/Cln2/Whi5 positive feedback loop was strongly nonlinear, and they suggested this could be due to cooperativity of multiple phosphorylation sites on Whi5 [70]. They also found that Cln2 leakage could activate the positive feedback loop without any external signal, explaining how the cell cycle can be started in the absence of Cln3. Cln3 has a short half-life, it is present in low amounts (less than 100 molecules), and it is localized on the cytoplasmic face of the Endoplasmic Reticulum (ER) in early G1 phase. In order to ensure that Cln3 is not released from the ER too early, Whi3 acts as a translational

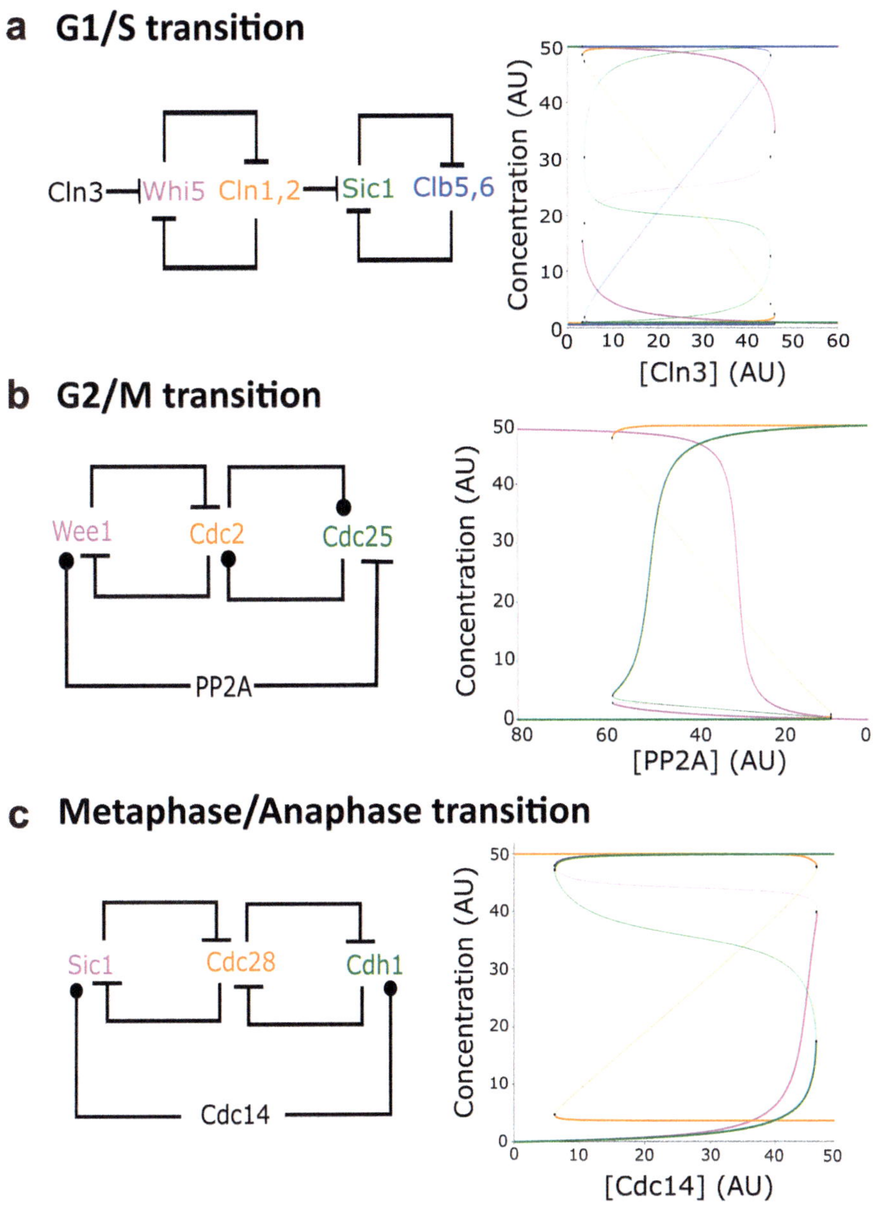

Fig. 3 Bistability in the simplified networks of the cell cycle transition regulatory networks. Left panels: the basic architecture of the regulatory networks controlling: (**a**) the G1–S transition in *Saccharomyces cerevisiae*; (**b**) the G2–M transition in *Schizosaccharomyces pombe*; and (**c**) the metaphase-to-anaphase transition in *Saccharomyces cerevisiae.* All these transitions include as their core a series of positive feedback loops. Point ended lines stand for activation, blunt ended lines for inhibitory effects. Right panels: bifurcation diagrams of the corresponding regulatory networks. Colors of plots correspond to the molecule names in the left panel, thick lines represent stable, thin lines, unstable steady states. On panel **b** we have the external input PP2A acting on the reverse scale as decrease in PP2A (and increase in total cyclin levels) can induce the G2–M transition

inhibitor for Cln3 [71–74], binding Cln3 mRNA to prevent the translation, and thus its accumulation on the protein level [71, 73]. In late G1, Cln3 is released from the ER by the chaperone Ydj1, and this causes the accumulation of Cln3 in the nucleus [75]. The role of this chaperone based regulation could be via the growth-rate of individual cells affecting chaperon availability, which then directly controls the cell cycle "Start" module through Cln3 [76].

One of the most recent and complete mathematical models of Start and the whole cell cycle regulation of yeast was provided by Adames et al. [77]. This model includes Ydj1, Whi3, and Whi5, as well as cyclins. It was optimized using 228 mutants and was able to reproduce the behavior of 214 of these. It was also used to predict the phenotype of 45 strains, which were not included for the parameter optimization, among these, 15 were newly constructed and experimentally validated against the model. The authors discuss in detail how the current model of the cell cycle needs revision based upon the failure to match certain mutants.

3.2 Models of the G2–M Transition

The G2–M transition occurs between interphase (including G1, S, and G2 phases) and mitosis. The checkpoint controlling this transition ensures that the hereditary material has been properly duplicated, and that all DNA damages have been repaired. There are many models with varying complexity that try to explain the most important features of this transition. However, the core of these models contains the same two positive feedback loops (Fig. 3b).

As in the other phases of the cell cycle, a specific Cdk/cyclin complex, formed by the Cdk Cdc2 and the cyclin Cdc13, promotes the exit from G2 and the entry into mitosis in fission yeast (Fig. 1). The transition occurs when the concentration of active Cdc2/Cdc13 exceeds a given threshold. Cdc2/Cdc13 active form induces the phosphorylation of its downstream substrates, controlling the events in early mitosis, like chromosome condensation and spindle assembly [8]. Cdc2/Cdc13 is the central complex of the G2–M transition core, so its activation state must be tightly controlled. Cdc2/Cdc13 activation at the G2–M transition is controlled by changes in its phosphorylation state. The most crucial of the kinases that control Cdc2/Cdc13 phosphorylation is Wee1, while the phosphatase Cdc25 can counteract this phosphorylation. The names of these regulators originate from research on the fission yeast *Schizosaccharomyces pombe*, which will be used here, although the orthologs of these also exist in budding yeast (Swe1 corresponds to Wee1 and Mih1 to Cdc25) [78].

In G2 phase, Wee1 kinase activity is high and Cdc25 phosphatase activity is low, keeping Cdc2/Cdc13 in the inactive phosphorylated form and maintaining the switch in the off state. When a small amount of Cdc25 becomes activated and/or some Cdc2/Cdc13 retains some residual activity [79, 80], the concentration of unphosphorylated active Cdc2/Cdc13 rises. This suppresses Wee1 activity and activates Cdc25, causing a swift activation of Cdc2/Cdc13, due to the positive feedback loops. This event initiates the transition from G2 to mitosis [79]. Wee1 and Cdc25 were the first molecules discovered to be regulating a crucial cell cycle transition [10], therefore they were incorporated in some of the earliest models of cell cycle regulation [81–83]. These positive feedbacks were also observed to be important in controlling the length of mitosis [84].

This main core can be extended to include indirect regulatory molecules, which interact with Cdc2/Cdc13 or counteract its kinase activity. The phosphatase PP2A bound to the regulatory subunit B55δ (PP2A from now on) counteracts many of the effects of Cdc2/Cdc13, providing thresholds for the phosphorylation reactions [85]. Furthermore, PP2A indirectly inhibits Cdc2/Cdc13 by the inhibition of Cdc25 and the activation of Wee1 [85]. The kinase Greatwall (Gwl) is directly activated by Cdc2/Cdc13. Once it is activated, Gwl indirectly inhibits PP2A via ENSA (low molecular-weight phosphatase inhibitor Endosulfine) and Arpp19 [86]. More recent models of the G2–M transition include some of these indirect players [78, 79, 87], while others tried to minimize the number of players without losing the dynamical features of the system [87–89].

3.3 Models of the Metaphase–Anaphase Transition

The metaphase-to-anaphase transition of the cell cycle, shown in Fig. 3c for the budding yeast *S. cerevisiae*, is also controlled by multiple positive feedback loops. The feedback between Cdk/CyclinB and its inhibitor Sic1, which acts a switch controlling G1/S transition, was also shown to affect the irreversibility of the M–G1 transition [90]. This study also provides a mathematical model of this system. Other models have investigated the mitotic focus of the spindle assembly checkpoint and how another positive feedback is important in inducing the initial step of this transition [91]. A further positive feedback occurs downstream of this, where the molecules that induce chromosome separation also activate the phosphatase Cdc14, which in turn helps their activation [92]. Despite the discoveries of these feedback loops, and the models that explain their behavior in isolation, an integrated model of the metaphase-to-anaphase transition is still lacking.

3.4 Common Features of Cell Cycle Transitions

The core of each cell cycle transition, as described above, is controlled by multiple positive feedback loops [38, 93]. The networks of Fig. 3 can be converted into a set of ODEs (Table 1) that can be directly investigated in software tools such as XPP-AUT (http://www.math.pitt.edu/~bard/xpp/xpp.html) or Oscill8 (http://oscill8.sourceforge.net/). These motifs produce a bistable switch that can be studied through bifurcation diagrams, as in our example in Fig. 2e. The bifurcation diagrams of the

Table 1
Equations and parameters for the models of Fig. 3

	Parameters/initial conditions	Equations
G1/S	*Initial conditions* init whi5on = siclon = 50 clnlon = clb5on = 0 *Parameters* param totwhi5 = totclnl = 50 totsicl = totclb5 = 50 Thresholdl = 50 Threshold2 = 25 Threshold3 = 50 Threshold4 = 20 *k*1–*k*8 = 1 *J*1–*J*8 = 1	dwhi5on/d*t* = *k*l × (totwhi5 - whi5on) × (Thresholdl + 0.1)/(*J*6 + (totwhi5 - whi5on)) − *k*2 × whi5on × (clnlon + cln3 + 0.1)/(*J*1 + whi5on) dclnlon/d*t* = *k*3 × (totclnl - clnlon) × (Threshold2 + 0.1)/(*J*2 + (tot-clnl − clnlon)) − *k*4 × clnlon × (whi5on + 0.1)/(*J*3 + clnlon)dsiclon/d*t* = *k*5 × (tot-sicl − siclon) × (Threshold3 + 0.1)/(*J*4 + (tot-sicl − siclon)) − *k*6 × siclon × (clnlon + clb5on + 0.1)/(*J*5 + siclon) dclb5on/d*t* = *k*7 × (totclb5 - clb5on) × (Threshold4 + 0.1)/(*J*7 + (tot-clb5 − clb5on)) − *k*8 × clb5on × (siclon + 0.1)/(*J*8 + clb5on)
G2/M	*Initial conditions* init cdc2on = cdc25on = 0, weelon = 50 *Parameters* param totweel = totcdc2 = 50, totcdc25 = 50, Threshold = 30, *k*l–*k*6 = 1, *J*1–*J*6 = 1	dweelon/d*t* = *k*l × (totweel − weelon) × (PP2A + 0.1)/(*J*6 + (totweel − weelon)) − *k*2 × weelon × (cdc2on + Thresh-old + 0.1)/(*J*1 + weelon) dcdc2on/d*t* = *k*3 × (totcdc2 − cdc2on) × (cdc25on + 0.1)/(*J*2 + (totcdc2 − cdc2on)) − *k*4 × cdc2on × (weelon + 0.1) /(*J*3 + cdc2on) dcdc25on/d*t* = *k*5 × (totcdc25 − cdc25on) × (cdc2on + 0.1) /(*J*4 + (totcdc25 − cdc25on)) − *k*6 × cdc25on × (PP2A + 0.1)/(*J*5 + cdc25on)
Metaphase/ anaphase	*Initial conditions* init cdc28on = 50, siclon = cdhlon = 0 *Parameters* param totsicl = totcdc28 = 50, totcdhl = 50, *k*l = 1.1, Threshold = 80, *k*2–*k*6 = 1, *J*1–*J*6 = 1	dsiclon/d*t* = *k*l × (tot-sicl − siclon) × (0.1 + cdc14)/(*J*6 + (tot-sicl − siclon)) − *k*2 × siclon × (0.1 + cdc28on)/(*J*1 + siclon)dcdc28on/d*t* = *k*3 × (totcdc28 − cdc28on) × (0.1)/(*J*2 + (totcdc28 − cdc28on)) − *k*4 × cdc28on × (siclon + cdhlon + 0.1)/(*J*3 + cdc28on) dcdhlon/d*t* = *k*5 × (totcdhl − cdhlon) × (0.1 + cdc14)/(*J*4 + (totcdhl − cdhlon)) − *k*6 × cdhlon × (0.1 + cdc28on)/ (*J*5 + cdhlon)

From the core components of each transition (Fig. 3), basic bifurcation diagrams can be built. All parameters not listed are equal to 1. These equations and parameters can be copied in a text file and directly be run in XPP-AUT (http://www.math.pitt.edu/~bard/xpp/xpp.html) or Oscill8 (http://oscill8.sourceforge.net/). Parameter values that are equal are lumped in this table, when copying to these tools, please split these and define each parameter individually three cell cycle switches share many common features (Fig. 3). Strikingly, each of the networks enables a large "jump" in steady state for the molecules that stabilize the next phase (Clb5,6, Cdc2and Cdh1/Sic1 respectively). Therefore, it is observed that the cell cycle relies on diverse positive feedback loops to ensure that molecules will rapidly become entirely active when the inactivation signal is suppressed [34–36, 38, 58, 84, 90, 92]. It is also interesting to note that in the case of the G1–S transition the two feedbacks are not directly coupled, rather the first switch initiates the second one. In this case the activity of Clb5,6 is almost zero right until the transition, so this may be significant in situations where the "target" molecule is required to stay at a very low concentration until the switch can flip.

4 Conclusion and Outlook

As our knowledge of the players of the cell cycle increases, the regulatory networks for all the transitions grow. Strikingly, the new players also interact in such a way that more positive feedback loops become apparent [94]. The question that arises from this observation is, why does each cell cycle transition have more than one positive feedback loop, when a single positive feedback loop will create a switch upon its own? Increasing the number of molecules implies a higher energetic cost (maintain the genetic copies, synthesize and process the molecules), which will only make evolutionary sense if it presents a clear advantage.

It was found by Brandman et al. [35] that interlinked fast and slow positive feedback loops have new characteristics that are not possible to obtain using a single feedback loop. These interconnected loops cause the switch to be rapidly inducible and resistant to noise. Both of these properties are desirable in cell cycle transitions. Once the decision to proceed in the cell cycle has been made, the transition should happen fast and the system should flip in a way that changes in the environment or other perturbations cannot flip it back. It has also been shown that larger networks, with multiple positive feedback loops can emulate the behavior of smaller networks [88, 89, 95]. In particular, the G2–M transition can be simplified to a population protocol called Approximate Majority (AM) [96], which has asymptotically optimal switching speed and a high resistance to noise. When investigating noise, the more complex systems were found to have a higher resistance to both intrinsic and extrinsic noise than smaller networks [97]. This high level of complexity and redundancy through the presence of multiple coupled feedback loops is responsible for the robustness of the cell cycle regulatory network. Unfortunately, the same robustness leads to resistance to many drugs, as alternative pathways and

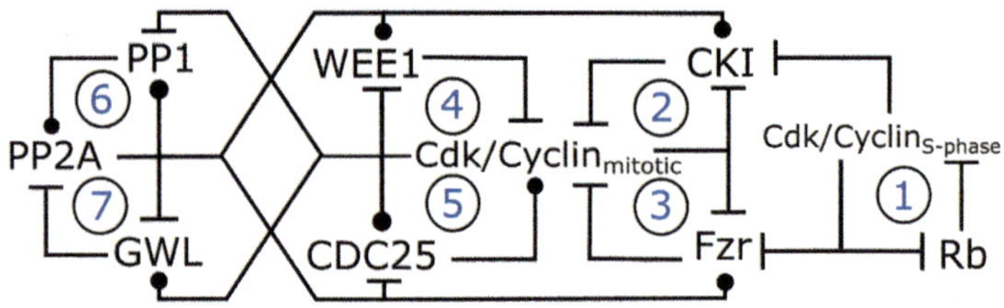

Fig. 4 Regulatory network combining multiple positive feedback loops controlling cell cycle transitions. Notation follows the naming convention in humans, highlighting how far these proteins are conserved. Loops 1, 2, and 3 with the Retinoblastoma protein (Rb) Cdk inhibitors (CKI–p27Kip1) and the Cdh1 homolog Fizzy related ubiquitin ligase are controlling various steps during the G1–S transition by inhibiting various Cdk/cyclin complexes. Loops 4 and 5 control the G2–M transition. Positive feedback loops 6 and 7 of PP1, PP2A, and Greatwall (GWL), together with loops 2 and 3 are important for mitotic exit

feedback loops could keep targeted cells alive and enable them to divide [98].

Cell cycle switches therefore utilize multiple feedback loops in order to obtain the speed and robustness that they require. But these cell cycle switches are dependent on each other and should happen in proper order. The cell cycle as a whole can be thought of as each of these transitions linked together as seen in Fig. 4. Here, seven direct positive feedback loops have been introduced, but the complete picture is much more complex with many more feedback loops [99]. It is our future task to develop methods to investigate such more complex models of cell cycle regulation.

We have now a comprehensive knowledge on the main controllers of the key cell cycle transitions. We know that most of these molecules and their interactions are strongly conserved among eukaryotes, highlighting that the key feature of the cell cycle transitions are their irreversibility and robustness. These are present thanks to the many positive feedback loops and the resulting multistability of the regulatory network. As more and more data becomes available on the genome-scale level and also on the individual cell level, we will need to update our current knowledge and mathematical modeling could be the tool to merge these data and test if our knowledge is indeed matching what can be found experimentally. In the future this will enable the development of cell-type specific cell cycle models, where incoming and outgoing signals, as well as variations in gene expression will be also considered.

Acknowledgment

R.H.B is supported by Microsoft Research through its PhD Scholarship Programme and K.J. is supported by the EPSRC Centre for Doctoral Training in Cross-Disciplinary Approaches to Non-Equilibrium Systems (CANES, EP/L015854/1).

References

1. Donnell MO, Langston L, Stillman B, Donnell MO, Langston L, Bell SP, Kaguni JM, Pfeiffer V, Lingner J, Cotmore SF, Tattersall P, Zielke N, Edgar BA, Melvin L (2013) Principles and concepts of DNA replication. Cold Spring Harb Perspect Biol 5:1–14
2. McAdams HH, Shapiro L (2003) A bacterial cell-cycle regulatory network operating in time and space. Science (80) 301(5641):1874–1877
3. Ausmees N, Jacobs-Wagner C (2003) Spatial and temporal control of differentiation and cell cycle progression in Caulobacter crescentus. Annu Rev Microbiol 57(1):225–247
4. Lundgren M, Bernander R (2005) Archaeal cell cycle progress. Curr Opin Microbiol 8(6):662–668
5. Lindås A-C, Bernander R (2013) The cell cycle of archaea. Nat Rev Microbiol 11(9):627–638
6. Kelman LM, Kelman Z (2014) Archaeal DNA replication. Annu Rev Genet 48:71–97
7. Csikász-Nagy A, Palmisano A, Zámborszky J (2011) Molecular network dynamics of cell cycle control: transitions to start and finish. Methods Mol Biol 761:277–291
8. Alberts B, Johnson A, Lewis J, Morgan D, Raff M, Roberts K (2015) Molecular biology of the cell, 6th edn. Garland Science, New York
9. Hartwell LH, Mortimer RK, Culotti J, Culotti M (1973) Genetic control of the cell division cycle in yeast: V. Genetic analysis of cdc mutants. Genetics 74(2):267–286
10. Nurse P (1975) Genetic control of cell size at cell division in yeast. Nature 256:547–551
11. Evans T, Rosenthal ET, Youngblom J, Distel D, Hunt T (1983) Cyclin: a protein specified by maternal mRNA in sea urchin eggs that is destroyed at each cleavage division. Cell 33(2):389–396
12. Tyson JJ, Novak B (2001) Regulation of the eukaryotic cell cycle: molecular antagonism, hysteresis, and irreversible transitions. J Theor Biol 210(2):249–263
13. Santos SDM, Ferrell JE (2008) Systems biology: on the cell cycle and its switches. Nature 454(7202):288–289
14. Hartvvell LH, Weinert T (1989) Checkpoints: controls that ensure the order of cell cycle events. Science 246(4930):629–634
15. Tyson JJ, Csikasz-Nagy A, Novak B (2002) The dynamics of cell cycle regulation. BioEssays 24(12):1095–1109
16. Musacchio A, Ciliberto A (2012) The spindle-assembly checkpoint and the beauty of self-destruction. Nat Struct Mol Biol 19(11):1059–1061
17. Gunawardena J (2014) Models in biology: 'accurate descriptions of our pathetic thinking'. BMC Biol 12:29
18. Tyson JJ, Novák B (2015) Models in biology: lessons from modeling regulation of the eukaryotic cell cycle. BMC Biol 13(1):1
19. Tyson JJ (1999) Models of cell cycle control in eukaryotes. J Biotechnol 71(1):239–244
20. Funahashi A, Morohashi M, Kitano H, Tanimura N (2003) CellDesigner: a process diagram editor for gene-regulatory and biochemical networks. Biosilico 1(5):159–162
21. Funahashi A, Matsuoka Y, Jouraku A, Morohashi M, Kikuchi N, Kitano H (2008) CellDesigner 3.5: a versatile modeling tool for biochemical networks. Proc IEEE 96(8):1254–1265
22. Kitano H, Funahashi A, Matsuoka Y, Oda K (2005) Using process diagrams for the graphical representation of biological networks. Nat Biotechnol 23(8):961–966
23. Alon U (2007) Network motifs: theory and experimental approaches. Nat Rev Genet 8(6):450–461
24. Griffith JS (1968) Mathematics of cellular control processes I. Negative feedback to one gene. J Theor Biol 20(2):202–208
25. R. Thomas On the relation between the logical structure of systems and their ability to generate multiple steady states or sustained oscillations In J Della, DJ Demongeot, B Lacolle Numerical methods in the study of critical phenomena, Springer Berlin, 1981, pp. 180–193
26. Thomas R, d'Ari R (1990) Biological feedback. CRC press, Boca Raton
27. Thomas R (2002) Laws for the dynamics of regulatory networks. Int J Dev Biol 42(3):479–485
28. Tyson JJ, Novák B (2010) Functional motifs in biochemical reaction networks. Annu Rev Phys Chem 61:219

29. Tyson JJ, Chen KC, Novak B (2003) Sniffers, buzzers, toggles and blinkers: dynamics of regulatory and signaling pathways in the cell. Curr Opin Cell Biol 15:221–231
30. Novák B, Tyson JJ (2008) Design principles of biochemical oscillators. Nat Rev Mol Cell Biol 9(12):981–991
31. Goldbeter A (1991) A minimal cascade model for the mitotic oscillator involving cyclin and cdc2 kinase. Proc Natl Acad Sci 88(20):9107–9111
32. Griffith JS (1968) Mathematics of cellular control processes II. Positive feedback to one gene. J Theor Biol 20(2):209–216
33. Xiong W, Ferrell JE (2003) A positive-feedback-based bistable 'memory module'that governs a cell fate decision. Nature 426(6965):460–465
34. Brandman O, Meyer T (2008) Feedback loops shape cellular signals in space and time. Science (80) 322(5900):390–395
35. Brandman O, Ferrell JE, Li R, Meyer T (2005) Interlinked fast and slow positive feedback loops drive reliable cell decisions. Science (80) 310(5747):496–498
36. Ferrell JE (2008) Feedback regulation of opposing enzymes generates robust, all-or-none bistable responses. Curr Biol 18(6):R244–R245
37. Angeli D, Ferrell JE, Sontag ED (2004) Detection of multistability, bifurcations, and hysteresis in a large class of biological positive-feedback systems. Proc Natl Acad Sci 101(7):1822–1827
38. Novak B, Tyson JJ, Gyorffy B, Csikasz-Nagy A (2007) Irreversible cell-cycle transitions are due to systems-level feedback. Nat Cell Biol 9(7):724–728
39. Goldenfeld N, Kadanoff LP (1999) Simple lessons from complexity. Science (80) 284(5411):87–89
40. Kauffman S (1969) Homeostasis and differentiation in random genetic control networks. Nature 224:177–178
41. Thomas R (1973) Boolean formalization of genetic control circuits. J Theor Biol 42(3):563–585
42. Davidich M, Bornholdt S (2008) The transition from differential equations to Boolean networks: a case study in simplifying a regulatory network model. J Theor Biol 255(3):269–277
43. Karlebach G, Shamir R (2008) Modelling and analysis of gene regulatory networks. Nat Rev Mol Cell Biol 9(10):770–780
44. Wang R-S, Saadatpour A, Albert R (2012) Boolean modeling in systems biology: an overview of methodology and applications. Phys Biol 9(5):55001
45. Mura I, Csikász-Nagy A (2008) Stochastic petri net extension of a yeast cell cycle model. J Theor Biol 254(4):850–860
46. Kar S, Baumann WT, Paul MR, Tyson JJ (2009) Exploring the roles of noise in the eukaryotic cell cycle. Proc Natl Acad Sci 106(16):6471–6476
47. Barik D, Baumann WT, Paul MR, Novak B, Tyson JJ (2010) A model of yeast cell-cycle regulation based on multisite phosphorylation. Mol Syst Biol 6(1):405
48. Goldbeter A, Koshland DE (1981) An amplified sensitivity arising from covalent modification in biological systems. Proc Natl Acad Sci 78(11):6840–6844
49. Rudorf S, Thommen M, Rodnina MV, Lipowsky R (2014) Deducing the kinetics of protein synthesis in vivo from the transition rates measured in vitro. PLoS Comput Biol 10(10):e1003909
50. Davidi D, Noor E, Liebermeister W, Bar-Even A, Flamholz A, Tummler K, Barenholz U, Goldenfeld M, Shlomi T, Milo R (2016) Global characterization of in vivo enzyme catalytic rates and their correspondence to in vitro kcat measurements. Proc Natl Acad Sci 113(12):3401–3406
51. Kuznetsov Y (2013) Elements of applied bifurcation theory. Springer, New York
52. Hoops S, Sahle S, Gauges R, Lee C, Pahle J, Simus N, Singhal M, Xu L, Mendes P, Kummer U (2006) COPASI—a complex pathway simulator. Bioinformatics 22(24):3067–3074
53. Zwolak JW, Tyson JJ, Watson LT (2005) Parameter estimation for a mathematical model of the cell cycle in frog eggs. J Comput Biol 12(1):48–63
54. Panning TD, Watson LT, Allen NA, Chen KC, Shaffer CA, Tyson JJ (2008) Deterministic parallel global parameter estimation for a model of the budding yeast cell cycle. J Glob Optim 40(4):719–738
55. Tsai TY-C, Choi YS, Ma W, Pomerening JR, Tang C, Ferrell JE (2008) Robust, tunable biological oscillations from interlinked positive and negative feedback loops. Science (80) 321(5885):126–129
56. Csikász-Nagy A (2009) Computational systems biology of the cell cycle. Brief Bioinform 10(4):424–434

57. Ferrell JE, Tsai TY-C, Yang Q (2011) Modeling the cell cycle: why do certain circuits oscillate? Cell 144(6):874–885
58. Skotheim JM, Di Talia S, Siggia ED, Cross FR (2008) Positive feedback of G1 cyclins ensures coherent cell cycle entry. Nature 454(7202):291–296
59. Bloom J, Cross FR (2007) Multiple levels of cyclin specificity in cell-cycle control. Nat Rev Mol Cell Biol 8(2):149–160
60. Nash P, Tang X, Orlicky S, Chen Q, Gertler FB, Mendenhall MD, Sicheri F, Pawson T, Tyers M (2001) Multisite phosphorylation of a CDK inhibitor sets a threshold for the onset of DNA replication. Nature 414(6863):514–521
61. Ferrell JE (2002) Self-perpetuating states in signal transduction: positive feedback, double-negative feedback and bistability. Curr Opin Cell Biol 14(2):140–148
62. Chen KC, Csikasz-Nagy A, Gyorffy B, Val J, Novak B, Tyson JJ (Jan. 2000) Kinetic analysis of a molecular model of the budding yeast cell cycle. Mol Biol Cell 11(1):369–391
63. Chen KC, Calzone L, Csikasz-Nagy A, Cross FR, Novak B, Tyson JJ (2004) Integrative analysis of cell cycle control in budding yeast. Mol Biol Cell 15(8):3841–3862
64. Barberis M, Klipp E (2007) Insights into the network controlling the G1/S transition in budding yeast. Genome Inform 18:85–99
65. Costanzo M, Nishikawa JL, Tang X, Millman JS, Schub O, Breitkreuz K, Dewar D, Rupes I, Andrews B, Tyers M (2004) CDK activity antagonizes Whi5, an inhibitor of G1/S transcription in yeast. Cell 117(7):899–913
66. Schmoller KM, Turner JJ, Kõivomägi M, Skotheim JM (2015) Dilution of the cell cycle inhibitor Whi5 controls budding-yeast cell size. Nature 526(7572):268–272
67. de Bruin RAM, McDonald WH, Kalashnikova TI, Yates J, Wittenberg C (2004) Cln3 activates G1-specific transcription via phosphorylation of the SBF bound repressor Whi5. Cell 117(7):887–898
68. Wagner MV, Smolka MB, de Bruin RAM, Zhou H, Wittenberg C, Dowdy SF (Jan. 2009) Whi5 regulation by site specific CDK-phosphorylation in Saccharomyces cerevisiae. PLoS One 4(1):e4300
69. Liu X et al (2015) Reliable cell cycle commitment in budding yeast is ensured by signal integration. Elife 4:1–19
70. Charvin G, Oikonomou C, Siggia ED, Cross FR (2010) Origin of irreversibility of cell cycle start in budding yeast. PLoS Biol 8(1):e1000284
71. Garí E, Volpe T, Wang H, Gallego C, Futcher B, Aldea M (2001) Whi3 binds the mRNA of the G1 cyclin CLN3 to modulate cell fate in budding yeast. Genes Dev 15(21):2803–2808
72. Wang H, Garí E, Vergés E, Gallego C, Aldea M (2004) Recruitment of Cdc28 by Whi3 restricts nuclear accumulation of the G1 cyclin-Cdk complex to late G1. EMBO J 23(1):180–190
73. Nash RS, Volpe T, Futcher B (2001) Isolation and characterization of WHI3, a size-control gene of Saccharomyces cerevisiae. Genetics 157(4):1469–1480
74. Mizunuma M, Tsubakiyama R, Ogawa T, Shitamukai A, Kobayashi Y, Inai T, Kume K, Hirata D (2013) Ras/cAMP-dependent protein kinase (PKA) regulates multiple aspects of cellular events by phosphorylating the Whi3 cell cycle regulator in budding yeast. J Biol Chem 288(15):10558–10566
75. Vergés E, Colomina N, Garí E, Gallego C, Aldea M (2007) Cyclin Cln3 is retained at the ER and released by the J chaperone Ydj1 in late G1 to trigger cell cycle entry. Mol Cell 26(5):649–662
76. Ferrezuelo F, Colomina N, Palmisano A, Garí E, Gallego C, Csikász-Nagy A, Aldea M (2012) The critical size is set at a single-cell level by growth rate to attain homeostasis and adaptation. Nat Commun 3:1012
77. Adames NR, Schuck PL, Chen KC, Murali TM, Tyson JJ, Peccoud J (2015) Experimental testing of a new integrated model of the budding yeast Start transition. Mol Biol Cell 26(22):3966–3984
78. Csikász-Nagy A, Battogtokh D, Chen KC, Novák B, Tyson JJ (2006) Analysis of a generic model of eukaryotic cell-cycle regulation. Biophys J 90(12):4361–4379
79. Domingo-Sananes MR, Kapuy O, Hunt T, Novak B (2011) Switches and latches: a biochemical tug-of-war between the kinases and phosphatases that control mitosis. Philos Trans R Soc B 366(1584):3584–3594
80. O'Farrell PH (2001) Triggering the all-or-nothing switch into mitosis. Trends Cell Biol 11(12):512–519
81. Norel R, Agur Z (1991) A model for the adjustment of the mitotic clock by cyclin and MPF levels. Science 251:1076
82. Tyson JJ (1991) Modeling the cell division cycle: cdc2 and cyclin interactions. Proc Natl Acad Sci 88(16):7328–7332

83. Novak B, Tyson JJ (1993) Numerical analysis of a comprehensive model of M-phase control in Xenopus oocyte extracts and intact embryos. J Cell Sci 106(4):1153–1168
84. Araujo AR, Gelens L, Sheriff RSM, Santos SDM (2016) Positive feedback keeps duration of mitosis temporally insulated from upstream cell-cycle events. Mol Cell 64(2):362–375
85. Krasinska L, Domingo-Sananes MR, Kapuy O, Parisis N, Harker B, Moorhead G, Rossignol M, Novak B, Fisher D (2011) Protein phosphatase 2A controls the order and dynamics of cell-cycle transitions. Mol Cell 44(3):437–450
86. Mochida S, Maslen SL, Skehel M, Hunt T (2010) Greatwall phosphorylates an inhibitor of protein phosphatase 2A that is essential for mitosis. Science (80) 330(6011):1670–1673
87. Gérard C, Tyson JJ, Coudreuse D, Novák B (2015) Cell cycle control by a minimal Cdk network. PLoS Comput Biol 11(2):e1004056
88. Cardelli L, Csikász-Nagy A (2012) The cell cycle switch computes approximate majority. Sci Rep 2:656
89. Cardelli L (2014) Morphisms of reaction networks that couple structure to function. BMC Syst Biol 8(1):84
90. López-Avilés S, Kapuy O, Novák B, Uhlmann F (2009) Irreversibility of mitotic exit is the consequence of systems-level feedback. Nature 459(7246):592–595
91. Simonetta M, Manzoni R, Mosca R, Mapelli M, Massimiliano L, Vink M, Novak B, Musacchio A, Ciliberto A (2009) The influence of catalysis on mad2 activation dynamics. PLoS Biol 7(1):e1000010
92. Holt LJ, Krutchinsky AN, Morgan DO (2008) Positive feedback sharpens the anaphase switch. Nature 454(7202):353–357
93. Romanel A, Jensen LJ, Cardelli L, Csikász-Nagy A (2012) Transcriptional regulation is a major controller of cell cycle transition dynamics. PLoS One 7(1):e29716
94. Fisher D, Krasinska L, Coudreuse D, Novák B (2012) Phosphorylation network dynamics in the control of cell cycle transitions. J Cell Sci 125(Pt 20):4703–4711
95. Cardelli L, Hernansaiz-Ballesteros RD, Dalchau N, Csikász-Nagy A (2017) Efficient Switches in Biology and Computer Science. PLoS Comput Biol 13(1):e1005100. https://doi.org/10.1371/journal.pcbi.1005100
96. Angluin D, Aspnes J, Eisenstat D (2008) A simple population protocol for fast robust approximate majority. Distrib Comput 21(2):87–102
97. Cardelli L, Csikász-Nagy A, Dalchau N, Tribastone M, Tschaikowski M (2016) Noise reduction in complex biological switches. Sci Rep 6:20214
98. Kitano H (2007) A robustness-based approach to systems-oriented drug design. Nat Rev Drug Discov 5(3):202–210
99. Kaizu K, Ghosh S, Matsuoka Y, Moriya H, Shimizu-Yoshida Y, Kitano H (2010) A comprehensive molecular interaction map of the budding yeast cell cycle. Mol Syst Biol 6(1):415

Chapter 15

Simultaneous Profiling of DNA Accessibility and Gene Expression Dynamics with ATAC-Seq and RNA-Seq

David G. Hendrickson, Ilya Soifer, Bernd J. Wranik, David Botstein, and R. Scott McIsaac

Abstract

This chapter describes sequencing-based methods for profiling dynamic changes in DNA accessibility and gene expression in *Saccharomyces cerevisiae*. Assay for transposase-accessible chromatin with high-throughput sequencing (ATAC-Seq) is a powerful technique for identifying nucleosome-free regions of the genome. Combining ATAC-Seq with RNA Sequencing (RNA-Seq) is a rapid approach for studying the relationship between genome structure and changes in global patterns of gene expression from a single experiment. A laboratory protocol is presented for these methods as well as examples of typical results and visualizations.

Key words Genomics, Chromatin, Gene regulation, Transcription, Transcription factors

1 Introduction

Eukaryotic genomes are organized and packaged into chromatin, a complex of DNA and protein that enables chromosomes to fit inside the nucleus. The local accessibility of chromatin is a hallmark of transcriptional activity. Numerous methodologies have been developed for measuring chromatin accessibility including DNase-Seq [1, 2], MNase-Seq [3], FAIRE-Seq [4], and most recently, ATAC-Seq [5]. Unlike other methods, ATAC-Seq allows the identification of accessible DNA (i.e., nucleosome-free regions of the genome) using a "tagmentation" reaction, in which a hyperactive Tn5 transposase cleaves and tags double-stranded DNA with universal priming sequences [6]. Since Tn5 preferentially inserts into DNA regions that are not bound by nucleosomes, PCR amplification of these samples followed by next-generation sequencing reveals a global picture of genomic structure. Those types of experiments can be performed with material from low cell number or even from single, cells [7, 8]. The speed, scalability,

Louise von Stechow and Alberto Santos Delgado (eds.), *Computational Cell Biology: Methods and Protocols*, Methods in Molecular Biology, vol. 1819, https://doi.org/10.1007/978-1-4939-8618-7_15,

and relative ease of processing samples with ATAC-Seq make it an appealing approach for exploring chromatin biology generally, and especially for studies with many samples such as time series experiments. The combination of ATAC-Seq with RNA-Seq offers a powerful approach for determining how information encoded in chromatin manifests itself with changes in gene expression levels. In this chapter, we present laboratory protocols for ATAC-Seq and RNA-Seq library preparation and informatic procedures for processing the resulting data.

2 Materials

These library preparation protocols are compatible with Illumina MiSeq and HiSeq platforms.

2.1 ATAC-Seq

2.1.1 Spheroplast and Transposition

1. SB (spheroplasting buffer) Buffer (40 mM HEPES [pH 7.5], 10 mM $MgCl_2$, 1 M sorbitol). Filter-sterilized and stored at 4 °C.
2. 10 mg/mL Zymolyase-100T in SB Buffer.
3. Tagmentation DNA Buffer (Nextera DNA Sample Prep Kit).
4. Nuclease-free water.
5. Nextera Tagment DNA Enzyme 1 (Nextera DNA Sample Prep Kit).
6. Centrifuge.
7. Eppendorf Thermoshaker.
8. DNA Clean & Concentrator™-5 kit.

2.1.2 Library Amplification via Polymerase Chain Reaction

1. NEBNext Hi-Fidelity 2× PCR Master Mix.
2. Universal Primer Ad1 (10 μM).
3. Barcoded reverse primer (Ad2.[1–24]) (10 μM).
4. Thermocycler.
5. HPLC-(high performance liquid chromatography) purified primers dissolved in water to 10 μM (See Table 1 for primer sequences).

2.1.3 Solid Phase Reversible Immobilization (SPRI) Cleanup and Library Size Selection: Removing Large DNA Fragments

1. Thermomixer (incubator blocks or water baths can also be used).
2. DynaMag-96 Side Skirted Magnet.
3. Ampure XP (SPRI) beads.

Table 1
Representative ATAC-Seq read alignment statistics

Sample name	Total sequencing reads	Mapped reads	rDNA reads (with mapping quality > = 30)	Mitochondrial Reads	Filtered genomic reads
GCN4_t0_S1_ L001_R1_001	36,068,090	34,794,314	586,347	7,472,833	24,044,497
GCN4_t05_S54_ L008_R1_001	149,081,020	144,008,053	2,523,997	30,349,380	99,810,124
GCN4_t10_S41_ L006_R1_001	34,165,863	33,039,145	645,286	7,119,619	22,749,141
GCN4_t15_S33_ L005_R1_001	46,232,603	44,692,553	680,719	9,144,703	31,392,244
GCN4_t20_S34_ L005_R1_001	23,013,433	22,259,985	269,781	4,523,163	15,741,796
GCN4_t30_S15_ L003_R1_001	21,944,289	21,197,299	406,904	4,531,947	14,623,289
GCN4_t90_S55_ L008_R1_001	183,787,064	177,685,093	2,260,802	38,617,000	123,032,428

2.1.4 SPRI Cleanup and Library Size Selection: Removing Small DNA Fragments

1. Thermomixer (incubator blocks or water baths can also be used).
2. Ampure XP (SPRI) beads DynaMag-96.
3. DynaMag-96 Side Skirted Magnet.
4. 80% ethanol.
5. DNA Clean & Concentrator™-5 kit.
6. Bioanalyzer or Qubit for molar concentration determination.

2.2 RNA-Seq

2.2.1 RiboZero Bead Preparation

1. Ribo-Zero Gold rRNA Removal Kit.
2. 1.5 mL tubes.
3. RiboGuard RNase inhibitor.

2.2.2 RiboZero Treatment of RNA (1–2.5 μg of Input RNA)

1. RiboZero Reaction Buffer.
2. RiboZero rRNA Removal Solution.
3. PCR plate spinner.
4. Thermomixer (Incubator blocks or water baths can also be used.).

2.2.3 Magnetic Bead Reaction to Remove rRNA

1. Axygen 96-Well Non-Skirted PCR Microplates.
2. DynaMag-96 Side Skirted Magnet.
3. MicroAmp Clear Adhesive Film.

2.2.4 RNA Cleanup

1. RNA Clean Ampure XP (SPRI) beads.
2. MicroAmp Clear Adhesive Film.
3. 80% ethanol.
4. Centrifuge.
5. Thermocycler.

2.2.5 cDNA Synthesis

1. TruSeq Stranded mRNA kit.
2. First Strand Synthesis Act D Mix.
3. Superscript III Reverse Transcriptase (Thermo Fisher).
4. Thermocycler.
5. Resuspension Buffer (TruSeq Kit).
6. Second Strand Marking Master Mix (TruSeq Kit).

2.2.6 cDNA Cleanup

1. DynaMag-96 Side Skirted Magnet.
2. 80% ethanol.
3. MicroAmp Clear Adhesive Film.
4. Dry beads at 37 °C for 4–5 min.
5. Resuspension Buffer (TruSeq kit).
6. Axygen 96-Well Non-Skirted PCR Microplates.

2.2.7 cDNA 3′ Adenylation

1. Ampure XP beads.
2. A-tailing Mix (TruSeq Kit).
3. Resuspension Buffer (TruSeq Kit).
4. Thermocycler.

2.2.8 Adapter Index Ligation

1. Resuspension Buffer (TruSeq Kit).
2. Ligation Mix (TruSeq Kit).
3. Unique RNA Adapter Index (as described in TruSeq Kit).
4. Thermocycler.
5. Stop Ligation Buffer (TruSeq Kit).

2.2.9 Double Cleanup

1. Ampure XP beads.
2. PCR Master Mix.
3. Primer Cocktail.
4. TruSeq Stranded LT Kit.
5. 80% ethanol.
6. DynaMag-96 Side Skirted Magnet.
7. MicroAmp Clear Adhesive Film.
8. Resuspension buffer (TruSeq Kit).

2.2.10 Library Amplification

1. PCR primer cocktail.
2. PCR Master Mix.
3. Thermocycler.

2.2.11 Final Cleanup

1. AMPure XP beads.
2. DynaMag-96 Side Skirted Magnet.
3. 80% ethanol.
4. Resuspension Buffer (TruSeq Kit).
5. Strip-tube(s).
6. Bioanalyzer or Qubit.

2.3 Bioinformatics

1. bcl2fastq software (Illumina, version 2.17).
2. bwa-mem (https://sourceforge.net/projects/bio-bwa/files/, version 0.7.15).
3. custom python script that uses the pysam library (version 0.10).
4. samtools (http://samtools.sourceforge.net/).
5. bedtools2 (http://bedtools.readthedocs.io/en/latest/).
6. bedGraphToBigWig from UCSC userApps (v326).
7. Tophat2 (https://ccb.jhu.edu/software/tophat/index.shtml).
8. Cuffdiff2 (http://cole-trapnell-lab.github.io/cufflinks/).

3 Methods

3.1 ATAC-Seq

3.1.1 Spheroplast and Transposition

1. Collect 1–5 million yeast cells by centrifugation at room temperature for 1 min at 4600 rpm (2000 relative centrifugal force, rcf).
2. Wash cells twice with 200 μL of spheroplasting buffer (SB) at room temperature.
3. Resuspend cells in 190 μL SB at room temperature + 10 μL of 10 mg/mL zymolyase dissolved in SB.
4. Incubate for 30 min at 30 °C with rocking or slow shaking.
5. Pellet cells by centrifugation for 1 min at 4600 rpm at room temperature.
6. Wash spheroplasted cells twice with 200 μL of SB at room temperature.
7. Resuspend pellet in 50 μL transposition mix (25 μL Tagment DNA Buffer, 22.5 μL nuclease-free water, 2.5 μL Nextera Tagment DNA Enzyme 1).

8. Incubate at 37 °C for exactly 30 min (use heat block or Eppendorf Thermomixer, but do not shake/agitate the samples).
9. Purify using DNA Clean & Concentrator™-5 Kit and elute in 11 μL H_2O

Samples can be frozen and stored at −20 °c until ready to proceed.

3.1.2 Library Amplification via Polymerase Chain Reaction (PCR) (See Table 2)

1. Set up PCR reaction for library amplification: 25 μL NEBNext Hi-Fidelity 2× PCR Master Mix, 7.5 μL water, 6.25 μL Universal Primer Ad1 (10 μM), 6.25 μL barcoded reverse primer (Ad2.[1–24]) (10 μM), 5 μL purified transposed DNA from Subheading 3.1.1.
2. Run PCR Amplification (1 cycle: 72 °C for 5 min; 1 cycle: 98 °C for 30 s; 8 cycles: 98 °C for 10 s, 63 °C for 30 s, 72 °C for 1 min; hold at 4 °C). (Note that for ATAC-Seq, eight amplification cycles is a reasonable starting point but not necessarily the optimal number of amplification cycles in all cases).

3.1.3 SPRI Cleanup and Library Size Selection: Removing Large DNA Fragments

Perform a 0.4× SPRI cleanup to remove large (>1000 bp) amplicons. Before starting, equilibrate the Ampure XP beads at room temperature and set the thermomixer to 37 °C.

1. Add 20 μL Ampure XP beads to each 50 μL PCR reaction and mix by pipetting.
2. Incubate at room temperature for 5 min.
3. Place on the DynaMag-96 magnet for 2–3 min at room temperature until solution is clear.
4. Transfer supernatant (65 μL) to new well (large amplicons will remain behind bound to the beads).

3.1.4 SPRI Cleanup and Library Size Selection: Removing Small DNA Fragments

Perform a 1.5× SPRI cleanup to capture amplicons >100 bp and remove small amplicons (<100 bp).

1. Add 57 μL Ampure XP beads to 65 μL supernatant from Subheading 3.1.3, and mix by pipetting.
2. Incubate for 5 min at room temperature.
3. Place on magnet for 2–3 min at room temperature until solution is clear.
4. Discard supernatant containing fragments <100 bp.
5. Wash beads 2× with 200 μL 80% ethanol.
6. Aspirate ethanol, cover and spin plate in a PCR plate spinner.
7. Place on magnet and remove residual ethanol.
8. Dry beads for 4–5 min at 37 °C using the thermomixer.
9. Resuspend beads in 22.5 μL H_2O.
10. Incubate at room temperature for 2 min.

Table 2
Primers for ATAC-Seq library preparation. Note that Ad1 is the forward primer, and reverse primers Ad2.1-Ad2.24 carry unique barcodes

Name	Barcode	Sequence (5′ → 3′)
Ad1		AATGATACGGCGACCACCGAGATCTACACTCGTCGGC AGCGTCAGATGTG
Ad2.1	TAAGGCGA	CAAGCAGAAGACGGCATACGAGATTCGCCTTAGTCTC GTGGGCTCGGAGATGT
Ad2.2	CGTACTAG	CAAGCAGAAGACGGCATACGAGATCTAGTACGGTCTC GTGGGCTCGGAGATGT
Ad2.3	AGGCAGAA	CAAGCAGAAGACGGCATACGAGATTTCTGCCTGTC TCGTGGGCTCGGAGATGT
Ad2.4	TCCTGAGC	CAAGCAGAAGACGGCATACGAGATGCTCAGGAGTC TCGTGGGCTCGGAGATGT
Ad2.5	GGACTCCT	CAAGCAGAAGACGGCATACGAGATAGGAGTCCGTCT CGTGGGCTCGGAGATGT
Ad2.6	TAGGCATG	CAAGCAGAAGACGGCATACGAGATCATGCCTAGTCT CGTGGGCTCGGAGATGT
Ad2.7	CTCTCTAC	CAAGCAGAAGACGGCATACGAGATGTAGAGAGGT CTCGTGGGCTCGGAGATGT
Ad2.8	CAGAGAGG	CAAGCAGAAGACGGCATACGAGATCCTCTCT GGTCTCGTGGGCTCGGAGATGT
Ad2.9	GCTACGCT	CAAGCAGAAGACGGCATACGAGATAGCGTAGCGTC TCGTGGGCTCGGAGATGT
Ad2.10	CGAGGCTG	CAAGCAGAAGACGGCATACGAGATCAGCCTCGGT CTCGTGGGCTCGGAGATGT
Ad2.11	AAGAGGCA	CAAGCAGAAGACGGCATACGAGATTGCCTCT TGTCTCGTGGGCTCGGAGATGT
Ad2.12	GTAGAGGA	CAAGCAGAAGACGGCATACGAGATTCCTCTAC GTCTCGTGGGCTCGGAGATGT
Ad2.13	GTCGTGAT	CAAGCAGAAGACGGCATACGAGATATCACGACGT CTCGTGGGCTCGGAGATGT
Ad2.14	ACCACTGT	CAAGCAGAAGACGGCATACGAGATACAGTGGTGT CTCGTGGGCTCGGAGATGT
Ad2.15	TGGATCTG	CAAGCAGAAGACGGCATACGAGATCAGATCCAG TCTCGTGGGCTCGGAGATGT
Ad2.16	CCGTTTGT	CAAGCAGAAGACGGCATACGAGATACAAACGGGTC TCGTGGGCTCGGAGATGT
Ad2.17	TGCTGGGT	CAAGCAGAAGACGGCATACGAGATACCCAGC AGTCTCGTGGGCTCGGAGATGT

(continued)

Table 2 (continued)

Name	Barcode	Sequence (5′ → 3′)
Ad2.20	GTGTGGTG	CAAGCAGAAGACGGCATACGAGATCACCACACG TCTCGTGGGCTCGGAGATGT
Ad2.21	TGGGTTTC	CAAGCAGAAGACGGCATACGAGATGAAACCCAGT CTCGTGGGCTCGGAGATGT
Ad2.22	TGGTCACA	CAAGCAGAAGACGGCATACGAGATTGTGACCAGT CTCGTGGGCTCGGAGATGT
Ad2.23	TTGACCCT	CAAGCAGAAGACGGCATACGAGATAGGGTCAAGT CTCGTGGGCTCGGAGATGT
Ad2.24	CCACTCCT	CAAGCAGAAGACGGCATACGAGATAGGAGTGGGTCT CGTGGGCTCGGAGATGT

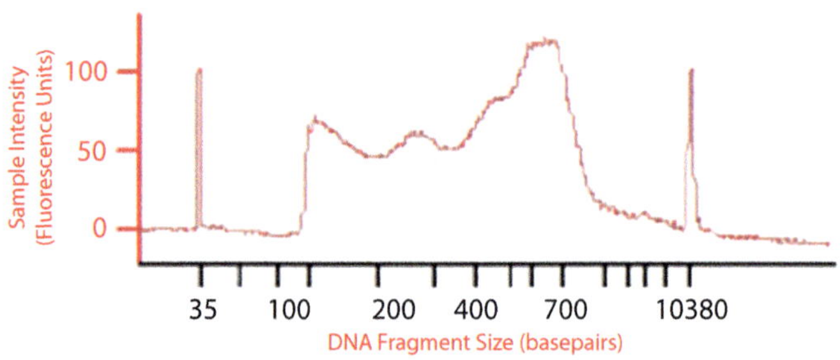

Fig. 1 Typical ATAC-Seq library trace obtained with Bioanalyzer. The periodicity of the fragment lengths is due to the periodicity of accessible regions imposed by nucleosome positioning in the genome

11. Place plate on magnet, allow solution to clear and collect supernatant containing DNA.
12. Purify using DNA Clean & Concentrator™ −5 Kit, elute in 22 μL H_2O and analyze on BioAnalyzer or Qubit (Fig. 1).

3.2 RNA-Seq

3.2.1 RiboZero Bead Preparation

1. Aliquot 225 μL RiboZero magnetic beads per reaction into a 1.5 mL tube (up to 5× reactions per tube).
2. Place tube on magnet, allow solution to clear and discard supernatant.
3. Wash beads 2× with water (equal volume), making sure to fully resuspend the beads with each wash.
4. Mix 60 μL Magnetic Bead Resuspension Buffer + 1 μL RiboGuard RNase inhibitor per reaction.
5. Resuspend beads, per reaction, in 61 μL of mix from step #4 and place into the wells of a 96-well plate. Leave the plate covered at room temperature until ready for use in Subheading 3.2.3.

3.2.2 RiboZero Treatment of RNA (1–2.5 μg of Input RNA)

1. For each reaction, mix (28 – X μL) RNase-free water, 4 μL RiboZero Reaction Buffer, X μL RNA, 8 μL RiboZero rRNA Removal Solution (Total volume = 40 μL).
2. Mix by pipetting up and down.
3. Incubate at 68 °C for 10 min, then spin for 10 s in a PCR plate spinner.
4. Incubate at room temperature for 5 min.

3.2.3 Magnetic Bead Reaction to Remove rRNA

1. Add 40 μL treated RNA (from Subheading 3.2.2) to 65 μL washed beads (from Subheading 3.2.1).
2. Mix thoroughly by pipetting, cover with clear film and vortex (medium) for 10 s.
3. Incubate at room temperature for 5 min, vortex (medium) for 5 s.
4. Incubate at 50 °C for 5 min (use heated lid to reduce condensation).
5. Place plate on magnet and allow solution to clear.
6. Remove supernatant (100 μL) and transfer to new well(s) in a 96-well plate.

3.2.4 RNA Cleanup

1. Add RNA Clean XP beads 1:1 (ca. 100 μL) to supernatant from **step 15** and mix by pipetting.
2. Incubate at room temperature for 15 min.
3. Place on magnet, allow solution to clear and discard supernatant.
4. Wash by adding 200 μL of 80% ethanol and incubating at room temperature for 30 s (leave on magnet while adding ethanol, do not resuspend beads).
5. Discard supernatant and repeat **step 4**.
6. Discard supernatant, seal plate and spin to remove any residual ethanol.
7. Dry beads at 37 °C for 4–5 min.
8. Resuspend in 19.5 μL of Fragment Prime Finish from the TruSeq Stranded mRNA kit, seal plate and run the following thermocycler program: 94 °C for 8 min, hold at 4 °C.

3.2.5 cDNA Synthesis

1. Thaw First Strand Synthesis Master Mix.
2. Place plate on magnet and allow solution to clear.
3. Transfer 17 μL of supernatant to a fresh well(s) of a 96-well plate.
4. Add 8 μL First Strand Master Mix plus Superscript III (for every new kit add 50 μL of Superscript III to one tube of First Strand Master Mix), mix by pipetting, and run the following thermocycler program: 25 °C for 10 min, 42 °C for 50 min, 70 °C for 15 min, hold at 4 °C.

5. Thaw Resuspension Buffer and Second Strand Master Mix.
6. Add 5 μL Resuspension Buffer to the reaction from **step 4**.
7. Add 20 μL Second Strand Master Mix and mix well by pipetting.
8. Use a thermocycler to incubate at 16 °C for 60 min, then hold at 25 °C.
9. Bring AMPure XP beads to room temperature.

3.2.6 cDNA Cleanup

1. Add 2× volume of beads to each sample (100 μL) and mix by pipetting.
2. Incubate 15 min at room temperature.
3. Place on magnet, allow solution to clear and discard supernatant.
4. Wash by adding 200 μL 80% ethanol and incubate at room temperature for 30 s (leave samples on the magnet, do not resuspend the beads).
5. Discard supernatant and repeat **step 4**.
6. Discard supernatant, seal plate and spin down, remove any residual ethanol.
7. Dry beads at 37 °C for 4–5 min.
8. Place on magnet and allow solution to clear.
9. Add 17.5 μL Resuspension Buffer, incubate at room temperature for 2 min.
10. Place on magnet and allow solution to clear.
11. Transfer 15 μL to fresh well(s) of a 96-well plate

    Samples can be frozen and stored at −20 °c until ready to proceed.

3.2.7 cDNA 3′ Adenylation

1. Ensure cDNA, Ampure XP beads are at room temperature before beginning; thaw A-tailing Mix and Resuspension Buffer.
2. Add 2.5 μL Resuspension Buffer and 12.5 μL A-tailing Mix to 15 μL of cDNA from Subheading 3.2.6.
3. Mix by pipetting, seal plate and run the following thermocycler program: 37 °C for 30 min, 70 °C for 5 min, hold at 4 °C.

3.2.8 Adapter Index Ligation

1. Add 2.5 μL Resuspension Buffer, 2.5 μL Ligation Mix and 2.5 μL of a unique RNA Adapter Index to each sample.
2. Mix well by pipetting and run the following thermocycler program: 30 °C for 10 min, hold at 25 °C.
3. Add 5 μL Stop Ligation Buffer and mix by pipetting.

3.2.9 Double Cleanup

1. Bring Ampure XP beads to room temperature, thaw PCR Master Mix and Primer Cocktail for TruSeq Stranded LT Kit

and place on ice; also prepare >800 μL of 80% ethanol per sample.

2. Add 42 μL AMPure XP beads to each sample, mix well and incubate at room temperature for 15 min.
3. Place on magnet and allow supernatant to clear, then discard supernatant.
4. Wash by adding 200 μL 80% ethanol and incubating at room temperature for 30 s (leave samples on the magnet, do not resuspend the beads).
5. Discard supernatant and repeat **step 4**.
6. Seal plate, spin down and remove residual ethanol.
7. Dry beads at 37 °C for 4–5 min.
8. Place on magnet and allow solution to clear.
9. Add 52.5 μL Resuspension Buffer, incubate at room temperature for 2 min.
10. Place on magnet and allow solution to clear.
11. Transfer 50 μL to fresh well(s) of a 96-well plate.
12. Add 50 μL of AMPure XP beads to each sample, mix well and incubate at room temperature for 15 min.
13. Repeat **steps 3–8**.
14. Add 22.5 μL Resuspension Buffer, incubate at room temperature for 2 min.
15. Place on magnet and allow solution to clear.
16. Transfer 20 μL to a fresh well in a 96-well plate

Samples can be frozen and stored at −20 °C until ready to proceed.

3.2.10 Library Amplification

1. Add 5 μL of PCR primer cocktail and 25 μL of PCR Master Mix and mix well by pipetting.
2. Run the following thermocycler program: 1 cycle: 98 °C for 30 s; 10 cycles: 98 °C for 10 s, 60 °C for 30 s, 72 °C for 30 s; 1 cycle: 72 °C for 5 min; hold at 4 °C.

3.2.11 Final Cleanup

1. Add 50 μL AMPure XP beads to each sample, mix well and incubate at room temperature for 15 min.
2. Place on magnet and allow supernatant to clear, and then discard supernatant.
3. Wash by adding 200 μL 80% ethanol and incubating at room temperature for 30 s (leave samples on the magnet, do not resuspend the beads).
4. Discard supernatant and repeat **step 3**.
5. Seal plate, spin down and remove residual ethanol.
6. Dry beads at 37 °C for 4–5 min.
7. Place on magnet and allow solution to clear.

8. Add 32.5 μL Resuspension Buffer, incubate at room temperature for 2 min.
9. Place on magnet, allow supernatant to clear and transfer 30 μL to fresh strip-tube(s). Analyze with Bioanalyzer or Qubit.

   Sequencing of libraries is performed using Illumina HiSeq or MiSeq platforms.

3.3 Bioinformatics

The aim of the analysis of ATAC-seq data is to produce the nucleotide level accessibility profile. In this profile, the value of every position in the genome is proportional to the probability of Tn5 insertion at this position. Code for performing the analysis below can be found at: https://github.com/calico/hendrickson_2017_supplement

3.3.1 Demultiplexing

Illumina libraries are demultiplexed using bcl2fastq software (version 2.17) with adapter trimming enabled.

Example command:

```
bcl2fastq \
--loading-threads 2 \
--demultiplexing-threads 6 \
--processing-threads 30 \
--writing-threads 2 \
-R /data/illumina/hiseq4000/160502_K00125_
0030_AH5GJTBBXX/ -o . --sample-sheet
SampleSheet.csv --use-bases-mask Y150,
I8,I8,Y150
```

3.3.2 Read Alignment

ATAC-seq libraries are aligned to the *S. cerevisiae* genome (SacCer3) using bwa-mem (version 0.7.15) with default parameters [9].

Example command:

```
bwa mem -M -t 16 SacCer3.fasta sample_R1.
fastq.gz sample_R2 | samtools view -bh -o
output.bam
```

3.3.3 Alignment Filtering

Resulting BAM files are first filtered using a custom python script that uses the pysam library (version 0.10):

1. The alignment must be primary (`not aln.is_secondary`).
2. Both read and its mate pair are mapped (`not aln.is_unmapped and not aln.mate_is_unmapped`).
3. The alignment has mapping quality of at least 30 (`aln.mapping_quality >= 30`).
4. Reads are mapped concordantly (`aln.is_reverse ^aln.mate_is_reverse`).
5. The direction of the pair is F1R2:

```
    if not aln.is_reverse and
                aln.reference_start
                <= aln.next_reference
                _start:
            return True
        elif aln.is_reverse and
                aln.reference_start
                >= aln.next_reference
                _start :
            return True
        return False
```

3.3.4 Visualizing Alignments

For visualization purposes, the resulting filtered BAM files are sorted and indexed using samtools.

Example command sequence:

```
samtools sort  -@ 16  -o output.
sort.bam -O bam -T tmp output.bam
samtools index output.sort.bam
```

3.3.5 Create Insertion Tracks

Filtered BAM files are then converted into the tracks of the number of insertions at each position in the genome (referred to as bedGraph files). For every aligned fragment the insertion locations are calculated by shifting the fragment ends of reads that align to "+" strand by four nucleotides and shifting the fragment ends of reads that align to "−"strand by five nucleotides in the 3′ direction to reflect the distance to the center of the transposase binding site [5].

3.3.6 Normalization

Since a significant and variable fraction of the reads (50%) are mitochondrial/ribosomal (Table 1), to compare numbers of insertions between samples we normalize the number of insertions in each location by the total number of insertions at nonmitochondrial, non-rDNA positions.

3.3.7 Visualizing Insertion Tracks

For visualization purposes, the normalized bedGraph files are converted to bigWig format using the utility bedGraphToBigWig from UCSC userApps (v326).

Example command:

```
bedGraphToBigWig output.bedGraph
SacCer3.fasta.len output.bigwig
```

3.3.8 RNA-Seq Analysis

For RNA-Seq, sequencing reads are aligned to the SacCer3 genome assembly accessed from using Tophat2 with the options -I 10000 -i 10 to constrain splice junction searching to >10 base pairs and <10,000 base pairs [10]. Estimates of transcript abundance and calls of differential expression are determined using Cuffdiff2 with a false discovery rate of 5% [11].

4 Illustrative Example

To demonstrate that DNA accessibility and gene expression can be simultaneously profiled, we performed a time series experiment in which a transcription factor, Gcn4, was specifically induced by the small molecule β-estradiol such that the cells begin with no Gcn4 activity and, as a function of time, accumulate the factor to levels well above those found normally [12–14].

Using the protocols provided in this chapter, we found that time-dependent Gcn4-mediated changes in DNA accessibility and gene expression patterns are clearly identifiable, such as at the *TMT1* locus where Gcn4 directly binds [15] (Fig. 2). Gcn4 activates the expression of *TMT1,* which encodes a metabolic enzyme that catalyzes the esterification of trans-aconitate [17], an inhibitor of the citric acid cycle, and of 3-isopropylmalate [18], a leucine biosynthetic pathway intermediate (Fig. 2). Increased accessibility at the site of Gcn4 binding, as measured by ATAC-Seq, is detected within 10 min of Gcn4 activation (Fig. 2), indicating that changes in ATAC-Seq signal provide a sensitive measure of transcription factor binding.

To further investigate the connection between ATAC-Seq and RNA-Seq, we calculated the correlation between accessibility and gene expression levels for every gene in the genome, and found

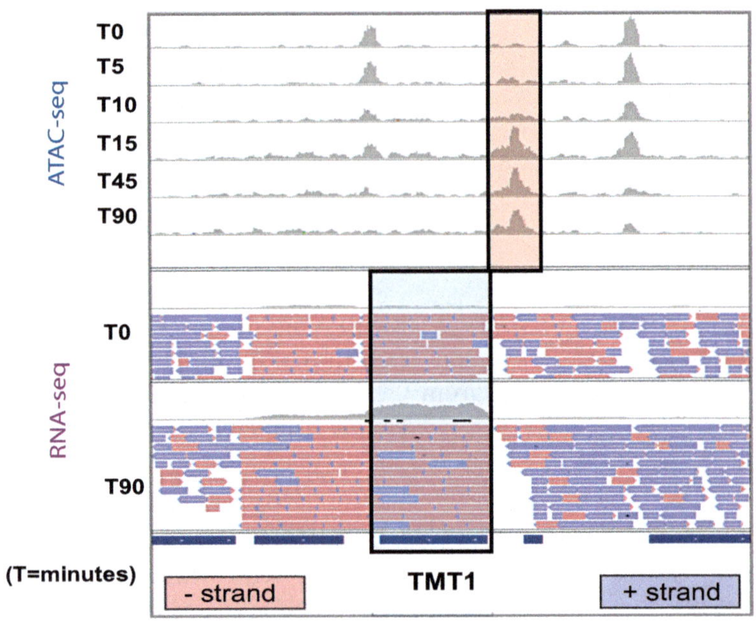

Fig. 2 Simultaneous profiling of DNA accessibility and gene expression dynamics. Levels of DNA accessibility and gene expression were measured following induction of GCN4 in strain DBY12423 [13]. Visualization of the TMT1 locus was performed using the Integrative Genomics Viewer (IGV) [16]

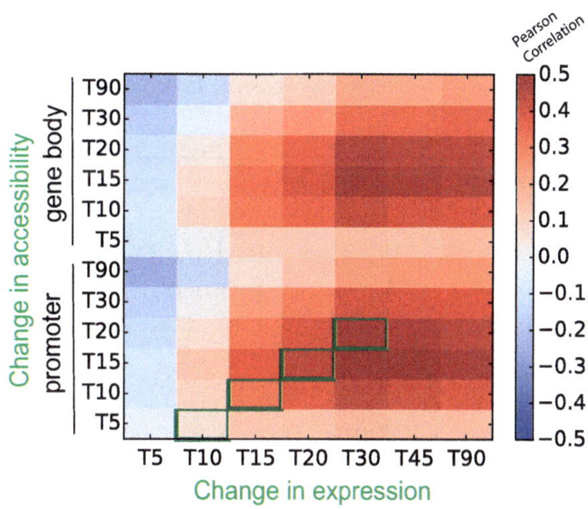

Fig. 3 Changes in DNA accessibility are correlated with future changes in gene expression. Pearson correlation matrix of changes in depth-normalized accessibility (*y*-axis) against changes in gene expression (*x*-axis) for all genes from 5 min to 90 min following GCN4 induction. Green boxes highlight that changes in accessibility precede changes in expression at early time points. Accessibility in promoter regions was calculated as a sum of normalized number of insertions from −200 ➔ 0 base pairs relative to the annotated transcription start site (TSS) from a transcription isoform sequencing (TIF-seq) dataset [19]. Accessibility in gene body was calculated by taking the sum of normalized insertion counts 0 ➔ 200 base pairs downstream of TSS

that accessibility in promoter regions is better correlated with gene expression levels than is accessibility in gene bodies. Additionally, *changes* in accessibility in gene bodies/promoter regions are better correlated with changes in gene expression than static measurements alone (Fig. 3). Since the highest observed correlation is 0.5, changes in gene expression are not necessarily reflected in changes of DNA accessibility, consistent with a previous report using MNase-Seq [20].

5 Conclusion and Outlook

The methods described above enable simultaneous profiling of gene expression in *S. cerevisiae* and can be broadly applied to other eukaryotic systems. With these tools in hand, we have the ability to track, with relatively few cells, how the genome responds to any perturbation of interest. ATAC-Seq and RNA-Seq provide complementary information on cellular states, and identifying functional connections between these two layers of biological organization will likely improve with recent computational advances in machine learning. Our expectation is that the rules that govern the fundamental interplay between chromatin and gene expression (to help give rise to homeostasis) will be deciphered to a deeper level as more time series experiments—in which DNA accessibility and gene expression levels are simultaneously profiled—are performed.

Acknowledgments

We thank Barbara Dunn for providing initial guidance on the ATAC-Seq protocol.

References

1. Song L, Crawford GE (2010) DNase-seq: a high-resolution technique for mapping active gene regulatory elements across the genome from mammalian cells. Cold Spring Harb Protoc 2010(2):pdb prot5384
2. Siersbaek R et al (2011) Extensive chromatin remodelling and establishment of transcription factor 'hotspots' during early adipogenesis. EMBO J 30(8):1459–1472
3. Meyer CA, Liu XS (2014) Identifying and mitigating bias in next-generation sequencing methods for chromatin biology. Nat Rev Genet 15(11):709–721
4. Simon JM et al (2012) Using formaldehyde-assisted isolation of regulatory elements (FAIRE) to isolate active regulatory DNA. Nat Protoc 7(2):256–267
5. Buenrostro JD et al (2013) Transposition of native chromatin for fast and sensitive epigenomic profiling of open chromatin, DNA-binding proteins and nucleosome position. Nat Methods 10(12):1213–1218
6. Adey A et al (2010) Rapid, low-input, low-bias construction of shotgun fragment libraries by high-density in vitro transposition. Genome Biol 11(12):R119
7. Buenrostro JD et al (2015) ATAC-seq: a method for assaying chromatin accessibility genome-wide. Curr Protoc Mol Biol 109:21 29 1–21 29 9
8. Buenrostro JD et al (2015) Single-cell chromatin accessibility reveals principles of regulatory variation. Nature 523(7561): 486–490
9. Li H (2013) Aligning sequence reads, clone sequences and assembly contigs with BWAMEM. eprint arXiv: 1303.3997
10. Kim D et al (2013) TopHat2: accurate alignment of transcriptomes in the presence of insertions, deletions and gene fusions. Genome Biol 14(4):R36
11. Trapnell C et al (2013) Differential analysis of gene regulation at transcript resolution with RNA-seq. Nat Biotechnol 31(1): 46–53
12. McIsaac RS et al (2014) Synthetic biology tools for programming gene expression without nutritional perturbations in Saccharomyces cerevisiae. Nucleic Acids Res 42(6):e48
13. McIsaac RS et al (2013) Synthetic gene expression perturbation systems with rapid, tunable, single-gene specificity in yeast. Nucleic Acids Res 41(4):e57
14. McIsaac RS et al (2012) Perturbation-based analysis and modeling of combinatorial regulation in the yeast sulfur assimilation pathway. Mol Biol Cell 23(15):2993–3007
15. Teixeira MC et al (2014) The YEASTRACT database: an upgraded information system for the analysis of gene and genomic transcription regulation in Saccharomyces cerevisiae. Nucleic Acids Res 42(Database issue):D161–D166
16. Thorvaldsdottir H, Robinson JT, Mesirov JP (2013) Integrative genomics viewer (IGV): high-performance genomics data visualization and exploration. Brief Bioinform 14(2): 178–192
17. Cai H et al (2001) Identification of the gene and characterization of the activity of the trans-aconitate methyltransferase from Saccharomyces cerevisiae. Biochemistry 40(45):13699–13709

18. Katz JE et al (2004) 3-Isopropylmalate is the major endogenous substrate of the Saccharomyces cerevisiae trans-aconitate methyltransferase. Biochemistry 43(20):5976–5986
19. Pelechano V, Wei W, Steinmetz LM (2013) Extensive transcriptional heterogeneity revealed by isoform profiling. Nature 497(7447): 127–131
20. Huebert DJ et al (2012) Dynamic changes in nucleosome occupancy are not predictive of gene expression dynamics but are linked to transcription and chromatin regulators. Mol Cell Biol 32(9):1645–1653

Chapter 16

Computational Network Analysis for Drug Toxicity Prediction

C. Hardt, C. Bauer, J. Schuchhardt, and R. Herwig

Abstract

The computational prediction of compound effects from molecular data is an important task in hazard and risk assessment and pivotal for judging the safety of any drug, chemical or cosmetic compound. In particular, the identification of such compound effects at the level of molecular interaction networks can be helpful for the construction of adverse outcome pathways (AOPs). AOPs emerged as a guiding concept for toxicity prediction, because of the inherent mechanistic information of such networks. In fact, integrating molecular interactions in transcriptome analysis and observing expression changes in closely interacting genes might allow identifying the key molecular initiating events of compound toxicity.

In this work we describe a computational approach that is suitable for the identification of such network modules from transcriptomics data, which is the major molecular readout of toxicogenomics studies. The approach is composed of different tools (1) for primary data analysis, i.e., the biostatistical quantification of the gene expression changes, (2) for functional annotation and prioritization of genes using literature mining, as well as (3) for the construction of an interaction network that consists of interactions with high confidence and the identification of predictive modules from these networks. We describe the different steps of the approach and demonstrate its performance with public data on drugs that induce hepatic and cardiac toxicity.

Key words Network analysis, Drug Toxicity, Toxicogenomics, Computational modeling, Literature mining, Molecular interactions

1 Introduction

Transcriptome analysis is the major read-out of toxicogenomics. Large data sets of gene expression measurements in response to treatments with different compounds are available for various in vivo and in vitro systems covering multiple target organs. While in many other research areas, for example cancer research, high-throughput sequencing has rapidly replaced microarray technology as the standard platform for transcriptome analysis and has given rise to the generation of large data repositories, the major resources of toxicogenomics data are still based on microarray technology as can be seen from publicly available data collections such as diXa (data infrastructure for chemical safety) [1]. In principle,

Louise von Stechow and Alberto Santos Delgado (eds.), *Computational Cell Biology: Methods and Protocols*, Methods in Molecular Biology, vol. 1819, https://doi.org/10.1007/978-1-4939-8618-7_16,

gene expression readouts from RNA-seq and microarrays lead to comparable signals in predicting phenotypical endpoints. However, the main advantage of RNA-seq over microarrays is that sequencing is not restricted to the number of reporter probes and, thus, allows a much deeper characterization of the gene expression state. Furthermore, the mapping of short reads to the genome allows identification of the different active splice variants as well as de novo exon prediction [2–4]. Nonetheless, considering the high number of transcriptomics datasets already generated with microarray-based technologies, as well as the fact that RNA-seq is being integrated more slowly into toxicogenomics than into other application domains, justifies the current focus on microarrays. We have built a pipeline for processing such data utilizing the R/Bioconductor software [5].

Although, it has been shown that transcriptome profiles can be used for prediction of toxicity [6, 7], it is not a straightforward task to extract mechanistic information from such experiments and thus an emerging problem of transcriptome analysis is to assess the functional content and the molecular interplay of the genes that show changes in gene expression upon compound treatment. Molecular interactions are key drivers of biological function and molecular interaction networks are reasonable descriptors of cellular action. A major source of molecular interactions is constituted by protein–protein interactions that have been measured by various experimental technologies in recent years such as yeast two-hybrid systems or mass spectrometry [8]. Subsequently, these interaction data sets have been collected and made available through various different databases [9]. In order to generate larger mappings of the interactome, attempts are ongoing to integrate these different resources in comprehensive maps of the human (and other organisms') interactomes [10]. In this context it is pivotal to gain a view of the interactome that is as comprehensive as possible, and to further combine and integrate the contents of the different resources that hold information on molecular interactions. One attempt in this direction is the ConsensusPathDB which currently integrates 530,000 interactions of different types agglomerated from 32 public resources [11] and which is used in this protocol as a resource for constructing the human interaction network.

The combination of gene expression information and molecular interaction networks with the aim to draw mechanistic conclusions from high-throughput data has been intensively studied including graph-theoretic, probabilistic, and clustering approaches [12]. In most cases these algorithms use gene expression, and/or other data for prioritizing the genes (e.g., literature context mining), in order to transform the initially unweighted interaction graph into a weighted graph where each node is weighted by its expression change and its relevance for the problem domain under analysis. The goal of these algorithms is then to compute subgraphs (network modules) that agglomerate many of the expression changes. The rationale is that these network modules point to mechanistic, biologically more relevant changes than results of gene-by-gene analyses because they take into account the biological relations of the genes. For network inference in this protocol we have used a graph-theoretic method, BioNet, that is based on minimal spanning tree computation and that addresses the mathematical problem of Steiner tree construction [13]. This algorithm has been adapted and implemented as an R/Bioconductor package [14] so that the computation of functional modules can be easily fit into bioinformatics workflows for identifying toxicity modules from transcriptome data.

In this protocol, we combine the above described components for human interaction network construction, gene node scoring and network inferences into a protocol for identifying network modules from drug response gene expression data (Fig. 1). The protocol is largely based on R/Bioconductor packages and public resources, thus assuring high reproducibility for users. Given a certain compound we generate two independent prioritization scores: the first one is based on microarray data and delivers for each gene a significance *p*-value for differential expression comparing the treatment vs. the control experiments; the second one is based on literature mining and delivers for each gene a significance *p*-value for the problem domain under analysis (here liver and heart toxicity). These scores are combined and mapped onto a comprehensive human protein–protein network constructed from the ConsensusPathDB. Finally, the BioNet algorithm is used to identify network modules from the weighted interaction graph.

2 Materials

2.1 R/Bioconductor Packages

Microarray data analysis and network analysis are done with the R/Bioconductor software packages. This can be run on a wide variety of UNIX platforms, MS Windows and MacOS. We recommend installing the latest version of R/Bioconductor freely available at www.r-project.org [5].

2.1.1 Remapping of Oligoprobes

As a preparatory step for the analysis of Affymetrix Gene Chips, we recommend re-defining the original probe sets, i.e., the sets of oligonucleotide sequences (oligoprobes) that are used by the provider as reporters for the genes. This procedure enables the assignment of a unique set of reporter oligoprobes to the respective gene and, thus, avoiding the cross-hybridization problem. It has been shown that remapping oligoprobes unambiguously to the latest genome-build increases performance of Affymetrix Gene Chip transcriptomics platforms [15]. Re-defined probe sets in the form of so-called custom CDF files are freely available for

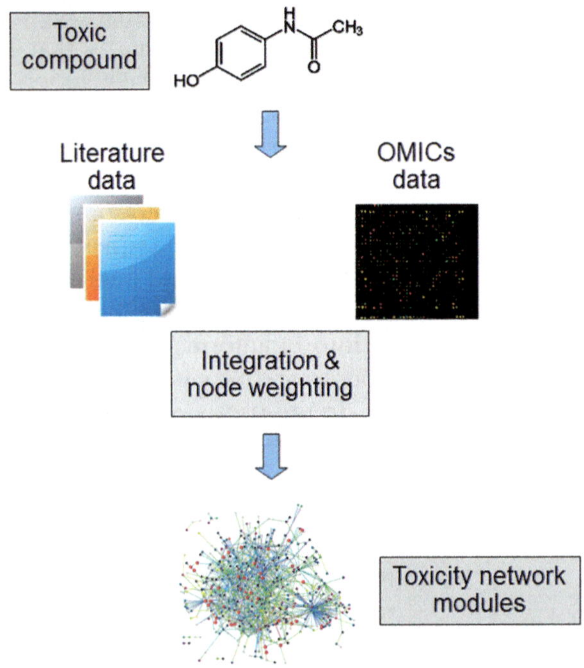

Fig. 1 Workflow for computing toxicity network modules. Compound action is characterized in two ways: Literature-based identification of related genes and proteins and omics-based identification of differentially expressed genes. As a further resource a large protein–protein interaction network is generated from public information. The nodes of the network (i.e., genes) are weighted according to the experimental and literature data and subsequently investigated by graph algorithms for the identification of densely connected subnetworks (hubs or cliques)

diverse chip platforms at http://brainarray.mbni.med.umich.edu/CustomCDF.

When downloading the "*CustomCDF*" R package and the custom CDFs, the corresponding versions of R and BioConductor should be used, for example, for CustomCDF Version 21 (the most current version) R 3.3 and BioC 3.3 should be used.

2.1.2 Normalization

Affymetrix array data suffer from microarray background signals (optical noise) as well as from nonspecific binding of the oligo-probes. For background correction and normalization the package "*gcrma*" (GC Robust Multi-array Average; [16]) has been used.

2.1.3 Differential Expression

The normalized microarray data can be fed into the package "*limma*" [17] in order to calculate differentially expressed genes, i.e., genes that are upregulated or downregulated significantly when comparing compound treatment against control experiments. It estimates fold-changes and standard errors by fitting a linear model to each gene profile and uses an empirical Bayesian approach to smoothen these errors.

2.1.4 Network Module Analysis Identification

In order to retrieve mechanistic information molecular networks are inferred. These networks provide causal links between the genes and proteins. The goal of such network inference is to identify functional modules, i.e., subnetworks that show a significant response to a certain drug treatment, in order to better describe the functional consequences of differential expression in the context of gene–protein interactions. In this work we have used the BioNet algorithm [14] available as the R package "*BioNet*" (see Subheading 2.4).

2.2 Literature Mining

Text mining is an information retrieval task aimed at discovering new information by automatically extracting it from different textual resources [18]. Probably the largest and most important textual resources for biomedical knowledge are papers that were published in peer reviewed journals. These publications are accessible via databases such as PubMed (https://www.ncbi.nlm.nih.gov/pubmed/) currently containing more than 26 million articles.

The most important step in the text-mining process is the extraction of terms or features. This step is called named entity recognition (NER). NER is a well-defined subtask of natural language processing (NLP) and aims to identify named entities—here known genes or proteins—in a larger context. NER of genes or compounds is a nontrivial task considering the different names and synonyms including systematic and nonsystematic nomenclatures. In addition, gene names or descriptions are often ambiguous referring to different gene identifiers. To increase specificity of the results, it is important to resolve these ambiguities (*see* Subheading 3.2).

2.3 ConsensusPathDB—Constructing an Interaction Network Resource

ConsensusPathDB is a meta-database for molecular interactions and pathways that currently integrates 32 public resources. With more than 530,000 different, experimentally derived molecular interactions, it represents the most comprehensive model of the human cellular interactome. The database is accessible through a web server (http://consensuspathDB.org) with functionality for characterizing genes, proteins, metabolites and other types of biomolecules at the level of pathways and interaction networks. ConsensusPathDB can be accessed via a web server or via web services. Any modern HTML5-enabled browser can be used.

Besides information on biological pathways the ConsensusPathDB holds an integrated network of molecular interactions that was constructed from 19 different databases comprising more than 300,000 binary protein–protein interactions (PPIs).

Since PPI networks can contain a large share of false positive interactions, the quality assessment of individual PPIs is particularly critical. In ConsensusPathDB we provide a quality check for PPIs consisting of the following methods:

2.3.1 Topology-Based Criteria (Unsupervised Methods)

CAPPIC [19]—applies Markov clustering and evaluates the confidence that a PPI belongs to a local cluster of PPIs; score range [0, 1].

Common neighbors [20]—evaluates common network neighborhoods of interacting nodes; score range [0, 1].

Geometric embedding [21]—embeds the network in a geometric space and weights edges according to the distance between their incident nodes in that space; score range [0, 1].

2.3.2 Annotation-Based Criteria (Supervised Methods)

Literature evidence [22]—number of publications that report the interaction; score range [0,∞).

Pathway co-occurrence [22]—indicates whether interaction partners are found in the same pathway; score range {0, 1}.

Semantic similarity [23]—similarity of interaction partners computed from the GO tree; score range [0, 1].

Based on these measures we have defined a meta-scoring approach, called IntScore, which combines the individual confidence scores of a PPI to an aggregated score that indicates whether the interaction is plausible or not [24].

2.4 Network Module Analysis

The input of network module analysis is typically a weighted graph where nodes represent genes/proteins and edges represent interactions between the genes/proteins. Nodes and/or edges are weighted according to the data. Given the PPI network (Subheading 2.3 above) and the *p*-values derived from microarray analysis (Subheading 2.1) and literature mining (Subheading 2.2), network analysis can be applied in order to compute the predictive toxicity networks modules from the large PPI network.

BioNet is available as an R-package that can easily be integrated in computational workflows. The distribution of *p*-values derived from literature mining and gene expression measurements is transformed into a mixture of two Beta-functions in order to separate signal (*B(a, 1)* distribution) from noise (*B(1, 1)* uniform distribution).

Such a function has the form:

$$f(x|a,\lambda) = \lambda + (1-\lambda)\,ax^{(a-1)} \quad \text{with} \quad 0 < x \leq 1, 0 < a < 1.$$

The function uses the initial *p*-values to derive estimates for *a* and λ.

In the next step, given a certain false-discovery rate, *t(FDR)*, this fit is used to derive for each node *x* a node score, *S(x)*, by:

$$S(x) = (a-1)\left(\log(x) - \log\left(t(FDR)\right)\right).$$

Once the nodes of the PPI network have been weighted according to these values the BioNet approach applies linear integer programming [25] in order to derive a high-scoring subnetwork by the following steps:

1. In the first step all positive connected nodes are aggregated into higher level nodes called meta-nodes.
2. Edges are scored based on the connected node's scores.
3. From these edge scores a minimum spanning tree (MST) is calculated [26].
4. All paths between positive meta-nodes are calculated based on the MST to obtain the negative nodes between the positives.
5. Upon these negative nodes again an MST is calculated from which the path with the highest score, which is based on node scores of negative nodes and the positive meta-nodes they connect, gives the resulting approximated module.

The result of the BioNet algorithm is a subnetwork that agglomerates highly significant nodes. The sizes of the modules are dependent on the algorithmic parameter, i.e., the false-discovery rate *t(FDR)*.

2.5 Data Sets for Toxicity Studies

2.5.1 Open TG-GATEs (http://toxico.nibiohn.go.jp)

Open TG-GATEs is a public toxicogenomics repository set up by the Japanese Toxicogenomics Project and Toxigenomics Informatics Project. It provides gene expression data for six different cell types (human in vitro, rat in vivo/in vitro with single/repeated dose) treated with 170 compounds of three dosages after three time points [27].

2.5.2 DrugMatrix (https://ntp.niehs.nih.gov/drugmatrix/)

The US toxicology database DrugMatrix offers gene expression data for 638 different drug treatments across six tissue types, ranging from 1 to 3 time points and dosages, respectively [28].

2.5.3 ToxDB (http://toxdb.molgen.mpg.de)

We have developed ToxDB [29] to integrate expression data from the above repositories (Open TG-GATEs and DrugMatrix) and combined them with pathway and PPI data from ConsensusPathDB. In total, 1152 treatment data sets are currently contained. ToxDB allows the user to browse through a drug treatment's effects on cellular pathway response. Vice versa, the user can browse through pathways and find out which drug treatments have the greatest effect on them. Additionally, ToxDB provides a differential pathway analysis, where the users can select their own two sets of drug treatments and retrieve pathways that differentiate best between the two sets.

2.5.4 Use Cases

To test the implemented pipeline, we assembled data sets for three hepatotoxic compounds (acetaminophen, azathioprine, and rifampin) and two cardiotoxic compounds (doxorubicin and cyclophosphamide). We retrieved gene expression data and literature data and performed the above mentioned analysis steps.

3 Methods

3.1 Primary Analysis of Microarray Data

After downloading and installing the correct customCDF file and starting the latest versions of R/BioConductor (R v3.4.0, Bioconductor v3.5), follow this standardized pipeline in R to normalize Affymetrix data and calculate differential gene expression between data sets (e.g., treatment vs control).

3.1.1 Normalize Affymetrix Data with "gcrma"

1. Load the "*gcrma*" package by typing library(gcrma) in the Bioconductor workspace.
2. Read in the CEL files using a custom CDF file (e.g., "*HS133A_HS_UG*") for remapping the probe sets by typing (*see* **Note 1**):

```
library(affy)
set ← ReadAffy(celfile.path=<CEL

directory>, cdfname = <custom CDF name>)
```

3. Normalize the data: nset ← gcrma(set).

3.1.2 Calculate Differential Expression Based on Treatment vs. Control Data with "limma"

1. Load the "*limma*" package: library(limma).
2. Fit probe set to linear model where "design" corresponds to the design matrix (dependent on setup of treatment, control and their replicates): fit ← lmFit(nset, design).
3. Perform empirical Bayes adjustment: fit ← eBayes(fit).
4. Show most differentially expressed genes, sorted by False Discovery Rate (FDR)- corrected *p*-value after Student's t-test (Table 1): topTable(fit).
5. It may be necessary to map probe set IDs to gene IDs, for example when different isoforms need to be condensed to a gene readout or if original CDFs are taken instead of custom CDFs. This mapping can be done, e.g., by taking the most significantly differentially expressed probe set per gene. Alternatives would be to take the geometric mean, median, or the minimum of expression fold change among probe sets per gene (ordered from less to more conservative approaches).
6. Check the respective manuals/tutorials of the packages for more info and additional options (*see* **Note 2**).

3.2 Literature Mining

In order to define the textual resource for text mining a PubMed query is required, which considers different aspects of the research question. We use AND, OR, and * (wildcard) operators for

Table 1
Exemplary output of *topTable(fit)* as part of the "*limma*" package

ID	Name	logFC	AveExpr	T	*P*Value	adj*P*Value
A_44_P271658	NM_053289	−4.652897	12.989191	−38.626454	1.46E-11	4.76E-07
A_44_P735711	TC524300	−4.360493	11.151226	−36.104656	2.72E-11	4.76E-07
A_43_P17697	NM_001006975	−4.367338	11.000957	−35.531527	3.16E-11	4.76E-07
A_42_P565621	XM_213932	−4.053834	11.297143	−33.342658	5.67E-11	5.28E-07
...	...	...	...	...	...	...

combining different query terms. Combining two queries with AND means matching of both queries, OR means matching of at least one query and * means anything following the query term:

1. Using synonyms for a given compound, e.g., paracetamol, by searching ChEMBL (https://www.ebi.ac.uk/chembl): APAP *OR* Paracetamol *OR* Acetaminophen *OR* Tylenol *OR* Phenaphen.
2. Defining a query to search for hepatotoxicity: hepatotoxic* *OR* (liver *AND* injury) *OR* (liver *AND* damage) *OR* (toxic* *AND* liver).
3. Definition of a species: human (Other organisms are not explicitly excluded).

All three fragments are connected with a logical *AND* to form a query. Using this query we identified more than 2500 relevant articles from PubMed (*see* **Note 3**). These articles can be downloaded from PubMed as an XML file (click on 'Send to' → File → Format: XML).

We then used the library GNAT (http://gnat.sourceforge.net/), which does not only perform NER but also gene mention normalization [30]. Gene mention normalization scans the whole paragraph for gene/protein related entities. The corresponding knowledge is managed with a MySQL database that comes with the installation of the tool. The database contains information on the relation between genes and species, GO-terms, chromosomal localization or diseases. Nearby occurrences of these relations are then used in a scoring function to remove false positive hits and to resolve ambiguities. NER for chemicals can be performed with the ChemSpot library ([31]; https://www.informatik.hu-berlin.de/de/forschung/gebiete/wbi/resources/chemspot/chemspot). Both tools include the required dictionaries and can be used either from command line or directly be integrated into Java source code. For example, to use ChemSpot in Java we first have to create a ChemSpot Object:

```
ChemSpot tagger = ChemSpotFactory.
 createChemSpot(
    chemSpotDir+"dict.zip",chemSpotDir
    +"ids.zip", chemSpotDir+"multiclass.bin");
```

This object can then be used to tag chemicals in a text and receive identifiers:

```
for (Mention mention : tagger.tag(text)) {
     mention.getText();
     mention.getCHID();
}
```

Having extracted genes and compounds, the next step is to assess their relevance (*see* **Note 3**). An obvious measure is the number of publications a gene appears in. But these raw counts may not be the best score since there are omnipresent genes (such as p53, TNF or albumin) and compounds (e.g., water or sodium), which are not specific for the question of interest and/or very well investigated and, thus, present with an overwhelming body of literature. To assess the relevance we can compare the question of interest with a reference set and calculate a *p*-value using Fisher's exact test. The reference set is created by downloading all publications from PubMed and perform the NER as described above (Table 2).

3.3 PPI Network Construction and Confidence Assessment

We have computed the aggregated confidence scores by applying IntScore to the original PPI network of ConsensusPathDB and retrieved a high-quality network consisting of 9533 nodes with 80,422 interactions (Fig. 2a). The node degree in the remaining network follows a power law distribution, which is postulated for biological (scale-free) networks [32], i.e., a network that has few hubs with many connections and many nodes with fewer connections (Fig. 2b). Additionally, the IntScore functionality was incorporated into ConsensusPathDB: The values resulting from the aggregated confidence scoring are displayed for each interaction so that the user can decide whether or not to work further on a particular PPI (Fig. 2c). The PPI network, along with the quality assessment scores, can be downloaded via http://cpdb.molgen.mpg.de/download/ConsensusPathDB_human_PPI.gz (*see* **Note 4**).

3.4 Network Module Prediction and Interpretation

Follow these steps to combine differential expression data with PPI network data to generate induced network modules:

1. Download the current PPI network from ConsensusPathDB and unpack it.
   - Optional: Filter for high-confidence interactions, e.g., confidence >0.5 to get a "high-quality" PPI network.

Table 2
Results of the literature mining approach for the compound rifampin (top twenty genes)

Symbol	Description	NTest	SizeTest	NRef	*P*Value
NR1I2	Nuclear receptor subfamily 1, group I, member 2	115	783	3919	3,6627E-266
CYP3A4	Cytochrome P450, family 3, subfamily A, polypeptide 4	122	783	6959	3,2083E-255
AADAC	Arylacetamide deacetylase	15	783	52	7,37685E-53
SLCO1B1	Solute carrier organic anion transporter family, member 1B1	21	783	1447	9,72969E-45
NR1I3	Nuclear receptor subfamily 1, group I, member 3	18	783	1299	1,76942E-38
SLC10A1	Solute carrier family 10 (sodium/bile acid cotransporter), member 1	13	783	519	2,49137E-32
ABCB11	ATP-binding cassette, subfamily B (MDR/TAP), member 11	15	783	1324	2,84992E-31
SLCO1B3	Solute carrier organic anion transporter family, member 1B3	12	783	706	5,6643E-28
GPT	Glutamic-pyruvate transaminase (alanine aminotransferase)	24	783	15,661	1,25013E-26
G6PC	Glucose-6-phosphatase, catalytic subunit	15	783	2793	3,84774E-26
CYP2B6	Cytochrome P450, family 2, subfamily B, polypeptide 6	16	783	4767	2,95472E-24
MIR34A	microRNA 34a	11	783	1151	5,65159E-23
CYP2E1	Cytochrome P450, family 2, subfamily E, polypeptide 1	12	783	2680	1,59033E-20
CYP1A2	Cytochrome P450, family 1, subfamily A, polypeptide 2	12	783	3178	1,41152E-19
SLCO1A2	Solute carrier organic anion transporter family, member 1A2	8	783	508	1,48373E-19
SGK3	Serum/glucocorticoid regulated kinase family, member 3	7	783	267	2,64929E-19
CYP3A5	Cytochrome P450, family 3, subfamily A, polypeptide 5	10	783	1757	9,65869E-19
NR1H4	Nuclear receptor subfamily 1, group H, member 4	10	783	2126	7,66336E-18
NCOR2	Nuclear receptor corepressor 2	8	783	976	4,98813E-17
ABCC3	ATP-binding cassette, subfamily C (CFTR/MRP), member 3	7	783	898	3,89841E-15
...	...	...	...	...	...

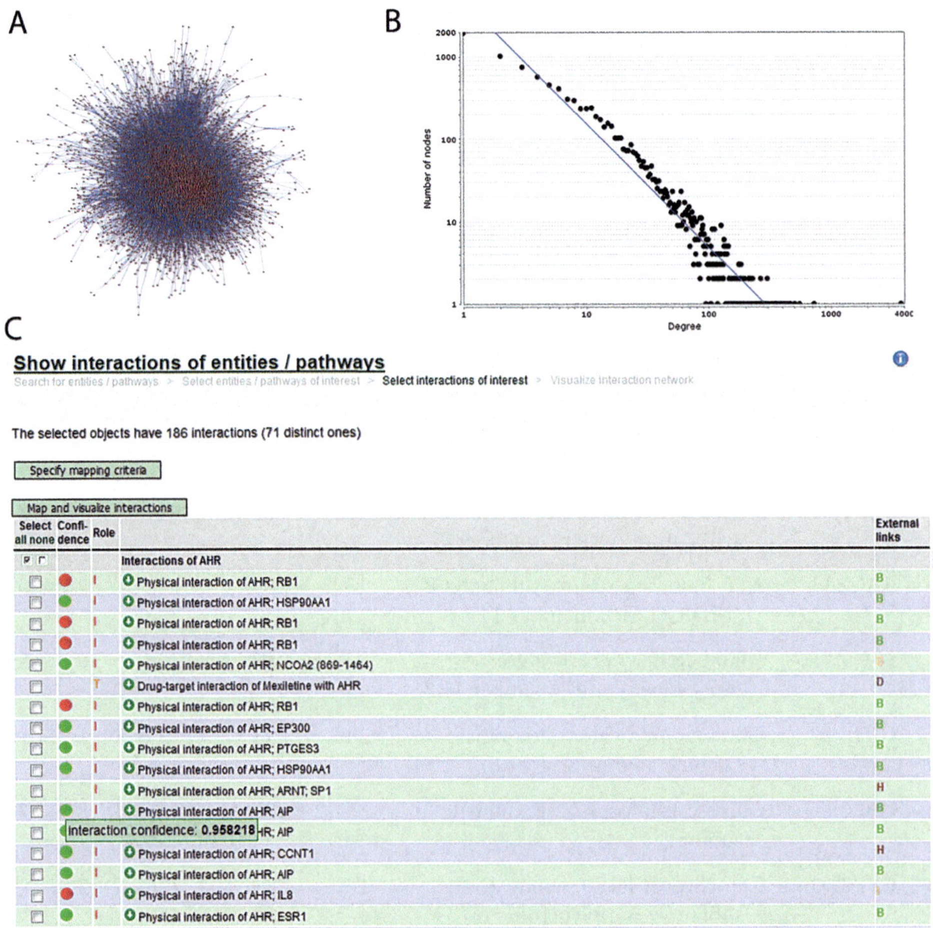

Fig. 2 PPI network (**a**) High-quality PPI network with 9533 nodes and 80,422 interactions derived from ConsensusPathDB after confidence scoring using IntScore. (**b**) Node degree distribution of the human PPI network. The node distribution follows a power law distribution $y = a^*x^b$ with parameters $a = 4896.5$ and $b = -1.504$. (**c**) Interaction confidence scores are integrated in the ConsensusPathDB as a color-coded aid for the user

2. Reformat the PPI file to standard interaction format (cf. Table 3).
3. Start R and load the "*BioNet*" package: library(BioNet).
4. Load the PPI file in standard interaction format:

```
ppi ← read.csv(<PPI file in sif>, sep="\t",
header=T)
```

5. Convert the PPI graph to 'GraphNEL' format:

```
ppiG ← ftM2graphNEL(ppi,edgemode=
"undirected")
```

Table 3
PPI interactions in *s*tandard *i*nteraction *f*ormat (*.sif*)

Protein 1	Protein 2
TSEN2	POLD1
CUL1	POLD1
CUL1	SNRPC
CUL1	FBXO17
...	...

6. Load into R a file containing fold changes and *p*-values of genes with the same IDs as in the PPI file (acquired as in Subheadings 3.1 and 3.2). Let pval be the vector of *p*-values.

7. Create a subnetwork with the genes that have been measured for differential expression (e.g., with gene identifier column "Gene ID"):

```
deData ← read.csv(<diffEx file
acquired in Method 3.1 step 8>,
sep="\t", header=T)
subnet ← subNetwork(deData$Gene.ID,
ppiG)
```

8. Fit a beta-uniform mixture model of the following form:

```
fb ← fitBumModel(pval, plot = FALSE)
```

   The mixture model concept is typically used to fit a bimodal distribution to the data, for example when one wishes to distinct noise from signal or subsets of data that follow different distributions.

9. Score every network node x using a user-defined false-discovery rate (e.g., FDR = 0.01) as a threshold:

```
scores ← scoreNodes(subnet, fb, fdr = 0.01)
```

   The FDR is the critical threshold of the algorithm because it influences the sizes of the resulting network modules (*see* **Note 5**).

10. Calculate the maximum scoring subnetwork (running the heuristic "*fastHeinz*"):

```
module ← runFastHeinz (subnet, scores)
```

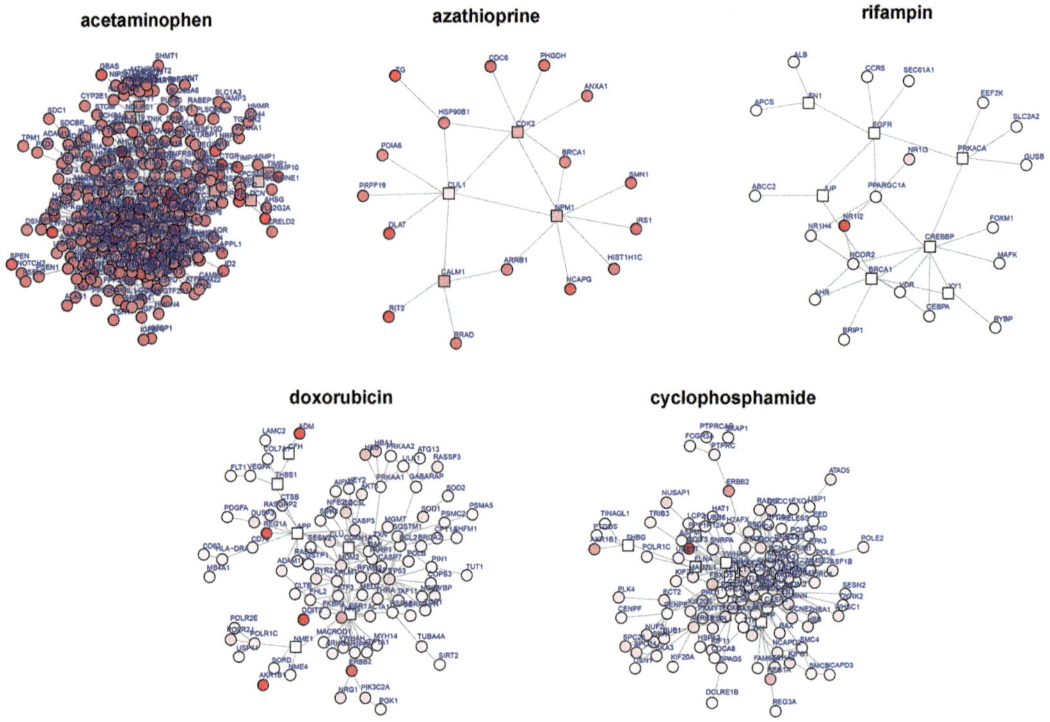

Fig. 3 Network modules. Combining PPI-information with the weighted top genes from literature mining and OMICS analysis key toxicity modules were computed using the BioNet algorithm (*t(FDR)* = 0.02). The five compounds acetaminophen, azathioprine, rifampin, doxorubicin, and cyclophosphamide were analyzed. Node coloring represents the relative weight of each gene. Nodes in dark red are identified with very high significance and represent key mediators of toxicity

11. Visualize the induced network module:

```
plotModule(module, scores = scores, diff
.expr = logFC)
```

We refer to the tutorial of BioNet [14] for a more detailed description of the method as well as additional options (*see* **Note 5**). The R-code used for Subheadings 3.1 and 3.4 is accessible under https://github.molgen.mpg.de/herwig/Hardt_et_al_MethodsMolBiol/.

Network modules were computed as described above for all five compounds under study (Fig. 3). Based on the different node weighing procedures derived from the *p*-values from literature mining and microarray analyses, the modules include different numbers of genes. Resulting gene lists contain 314 genes for acetaminophen, 20 for azathioprine, 27 for rifampin, 98 for doxorubicin, and 125 for cyclophosphamide.

We exemplify the results for the hepatotoxic compound rifampin. Rifampin is an antibiotic that inhibits DNA-dependent RNA polymerase activity in susceptible cells. Specifically, it interacts with bacterial RNA polymerase but does not inhibit the mammalian enzyme. It is bactericidal and has a very broad spectrum of activity against most gram-positive and gram-negative organisms and specifically *Mycobacterium tuberculosis* [33]. Because of rapid emergence of resistant bacteria through mutations in the rpoB gene, use is limited and it was suggested that rifampin could be used in combination with other antimicrobial agents [34]. Rifampin is metabolized in the liver and eliminated in bile and, to a much lesser extent, in urine. The mechanism of rifampin hepatotoxicity is not well understood. However, Rifampin is extensively metabolized by the liver and induces multiple hepatic enzymes including CYP3A4 [35]. Patients with preexisting liver disease and cirrhosis are particularly likely to develop jaundice on rifampin therapy (https://livertox.nlm.nih.gov/Rifampin.htm).

We uploaded the summarized *p*-values from literature mining and microarray analysis to the PPI network. The BioNet node weighting yields a transformation of the original *p*-value distribution into a Beta-distribution mixture (Fig. 4a), which results in a reasonable fit (Fig. 4b). Scoring with FDR threshold ($q = 0.02$) yields a module consisting of 27 genes that is highly responsive for rifampin treatment (Fig. 4c).

The selected response genes enrich important pathways of rifampin metabolization and liver toxicity such as the bile acid pathway, the role of nuclear receptors in lipid metabolism and toxicity (Fig. 5). The different pathway concepts are highly overlapping as can be seen from Fig. 6.

The most reactive protein in the computed network module for rifampin is NR1I2. This gene product belongs to the nuclear receptor superfamily, members of which are transcription factors characterized by a ligand-binding domain and a DNA-binding domain. The encoded protein is a transcriptional regulator of the cytochrome P450 gene CYP3A4, binding to the CYP3A4 promoter as a heterodimer with the 9-cis retinoic acid receptor RXR. It is activated by a range of compounds that induce CYP3A4, including dexamethasone and rifampicin [36].

Thus, we conclude that the combined approach conducted here generates network modules that are highly predictive of compound toxicity. These modules can be the basis for molecular models and be further used for toxicity studies and the construction of AOPs.

3.5 Conclusion and Outlook

The response of cells to drugs, chemicals, or other compounds cannot be described exhaustively at the level of single genes or proteins. Thus, network approaches that take into account the interactions of the various genes and proteins and provide a mechanistic view on the transcriptome results are better suited for describing the molecular effects of compounds. The approach presented here interrogates the global protein–protein interaction

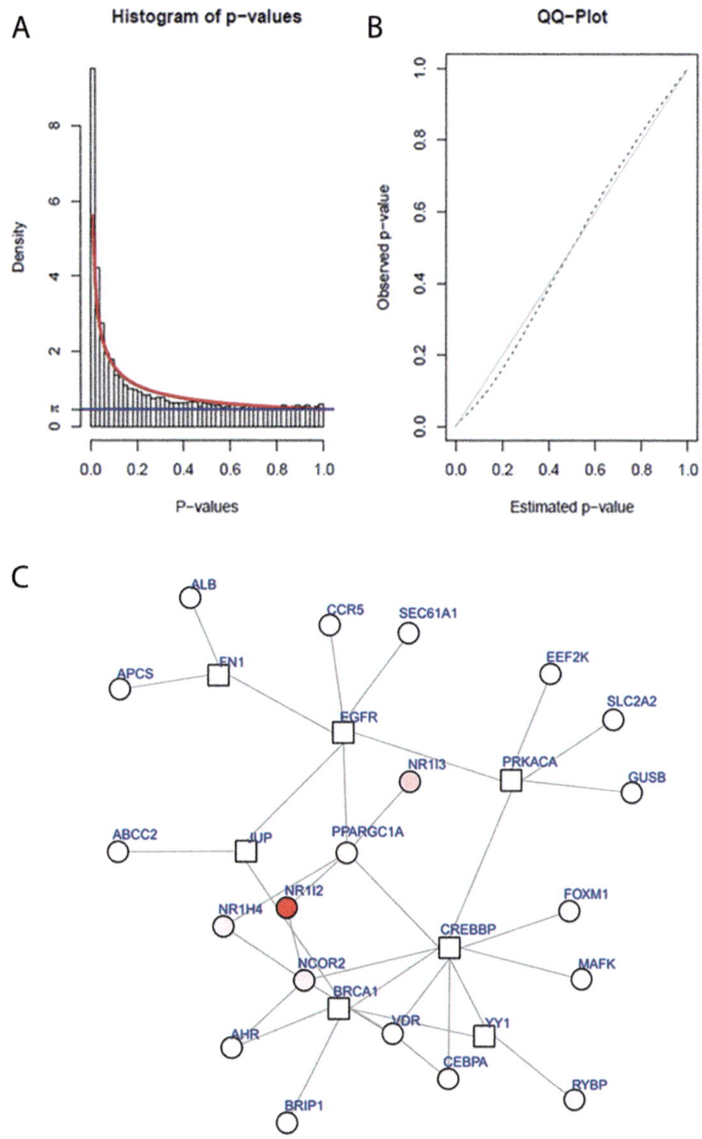

Fig. 4 BioNet network module computation. (**a**) Distribution of *p*-values from rifampin. Fitted curves correspond to the estimated Beta functions for signal and noise that are subsequently being used to weight the network nodes. (**b**) QQ-plot of fitted (*X*-axis) vs. experimental (*Y*-axis) *p*-values. (**c**) Computed module ($t(FDR) = 0.02$) with 27 response genes (*BRCA1, RYBP, PRKACA, NCOR2, VDR, YY1, FOXM1, APCS, BRIP1, AHR, CEBPA, NR1H4, PPARGC1A, NR1I2, CCR5, EEF2K, MAFK, SLC2A2, SEC61A1, FN1, GUSB, EGFR, NR1I3, JUP, CREBBP, ALB, ABCC2*)

network and aims at identifying high-scoring subnetworks that agglomerate a large amount of the observed expression changes by mapping gene expression data onto the network and applying graph theoretical algorithms for network inference.

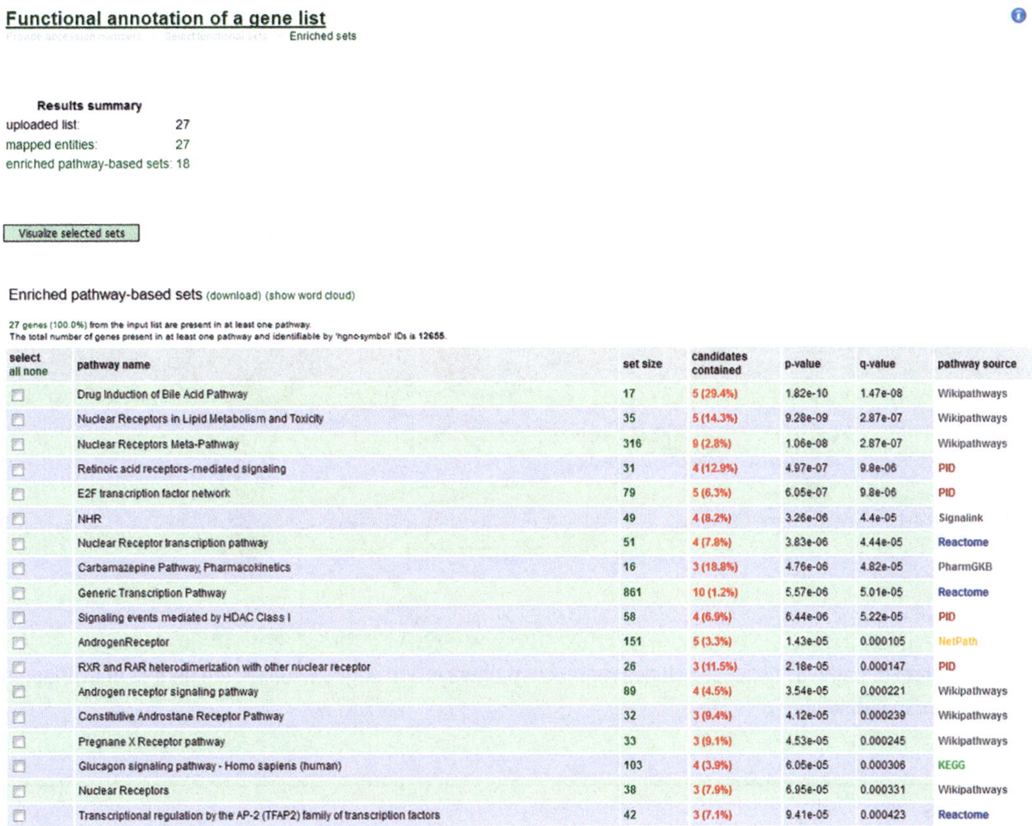

Functional annotation of a gene list

Enriched sets

Results summary

uploaded list: 27
mapped entities: 27
enriched pathway-based sets: 18

Visualize selected sets

Enriched pathway-based sets (download) (show word cloud)

27 genes (100.0%) from the input list are present in at least one pathway.
The total number of genes present in at least one pathway and identifiable by 'hgnc-symbol' IDs is 12655.

select all none	pathway name	set size	candidates contained	p-value	q-value	pathway source
☐	Drug Induction of Bile Acid Pathway	17	5 (29.4%)	1.82e-10	1.47e-08	Wikipathways
☐	Nuclear Receptors in Lipid Metabolism and Toxicity	35	5 (14.3%)	9.28e-09	2.87e-07	Wikipathways
☐	Nuclear Receptors Meta-Pathway	316	9 (2.8%)	1.06e-08	2.87e-07	Wikipathways
☐	Retinoic acid receptors-mediated signaling	31	4 (12.9%)	4.97e-07	9.8e-06	PID
☐	E2F transcription factor network	79	5 (6.3%)	6.05e-07	9.8e-06	PID
☐	NHR	49	4 (8.2%)	3.26e-06	4.4e-05	Signalink
☐	Nuclear Receptor transcription pathway	51	4 (7.8%)	3.83e-06	4.44e-05	Reactome
☐	Carbamazepine Pathway, Pharmacokinetics	16	3 (18.8%)	4.76e-06	4.82e-05	PharmGKB
☐	Generic Transcription Pathway	861	10 (1.2%)	5.57e-06	5.01e-05	Reactome
☐	Signaling events mediated by HDAC Class I	58	4 (6.9%)	6.44e-06	5.22e-05	PID
☐	AndrogenReceptor	151	5 (3.3%)	1.43e-05	0.000105	NetPath
☐	RXR and RAR heterodimerization with other nuclear receptor	26	3 (11.5%)	2.18e-05	0.000147	PID
☐	Androgen receptor signaling pathway	89	4 (4.5%)	3.54e-05	0.000221	Wikipathways
☐	Constitutive Androstane Receptor Pathway	32	3 (9.4%)	4.12e-05	0.000239	Wikipathways
☐	Pregnane X Receptor pathway	33	3 (9.1%)	4.53e-05	0.000245	Wikipathways
☐	Glucagon signaling pathway - Homo sapiens (human)	103	4 (3.9%)	6.05e-05	0.000306	KEGG
☐	Nuclear Receptors	38	3 (7.9%)	6.95e-05	0.000331	Wikipathways
☐	Transcriptional regulation by the AP-2 (TFAP2) family of transcription factors	42	3 (7.1%)	9.41e-05	0.000423	Reactome

Fig. 5 Overrepresentation analysis with module genes. Overrepresentation analysis was done within the ConsensusPathDB inserting the 27 proteins of the computed network module

The presented approach has three main components: (1) the underlying network to be inferred, (2) the weighting of the nodes, and (3) the inference algorithm. It should be noted that there exist many alternative approaches and that there is no consensus on any of these three components. For example, a network of interest could consist of other interactions, for example gene regulatory interactions or molecular signaling pathways or might be derived by experimental data (coexpression networks); the weighting of the nodes could be done alternatively by taking into account other types of experimental data, for example mutations, protein expression, methylation scores or combinations thereof; finally, the inference could be based on other algorithms, for example diffusion models and random walks. Many different approaches exist (for a recent review see [37]) in parallel and a comparison of methods with a controlled network is still missing. Thus, the computation of network modules that link genotype with phenotype remains a challenge.

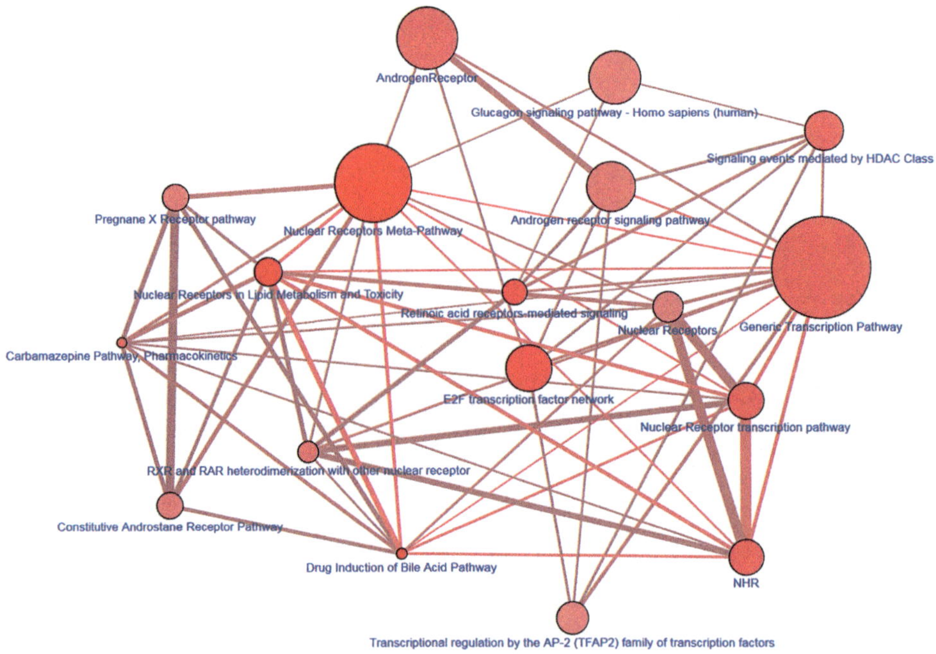

Fig. 6 Overlap of pathway concepts. Nodes represent pathways and edges visualize the degree of overlapping genes between pathways

4 Notes

1. The remapping of oligoprobes prior to primary data analysis results in a redefinition of the original oligonucleotide probe sets and the results are not straightforwardly compatible with standard Affymetrix data analysis results. In particular it should be noted that after redefinition probes match a unique gene locus avoiding the cross-hybridization problem, probe sets per gene are no longer uniform in size (i.e., 11 probes per probe set) but have a varying number of probes (≥ 3), and the resulting number of probe sets/genes differs. A recent comparison of both protocols in a toxicogenomics setting can be found in [38].
2. Transcriptome data can be processed in different ways even using a single software platform such as Bioconductor. There are multiple alternatives to the GCRMA normalization and the detection of differentially expressed genes. A comparative study has been published with the MAQC-II study [39]. The main difficulty for biostatistical tools and models is to correctly quantify expression data under the null hypothesis, i.e., the assumption that no change of expression is observed between the treatment and control conditions. Models cover Bayesian approaches, linear models, generalized linear models among others. We have used limma, a linear model approach, since

this is a well-established and widely used method for detecting differentially expressed genes.

3. Most of the articles listed in PubMed are not freely available in full text. Even if some articles are available in full text as in PMC (https://www.ncbi.nlm.nih.gov/pmc/), important information is often presented in tables or images, making automated interpretation and extraction of information highly complex. By performing text-mining only on freely available abstracts, important information in the full text article may be missed. On the other hand, the most important information in the paper is usually given in the abstract.

   The presented approach searches for co-occurrence of factors. The co-occurrence does not imply a physical interaction between the factors. Instead, it identifies factors that are significantly reported together which may hint for a hidden connection. To identify physical interactions the literature mining needs to consider the semantic properties of the sentences, e.g., by searching for interaction keywords (such as "*bind*," "*interact*," and "*inhibit*") typically located between the factors.

   Another problem is the correlation of information in literature. Information is copied and passed on especially in a concrete research field. Due to this effect even wrong information may occur repetitively and can therefore appear to be significantly associated with the question of interest. However, it can be very helpful and powerful to uncover hidden relations, such as patterns or trends.

4. The charted interactome of most organisms is still incomplete and the prediction of interactions is highly error-prone. This necessitates a careful quality control of the interactions in particular when interaction graphs are composed from different heterogeneous resources. In ConsensusPathDB we therefore use an aggregated score for PPI quality checking, which consists of the methods described in Subheading 2.3. In addition, it should be noted that interactions are tissue and time-dependent what could not be resolved easily. Thus, there remain some ambiguities when using PPIs.

5. There is currently no consensus about choosing between the different strategies of network module identification and new methods are constantly being developed. We have selected BioNet as one particular method, which is easy to integrate into R workflows. The critical parameter of the BioNet network inference algorithm is the FDR level. This level determines the size of the output module: The lower the FDR threshold is set, the smaller the resulting network module will be, which makes a bit of tinkering advisable in order to receive an interpretable network module. Besides this method, there are multiple other ones [40]. Critical issues of the methods are typically the size

of the PPI network as well as the proper weighting of the nodes. In our examples we have used a combined scoring with significance *p*-values derived from gene expression data and literature mining, which could be interpreted as a combination of unsupervised and supervised weights. A few studies have been carried out to compare the performance of the different algorithms (for example [41]) with BioNet having a very good precision rate. However, it has been shown that no single method outperforms others overall so that the selection of the network inference method should be based on the concrete purpose and biological question of the study.

Acknowledgment

This work was supported by the EC's 7th Framework Programme with the grant HeCaToS (602156) and by the Max Planck Society.

References

1. Hendrickx DM, Aerts HJ, Caiment F et al (2015) diXa: a data infrastructure for chemical safety. Bioinformatics 31:1505–1507
2. vanDelft J, Gaj J, Lienhard M et al (2012) RNA-Seq provides new insights in the transcriptome responses induced by the carcinogen benzo[a]pyrene. Toxicol Sci 130:427–439
3. Zhang W, Yu Y, Hertwig F et al (2015) Comparison of RNA-seq and microarray-based models for clinical endpoint prediction. Genome Biol 16:133
4. Rasche A, Lienhard M, Yaspo ML, Lehrach H, Herwig R (2014) AHR-seq: identification of differential splicing in RNA-seq data. Nucleic Acids Res 42:e110
5. Gentleman R (2008) R programming for bioinformatics. CRC Press, Boca Raton FL
6. Yildirimman R, Brolén G, Vilardell M et al (2011) Human embryonic stem cell derived hepatocyte-like cells as a tool for in vitro hazard assessment of chemical carcinogenesis. Toxicol Sci 124:278–290
7. Gusenleitner D, Auerbach SS, Melia T, Gómez HF, Sherr DH, Monti S (2014) Genomic models of short-term exposure accurately predict long-term chemical carcinogenicity and identify putative mechanisms of action. PLoS One 9:e102579
8. Vidal M, Cusick ME, Barabasi AL (2011) Interactome networks and human disease. Cell 144:986–998
9. Bader GD, Cary MP, Sander C (2006) Pathguide: a pathway resource list. Nucleic Acids Res 34(Database issue):D504–D506
10. Szklarczyk D, Franceschini A, Wyder S et al (2015) STRING v10: protein-protein interaction networks, integrated over the tree of life. Nucleic Acids Res 43:D447–D452
11. Herwig R, Hardt C, Lienhard M, Kamburov A (2016) Analyzing and interpreting genome data at the network level with ConsensusPathDB. Nat Protoc 11:1889–1907
12. Sharan R, Ulitsky I, Shamir R (2007) Network-based prediction of protein function. Mol Syst Biol 3:88
13. Ljubic I, Weiskircher R, Pferschy U et al (2006) An algorithmic framework for the exact solution of the prize-collecting Steiner tree problem. Math Progr Series B 105:427–449
14. Beisser D, Klau GW, Dandekar T, Müller T, Dittrich MT (2010) BioNet: an R-package for the functional analysis of biological networks. Bioinformatics 26:1129–1130
15. Dai M, Wang P, Boyd AD et al (2005) Evolving gene/transcript definitions significantly alter the interpretation of GeneChip data. Nucleic Acids Res 33:e175
16. Wu J, Gentry RI (2016) gcrma: Background adjustment using sequence information. R package version 2.46.0
17. Ritchie ME, Phipson B, Wu D et al (2015) Limma powers differential expressionanalyses

for RNA-sequencing and microarray studies. Nucleic Acids Res 43:E47
18. Manconi A, Vargiu E, Armano G, Milanesi L (2012) Literature retrieval and mining in bioinformatics: state of the art and challenges. Adv Bioinforma 2012:573846
19. Kamburov A, Grossmann A, Herwig R, Stelzl U (2012) Cluster-based assessment of protein-protein interaction confidence. BMC Bioinformatics 13:262
20. Goldberg DS, Roth FP (2003) Assessing experimentally derived interactions in a small world. Proc Natl Acad Sci U S A 100:4372–4376
21. Kuchaiev O, Rasajski M, Higham DJ, Przulj N (2009) Geometric de-noising of protein-protein interaction networks. PLoS Comp Biol 5:e1000454
22. Kamburov A, Stelzl U, Lehrach H, Herwig R (2013) The ConsensusPathDB interaction database: 2013 update. Nucleic Acids Res 41:D793–D800
23. Yu G, Fei L, Qin Y, Bo X, Wu Y, Wang S (2010) GOSemSim: an R package for measuring semantic similarity among GO terms and gene products. Bioinformatics 26:976–978
24. Kamburov A, Stelzl U, Herwig R (2012) IntScore–a web tool for confidence scoring of biological interactions. Nucleic Acids Res 40:W140–W146
25. Schrijver A (1998) Theory of linear and integer programming. John Wiley and Sons, Hoboken, New Jersey
26. Cormen TH, Leiserson CE, Rivest RL, Stein C (2001) Introduction to algorithms, 2nd edn. MIT Press, Cambridge MA
27. Uehara T, Ono A, Maruyama T et al (2010) The Japanese toxicogenomics project: application of toxicogenomics. Mol Nutr Food Res 54:218–227
28. Ganter B, Tugendreich S, Pearson CI et al (2005) Development of a large-scale chemogenomics database to improve drug candidate selection and to understand mechanisms of chemical toxicity and action. J Biotechnol 119:219–244
29. Hardt C, Beber M, Rasche A et al (2016) ToxDB: pathway-level interpretation of drug-treatment data. Database (Oxford) 2016:baw052
30. Hakenberg J, Gerner M, Haeussler M et al (2011) The GNAT library for local and remote gene mention normalization. Bioinformatics 27:2769–2771
31. Rocktaschel T, Weidlich M, Leser U (2012) ChemSpot: a hybrid system for chemical named entity recognition. Bioinformatics 28:1633–1640
32. Barabasi AL, Oltvai ZN (2004) Network biology: understanding the cell's functional organization. Nat Rev Genet 5:101–113
33. Campbell EA, Korzheva N, Mustaev A et al (2001) Structural mechanism for rifampicin inhibition of bacterial RNA polymerase. Cell 104:901–912
34. Cassir N, Rolain JM, Brouqui P (2014) A new strategy to fight antimicrobial resistance: the revival of old antibiotics. Front Microbiol 5:551
35. Li T, Chiang JY (2006) Rifampicin induction of CYP3A4 requires pregnane X receptor cross talk with hepatocyte nuclear factor 4alpha and coactivators, and suppression of small heterodimer partner gene expression. Drug Metab Dispos 34:756–764
36. Ihunnah CA, Jiang M, Xie W (2011) Nuclear receptor PXR, transcriptional circuits and metabolic relevance. Biochim Biophys Acta 1812:956–963
37. Kim YA, Cho DY, Przytycka TM (2016) Understanding genotype-phenotype effects in cancer via network approaches. PLoS Comput Biol 12:e1004747
38. Herwig R, Gmuender H, Corvi R et al (2016) Inter-laboratory study of human in vitro toxicogenomics-based tests as alternative methods for evaluating chemical carcinogenicity: a bioinformatics perspective. Arch Toxicol 90:2215–2229
39. MAQC Consortium (2010) The MicroArray quality control (MAQC)-II study of common practices for the development and validation of microarray-based predictive models. Nat Biotechnol 28:827–838
40. Creixell P, Reimand J, Haider S et al (2015) Pathway and network analysis of cancer genomes. Nat Methods 12:615–621
41. He H, Lin D, Zhang J, Wang YP, Deng HW (2017) Comparison of statistical methods for subnetwork detection in the integration of gene expression and protein interaction network. BMC Bioinformatics 18:149

Chapter 17

Modeling the Epigenetic Landscape in Plant Development

Jose Davila-Velderrain, Jose Luis Caldu-Primo, Juan Carlos Martinez-Garcia, and Elena R. Alvarez-Buylla

Abstract

Computational mechanistic models enable a systems-level understanding of plant development by integrating available molecular experimental data and simulating their collective dynamical behavior. Boolean gene regulatory network dynamical models have been extensively used as a qualitative modeling framework for such purpose. More recently, network modeling protocols have been extended to model the epigenetic landscape associated with gene regulatory networks. In addition to understanding the concerted action of interconnected genes, epigenetic landscape models aim to uncover the patterns of cell state transition events that emerge under diverse genetic and environmental background conditions. In this chapter we present simple protocols that naturally extend gene regulatory network modeling and demonstrate their use in modeling plant developmental processes under the epigenetic landscape framework. We focus on conceptual clarity and practical implementation, providing directions to the corresponding technical literature. The protocols presented here can be applied to any well-characterized gene regulatory network in plants, animals, or human disease.

Key words Epigenetic landscape, Gene regulatory networks, Dynamical systems, Systems biology, Cell differentiation, Attractors, Morphogenesis, Development

1 Introduction

Twenty-first century biology is largely an interdisciplinary endeavor. The ever-increasing accumulation of molecular experimental data point toward the need of encompassing integrative approaches to better understand observations and to unravel underlying mechanisms [1, 2]. The development of a multicellular organism is a complex process involving the dynamical interplay of multiple genetic and nongenetic factors at different spatial and temporal scales [3]. Given the complexity of such intricate systems, computational modeling has been a natural approach for the study of developmental dynamics. In particular, systems-level mechanistic models have been instrumental to simulate and better understand

Louise von Stechow and Alberto Santos Delgado (eds.), *Computational Cell Biology: Methods and Protocols*, Methods in Molecular Biology, vol. 1819, https://doi.org/10.1007/978-1-4939-8618-7_17,

the forces driving cellular differentiation and morphogenesis in plant development [4].

One of the first attempts to understand the emergence of development as a dynamical system driven by complex interactions—as well as its association with genetics, the environment, and evolution—was the conceptual model of the Epigenetic Landscape (EL) proposed by C.H. Waddington [5]. In its original form, the EL depicts the development of a body part as a series of bifurcating trajectories undergone by balls rolling down a surface with hills and valleys, until reaching terminally differentiated states. The global shape of the landscape is assumed to be determined by underlying systems of complex genetic interactions. More recently, the EL has been grounded in the framework of gene regulatory networks (GRN), where cell states undergo differentiation transitions driven by the dynamics emerging from intracellular regulatory constraints [6]. In this chapter we focus on demonstrating how well established GRN modeling protocols can be extended in order to develop EL models for the study of cell differentiation and morphogenesis. We illustrate this approach on plant development examples as study systems.

GRN modeling has been successfully used for the theoretical study of developmental processes [4, 7–10]. We have previously presented general protocols for the implementation of GRN dynamical models and discussed their interpretation (*see* for example refs. 1 and 11). Given that GRN models are the basis to develop epigenetic landscape frameworks, we will include here a brief, nontechnical explanation of key concepts and methods for GRN modeling. We point the reader to our previous methods papers for a more in-depth exposition [7, 10, 11].

A GRN dynamical model is a formal representation of fragmentary knowledge about molecular regulatory mechanisms, which is integrated using the tools of dynamical systems theory [1, 2]. A GRN consists of a set of nodes and their mutual interactions (*see* **Note 1**). Both nodes and interactions are formalized in a set of rules that specify how the activity of each node affects the behavior of its interacting partners through time. In these models the activity of each node commonly represents the expression of a gene, and the goal of the model is to simulate the coordinated behavior of the complete gene expression profile. Because of the regulatory constraints specified in the dynamical rules, genes will display specific patterns of expression. For example, a regulatory constraint could simply indicate that an active gene can no longer remain expressed because its inhibiting partners become active (Fig. 1).

The power of the integrative GRN resides in its inherent ability to consider all such constraints in parallel and for each time step. The global consequence of the regulatory constraints

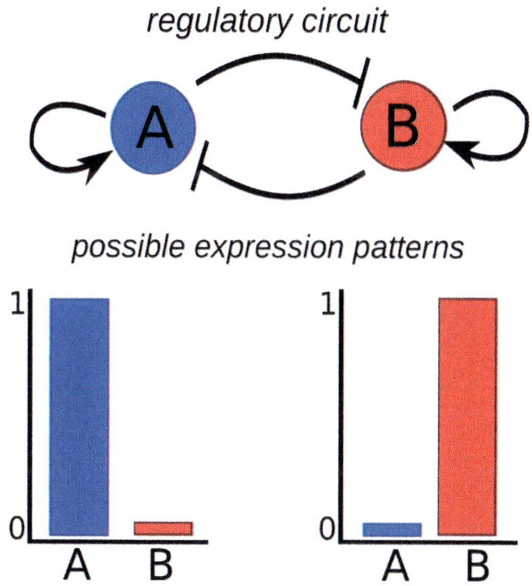

Fig. 1 Regulatory constraints. A classic example of a regulatory circuit that imposes a regulatory constraint is the bistable switch shown in the Fig. A simple circuit is formed by two mutually repressing genes. This regulatory structure causes the system to have only three stable expression patterns: a trivial pattern in which none of the genes is expressed and two opposing behaviors in which either one of the two genes is expressed and the other gene is silent

is that only certain activity configurations (expression profiles) will be consistent with the regulatory logic and thus stably maintained in time. Such configurations are called attractors in the theory of dynamical systems and represent specific expression profiles in which the state of each gene does not change anymore (*see* **Note 2**).

Because of their robustness—(quasi-)stationary and stable—attractor states likely represent behaviors we would expect to observe in nature in terms of the gene activity characteristic of each cell type. This is a consequence of the information integrated in the GRN being experimentally grounded (for details, *see* ref. 6 and 11). For practical purposes, the attractors recovered using GRN dynamical models (*see* Subheading 3 below) correspond to the cell states observed in real developmental systems and are empirically recovered by gene expression profiles.

The dynamical rules specifying the GRN can be modeled with different types of equations or procedural algorithms. Here, we will focus on models with discrete state and time variables, which are modeled with discrete map equations; we will focus on the simplest type of such models: the Boolean network model [12]. Previous studies have shown that such qualitative models capture the systemic behavior of GRNs underlying cell differentiation ([7, 9] and references therein). The capacity of Boolean models to recover systemic behaviors, despite their qualitative nature, evidences that biological behaviors are largely guided by the overall structure of the networks rather than the specific form of the kinetic functions [13].

In addition to successfully modeling cellular differentiation in diverse plant developmental processes [14–16], Boolean GRN models have also been used to operationalize the otherwise metaphorical model of the EL [6, 17–19]. In this context, as we shown below, Boolean GRNs are extended by the introduction of stochasticity in order to simulate cell state transitions (i.e., differentiation events), enabling the discovery of regular patterns of such transitions. These ideas were first applied in plant development in the study of the morphogenetic patterns observed during early flower development in *Arabidopsis* [17]. Based on this original study, as well as on more recent work proposing a general framework for modeling the EL associated with Boolean GRNs [6, 17], we recently implemented previously dispersed algorithms in a common modeling framework with the goal of facilitating the overall use of EL models [18]. In this chapter, we make use of some of these tools in order to provide a general protocol for modeling the EL associated with Boolean networks, a methodological standpoint that has been applied successfully in the study of several plant developmental processes [17, 20].

Given that the EL methods considered here constitute extensions to well-established protocols of GRN modeling and in order to provide a comprehensive overview, we also include a brief protocol for modeling Boolean GRNs. For all the examples we consider a well-studied developmental GRN: the root stem cell niche GRN (root SCN-GRN) of *Arabidopsis thaliana* [15].

The *A. thaliana* root stem cell niche (SCN) is part of the root apical meristem at the acropetal end of the primary root. The root SCN is well described at the anatomical level. It is formed by an inner core of four cells that rarely divide and constitute the quiescent center (QC), surrounded by three sets of initial cells that give rise to the different types of differentiated cells in the root: vascular initials (VI), cortex–endodermis initials (CEI), and columella–epidermis–lateral root cap initials (CEpI) (Fig. 2a) [15]. Some of the molecular components implicated in root SCN cellular patterning and maintenance have been uncovered and characterized. Three main pathways are involved. One component is the module of *SHORT-ROOT* (*SHR*) and its target gene *SCARECROW* (*SCR*); both genes regulate the transcription of several genes involved in the root cellular organization [21, 22]. Another component involves the *PLETHORA* (*PLT*) genes, which are transcribed in response to auxin accumulation [23]. Finally, *WUSCHEL RELATED HOMEOBOX 5* (*WOX5*) is a gene expressed exclusively in the QC, and it has been implicated in maintaining stable gene expression and keeping the distal SC

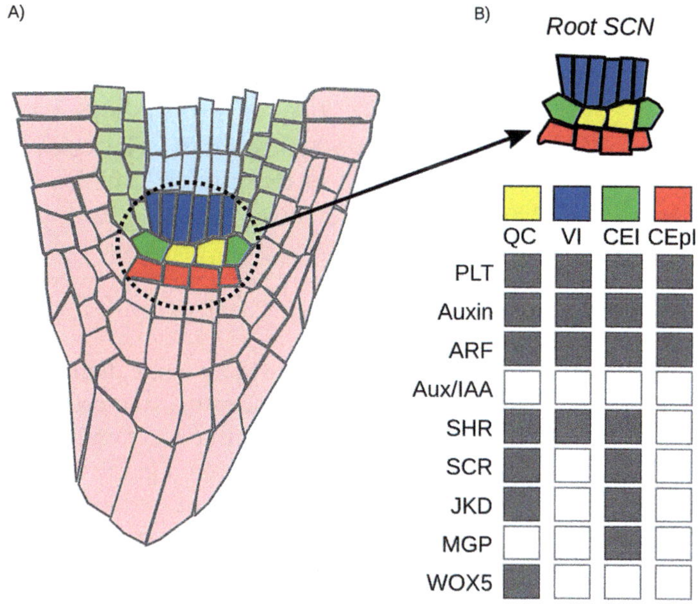

Fig. 2 Simplified cellular pattern of the root stem cell niche. (**a**) Root tip of Arabidopsis thaliana colorized to show the corresponding cell types, highlighted is the root SCN. The four cell types within the SCN are: quiescent center (yellow), cortex–endodermis initials (green), vascular initials (blue), and columella–epidermis–lateral root cap initials (red). Root cells formed from the SCN are shadowed with the color of the corresponding SCN cell type. (**b**) Gene expression configuration in the cell types of the SCN. Gene activity is coded in squares: gray for active genes and white for inactive genes

undifferentiated [24]. Work in our group integrated the gene interactions recovered through molecular evidence and proposed a GRN that recovers the expected gene configurations of the different root SCN cell types [15] (Fig. 2b). In what follows, we will reanalyze such a network following the epigenetic landscape framework.

2 Materials

2.1 Data

The input to all the analyses described here is any previously specified Boolean GRN dynamical model (for example the model in [15]), which consists of a list of state variables (i.e., molecules of interest) and a corresponding set of Boolean logical propositions describing their regulatory interactions. For the root SCN example, we consider the experimental data in Table 4 and the Boolean propositions in Table 5. In addition, for validation and prediction purposes, observable wild-type cellular phenotypes and mutant (loss- and gain-of-function) gene activation configurations need to be previously defined from literature. We consider the gene expression profiles extracted from Azpeitia and collaborators [15].

2.2 Software

The R statistical programming environment (www.R-project.org) is used to implement all the analyses presented here, making use of the functions from the contributed package *BoolNet* [25], as well as those kept at https://github.com/JoseDDesoj/Epigenetic-Attractors-Landscape-R. The complete protocol for the analyses presented in this chapter can also be found in the Github folder mentioned above.

3 Methods

3.1 Definitions

We start by providing some definitions that are key to understanding how the conceptual framework of the epigenetic landscape is enclosed within the context of GRN dynamics.

A GRN is a directed graph composed of *nodes* and *edges*. Each node represents a gene (or another molecular type), and a state variable is assigned to each node to specify a *gene state*. Nodes are connected by edges that represent regulatory interactions among genes. In addition to the graph, which is simply represented as a wiring diagram (Fig. 3), the incoming regulatory interactions of each node are formalized as *updating rules* that determine the temporal activity of each node. In a Boolean GRN, gene activity changes are assumed in discrete time units. The gene state variables take only one of two values (0 when the gene is inactive, and 1 when the gene is active). The updating rules are logical propositions that determine the state of a given node in the next time step given the state of its incoming regulatory nodes (*see* Tables 3 and 5 for examples). The GRN is completely specified by the set of nodes and their corresponding updating rules [1]. Both the set of nodes and the updating rules are proposed based on experimental molecular data.

A gene configuration of the complete network (*network state*) is specified at each time step by the state of all the genes constituting the GRN. Given that gene states take only two values (0 or 1), a GRN with n genes can only take state values from a finite *state space* of 2^n possible states. Starting from a given network state, the parallel implementation of the updating rules will determine a specific future network state. By the iterative application of the updating rules starting from a given network state it is possible for the network to reach either a stationary state that does not change after applying the updating rules (*fixed point attractor*). Alternatively, the networks can reach a close set of states from where the updating rules map to themselves (*cyclic attractor*). These stationary states are called *attractors*. If the updating rules are applied exhaustively starting from every possible network state, subsets of states converging to the same attractor can be determined. The regions in the state space converging to the same attractor are called *basins of attraction*. Importantly,

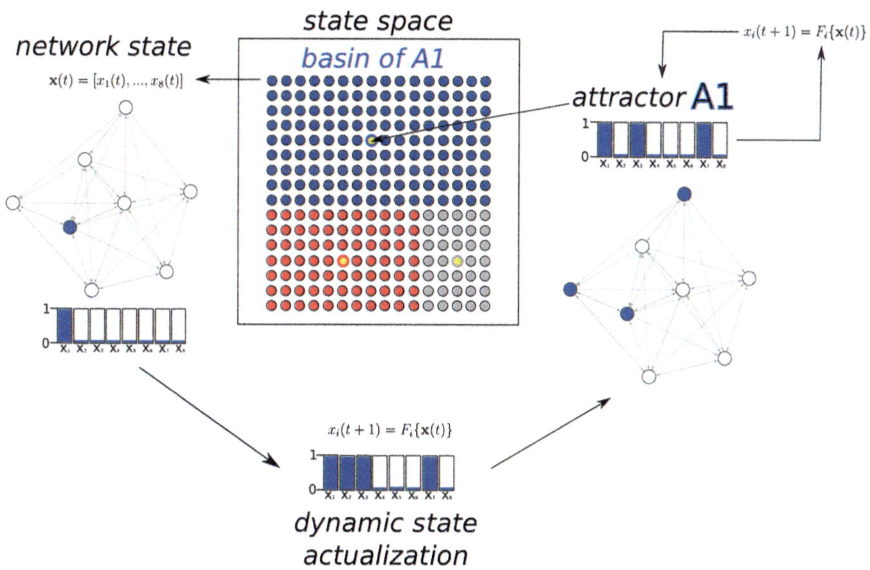

Fig. 3 Schematic representation of an arbitrary GRN dynamical model. This diagram illustrates the general protocol of a dynamical analysis of a Boolean network. The state space contains all of the possible network states. Application of the logical functions to every network state eventually leads to an attractor. All the network states that lead to a particular attractor constitute its basin of attraction. The network state space is then structured by the basins of attraction of the different attractors

attractors represent activity configurations that are consistent with the regulatory logic specified in the updating functions—similar to the way gene activity profiles observed in vivo maintain specific cellular phenotypes. Thus, the attractor states of GRN models are expected to correspond to the cellular phenotypes observed in the real developmental systems being modeled. Interestingly, it has been shown in multiple developmental processes that this is indeed the case [14–16, 20].

As a result of the GRN dynamics, the state space is structured in a specific manner, which is reflected in the way the basins of attraction are organized. We refer to such structures as the *attractor landscapes*. The characterization of *attractor landscapes* is what allows us to explore the EL associated with the underlying GRN. The basins of attraction correspond to distinct differentiation states having different relative stability properties—akin to hierarchically organized valleys in the EL. Given the attractor landscape, the developmental pathways followed by a developmental system under normal or perturbed conditions can be explored by estimating the likelihood of specific single or sequential transitions among the basins of attraction (see below).

In Fig. 3 we schematically show the key concepts involved in modeling GRN dynamics. The implementation of the updating rules to a given network state will determine a specific network state for the next time step. Then, if one applies the updating functions

Table 1
General modeling concepts

Descriptive modeling	
Model	A mathematical expression or computer algorithm that relates the values of one or more *responsive* (dependent) variables with the values of a set of *predictor* (independent) variables
Prediction	Calculated values of the responsive variables, which are assessed by taking specific values of the predictor variables as input to the model
Explanation	A predictor variable x is said to *explain* a responsive variable y if the predicted values for y are in agreement (to a certain degree) with the observed values in a particular dataset comprising empirical values of x and y
Validation	The practice of testing the performance of a model by testing its predictive power using an independent dataset
Causal attribution	It is not possible to postulate the reasons why a certain quantitative relationship embedded in the model is able to *explain* one variable in terms of the other—"*correlation does not imply causation*"
Mechanistic modeling	
Model	Set of equations or computer code that describe how simplified properties of a real-world entity (system) change over time as a result of specific underlying processes
Prediction	Forecasting the future properties of the system or their long-term behavior
Explanation	The processes considered in the model account for the observed system behavior
Validation	The practice of contrasting model predictions with experimental observations of the real-world entity
Causal attribution	The predicted behavior results from the underlying *causal* processes considered in the model. The model is built by explicitly considering the processes that produce our observations

iteratively to this latter state, the network will eventually end up in an attractor state. In order to characterize the attractor landscape, we need to obtain all the attractors and their basins of attraction. After identifying the network's attractors, we need to characterize their corresponding attraction basin. To further describe the attraction basins, we can retrieve the complete mapping structure linking every possible initial state to its corresponding attractor, which is known as the network's transition table. Tables 1 and 2 summarize the concepts discussed in this section. In the following sections, we explain how to apply these concepts for systems biology modeling.

3.2 Dynamical Analysis of Boolean GRNs

Before presenting the real case study of plant development, we first show how the dynamics of an arbitrary Boolean GRN can be

Table 2
GRN dynamical model concepts

Concept	Definition
Node	Representation of a molecular species (gene, protein,...)
Edge	Representation of a given regulatory interaction
Node state (variable)	Expression value that a node takes at a certain time
Network state	Ordered set of node expression values at a certain time
State space	Set comprising all possible network states
Attractor	Stable and stationary (time-invariant) network states
Transitory state	Network states that are not (do not form) part of an attractor (attractor's basin)
Basin of attraction	Set comprising all the initial network states that eventually lead to a particular attractor
Biologically, observable attractor	Gene expression profiles (gene configurations) that have been obtained from experimental assays and reported in the scientific literature for particular cell types

analyzed. Here, the modeler interested in a specific developmental system builds a network by choosing the set of genes and their regulatory interactions based on prior knowledge extracted from literature or directly from experiments (*see* **Note 3**). Interactions among nodes are encoded as logical propositions constituting the updating rules. For demonstrative purposes, here we will start by generating a random network as an example.

We follow a simplified protocol for constructing and analyzing a Boolean GRN model considering only the essential steps. For an in-depth modeling protocol that outlines how GRNs are proposed from scratch using experimental molecular data *see* Refs. 1 and 11.

The present protocol includes only the next two essential steps:

1. Defining a set of nodes and their associated updating rules (*see* **Note 3**).
2. Characterizing the state space by identifying the attractors and their basins of attraction (*see* **Note 4**).

3.2.1 Building and Analyzing a Random Boolean GRN Model

1. *Defining the set of genes and updating functions:* Artificial random Boolean GRN models with specific properties can be easily generated using the software *BoolNet*. *BoolNet* includes the function `generateRandomNKNetwork(n, k)`, which generates a random network with *n* genes, each with an updating rule having *k* input genes. Using this

Table 3
Arbitrary Boolean GRN model

List of state variables
X = [Gene1, Gene2, Gene3, Gene4, Gene5, Gene6, Gene7, Gene8]
Boolean functions
Gene1, (!Gene1 & Gene8) | (Gene1 & !Gene8)
Gene2, (!Gene5 & !Gene3) | (Gene5 & Gene3)
Gene3, (!Gene3) | (!Gene8)
Gene4, (!Gene1 & Gene4)
Gene5, (Gene4 & !Gene5)
Gene6, (Gene4) | (Gene2)
Gene7, (!Gene8) | (!Gene7)
Gene8, (Gene5 & Gene1)

function, we created a network with eight genes, whose regulatory logic is determined by two input genes in each case. Note that the output network of this function will be different each time. The set of nodes and updating functions of the specific random network are included in Table 3. Alternatively, the updating rules can be loaded directly from a text file using the function `loadNetwork()`. To load a network in *BoolNet,* these rules have to be written using specific symbols: AND (&), OR (|), and NOT (!), in a syntax specified in the manual [25].

2. *Characterizing the state space: identifying the attractors and their basins of attraction*
   BoolNet package is implemented with functions that make the dynamical analysis of Boolean networks straightforward. Using *BoolNet*, the attractors of a given network can be easily obtained through the function `getAttractors(network)`. The default settings of this function recover the attractors of the given network by an exhaustive synchronous search. Intuitively, this means that it takes each of the network's 2^n possible states as a starting point and follows their transitions until they reach an attractor. Specialized algorithms grant efficient exhaustive search in networks with up to 29 nodes. In bigger networks, heuristic search algorithms perform well by specifying the corresponding arguments to the same function [25, 26]. Once the attractors have been identified, a graphical depiction of how the state space is structured can be obtained with the function `plotAttractors()`.

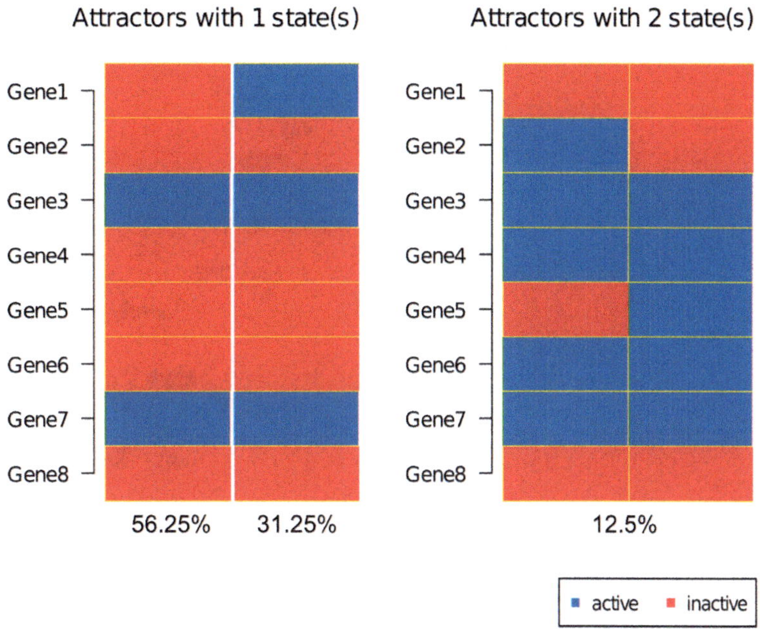

Fig. 4 Obtained attractors of the random GRN. The GRN recovered two fixed-point attractors and one cyclic attractor with two states. In the graph, blue color indicates expression or gene activation (1), while red color indicates no expression or inactivation (0)

This function outputs a graph as the one shown in Fig. 4, where rows correspond to genes and columns correspond to attractors. The activation state of each gene in the different attractors is color-coded (blue(red) cells represent active (inactive) genes). We can see that this network has an attractor landscape of three attractors, two of them are a fixed point and the third one is a cyclic attractor that oscillates between two states. The relative size of each attraction basin is shown under each attractor configuration (Fig. 4). This number represents the number of initial states that reach the given attractor divided by all the possible initial states.

To further describe the attraction basins, we recover the transition table, graphically represented as a state transition graph—i.e., a graph in which nodes are network states and arrows represent transition between nodes that result from the application of the updating rules. The *BoolNet* function `plotStateGraph()` generates a state transition graph, taking the computed attractors as input. Figure 5 shows the state transition graph for our example network. A different color is used for each attraction basin and attractors correspond to nodes having either only incoming edges (fixed-point attractor) or forming a loop (cyclic attractors with more than one state). The number of nodes in each basin of attraction defines its size.

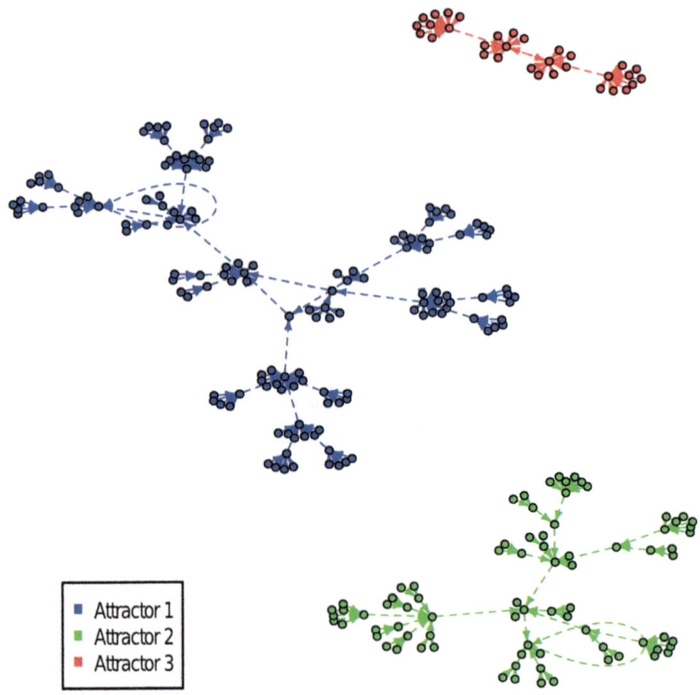

Fig. 5 State transition graph of the random GRN. Each node in this graph represents a network state connected to the state it takes after applying the updating rules. Each component of the graph corresponds to an attraction basin and is plotted in a different color

3.2.2 Building and Analyzing a 'Real' Boolean GRN Model from Experimental Data

We now turn to the *Arabidopsis* root SCN-GRN as an example of a dynamical analysis applied to a gene network based on experimental data [15]. As explained above, in order to model a developmental process as a Boolean GRN, the first step is to define the genes involved in the process and the regulatory interactions among them (Table 4). Azpeitia and collaborators proposed a Boolean network based on a thorough review of scientific literature (*see* **Note 3**). The obtained network recovers four attractors corresponding to the main cell types in the *Arabidopsis* root SCN [15].

It is important to recall that the four cell types of the root SCN can be distinguished by their expression pattern of the proposed set of genes (Fig. 2b). Defining the expected set of attractors is an indispensable step when building the GRN model because the attractors are used to validate the GRN (*see* **Note 3**). To further validate a proposed network model, it is important to simulate gene knockdown or mutations causing gene overexpression on the network and compare the results with experimental evidence (*see* **Note 5**). Azpeitia and collaborators followed this approach and showed that most predicted alterations to the stable configurations caused by mutant simulations are consistent with known empirical observations [15].

Table 4
Experimentally supported (real) interactions set for the root SCN-GRN, taken from [15]

Interactions	Experimental evidence
SHR -> SCR	The expression of *SCR* is reduced in shr mutants. ChIP-QRTPCR experiments show that *SHR* directly binds in vivo to the regulatory sequences of *SCR* and positively regulates its transcription
SCR -> SCR	In the scr mutant background promoter activity of SCR is absent in the QC and CEI. A ChIP-PCR assay confirmed that *SCR* directly binds to its own promoter and directs its own expression
JKD -> SCR	*SCR* mRNA expression is lost in QC and CEI cells in *jkd* mutants from the early plantule stage onward
MGP–|SCR	The double mutant *jkd/mgp* rescues the expression of *SCR* in QC and CEI cells, which is lost in the jkd single mutant
SHR -> MGP	The expression of *MGP* is severely reduced in the SHR background. *SHR* binding to *MGP* regulatory region has been confirmed by ChIP-PCR
SCR -> MGP	*SCR* directly binds to the *MGP* promoter, and *MGP* expression is reduced in the scr mutant background
SHR -> JKD	The post-embryonic expression of *JKD* is reduced in *shr* mutant roots
SCR -> JKD	The post-embryonic expression of *JKD* is reduced in *scr* mutant roots
SCR -> WOX5	*WOX5* is not expressed in *scr* mutants
SHR -> WOX5	*WOX5* expression is reduced in *shr* mutants
ARF(MP)-> WOX5	*WOX5* expression is rarely detected in *arf(mp)* or arf(*bdl)* mutants
ARF-> PLT	*PLT1* mRNA is overexpressed under ectopic auxin addition. *PLT1&2* mRNAs are absent in the majority of *arf(mp)* embryos
Aux/IAA–|ARF	Overexpression of *Aux/IAA* genes represses the expression of *ARF(DR5)* both in the presence and absence of auxin. Domains III & IV of *Aux/IAA* genes interact with domains III & IV of *ARF* stabilizing the dimerization that represses *ARF* transcriptional activity
Auxin–| aux/IAA	Auxin application destabilizes *Aux/IAA* proteins. *Aux/IAA* proteins are targets of ubiquitin-mediated auxin-dependent degradation

3.2.3 Dynamical Analysis with BoolNet

We use the logical rules proposed by Azpeitia and collaborators (summarized in Table 5) to perform the analysis in *BoolNet*. The rules are loaded by simply using the function `loadNetwork("root_SCN.txt")`, where root_SCN.txt is a text file containing the logical rules, which should be saved in the current R working directory. Figure 6 shows the wiring diagram of the root SCN-GRN, a graphical representation of the Boolean rules used to define the network.

Table 5
***Arabidopsis* root SCN-GRN Boolean functions**

List of state variables
X = [SCR, PLT, ARF, AUXIAA, AUX, SHR, JKD, MGP, WOX5]
Boolean functions
SCR, (SHR & SCR & !JKD & !MGP) | (SHR & SCR & JKD & MGP) | (SHR & SCR & JKD & !MGP)
PLT, ARF
ARF, !AUXIAA
AUXIAA, !AUX
AUX, AUX | !AUX
SHR, SHR
JKD, SHR & SCR
MGP, SHR & SCR & !WOX5
WOX5, (ARF & SHR & SCR & !MGP & !WOX5) | (ARF & SHR & SCR & !MGP & WOX5) | (ARF & SHR & SCR & MGP & WOX5)

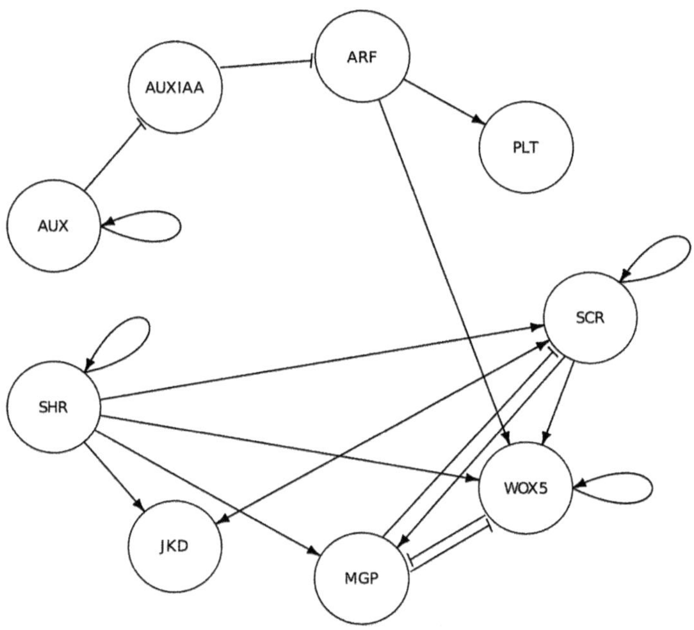

Fig. 6 Wiring diagram of the Arabidopsis root SCN-GRN. Single-cell root SCN GRNs proposed in [15]. Nodes represent genes and edges represent regulatory interactions among them. The effect of the interaction is symbolized with arrows and flat arrows for activating or repressing interactions, respectively. Note that when using *BoolNet*'s function plotNetworkWiring the resulting plot does not distinguish between activating and repressing interactions

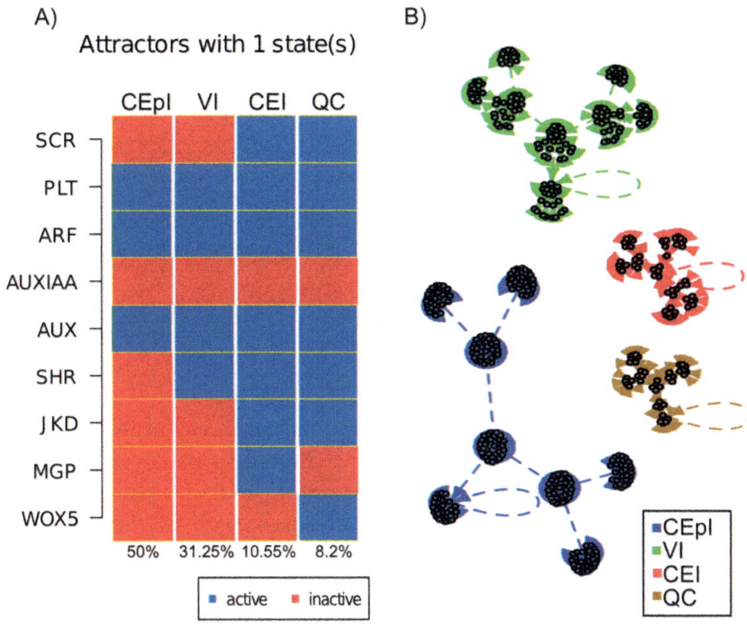

Fig. 7 Attractors and state transition graph of the root SCN-GRN. (**a**) Genetic configuration of the four attractors obtained from the root SCN-GRN. These attractors correspond to the experimental gene expression profiles of the cell/types found at the root SCN. (**b**) State transition graph, representing each attractor's basin of attraction

Once the network is loaded into *BoolNet*, we follow the same dynamical analysis presented above for the random network. In summary, we find the network's attractors and describe their corresponding attraction basins. The output of the functions `getAttractors()` and `plotAttractors()` indicate that the network converges to four attractors (Fig. 7a). The recovered attractors correspond to the expression profiles of the four cell types found in the *Arabidopsis* root SCN (Table 6):VI,CEI, QC and CEpI [15]. This suggests that cell-type gene-expression patterns in the root SCN result from the restrictions imposed by the uncovered GRN developmental module. As explained above, each attractor's basin of attraction can be graphically represented using the function `plotStateGraph()` (Fig. 7b).

3.3 Modeling the EL

In a systems biology conceptual framework, a cell is represented as a dynamical system governed by an underlying GRN. The trajectories leading to the attractors can be naturally associated to the valleys depicted in the EL metaphor proposed by Waddington [27]. Although the idea underlying the EL metaphor is intuitively easy to understand, it cannot be directly operationalized with the conventional Boolean GRN formalism. GRN attractors are steady states, this implies that once a network reaches an attractor state it

Table 6
Gene expression profiles of the root SCN cell types (expected attractors)

Cell type	PLT	Auxin	ARF	Aux/IAA	SHR	SCR	JKD	MGP	WOX5
QC	1	1	1	0	1	1	1	0	1
VI	1	1	1	0	1	0	0	0	0
CEI	1	1	1	0	1	1	1	1	0
CEpl	1	1	1	0	0	0	0	0	0

will stay there indefinitely unless an external force moves the system out of such a state. In real developmental processes, however, cells transit from one cell state (attractor) to another. A cell can change its state by two nonexclusive mechanisms. On the one hand, cell state changes can be driven by intrinsic stochastic fluctuations of the molecular system. On the other hand, cells can change state in response to extrinsic signaling factors [28, 29]. Both mechanisms can be modeled within GRN dynamics. Stochastic noise can be introduced to the model causing cells to jump around the landscape without requiring any parameter changes. The effect of extrinsic factors can be modeled by altering the parameters of the network structure and thus changing the attractors landscape [30]. Thus, in order to characterize the EL associated with a Boolean GRN, it is necessary to extend the discrete network model in order to explore transitions between attractors.

Here we consider a modeling extension to include stochastic intrinsic perturbations. This is achieved by randomly perturbing the state of the genes in the network causing "jumps" among attractors. This scenario assumes that developmental transitions are the natural consequence of the regulatory restrictions themselves and not of the signaling mechanisms. The fact that genetic regulatory interactions represented in a GRN are biochemical reactions subject to stochastic fluctuations makes the inclusion of stochasticity to any proposed GRN model a valid assumption. Given the nonlinear restrictions imposed by the underlying regulatory network, the potential jumping patterns among network states will not be equally likely, in spite of unbiased stochasticity. This interacting effect between nonlinearity and stochasticity enables the discovery of nontrivial, robust patterns of transitions.

Based on the considerations above, we proceed by introducing a general strategy for the practical extension of a validated Boolean GRN in order to produce an EL model. The framework is exemplified in the next section, and it comprises three steps:

1. Computational simulation of cell state changes in response to perturbations generated by introducing stochasticity into the Boolean dynamics.
2. Analysis of the prevailing paths of cell fate change for estimating an interattractor transition probability matrix.
3. Characterization of the temporal evolution of the probability distribution over attractor states.

3.3.1 Introducing Stochasticity to the Boolean Dynamics

Stochastic noise can be introduced to a GRN in different ways (*see* **Note 6**). Here we use the model known as stochasticity in nodes (SIN) [31], which introduces stochasticity by considering a fixed probability for a gene to disobey its updating rule. In other words, under the SIN model, even if the updating rules of a gene X imply that it should be active (or inactive) in the next time step, there is a certain probability of the contrary output to occur. Under this stochastic dynamics, a given initial configuration will no longer converge to the same attractor every time. The probability of the network to pass from one network state to another can be estimated by iterating the stochastic rule a large number of times (at least 1000 times) and estimating the frequency of the interstate transitions. The estimated transition probabilities can then be used to study the behavior of the system and to make statistical predictions of cell state transitions.

3.3.2 Building the Interattractor Transition Matrix

The state transition probabilities we are interested in are those that occur between network states belonging to different basins of attraction. The probabilities of passing from any attractor to any another are arranged into an interattractor transition matrix (IATM). In order to estimate the IATM, we first introduce stochasticity to the GRN and iterate the stochastic functions for every network state a large number of times. For every iteration, we must store the basin of attraction the original state belonged to, and the basin of attraction it reaches after introducing stochasticity and applying the updating rules. Using this simulated state transition, the interattractor transition probabilities are calculated as the relative number of times the network "jumps" to every possible attraction basin starting from the one the initial state belongs to.

3.3.3 Downstream Analyses to Characterize the EL

Once the IATM has been computed, downstream analyses can be performed in order to uncover the underlying EL structure emerging from the regulatory restrictions. Some of these analyses include computing the temporal order of attractor attainment and the attractor relative stability and global ordering. We will explain the basic ideas underlying these analyses.

3.3.4 Temporal Sequence of Attractor Attainment

Computing the temporal sequence of attractor attainment is a basic approach to developmental phenomena that involves uncovering the regulatory basis for the typical temporal sequence of cell-type acquisition. To perform it, it is necessary to have a hypothesis about the initial distribution of cell-types. This means that a given process begins with a population of cells distributed across the available cell-types. Further throughout the process, some of those cell types differentiate into other cell types. In the present model, the changes between cell-types are encoded in the IATM, and the process of cell population differentiation is simulated by multiplying an attractor distribution vector times the IATM. In order to perform this analysis, an initial cell-type (attractor) distribution vector must be defined.

This is a vector $P_X(t_0) = (p_1(t_0), p_2(t_0), \ldots, p_K(t_0))$, where $p_i(t_0)$ represents the probability of the network being in attractor i at the initial time $t = 0$. From a cell population point of view, the probability of attractor i at any time is interpreted as the percentage of cell type i in the population at that time. Having the initial distribution vector, the dynamics of each attractor's probability in time is simulated by iteratively multiplying the distribution vector times the IATM a certain number of time steps. As proposed in [17], the succession of attractors' probability *maxima* then corresponds to an intrinsic explanation for the emerging temporal order observed during a developmental process.

3.3.5 Relative Attractor Stability and Global Ordering

During development, the zygote differentiates into different cell types in a defined and robust order, which makes most steps in the cell differentiation process irreversible. GRNs, proposed as a representation of the genetic mechanism underlying cell differentiation, are expected to recover the observed sequence of attractors in the presence of noise. Zhou and collaborators proposed a method to calculate the global attractor ordering of a given GRN based on an attractor's relative stability [19]. Relative stability of attractors reflects the relative ease for transitioning from an attractor (A) to another attractor (B) given a certain degree of stochastic noise, i.e., the probability of passing from attractor A to attractor B. The relative stabilities in a GRN are expected to be asymmetric between any pair of attractors, which gives the attractor ordering directionality (i.e., irreversibility during cell differentiation).

Relative stabilities of a GRN attractor can be calculated by the mean first passage time (MFPT) from a given IATM as proposed by Zhou and collaborators [17]. The MFPT between attractors A and B is the expected number of time steps until reaching attractor B starting from attractor A. MFPTs can be calculated either by implementing the matrix-based algorithm proposed in [32] or by means of numerical simulation. After defining the MFPT among every attractor pair, a net transition rate ($d_{i,j}$) between attractor i and j is defined in terms of the MFPT as follows:

$$d_{i,j} = \frac{1}{\mathrm{MFPY}_{i,j}} - \frac{1}{\mathrm{MFPT}_{j,i}}$$

The attractor global ordering can be obtained by calculating the transition rate among all the attractors. The consistent global ordering of the attractors is given by the attractor permutation in which all transitory net transition rates from an initial attractor to a final attractor are positive, as proposed in [19]. This can be illustrated by constructing a network using the transition rates matrix as an adjacency matrix and highlighting the positive transition rates. In the global ordering network nodes are the attractors and the arrows connecting them are the transition rates among them. From such a network one can recover the global ordering looking for the path that connects all the attractors through positive transition rates.

We have implemented all the modeling extensions introduced in this section so that they can be applied directly to the output of the dynamical analyses presented in the previous sections. In what follows we exemplify their use.

3.3.6 Implementing the EL Protocol

We have coded functions in R for a practical implementation of the complete framework of EL modeling, applicable to a dynamically analyzed Boolean GRN (*see* Subheading 3.2). These functions recover the steps explained above for EL modeling, these are: calculation of the IATM, temporal sequence of attractor attainment, and attractors global ordering. We apply these functions to the *Arabidopsis* root SCN-GRN analyzed above, and interpret the results.

3.3.7 Calculating the IATM

The first step is the calculation of the IATM. For this purpose we coded the function `Implicit.InterAttractor.Simulation (Network, P.error, Nreps)`, which estimates the IATM by simulating stochastic node perturbations in the network dynamics. It takes as inputs a Boolean GRN, an error probability (*see* **Note 7**) and the desired number of iterations (1000 or more) to be performed over each network state. The function works as follows:

1. Recovers the state space of the network and the attraction basin each state belongs to.
2. Computes the state transition for every initial state.
3. Changes the state of every gene considering the given error probability.
4. Identifies the attraction basin of the resultant perturbed transitory state.
5. Records every change of attraction basin.

The steps are repeated the indicated number of times and the frequency of observed transitions among all the attractor basins is estimated (for details, *see* [18]). Using this function, we estimate the IATM for the root SCN-GRN considering an error probability

Table 7
Interattraction probability matrix for the root SCN-GRN, with an error probability of 0.05

	CepI	VI	CEI	QC
CepI	0.9498	0.0479	0.0011	0.0012
VI	0.0513	0.9032	0.0239	0.0216
CEI	0.0509	0.0745	0.8141	0.0605
QC	0.0502	0.0608	0.0796	0.8094

of 0.05 and 1000 repetitions. The resulting IATM is shown in Table 7. Note that due to stochasticity interattractor probabilities in different simulations will vary slightly.

3.3.8 Estimating the Attractor Temporal Evolution

Once having calculated the IATM, the temporal sequence of attractor attainment can be easily estimated. For a practical implementation, we coded the function.

`Plot.Probability.Evolution(TPM, Initial, Attrs Names, timeF)`. It takes as inputs an IATM, the name of the initial attractor, the names of all the network's attractors in the same order as they are obtained by `getAttractors`, and the number of times to iterate the process. As mentioned above, we need to provide an initial attractor distribution to calculate the temporal evolution. In this function, the initial attractor provided by the user is assumed to be the only cell type present in the initial distribution. The result of the function is a matrix with the attractor's probability distribution corresponding to each time-step, and a plot that illustrates the dynamics of each attractor's probability and their temporal attainment order.

We calculated the temporal order of cell-type attainment in the root SCN, using as input the IATM calculated above with an initial distribution of quiescent center cells for 50 time-steps: `Plot.Probability.Evolution (IAT_5, Initial="QC", AttrsNames=c ("CEpI","VI","CEI", "QC"), timeF = 50)`. In the resulting plot (Fig. 8) the obtained temporal order of cell types follows sequentially: QC, CEI, VI, and CEpI. The biological interpretation of this result is not straightforward since the root SCN has a continuous production of cells and the characteristic cell-types follow a topological order rather than a temporal one. Alvarez-Buylla and collaborators proposed one of the first methodological frameworks for the exploration of the EL. They proposed a GRN underlying the cell-types of early flowering in *A. thaliana*, and through an IAT approach they recovered the temporal sequence of cell types observed during flower development [17].

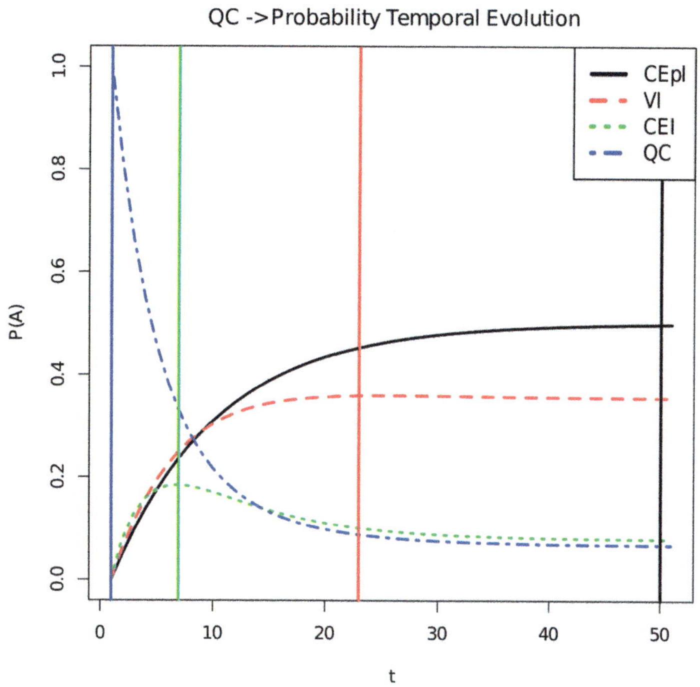

Fig. 8 Temporal sequence of cell–fate attainment pattern in the root SCN-GRN, starting with quiescent center cells. Each line in the plot corresponds to each attractor's probability of occurrence through time, the vertical lines indicate the time of the maximum probability for the corresponding color-coded attractor

The last analysis we present here is the global ordering of attractors, for which it is necessary to calculate the MPFT and transition rate matrices. We have created the functions `Calculate.MFPT.Matrix` and `MFPT.Transition.Rates`, which calculate the MFPT matrix among attractors of a BN and their corresponding transition rates, respectively. The former function calculates the MFPT taking as inputs an IATM and the names of the obtained attractors. The latter function takes as input the calculated MFPT. Once we have calculated the inter attractor transition rates, we can obtain the attractor global ordering. We created the function `Plot.Attractor.Global.Ordering`, which takes the transition rates matrix as input and creates a network from the attractors' transition rates that highlights the positive transition rates as red arrows. As mentioned above, the attractors global ordering is the path in the network that passes through every attractor by positive transition rates.

Using these functions and the IATM calculated above, we calculated the MFPT matrix (Table 8) and the transition rates matrix (Table 9) of the root SCN-GRN. Finally, we plotted the attractors' global ordering network (Fig. 9a) using the latter matrix as input. In Fig. 9, we can see that the temporal flow of the root

Table 8
Mean first passage time among attractors of the root SCN-GRN

	CEpI	VI	CEI	QC
CEpI	0	20.52	80.14	90.49
VI	20.38	0	63.26	72.76
CEI	20.40	16.34	0	59.53
QC	20.45	17.41	47.48	0

Table 9
Transition rates among attractors of the root SCN-GRN

	CEpI	VI	CEI	QC
CEpI	0	−0.0027	−0.0387	−0.0406
VI	0.0027	0	−0.0455	−0.0432
CEI	0.0387	0.0455	0	−0.0040
QC	0.0406	0.0432	0.0040	0

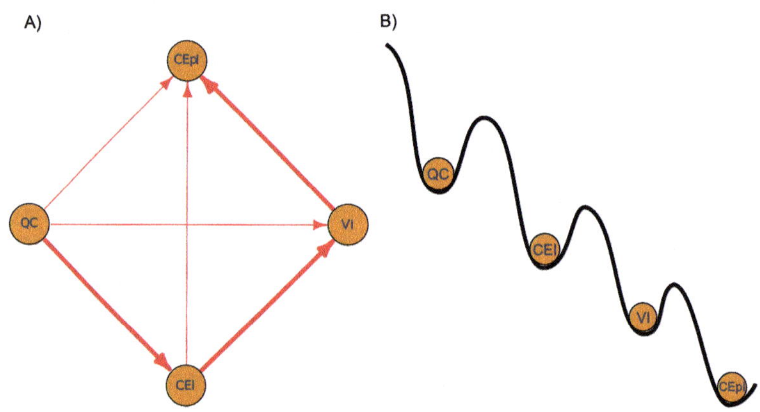

Fig. 9 Global order of the root SCN-GRN attractors. (**a**) Global order network of the SCN attractors. The global order corresponds to the path of positive interactions passing through all of the attractors, the corresponding path is highlighted in the image. (**b**) Graphical interpretation of the landscape inferred from the attractor's global order

SCN attractors follows the order: QC -> CEI -> VI-> CepI, which corresponds to the probability evolution of attractor attainment starting from QC. From the attractor global ordering, we can infer a representation of the EL as rifts and valleys as shown in Fig. 9b.

The transitions uncovered by the EL extension indicate that progenitor cells in the root stem cell niche are prone to differentiate into specific cell type given their current state, and as a natural consequence of the regulatory constraints. For example, our results would suggest that cells from the QC have a natural tendency to transit to the CEI state. Such prediction could be an interesting starting point to propose hypotheses for the analysis and interpretation of, for example, single-cell expression data, which is becoming more available [33, 34]. The predictions can also be relevant for in vitro differentiation experiments, given that the current model is not considering any additional, tissue-level restriction to the intrinsic regulatory interactions. Ultimately, the modeling protocols presented here should ideally be integrated with experiments. Interdisciplinary systems biology approaches considering the interplay among model building, analysis/simulation, and experimentation enable the discovery and interpretations of interesting and counterintuitive observed behaviors (see, for example 20).

4 Conclusion and Outlook

Computational modeling is a useful approach for understanding biological developmental processes. Integration of molecular genetic and genomic information in GRNs, allows the postulation of computer models that explore the epigenetic landscape, as theoretically proposed by C.H. Waddington [5]. Cross talk between experimental, theoretical, and computational approaches to biological systems is necessary for an integral comprehension of developmental processes.

In this chapter we present a framework for the analysis of the epigenetic landscape associated with a GRN. The proposed methodology is based on the Boolean modeling of GRNs, a useful scheme for explaining the different gene expression patterns among cell types as attractors of an underlying GRN. Moving forward from the Boolean dynamical analysis, the epigenetic landscape associated with a GRN is characterized by introducing stochasticity to the model and by measuring the probability of transitioning among attractors. The introduction of stochasticity makes possible the exploration of the epigenetic landscape implicit in the proposed GRN. We include a comprehensive exposition of our proposed methodology, together with its computational implementation. We expect the toolkit and conceptual interpretation put forward here to be useful resources for the systems biology community interested in modeling plant developmental processes.

5 Notes

1. A node in a GRN represents a variable describing the system behavior. In a general dynamical model variables are chosen both for practical and fundamental considerations. On the one hand, it is preferable to choose a variable whose value is easy to approximate experimentally. On the other hand, the modeler is commonly interested in finding variables that reflect relevant characteristics about the functional behavior of the system, and which are involved in the mechanistic basis underlying the latter. Although in a GRN, nodes are generically referred to as genes, these can represent any cellular element whose activity has a strong influence on the cellular phenotype. These elements commonly are proteins or protein complexes, signaling molecules, miRNAs, or groups of various elements representing a process. Due to the generality of the underlying modeling apparatus (i.e., a dynamical system), any kind of variable that changes with time can be considered a node in the network.
2. For a given GRN, an attractor is a network state which, if taken as an input for the Boolean functions, either does not move the network to another state (steady state attractor), or only moves through the same set of states (cyclic attractor). In other words, it is a network state (or sequence of states) that does not change in time. Every network state not belonging to an attractor is a transitory state that eventually leads to an attractor.
3. Boolean GRN model construction is an intuitive process in which experimental regulatory interactions are transferred into Boolean syntax. This process involves the integration of large amounts of experimental evidence obtained from the literature to be formalized as Boolean rules or truth tables. Experimental natural-language expressions can be stated as Boolean functions in a straightforward manner, as shown in Table 4. We will use an example from the root SCN-GRN to illustrate this process:

   Roots with *shr* or *scr* knockdown show reduced expression levels of JKD. This suggests that: SHR and SCR are positive regulators of JKD.

   The latter statement can be transformed easily into a Boolean function such as:

   JKD = SCR AND SHR
4. The dynamical analysis of a Boolean network recovers the network's steady states (attractors). In GRN models of developmental modules, attractors are considered as an abstract

representation of cellular phenotypes, experimentally accessible through gene expression profiling. In order to validate a proposed GRN, the expected attractors have to be defined, i.e., the expression profiles of the cell-types of interest must be identified from experimental evidence and coded into a Boolean vector. For a GRN to be experimentally validated, the attractors it recovers must match the expected attractors. Nevertheless, it is important to keep in mind that different networks converge to the same set of attractors. Experimental evidence should weight the evaluation process of competing models.

5. The BoolNet package has straightforward ways to implement knockout and overexpression simulation experiments. Specifically, the genes within the network can be set to a fixed value (0 for knockout, and 1 for overexpression). Calculations can then be performed on the modified network, with the only difference that the assigned value, and not the one generated by corresponding transition function, will be used through the simulation. The function *fixGenes()* takes as input the network, the name of the gene to be perturbed, and the value to be fixed (0 or 1). Then all the other dynamical analysis, such as attractor identification, can be performed on this new-perturbed network.
6. When working with Boolean GRNs, stochasticity can be introduced either by the SIN method (*see* Subheading 3.3.1) or by the stochasticity in function (SIF) method. In the SIF method, stochasticity is modeled at the level of biological functions (i.e., Boolean functions in the GRN), i.e., implicitly behaving contrary to what the Boolean function indicates and not just flipping the state of a gene as in the SIN model (for details *see* refs. 26, 31).
7. The level of noise (error probability) used in a stochastic model determines the behavior that will be recovered. When introducing stochasticity to a Boolean network, very small levels of noise are not strong enough to make the system leave an attractor so no state transitions will be recovered. On the other hand, high levels of noise cause the system to jump among attractors completely randomly losing the information contained in the network. An appropriate noise level shows a nontrivial behavior in which there are state changes following the logic of the network. Levels of error probability used for Boolean GRNs range normally form 0.01 to 0.1 [16, 17], but different values should be tested.

References

1. Davila-Velderrain J, Martinez-Garcia JC, Alvarez-Buylla ER (2015) Descriptive vs. mechanistic network models in plant development in the post-genomic era. Methods Mol Biol 1284:455–479
2. Álvarez-Buylla ER, Dávila-Velderrain J, Martínez-García JC (2016) Systems biology approaches to development beyond bioinformatics: nonlinear mechanistic models using plant systems. Bioscience 66(5):371–383
3. Forgacs G, Newman SA (2005) Biological physics of the developing embryo. Cambridge University Press, Cambridge
4. Alvarez-Buylla ER, Azpeitia E, Barrio R, Benítez M, Padilla-Longoria P (2010) From ABC genes to regulatory networks, epigenetic landscapes and flower morphogenesis: making biological sense of theoretical approaches. Semin Cell Dev Biol 21:108–117
5. Waddington CH (1957) The strategy of genes. George Allen & Unwin, Ltd., London
6. Davila-Velderrain J, Martinez-Garcia JC, Alvarez-Buylla ER (2015) Modeling the epigenetic attractors landscape: toward a post-genomic mechanistic understanding of development. Front Genet 6:160
7. Alvarez-Buylla ER, Balleza E, Benítez M et al (2008) Gene regulatory network models: a dynamic and integrative approach to development. SEB Exp Biol Ser 61:113
8. Kauffman SA (1969) Metabolic stability and epigenesis in randomly constructed genetic nets. J Theor Biol 22(3):437–467
9. Alvarez-Buylla ER, Benítez M, Davila EB et al (2007) Gene regulatory network models for plant development. Curr Opin Plant Biol 10(1):83–91
10. Davila-Velderrain J, Martinez-Garcia JC, Alvarez-Buylla ER (2016) Dynamic network modelling to understand flowering transition and floral patterning. J Exp Bot 67(9):2565–2572
11. Azpeitia E, Davila-Velderrain J, Villarreal C, Alvarez-Buylla ER (2014) Gene regulatory network models for floral organ determination. Methods Mol Biol 1110:441
12. Kaplan D, Glass L (2012) Understanding nonlinear dynamics. Springer, New York
13. Glass L, Kauffman SA (1973) The logical analysis of continuous, non-linear biochemical control networks. J Theor Biol 39(1):103–129
14. Espinosa-Soto C, Padilla-Longoria P, Alvarez-Buylla ER (2004) A gene regulatory network model for cell-fate determination during Arabidopsis thaliana flower development that is robust and recovers experimental gene expression profiles. Plant Cell 16(11):2923–2939
15. Azpeitia E, Benítez M, Vega I, Villarreal C, Alvarez-Buylla ER (2010) Single-cell and coupled GRN models of cell patterning in the Arabidopsis thaliana root stem cell niche. BMC Syst Biol 4(1):1
16. Benítez M, Espinosa-Soto C, Padilla-Longoria P, Alvarez-Buylla ER (2008) Interlinked nonlinear subnetworks underlie the formation of robust cellular patterns in Arabidopsis epidermis: a dynamic spatial model. BMC Syst Biol 2(1):1
17. Álvarez-Buylla ER, Chaos Á, Aldana M et al (2008) Floral morphogenesis: stochastic explorations of a gene network epigenetic landscape. PLoS One 3(11):e3626
18. Davila-Velderrain J, Juarez-Ramiro L, Martinez-Garcia JC, Alvarez-Buylla ER (2015) Methods for characterizing the epigenetic attractors landscape associated with Boolean gene regulatory networks. arXiv preprint arXiv:1510.04230
19. Zhou JX, Samal A, d'Hérouël AF, Price ND, Huang S (2016) Relative stability of network states in Boolean network models of gene regulation in development. Biosystems 142:15–24
20. Pérez-Ruiz RV, García-Ponce B, Marsch-Martínez N et al (2015) XAANTAL2 (AGL14) is an important component of the complex gene regulatory network that underlies arabidopsis shoot apical meristem transitions. Mol Plant 8(5):796–813
21. Cui H, Levesque MP, Vernoux T, Jung JW et al (2007) An evolutionarily conserved mechanism delimiting SHR movement defines a single layer of endodermis in plants. Science 316: 421–425
22. Levesque MP, Vernoux T, Busch W, Cui H et al (2006) Whole- genome analysis of the SHORT-ROOT developmental pathway in Arabidops. PLoS Biol 4:e143
23. Sarkar AK, Luijten M, Miyashima S, Lenhard M, Hashimoto T, Nakajima K et al (2007) Conserved factors regulate signalling in Arabidopsis thaliana shoot and root stem cell organizers. Nature 446:811–814
24. Stahl Y, Wink RH, Ingram GC, Simon R (2009) A signaling module controlling the stem cell niche in Arabidopsis root meristems. Curr Biol 19:909–914

25. Müssel C, Hopfensitz M, Kestler HA (2010) BoolNet—an R package for generation, reconstruction and analysis of Boolean networks. Bioinformatics 26(10):1378–1380
26. Garg A, Mohanram K, De Micheli G, Xenarios I (2012) Implicit methods for qualitative modeling of gene regulatory networks. Methods Mol Biol 786:397–443
27. Bhattacharya S, Zhang Q, Andersen ME (2011) A deterministic map of Waddington's epigenetic landscape for cell fate specification. BMC Syst Biol 5:85
28. Moris N, Pina C, Arias AM (2016) Transition states and cell fate decisions in epigenetic landscapes. Nat Rev Genet 17(11):693–703
29. Martinez-Sanchez ME, Mendoza L, Villarreal C, Álvarez-Buylla ER (2015) A minimal regulatory network of extrinsic and intrinsic factors recovers observed patterns of CD4+ T cell differentiation and plasticity. PLoS Comput Biol 11:e1004324
30. Davila-Velderrain J, Villarreal C, Alvarez-Buylla ER (2015) Reshaping the epigenetic landscape during early flower development: induction of attractor transitions by relative differences in gene decay rates. BMC Syst Biol 9(1):20
31. Garg A, Mohanram K, Di Cara A, De Micheli G, Xenarios I (2009) Modeling stochasticity and robustness in gene regulatory networks. Bioinformatics 25(12):i101–i109
32. Sheskin TJ (1995) Computing mean first passage times for a Markov chain. Int J Math Educ Sci Technol 26(5):729–735
33. Brennecke P, Anders S, Kim JK, Kołodziejczyk AA, Zhang X, Proserpio V, Baying B, Benes V, Teichmann SA, Marioni JC, Heisler MG (2013) Accounting for technical noise in single-cell RNA-seq experiments. Nat Methods 10(11):1093–1095. https://doi.org/10.1038/nmeth.2645
34. Efroni I, Ip P-L, Nawy T, Mello A, Birnbaum KD (2015) Quantification of cell identity from single-cell gene expression profiles. Genome Biol 16(1):9. https://doi.org/10.1186/s13059-015-0580-x

Chapter 18

Developing Network Models of Multiscale Host Responses Involved in Infections and Diseases

Rohith Palli and Juilee Thakar

Abstract

Complex interactions involved in host response to infections and diseases require advanced analytical tools to infer drivers of the response in order to develop strategies for intervention. This chapter discusses approaches to assemble interactions ranging from molecular to cellular levels and their analysis to investigate the cross talk between immune pathways. Particularly, construction of immune networks by either data-driven or literature-driven methods is explained. Next, graph theoretic approaches for probing static network properties as well as visualization of networks are discussed. Finally, development of Boolean models for simulation of network dynamics to investigate cross talk and emergent properties are considered along with Boolean-like models that may compensate for some of the limitations encountered in Boolean simulations. In conclusion, the chapter will allow readers to construct and analyze multiscale networks involved in immune responses.

Key words Boolean modeling, Biological networks, Gene networks, Graph theory, Systems immunology

1 Introduction

Molecular changes associated with diseases and infections are complex and encompass a large range of temporal- (millisecond to years) and spatial (molecular to whole body) scales [1]. Investigation of these molecular changes is necessary in order to identify determinants of disease/infection outcome. As technological advances are facilitating measurements of thousands of genes (RNA-sequencing), proteins (proteomics), and cells (multidimensional flow cytometry) over time, the requirement of advanced computational techniques to identify the determinant of the disease/infection outcome is becoming increasingly clear [2–4]. Network analysis allows investigation of the molecular and cellular level changes and association between those by depicting discrete biological entities (genes, proteins and cells) in a mathematical format to study molecular pathogenesis [5–7]. Typically, a

Louise von Stechow and Alberto Santos Delgado (eds.), *Computational Cell Biology: Methods and Protocols*,
Methods in Molecular Biology, vol. 1819, https://doi.org/10.1007/978-1-4939-8618-7_18,

network includes interactions between nodes representing biological entities such as genes, proteins, or cells, which are connected by edges. The definition of nodes depends on underlying questions; for example, nodes could represent specific molecules (e.g., genes or proteins) or multiple types of molecules as in a multilayered network encompassing mRNAs, transcription factors, and signaling proteins [8, 9]. Similarly, edges can describe a wide range of interactions, including physical binding, transcriptional or translational regulation, and in addition to other functional relationships, such as phosphorylation or catalysis [10, 11].

Defining nodes and edges in a context-specific manner allows investigation of regulatory principles, which underlie the observed disease or infection outcome [12]. Investigation of regulatory principles can identify cross talk between signaling pathways that is not reflected in linear examinations of pathways or traditional single-element perturbation experiments. Furthermore, computational perturbation of nodes and edges provides a low cost method for probing system-level dynamics. For example, one could investigate response to infections upon perturbation of signaling molecules by drugs. Thus, context-specific network modeling allows investigation of complex multiscale host responses.

Molecular level networks frequently investigate transcriptional and posttranscriptional regulation whereas cellular networks consider interactions between cells and cytokines in the context of infections and diseases [7, 13]. Explosion of RNA-sequencing and microarray data has facilitated network modeling of transcriptional networks. These networks can define states of specific immune cells or of mixtures of cells, such as peripheral blood mononuclear cells (PBMCs), and have been instrumental in understanding responses to various infections and vaccines [14, 15]. Cellular networks are much less studied at the systems level due to limitations in our ability to measure the cells and cytokines at large scale. However, recent development in multidimensional cytometry techniques, such as CyTOF, will facilitate large scale development of cell-cytokine networks and their modeling [16, 17]. In the meanwhile, manually curated cell-cytokines' networks at smaller scale have been developed [13] and have revealed cytokines involved in cross talks and modulating the infection outcome [18]. Thus, molecular and cellular networks have been instrumental in organizing and analyzing complex multitype datasets.

In this chapter, we discuss the construction of molecular networks, methods for analysis of static networks, methods for simulation of signals over biological networks, and available computational resources.

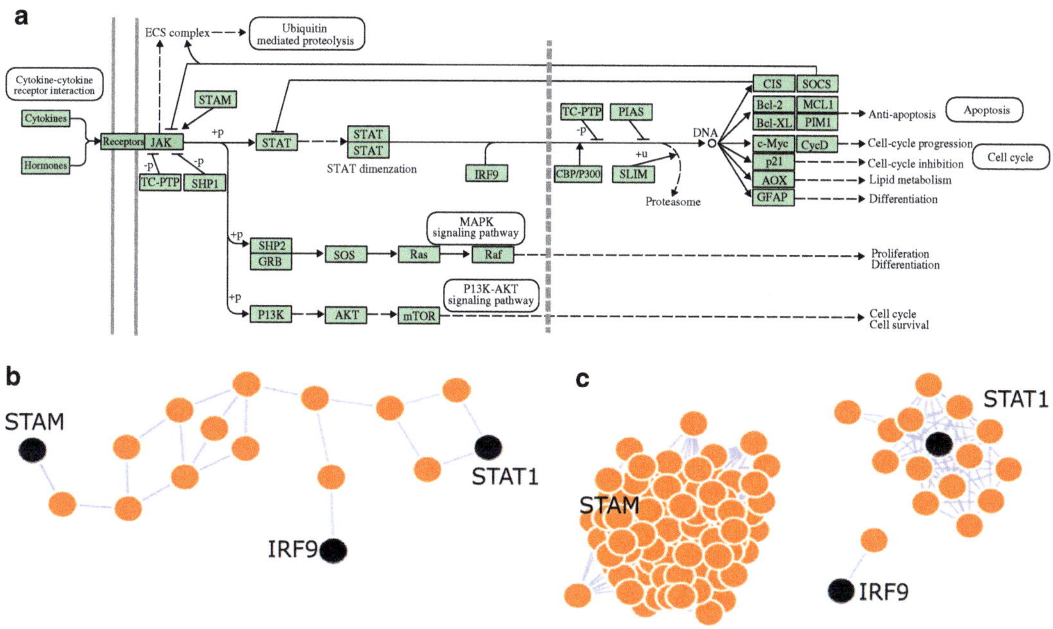

Fig. 1 Examples of data-driven and knowledge-driven antiviral networks: (**a**) JAK-STAT pathway from KEGG which is frequently enriched in transcriptomic datasets measuring response to several strains of influenza infections [15]. (**b**) Shortest path and (**c**) nearest neighbors of three genes (STAT1, IRF9, and STAM) from KEGG JAK-STAT pathway in Influenza responsive network inferred from transcriptomic data from Epithelial cell (A549) cells [20]

2 Construction of Interaction Networks at Molecular Level

Construction of interaction networks at molecular level can be performed using data-driven approaches, which infer linkages from high-throughput data, or on knowledge-driven approaches, which generally utilize either expert-curated interaction databases or text mining of literature [19].

Biological knowledge repositories serve as a starting point for knowledge-driven network construction. Some repositories specialize in particular types of pathways or networks whereas others attempt to create centralized, general-purpose network storehouses (Fig. 1). In general, such repositories fall into three categories (Table 1): (1) Storehouses of interaction data such as physical interactions, phosphorylation and genetics interactions, (2) Pathway repositories, which list predefined pathways and might allow the users to overlay data on top of those pathways and (3) Pathway aggregator tools, which mine the repositories of molecular interaction to display integrated information from independent sources.

The tools in the third category lead to the most robust and complete description of biological knowledge. One popular aggregator of pathway databases is Pathway Commons [33], which

Table 1
Three types of storehouses of molecular interactions: *I for interaction database, P for pathway repository, or N for network tool

Name	Type*	Description	Reference
PhosphoSitePlus	I	Protein phosphorylation networks	Hornbeck et al. [21]
Human Protein Reference Database	I	Protein-protein interaction networks	Prasad et al. [22]
Database of Interacting Proteins	I	Protein-protein interaction networks	Salwinski et al. [23]
The Kyoto Encyclopedia of Genes and Genomes (KEGG)	P	Wide variety of pathways, particularly signaling cascades and metabolic pathways	Kanehisa et al. [10]
Small Molecule Pathway Database	P	Small molecule pathways	Jewison et al. [24]
NetPath	P	Signaling pathways	Kandasamy et al. [25]
WikiPathways	N	Wikipedia-like pathway aggregator	Kutmon et al. [26]
Reactome	N	Forum for biological network storage and retrieval	Joshi-Tope et al. [27], Croft et al. [28], and Fabregat et al. [29]
HumanCyc	N	Forum for biological network storage and retrieval	Romero et al. [30]
BioGRID	N	Forum for biological network storage and retrieval	Chatr-Aryamontri et al. [31] and [32]
PathwayCommons	N	Forum for biological network storage and retrieval	Cerami et al. [33]

combines many databases into a unified resource. Furthermore, standardized file formats have been proposed to facilitate use of the networks, foster collaboration across disciplines, and ensure interoperability of databases and community code. Three notable file formats are Simple Interaction Format (SIF), Systems Biology Markup Language (SMBL), and Biological Pathways Exchange (BIOPAX) [34, 35]. SIF is the simplest way to represent networks. The SIF format represents biological interactions as a list of the interactions (edges) in the network along with interaction types [36]. SMBL and BIOPAX are both based on XML, a markup language that allows increased flexibility in defining data structure. Both formats are under active development as standards for the biological/ bioinformatics community. Of the two formats, BIOPAX includes more detailed metadata, but this is at the cost of being more complex [34, 35].

Apart from the three types of storehouses knowledge-driven networks can also be generated by mining the scientific literature. Literature mining must proceed in a stepwise manner through information (or literature) retrieval, entity (or node) recognition, information (or edge) extraction, text (rule) mining, and finally integration (of data) [37]. This information has been used to provide confidence intervals for the edges in molecular networks such as gene-gene association networks [38, 39]. However, frequently literature mining is the only method used for assembly of cellular networks because interactions at cellular level, e.g., between cells and cytokines are not typically stored in structured databases.

After identification of relevant papers, data mining algorithms start with entity recognition algorithms to determine nodes in the network. These algorithms either use machine learning approaches to identify the context of references to nodes in the literature or tag nodes based on dictionaries of equivalent gene names. The edge inference is then performed in a process called information extraction by either Natural Language Processing (NLP) or co-occurrence [37]. Co-occurrence methods yield edges when two terms occur close to each other while NLP methods are able to yield more specific rules by extracting information from the structure of language [37]. While co-occurrence networks are more likely to capture extant edges, NLP algorithms can extract more meaningful information. One example of this approach is the use of SciMiner to create gene-gene and vaccine-gene (vaccine configurations, genes as nodes and interactions as edges) interaction networks, and compare them with fever-associated genes in order to hypothesize a set of genes that mediate vaccine-associated fever outcomes [40]. Additionally, FunCoup [41–43] and STRING [39, 44] provide an integrated approach, combining literature-based information with the interaction databases.

Data-driven approaches, on the other hand, seek to leverage large data sets to determine network connectivity, and their development has been closely linked to the advent of high throughput technologies such as RNA-sequencing, proteomics, and metabolomics [20, 45–55]. Data-driven approaches measure associations between experimental measurements across time points, biologically independent replicates, or experimental perturbations of particular nodes (via knockout or overexpression of a gene for example) to determine linkages [55]. Associations can be determined by several metrics such as Pearson correlation and mutual information [18, 56]. However, these data-driven networks are dense and have many more nodes compared to the knowledge-driven networks. The associations could be observed due to primary or secondary regulations leading to higher frequency of false positives which do not inform about causality. Hence, to generate higher confidence data-driven networks, stringent cutoffs assessing strength of association and methods for integration of multiple data-sets should be used [20].

Construction of data-driven networks allows investigation of regulatory principles in a context-specific manner. We recently developed cell-specific influenza-responsive networks to identify specific antiviral responses generated by cells implicated in innate immune responses [20]. Our studies revealed very small overlap between antiviral responses across cell-types. Furthermore, analysis of patients' blood transcriptional profiles under various disease states have identified transcriptional modules that can act as disease-specific transcriptional fingerprints and biomarkers, demonstrating the ability of network-based analysis to assist in distinguishing disease states [57]. Other algorithms to infer gene association networks are ARACNE [45] and weighted gene coexpression network analysis (WGCNA) [58]. Notably, these data-driven approaches almost exclusively generate undirected networks (Fig. 1a). Moreover, confidence in the inferred edges can be increased by integrating data-driven networks and knowledge-driven approaches [19]. These integrated approaches also enable inference of directionality [19].

The size and density of biological networks has increased with the amount of data generated by high-throughput experiments. Large biological networks are difficult to explore manually and analyzing them computationally is costly. Reducing network complexity can often improve the interpretability and computability of the networks. Many approaches can efficiently reduce the complexity of networks while preserving functional properties. Liu et al. followed a structural approach of trimming extraneous nodes from the network and collapsing chains of single links [19]. Wang and Albert, on the other hand, took a structural approach by finding networks that recapitulate the functional properties of the original network with a simpler structure [59]. Software packages for generating simple networks that explain evidence include NET-SYNTHESIS, a stand-alone program for construction and reduction of biological networks; Cellular Network Optimizer (CellNOptR), an R package for generation and analysis of Boolean networks that can also reduce networks; and BoolNet, an R package for simulation and analysis of Boolean networks that includes network reduction functionality [60–62].

2.1 Static Network Analysis

After network construction, graph-theoretical approaches are used to investigate the topological characteristics of a network to identify key regulatory nodes and interactions. Here we will briefly review commonly used graph-theoretical measures (please refer to Chistensen et al. [12] for additional details). Degree and centrality are related graph-theoretic properties which reflect the importance of a node in determining behavior of the network as a whole. Network connectivity denotes the robustness of the network to perturbations of individual nodes or edges. The degree of a node is the number of interactions in which the node participates. The node degree can further be segregated (if the network is directed) into in-degree and out-degree for incoming and outgoing edges. The out-degrees in many molecular networks are distributed closely to the power law in a scale-free manner [63–65]. This means that there are relatively few nodes with high degree (known as hubs) and many nodes with low degree [7, 66]. Typically, transcription factor nodes have high out-degree, which could be functionally linked to one transcription factor regulating a number of genes. Centrality reflects the importance of each node in signal-transduction through the network and is an area of active research. We have developed novel measures of centrality to estimate inhibitory effects of the nodes in the cell-cytokine networks, revealing stimulus dependent activation of the immune cells [18]. Betweenness centrality, another measure of centrality, counts the number of shortest paths between two network nodes that pass through a particular node. Nodes with high betweenness centrality are typically important in cross talk among signaling cascades. Moreover, connectivity measures allow estimation of accessibility between parts of a network, for example k-connectivity measures the connectivity of a subset of components in the network so that removal of nodes $<k$ retrieves a connected graph. These strongly connected components often map to protein complexes and highly regulated biological pathways [67, 68]. Thus, graph theoretical analysis allows identification of key regulatory features of a biological network.

The next step in examining any network is to visualize it. Proper visualization highlights the nodes with specific regulatory role in the network. Though trivial on the surface, the useful arrangement of nodes on a 2-D surface can be difficult to achieve. Tools and techniques available for network layout have been extensively reviewed elsewhere [36, 69]. Broadly, network visualization techniques are grouped into force-based and hierarchical techniques [36, 69]. Force-directed graph layouts treat edges as springs and compute a minimal energy layout of nodes. They offer a simple, fast approach that, in undirected graphs, spreads nodes roughly evenly, while centering nodes with higher betweenness centrality [70, 71]. In contrast, a popular hierarchical algorithm, the Sugiyama framework, first removes all cycles from a network, turning it into a directed acyclic graph (DAG), which can be represented as a tree, then draws the edges removed to create the DAG [36, 69]. The Sugiyama framework easily handles networks that are hierarchical (i.e., with a clear "top" and "bottom") but falters when there are many-cycle in the networks. Similar hierarchical methods begin by placing high degree nodes, high centrality nodes, or sources as the root of a tree to which they add the remaining nodes in successive layers. Many of these approaches are implemented in Cytoscape [72], which has an easy-to-use GUI for manipulating graphs. Other software packages for visualizing graphs include IPath [73, 74], an interactive web explorer; iGraph, a general-purpose network software package in C, R, and Python; RedeR [70], a package for biological network visualization in R; and GraphViz [75], a stand-alone application that is compatible with Python via pyGraphViz.

Further, examination of recurrent patterns within networks yields insight into how nodes are regulated [76]. These recurrent patterns, the discrete functional units of networks, are referred to as network motifs [77, 78]. Though there has been much debate about the role of motif analysis for distinguishing the functional connectivity of networks, motif analysis may reveal how individual links coalesce to form functional units of regulation within networks [79]. Many groups have used such motifs to understand the biological significance of gene subnetworks [80–88]. We have shown that an abundance of well-known motifs such as bi-fans, feed-forward loops, negative and positive biparallel segments (both with and without feedback) are present in the cellular network of immune cells and cytokines (Fig. 2) [20]. Algorithms such as that described by Kashatan and colleagues perform efficient sampling of motif numbers rather than exhaustive detection because motif counts can provide insights into the overall topology of a network [89].

Several software packages exist for manipulation of networks in silico and computation of graph theoretic measures. IGraph [90] is a popular package for manipulation and computation of graphs in R. In Python, Networkx [91] is a particularly fast graph implementations method, due to its use of a RAM-intensive dictionary-based storage technique for data management. Networkx has similar functionality to IGraph. Similarly, the built-in graph and digraph data types in both MatLab [92] and Mathematica [93] are powerful and can achieve much of the same objectives as Networkx or IGraph [90, 91]. Finally, extensions of Cytoscape allow user-friendly, GUI-based approaches to calculate graph measures [72].

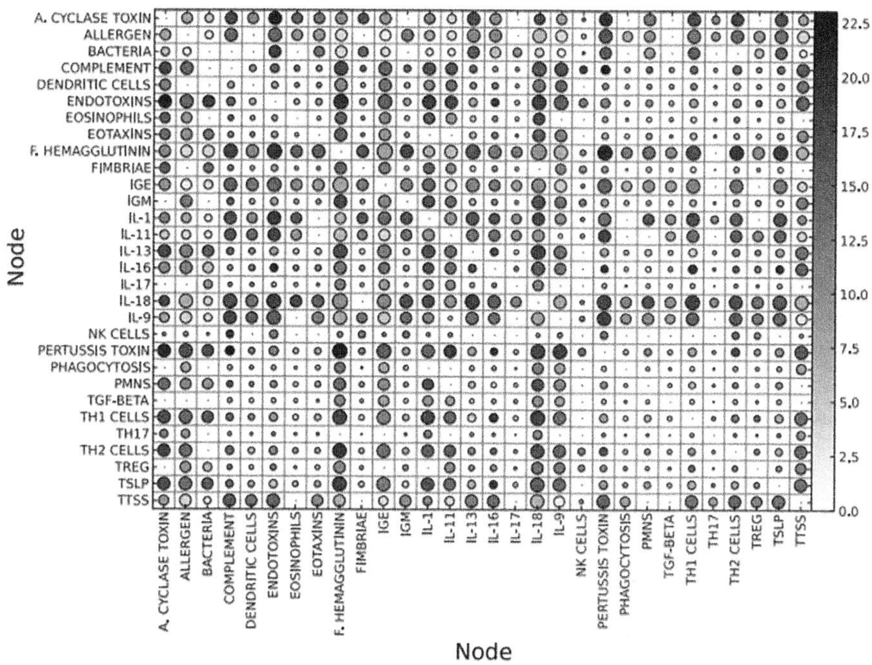

Fig. 2 Motif abundance analysis of cell-cytokine network: Coexistence (Z score > 2) of immune cells and cytokines in network motifs represent regulation of cells and cytokines by each other. Circle size corresponds to the raw number of observed motifs (maximum = 152), and grey gradation corresponds to the number of unique motif types. The figure reproduced from Campbell et al. [18]

3 Dynamics of the Networks Using Boolean Models

Dynamic analysis of networks allows investigation of emergent properties, which can be mapped to clinical symptoms or observed phenotypes such as outcome of infections (severe or mild) or cell fate (dead or alive). Investigation of network dynamics also allows prediction of the effect of node perturbations by experimental intervention such as RNA interference (RNAi) or by drugs on emergent properties. Translation of static networks into dynamic models, however, is not trivial. Mathematical models describing interactions between each node of the network by ordinary differential equations (ODEs) provide detailed descriptions of the behavior of the system. However, ODEs need a great deal of parametrization, which is frequently unachievable for biological systems [94]. In contrast, increasing evidence supports a switch-like approximation of signaling cascades enabling development of network models with less parametrization [94]. Boolean models, a type of discrete dynamic models, can help understand the effect of network topologies on signal transduction. The nodes in these models have discrete 'on' or 'off' states based on whether abundance (e.g., concentration, or activity) of a node (molecular

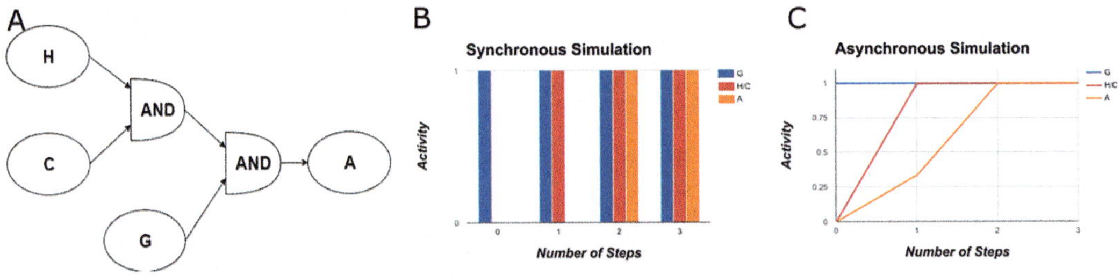

Fig. 3 Toy example of Boolean modeling: (**a**) circuit diagram representing regulation of A by C, H and G by the given logic rules. (**b**) results of a synchronous and (**c**) asynchronous simulation in which initial condition is G = ON and A=C=H=OFF. Activities between 0 and 1 represent the probability of the node being in an ON state at a given time-step and is calculated from 10,000 simulations. In b and c G is in blue, H and C in red, and A in yellow

species) is sufficient to induce downstream signaling (Fig. 3). The system is updated using rules that define how upstream nodes (nodes with outgoing edges to the updated node) regulate each node. Considering a network in which nodes A and B have edges going to node C, updating rules must specify whether A and B are necessary or sufficient for the activation of C (Fig. 3a). Rule specification is relatively straightforward for nodes with a single predecessor. However, complex logic rules using standard Boolean connectives may be required for nodes with higher in-degrees. If detailed knowledge of the system exists, such rules can be written by experts. However, if the prior knowledge does not exist, multiple rules may be tested in a systematic manner to determine which rule set reproduces known experimental behavior. This inference of the logic-rules requires perturbation or time series data so that robust rule selection can be performed [19, 60, 62, 95].

The emergent properties of the network can be studied by analyzing dynamics of the networks across all possible starting states/initial conditions [96]. The state of the network is then mapped as signal transfers through the network leading to phase transition maps. In other words, phase transition maps specify the state of every node in the system and indicate, which state each node will enter upon update; thus, illustrating the causal relationship between network states. However, simulation of all initial conditions is computationally expensive since the number of possible starting states grow at a rate of 2^n where n is the number of nodes in the network (because each node can either be on or off). In large systems where checking all states is not possible, the initial conditions of a simulation could be set to those of the system under steady state or resting conditions in order to minimize necessary computation. Care in interpretation is critical because the meaning of one and zero can vary significantly depending on the type of data. For example, zero could represent different levels of expression for different genes since 'inactive' genes (not transcriptionally induced upon stimulation) will often have different basal transcript levels [97].

To simulate the Boolean models, update rules can be applied simultaneously to all nodes (Synchronous update models) or in random order (asynchronous update models). On the one hand, asynchronous model frameworks facilitate definition of node-specific time-scales, improving the description of biological systems. On the other hand, asynchronous models can lead to increased parameterization. Synchronous update models can induce artificial steady states, and especially limit cycles which might not exist in real system's level dynamics [98]. Several models of asynchronous update have been proposed to account for varying time-scales of immunological interactions [97, 99–101]. In one popular asynchronous update method, nodes are updated in a random order and the current values of all nodes are used in all calculations [95]. This has the advantage of a limited number of parameters. However, the simulation must be run many times and needs to be averaged to avoid artifacts from particular randomly chosen update orders. Other asynchronous models allow for the incorporation of known information about a system; for example, faster processes can be updated first [6]. We have developed network models to identify representative mediators, to identify functions that could capture the essence of the host immune response as a whole, and to assess how their relative contribution dynamically changed over time during single infections, coinfections, and allergic responses (*see* Fig. 4) [99, 100]. Such models can be used to predict outcomes of coinfection from the host's response to single infections. Thus, Boolean models can be used to investigate dynamic response of immune interactions, to make predictions about the role of key regulatory cells and cytokines in the infections, and to investigate strategies for network interventions.

In an attempt to capture the elegance of Boolean networks while preserving the resolution of continuous models such as those based on ODEs, hybrid Boolean models were developed. These systems are minimally parameterized to functionally limit assumptions being made about the system, but nonetheless attempt to estimate the shape of response curves [102]. Piecewise linear formalism, the oldest such model, uses a continuous variable and an activity-like discrete variable regulated by differential equations and an activation threshold [96]. We have used piecewise linear formalism to model the dynamic outcome of the interplay between host immune components and pathogenic bacteria *Bordetellae bronchiseptica* [101]. We were able to determine activity thresholds for various signaling molecules necessary to establish the *B. bronchiseptica*-optimized cellular response. Thus, hybrid Boolean models offer an elegant way to study dynamic properties of the

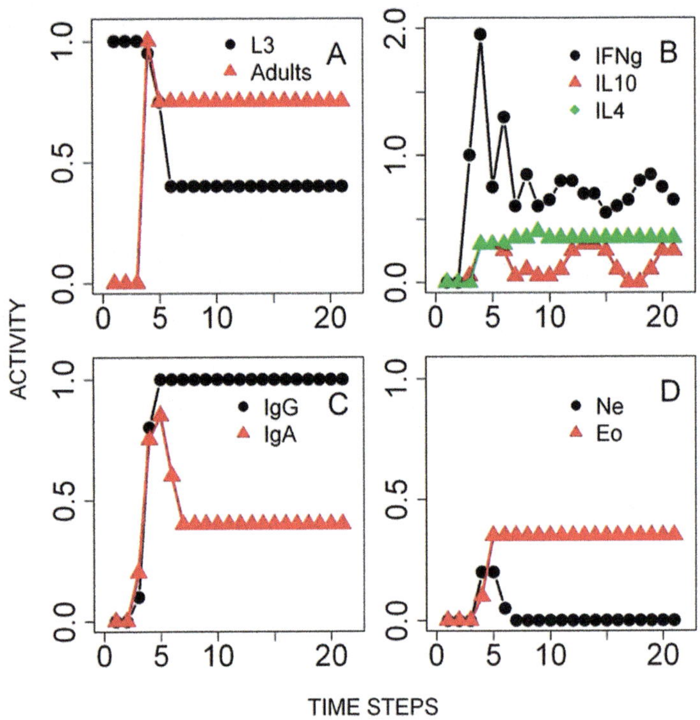

Fig. 4 An example of Boolean model developed to study immune response to *Trichostrongylus retortaeformis* infection: Activity profiles (the probability of the node being in an ON state at a given time-step) measured by asynchronous Boolean model of (**a**) Third stage infective larvae (L3) and adults. (**b**) Cytokines, IFNγ, IL4, and IL10 in the duodenum. (**c**) Mucus antibodies against helminth adult parasites. (**d**) Peripheral eosinophils and neutrophils. Note that the IFNγ concentration range is between 0 and 2 to describe additional non-immune mediated activation of that node by the tissue damage (refer to Thakar et al. [99] for more details and figure reproduced from Thakar et al. [99])

network with limited knowledge about the system's parameterization.

There have been many successful implementations of Boolean models. Software packages for Boolean Simulation include BoolNet, BooleanNet, CellNOpt, and CellNOptR [61, 62, 95, 103] in R, Python, Matlab, and R, respectively. BooleanNet, implemented in python, is particularly easy and flexible, allowing users to employ synchronous, asynchronous, and hybrid simulation modalities [95].

3.1 Beyond the Boolean Model

Boolean networks may not always be able to account for variability arising in biological systems. The switch-like assumption underlying Boolean models might not be able to capture the stochastic nature of transcriptional regulations seen in transcriptomics data that are usually based on a mixture of cells, including different cell types and different cell cycle stages. Moreover, transcriptional regulation is not the only level of regulation that might account for observed changes as epigenetic changes and post-transcriptional regulation may play a role.

A number of methods can aid to better capture a range of abundances, while maintaining the simplicity of rule-based models. The most thoroughly investigated among these methods are fuzzy logic models [104] which have also been applied in an immune context [19, 105, 106]. In fuzzy logic models, nodes can be assigned more than two (on/off) states. Fuzzy logic models attempt to minimize the impact of the switch-like assumption by ascribing a proportion of activation to each node. This proportional activation is then propagated by treating 'and' rules as min functions while treating 'or' rules as max functions. For example, if node Q has the update function Q = (A and B) or C, then, using the values from the specified time step (either previous or current depending on whether the simulation is synchronous or asynchronous), Q = max(min(A,B),C). This formalism captures a portion of the "fuzziness" or variability of the way nodes are turned on and off without sacrificing the simplicity of defining Boolean update functions. BoolNet and CellNOptR both have implementations of fuzzy logic network simulations built in [61, 62].

Probabilistic Boolean networks (PBNs) represent another approach to addressing variability of transcriptomic data. Rather than assuming that variability exists in node activation, PBNs assume variability arises from ambiguity in the update rules employed [107]. Thus, each node has multiple update functions from which one is chosen randomly before each update. PBNs introduce variability and account for non-switch-like behavior while preserving strict dependence of future states exclusively on current and not past states. PBNs also allow most analytic techniques from Boolean networks, such as derivatives and influence calculations, to be applied [107]. In conclusion, discrete dynamic modeling provides a feasible framework for whole-network simulations and can reveal key interaction mapping to clinical observations.

4 Limitations

Despite their evident usefulness, there are a number of limitations to Boolean and Boolean-like models. The first, and likely most important limitation, is the assumption about systemic variance. Boolean models mostly assume there is little stochastic variability. In contrast, fuzzy models ascribe variability to partial activation and PBNs ascribe variance to use of different wiring at random. Since these assumptions undergird the propagation of signal through

Boolean and Boolean-like models, they are critical for a careful interpretation of results. Further, simulation models can only be as accurate as the input provided; if a data-driven approach is initiated with biased data or missing pieces of the model, the results of the simulation are likely to be inaccurate (even if they are precise). Finally, since there is no guarantee that even the most carefully run simulation will recapitulate reality, models should be validated by experimental approaches.

5 Summary and Outlook

Integration of increasing amount of omics data requires the development of advanced analytical tools in order to assess relationships between genes, proteins and cells. Those tools can further be employed to identify signatures of biological entities, such as genes, proteins or cells that define or predict disease outcomes. Here we discussed approaches to assemble interaction networks ranging from molecular to cellular levels, via either data-driven or knowledge-driven methodologies, and their analysis to investigate the cross talk between signaling cascades. Specifically, we described static and dynamic approaches, including Boolean modeling, Fuzzy logic modeling, and Probabilistic Boolean Network modeling which enable identification of key regulatory components, prioritization of experimental tests and study of emergent properties. Thus, the chapter provides introductory material about construction and analysis of multiscale networks involved in immune responses alongside their limitations and uses.

References

1. Thakar J, Christensen C, Albert R (2008) Toward understanding the structure and function of cellular interaction networks. Bolyai Soc Math Stud 18:239–275
2. Saeys Y, Van Gassen S, Lambrecht BN (2016) Computational flow cytometry: helping to make sense of high-dimensional immunology data. Nat Rev Immunol 16:449–462
3. Noble WS, MacCoss MJ (2012) Computational and statistical analysis of protein mass spectrometry data. PLoS Comput Biol 8(1):e1002296
4. Anafi RC, Francey LJ, Hogenesch JB et al (2017) CYCLOPS reveals human transcriptional rhythms in health and disease. Proc Natl Acad Sci 114:201619320
5. Ma'ayan A (2011) Introduction to network analysis in systems biology. Sci Signal 4
6. Thakar J, Pilione M, Kirimanjeswara G et al (2007) Modeling systems-level regulation of host immune responses. PLoS Comput Biol 3:1022–1039
7. Thakar J, Christensen C, Albert R (2008) Toward understanding the structure and function of cellular interaction networks. Bolyai Soc Math Stud 18:239–275
8. Prescott TP, Papachristodoulou A (2014) Layered decomposition for the model order reduction of timescale separated biochemical reaction networks. J Theor Biol 356:113–122
9. Berenstein AJ, Magariños MP, Chernomoretz A et al (2016) A multilayer network approach

for guiding drug repositioning in neglected diseases. PLoS Negl Trop Dis 10:e0004300
10. Kanehisa M, Furumichi M, Tanabe M et al (2016) KEGG: new perspectives on genomes, pathways, diseases and drugs. Nucleic Acids Res 45:353–361
11. Wrzodek C, Büchel F, Ruff M et al (2013) Precise generation of systems biology models from KEGG pathways. BMC Syst Biol 7:15
12. Christensen C, Thakar J, Albert R (2007) Systems-level insights into cellular regulation: inferring, analysing, and modelling intracellular networks. IET Syst Biol 1:61–67
13. Shen-orr SS, Goldberger O, Garten Y et al (2009) Towards a cytokine-cell interaction knowledgebase of the adaptive immune system. Pac Symp Biocomput 2009:439–450
14. Thakar J, Hartmann BM, Marjanovic N et al (2015) Comparative analysis of anti-viral transcriptomics reveals novel effects of influenza immune antagonism. BMC Immunol 16:46
15. Hartmann BM, Thakar J, Albrecht RA et al (2015) Human dendritic cell response signatures distinguish 1918, pandemic, and seasonal H1N1 influenza viruses. J Virol 89:10190–10205
16. Bjornson ZB, Nolan GP, Fantl WJ (2013) Single-cell mass cytometry for analysis of immune system functional states. Curr Opin Immunol 25:484–494
17. Brodin P, Jojic V, Gao T et al (2015) Variation in the human immune system is largely driven by non-heritable influences. Cell 160:37–47
18. Campbell C, Thakar J, Albert RR (2011) Network analysis reveals cross-links of the immune pathways activated by bacteria and allergen. Phys Rev E Stat Nonlinear Soft Matter Phys 84:1–12
19. Liu H, Zhang F, Mishra SK et al (2016) Knowledge-guided fuzzy logic modeling to infer cellular signaling networks from proteomic data. Sci Rep 6:35652
20. Katanic D, Khan A, Thakar J (2016) PathCellNet: cell-type specific pathogen-response network explorer. J Immunol Methods 439: 15–22
21. Hornbeck PV, Zhang B, Murray B et al (2015) PhosphoSitePlus, 2014: mutations, PTMs and recalibrations. Nucleic Acids Res 43:D512–D520
22. Prasad TSK, Goel R, Kandasamy K et al (2009) Human protein reference database — 2009 update. Nucleic Acids Res 37: 767–772
23. Salwinski L, Miller CS, Smith AJ et al (2004) The database of interacting proteins: 2004 update. Nucleic Acids Res 32:449–451
24. Jewison T, Su Y, Disfany FM et al (2014) SMPDB 2.0: big improvements to the small molecule pathway database. Nucleic Acids Res 42:478–484
25. Kandasamy K, Mohan SS, Raju R et al (2010) NetPath: a public resource of curated signal transduction pathways. Genome Biol 11:R3
26. Kutmon M, Riutta A, Nunes N et al (2016) WikiPathways: capturing the full diversity of pathway knowledge. Nucleic Acids Res 44:D488–D494
27. Joshi-Tope G, Gillespie M, Vastrik I et al (2005) Reactome: a knowledgebase of biological pathways. Nucleic Acids Res 33
28. Croft D, Mundo A, Haw R et al (2014) The reactome pathway knowledgebase. Nucleic acids 42:D472–D477
29. Fabregat A, Sidiropoulos K, Garapati P et al (2016) The reactome pathway knowledgebase. Nucleic Acids Res 44:D481–D487
30. Romero P, Wagg J, Green ML et al (2005) Computational prediction of human metabolic pathways from the complete human genome. Genome Biol 6:R2
31. Chatr-aryamontri A, Oughtred R, Boucher L et al (2017) The BioGRID interaction database: 2017 update. Nucleic Acids Res 45:369–379
32. Stark C (2006) BioGRID: a general repository for interaction datasets. Nucleic Acids Res 34:D535–D539
33. Cerami EG, Gross BE, Demir E et al (2011) Pathway commons, a web resource for biological pathway data. Nucleic Acids Res 39: 685–690
34. Keating SM, Le Novère N (2013) Supporting SBML as a model exchange format in software applications. Methods Mol Biol 1021: 201–225
35. Demir E, Cary MP, Paley S et al (2010) The BioPAX community standard for pathway data sharing. Nat Biotechnol 28:935–942
36. Habermann B, Villaveces J, Koti P (2015) Tools for visualization and analysis of molecular networks, pathways, and -omics data. Adv Appl Bioinforma Chem 8:11
37. Jensen LJ, Saric J, Bork P (2006) Literature mining for the biologist: from information retrieval to biological discovery. Nat Rev Genet 7:119–129

38. Van Landeghem S, De Bodt S, Drebert ZJ et al (2013) The potential of text mining in data integration and network biology for plant research: a case study on Arabidopsis. Plant Cell 25:794–807
39. Snel B, Lehmann G, Bork P et al (2000) STRING: a web-server to retrieve and display the repeatedly occurring neighbourhood of a gene. Nucleic Acids Res 28: 3442–3444
40. Hur J, Ozgür A, Xiang Z et al (2012) Identification of fever and vaccine-associated gene interaction networks using ontology-based literature mining. J Biomed Semant 3:18
41. Alexeyenko A, Sonnhammer ELL (2009) Global networks of functional coupling in eukaryotes from comprehensive data integration. Genome Res 19:1107–1116
42. Schmitt T, Ogris C, Sonnhammer ELL (2014) FunCoup 3.0: database of genome-wide functional coupling networks. Nucleic Acids Res 42:D380–D388
43. Studham ME, Tjärnberg A, Nordling TEM et al (2014) Functional association networks as priors for gene regulatory network inference. Bioinformatics 30:130–138
44. Szklarczyk D, Morris JH, Cook H et al (2017) The STRING database in 2017: quality-controlled protein–protein association networks, made broadly accessible. Nucleic Acids Res 45:D362–D368
45. Margolin AA, Nemenman I, Basso K et al (2006) ARACNE: an algorithm for the reconstruction of gene regulatory networks in a mammalian cellular context. BMC Bioinformatics 7:S7
46. Bonneau R, Reiss DJ, Shannon P et al (2006) The Inferelator: an algorithm for learning parsimonious regulatory networks from systems-biology data sets de novo. Genome Biol 7:1
47. Faith JJ, Hayete B, Thaden JT et al (2007) Large-scale mapping and validation of Escherichia coli transcriptional regulation from a compendium of expression profiles. PLoS Biol e8:5
48. Marbach D, Mattiussi C, Floreano D (2009) Replaying the evolutionary tape: biomimetic reverse engineering of gene networks. Ann N Y Acad Sci 1158:234–245
49. Wise A, Bar-Joseph Z (2015) cDREM: inferring dynamic combinatorial gene regulation. J Comput Biol 22:324–333
50. Mitrea C, Taghavi Z, Bokanizad B et al (2013) Methods and approaches in the topology-based analysis of biological pathways. Front Physiol 4:278
51. Merico D, Isserlin R, Stueker O et al (2010) Enrichment map: a network-based method for gene-set enrichment visualization and interpretation. PLoS One 5(11):e13984
52. Ernst M, Du Y, Warsow G et al (2017) FocusHeuristics – expression-data-driven network optimization and disease gene prediction. Sci Rep 7:42638
53. Luo W, Friedman MS, Shedden K et al (2009) GAGE: generally applicable gene set enrichment for pathway analysis. BMC bioinformatics 10:161
54. Doncheva NT, Assenov Y, Domingues FS et al (2012) Topological analysis and interactive visualization of biological networks and protein structures. Nat Protoc 7:670–685
55. Marbach D, Prill RJ, Schaffter T et al (2010) Revealing strengths and weaknesses of methods for gene network inference. Proc Natl Acad Sci U S A 107:6286–6291
56. Song L, Langfelder P, Horvath S (2012) Comparison of co-expression measures: mutual information, correlation, and model based indices. BMC Bioinformatics. 13:328
57. Chaussabel D, Quinn C, Shen J et al (2009) A modular analysis framework for blood genomics studies: application to systemic lupus erythematosus. Immunity 29:150–164
58. Langfelder P, Horvath S (2008) WGCNA: an R package for weighted correlation network analysis. BMC bioinformatics 9:559
59. Wang R-S, Albert R (2011) Elementary signaling modes predict the essentiality of signal transduction network components. BMC Syst Biol 5:44
60. Kachalo S, Zhang R, Sontag E et al (2008) NET-SYNTHESIS: a software for synthesis, inference and simplification of signal transduction networks. Bioinformatics 24:293–295
61. Terfve C, Cokelaer T, Henriques D et al (2012) CellNOptR: a flexible toolkit to train protein signaling networks to data using multiple logic formalisms. BMC Syst Biol 6:133
62. Müssel C, Hopfensitz M, Kestler HA (2010) BoolNet-an R package for generation, reconstruction and analysis of Boolean networks. Bioinformatics 26:1378–1380
63. Jeong H, Mason SP, Barabási A-L et al (2001) Lethality and centrality in protein networks. Nature 411:41–42
64. Yook S-H, Oltvai ZN, Barabási A-L (2004) Functional and topological characterization

of protein interaction networks. Proteomics 4:928–942
65. Oltvai ZN, Barabási A-L, Jeong H et al (2000) The large-scale organization of metabolic networks. Nature 407:651–654
66. Bollobás B Riordan O (2002) Mathematical results on scale-free random graphs, Handbook of Graphs and Networks: from the Genome to the Internet pp 1–38
67. Brohée S, van Helden J, Wong L et al (2006) Protein complex prediction based on k - connected subgraphs in protein interaction network. BMC Bioinformatics 7:488
68. Zhang B, Horvath S (2005) A general framework for weighted gene co-expression network analysis. Stat Appl Genet Mol Biol 4:Article17
69. Pavlopoulos G, Wegener A-L, Schneider R (2008) A survey of visualization tools for biological network analysis. BioData Mining 1:12
70. Castro MA, Wang X, Fletcher MNC et al (2012) RedeR: R/bioconductor package for representing modular structures, nested networks and multiple levels of hierarchical associations. Genome Biol 13:R29
71. Fruchterman TMJ, Reingold EM (1991) Graph drawing by force-directed placement. Software-Practice & Experience 21:1129–1164
72. Shannon P, Markiel A, Ozier O et al (2003) Cytoscape: a software environment for integrated models of biomolecular interaction networks. Genome Res 13:2498–2504
73. Yamada T, Letunic I, Okuda S et al (2011) IPath2.0: interactive pathway explorer. Nucleic Acids Res 39:412–415
74. Letunic I, Yamada T, and Kanehisa M et al (2008) iPath: interactive exploration of biochemical pathways and networks
75. Ellson J, Gansner ER, Koutsofios E et al (2004) Graphviz and Dynagraph – static and dynamic graph drawing tools. In: Jünger M, Mutzel P (eds) Graph Drawing Software. Springer, Berlin, Heidelberg, pp 127–148
76. Milo R, Shen-Orr S, Itzkovitz S et al (2002) Network motifs: simple building blocks of complex networks. Science 298: 824–827
77. Shoval O, Alon U (2010) SnapShot: network motifs. Cell 143:326–326.e1
78. Alon U (2007) Network motifs: theory and experimental approaches. Nat Rev Genet 8:450–461
79. Shen-Orr SS, Milo R, Mangan S et al (2002) Network motifs in the transcriptional regulation network of Escherichia coli. Nat Genet 31:64–68
80. Eiben AE, Smith J, Albert R et al (2015) From evolutionary computation to the evolution of things. Nature 521:476–482
81. Collier JH, Allison L, Lesk AM et al (2014) A new statistical framework to assess structural alignment quality using information compression. Bioinformatics 30:i512–i518
82. Konagurthu AS, Allison L, Stuckey PJ et al (2011) Piecewise linear approximation of protein structures using the principle of minimum message length. Bioinformatics 27:i43–i51
83. Cohen AA, Kalisky T, Mayo A et al (2009) Protein dynamics in individual human cells: experiment and theory. PLoS One 4:e4901
84. Cohen AA, Geva-Zatorsky N, Eden E et al (2008) Dynamic proteomics of individual cancer cells in response to a drug. Science 322:1511–1516
85. Konagurthu AS, Lesk AM (2008) On the origin of distribution patterns of motifs in biological networks. BMC Syst Biol 2:73
86. Friedman EJ, Young K, Tremper G et al (2015) Directed network motifs in Alzheimer's disease and mild cognitive impairment. PLoS One 10:e0124453
87. Kalir S, Mangan S, Alon U (2005) A coherent feed-forward loop with a SUM input function prolongs flagella expression in Escherichia coli. Mol Syst Biol 1(2005):0006
88. Yeger-Lotem E, Sattath S, Kashtan N et al (2004) Network motifs in integrated cellular networks of transcription-regulation and protein-protein interaction. Proc Natl Acad Sci U S A 101:5934–5939
89. Kashtan N, Itzkovitz S, Milo R et al (2004) Efficient sampling algorithm for estimating subgraph concentrations and detecting network motifs. Bioinformatics 20: 1746–1758
90. Csardi G, Nepusz T (2006) The igraph software package for complex network research. InterJournal Complex Sy 1695:1–9
91. Hagberg AA, Schult DA, Swart PJ (2008) Exploring network structure, dynamics, and function using NetworkX. Proceedings of the 7th Python in Science Conference (SciPy) 2008:11–15
92. Ibrahim M, Jassim S, Cawthorne M et al (2014) A MATLAB tool for pathway

enrichment using a topology-based pathway regulation score. BMC Bioinformatics 15:358
93. Graph (2017) Wolfram Language and System Documentation Center
94. Thakar J, Poss M, Albert R et al (2010) Dynamic models of immune responses: what is the ideal level of detail? Theor Biol Med Model 7:35
95. Albert I, Thakar J, Li S et al (2008) Boolean network simulations for life scientists. Source Code Biol Med 3:16
96. Glass L, Kauffman SA (1973) The logical analysis of continuous, non-linear biochemical control networks. J Theor Biol 39:103–129
97. Anderson CS, DeDiego ML, Topham DJ et al (2016) Boolean modeling of cellular and molecular pathways involved in influenza infection. Comput Math Methods Med 2016:1–11
98. Saadatpour A, Albert I, Albert R (2010) Attractor analysis of asynchronous Boolean models of signal transduction networks. J Theor Biol 266:641–656
99. Thakar J, Pathak AK, Murphy L et al (2012) Network model of immune responses reveals key effectors to single and co-infection dynamics by a respiratory bacterium and a gastrointestinal helminth. PLoS Comput Biol 8(1):e1002345
100. Walsh ER, Thakar J, Stokes K et al (2011) Computational and experimental analysis reveals a requirement for eosinophil-derived IL-13 for the development of allergic airway responses in C57BL/6 mice. J Immunol 186:2936–2949
101. Thakar J, Saadatpour-Moghaddam A, Harvill ET et al (2009) Constraint-based network model of pathogen-immune system interactions. J R Soc Interface 6: 599–612
102. Wittmann DM, Krumsiek J, Saez-Rodriguez J et al (2009) Transforming Boolean models to continuous models: methodology and application to T-cell receptor signaling. BMC Syst Biol 3:98
103. Morris MK, Melas I, Saez-Rodriguez J (2013) Construction of cell type-specific logic models of signaling networks using CellNOpt. Methods Mol Biol 930:179–214
104. Morris MK, Saez-Rodriguez J, Sorger PK et al (2010) Logic-based models for the analysis of cell signaling networks. Biochemistry 49:3216–3224
105. Aldridge BB, Saez-Rodriguez J, Muhlich JL et al (2009) Fuzzy logic analysis of kinase pathway crosstalk in TNF/EGF/insulin-induced signaling. PLoS Comput Biol 5(4): e1000340
106. Schivo S, Scholma J, van der Vet PE et al (2016) Modelling with ANIMO: between fuzzy logic and differential equations. BMC Syst Biol 10:56
107. Shmulevich I, Dougherty ER, Kim S et al (2002) Probabilistic Boolean networks: a rule-based uncertainty model for gene regulatory networks. Bioinformatics (Oxford, England) 18:261–274

Part V

Computational Analyses of Heterogenous Cell Populations

Chapter 19

Exploring Dynamics and Noise in Gonadotropin-Releasing Hormone (GnRH) Signaling

Margaritis Voliotis, Kathryn L. Garner, Hussah Alobaid, Krasimira Tsaneva-Atanasova, and Craig A. McArdle

Abstract

Gonadotropin-releasing hormone (GnRH) acts via G-protein coupled receptors on pituitary gonadotropes. These are G_q-coupled receptors that mediate acute effects of GnRH on the exocytotic secretion of luteinizing hormone (LH) and follicle-stimulating hormone (FSH), as well as the chronic regulation of their synthesis. FSH and LH control steroidogenesis and gametogenesis in the gonads so GnRH mediates control of reproduction by the central nervous system. GnRH is secreted in short pulses and the effects of GnRH on its target cells are dependent on the dynamics of these pulses. Here we provide a brief overview of the signaling network activated by GnRH with emphasis on the use of high content imaging for their examination. We also describe computational approaches that we have used to simulate GnRH signaling in order to explore dynamics, noise, and information transfer in this system.

Key words GnRH, G-protein coupled receptor, NFAT, ERK, Mathematical modeling, Mutual information

1 Introduction

1.1 An Overview of GnRH Signaling

GnRH is a neuropeptide hormone that mediates the central control of reproduction. It does so by activating GnRH receptors (GnRHRs) on pituitary gonadotropes to control the synthesis and secretion of luteinizing hormone (LH) and follicle-stimulating hormone (FSH), pituitary gonadotropin hormones that are secreted exocytotically to control gametogenesis and steroidogenesis in the gonads [1–6]. The immediate effect of GnRH is to increase secretory vesicle fusion with the plasma membrane. In the long term GnRH increases gonadotropin synthesis, thereby controlling vesicle content. GnRHRs are members of the rhodopsin-like G-protein coupled receptor (GPCR) family. They

Margaritis Voliotis and Kathryn L. Garner contributed equally to this work.

Louise von Stechow and Alberto Santos Delgado (eds.), *Computational Cell Biology: Methods and Protocols*, Methods in Molecular Biology, vol. 1819, https://doi.org/10.1007/978-1-4939-8618-7_19,

signal via G_q, which leads to generation of the second messengers IP_3 (inositol 1,4,5 trisphosphate) and DAG (diacylglycerol) gonads [1–7]. IP_3 mobilizes Ca^{2+} from intracellular stores and this is followed by Ca^{2+} influx via L-type voltage-gated Ca^{2+} channels. Ca^{2+} then drives the regulated exocytotic secretion of LH and FSH [8–10]. Like many other GPCRs, GnRHRs mediate activation of mitogen-activated protein kinase (MAPK) cascades. Notably, GnRH activates the extracellular signal regulated kinase (ERK, here used to mean ERK1 and/or ERK2) and this is largely mediated by protein kinase C (PKC) in gonadotrope-derived cell lines [5, 11]. In some models GnRH also activates c-Jun N-terminal kinases (JNKs) [5, 12], p38 MAPK [13, 14] and/or ERK5 [15]. In gonadotropes, GnRH influences expression of many genes [16, 17] but emphasis has been on transcription of the gonadotrope signature genes (i.e., the genes encoding the common gonadotropin α-subunit, LHβ, FSHβ, and the GnRHR) which are all increased by GnRH [4]. PKC and the MAPKs above can all influence gonadotropin signature gene expression [2, 4, 18, 19]. Several Ca^{2+}-regulated proteins are known to mediate transcriptional effects of GnRH. These include calmodulin (CaM), calmodulin-dependent protein kinases, the calmodulin-dependent phosphatase calcineurin (Cn), and the Ca^{2+}-dependent transcription factor NFAT (nuclear factor of activated T-cells) [4, 20, 21].

1.2 Dynamics and Noise in GnRH Signaling

GnRH secretion is pulsatile and this is essential for reproduction because GnRH effects are dependent on pulse frequency [22, 23] with responses to GnRH often maximal at submaximal pulse frequency [2, 12, 19–21, 24–27]. In humans, GnRH pulses have durations of a few minutes and intervals of 30 min to several hours. The pulse frequency moreover differs under different physiological conditions with changes in frequency driving changes in reproductive status during development, through the menstrual cycle and with aging [25, 28, 29]. Stimulus dynamics are also crucial for therapeutic targeting of the system as pulses of agonists can increase or maintain circulating gonadotropin levels whereas sustained agonist treatment initially increases, and then reduces, them. Consequently, sustained agonists ultimately cause chemical castration and this effect is exploited to treat breast cancer, prostate cancer and other hormone-dependent conditions [1, 25, 28]. Given the physiological and pharmacological importance of GnRH dynamics, there is considerable interest in cellular mechanisms for decoding GnRH pulses. The desensitization of GnRH-stimulated gonadotropin secretion by constant agonist treatment is often attributed to receptor desensitization in spite of the fact that type I GnRHR (unlike other GPCRs) does not undergo rapid homologous desensitization [4, 30–32]. Desensitization of GnRH-stimulated gonadotropin secretion is therefore indicative of post-receptor adaptive mechanisms [4, 30–32]. Virtually all biological processes involve relationships between changing quantities. Rates of change are represented mathematically by derivatives that describe the evolution (in time) of some quantity of interest and constitute the nature of ordinary differential equation (ODE)-based modeling. Therefore, ODE models are ideally suited to mathematically/computationally describe and study dynamical biological systems, such as pulsatile GnRH signaling. To explore this further we developed mechanistic (ODE)-based models of simplified GnRH signaling networks and used these to show how pulsatile stimulation can increase efficiency and confer specificity on signaling [31, 33]. With models trained against wet-lab measures of GnRH signaling to ERK and NFAT [34, 35], we also showed how parallel pathways activated by GnRH pulses might converge and cooperate to generate the kind of nonmonotonic frequency–response relationships (i.e., maximal effects at sub-maximal pulse frequency) that are characteristic of this system [33].

Although dynamics have long been known to be important for GnRH signaling much less is known about the relevance of noise. Very often, experiments are undertaken assuming that all cells of a given type are identical. However, results from single cell measurements invariably reveal marked cell–cell variation [36–41]. Such heterogeneity is inevitable because of the stochasticity of biological processes. Most importantly, heterogeneity can also drive the health and function of cell populations because it is individual cells that have to sense their environment and make appropriate decisions. For GnRH signaling, high levels of cell–cell variability have been reported for effects on cytoplasmic Ca^{2+} concentration, gonadotropin secretion, effector activation, and gene expression [6, 16, 34, 42–46]. Our recent work has moreover sought to explore the effect of such heterogeneity on information transfer [43, 47, 48]. More generally, information theory, which was initially developed to analyze electronic communication, is increasingly used to measure how reliably biological signaling systems transfer information [36, 38–40, 48–50]. Here "information" refers to the uncertainty about the environment that is reduced by signaling, it can be quantified by means of Mutual Information (MI) between system inputs and outputs [36]. MI is measured in Bits with an MI of 1 Bit meaning that the system can unambiguously distinguish between two equally probable states of the environment. For cell signaling studies, the signal could be set to the concentration of stimulus and the response could be the measured amount of activated effector in an individual cell.

Where information theoretic approaches are used to analyze cell signaling pathways, they are effectively treated as noisy communication channels and MI can be used as a measure of the amount of information that they carry. We have used this approach to explore information transfer in HeLa cells or in LβT2 cells [43] as illustrated in Fig. 1. LβT2 cells are a murine gonadotrope-derived cell line with endogenous GnRHR, whereas HeLa cells are derived from a human cervical cancer and do not express endogenous GnRHR but can be transduced with recombinant Adenovirus (Ad) for GnRH expression facilitating (for example) comparative studies with different GnRH from different species [51, 52]. These cells can also be transduced with Ad for expression of ERK2-GFP (green fluorescent protein) or NFAT1c-EFP (emerald fluorescent protein), both of which translocate from the cytoplasm to the nucleus providing readouts for GnRH-mediated activation of the rapidly accelerated fibrosarcoma-kinase (Raf)/MAPK/ERK kinase (MEK)/ERK and Ca^{2+}/CaM/Cn pathways, respectively [34, 35, 53]. Alternatively, the cells can be fixed and stained for the dual phosphorylated/activated form of ERK (ppERK), and transcriptional readouts for the same pathways can be obtained by transduction with Ad for reporters in which early growth response protein-1 (Egr1)- or NFAT response elements drive expression of asRED or zsGREEN fluorophores respectively [34, 35, 42, 43, 53, 54]. Here, Egr1-driven fluorophore expression is used because GnRH causes a pronounced ERK-mediated increase in Egr1 expression [55, 56]. When fluorescence immunocytochemistry is used to stain for ppERK, the cells must be fixed and permeabilized but for the other readouts, live cell imaging is also possible. When live cell imaging is combined with cell tracking, cell–cell variability in response trajectories can also be considered. In either case, we use a high content imaging platform (automated fluorescence microscopy) to obtain measures of signaling in large numbers of individual cells. Combining high content imaging with stochastic modeling we explore the amount of information transferred via GnRHR to distinct effectors. We can further identify mechanisms by which loss of information through signaling can be mitigated. Here we describe some of the experimental and computational methods along with some of the key findings of the work.

2 Materials

2.1 Cell Culture, Adenovirus Transduction, and Reverse Transfection with siRNA

1. Cells: HeLa cells were from the European Collection of Cell Cultures and LβT2 cells were a gift form Prof. Pamela Mellon, UCSD, San Diego, USA.
2. Dulbecco's modified Eagle medium (DMEM).
3. OptiMEM.
4. Phosphate-buffered saline (PBS).
5. Fetal bovine serum (FBS).
6. Penicillin-streptomycin.
7. Trypsin solution.

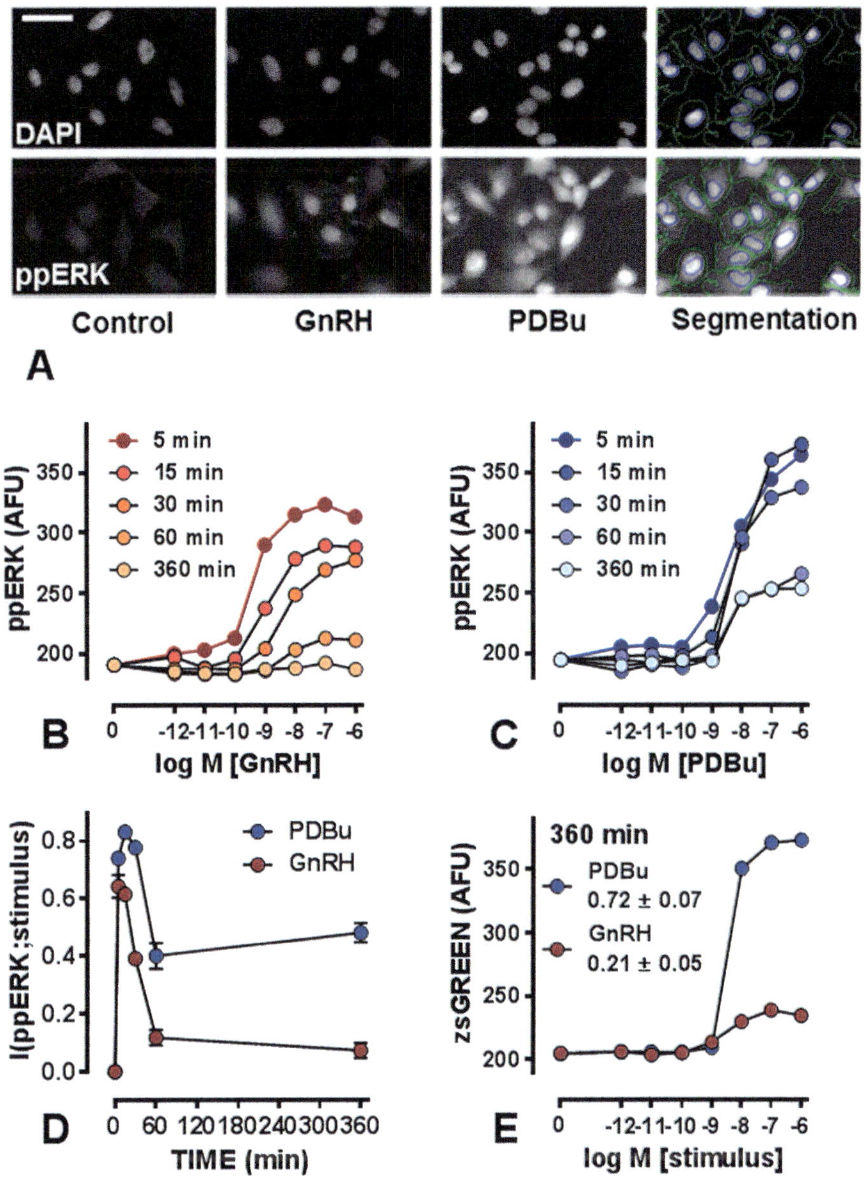

Fig. 1 Quantifying GnRHR-mediated ERK signaling. (**a**) The images show representative DAPI- and ppERK-stained cells cultured under control conditions or stimulated 5 min with 10^{-7} M GnRH or PDBu as indicated. The horizontal bar is approximately 20 μm and the right hand images show an example of the automated image segmentation used to define perimeters of the nuclei and cells (perimeters superimposed over PDBu-treated cells). Each image shows <1% of the area captured to generate the X-Y plots. Data from [43]. Panels (**b**) and (**c**) show population-averaged nuclear ppERK in HeLa cells transduced with Ad-GnRHR and Ad Egr1-zsGREEN and stimulated 5, 15, 30, 60, or 360 min with 0 or 10^{-12}–10^{-6} M GnRH, as indicated. Cells were fixed and stained (DAPI and ppERK) before image capture and analysis. Background values (without fluorophore) were 120–150 AFU and were not subtracted. Individual cell measures from these experiments were used to calculate I(ppERK;stimulus) and are shown in panel (**d**). Panel (**e**) shows population averaged zsGREEN values for cells stimulated for 6 h with the indicated concentration of GnRH or PDBu and the corresponding I(Egr1-zsGREEN;stimulus)

8. Normal goat serum (NGS).
9. Black-walled 96-well plates for imaging.
10. Matrigel basement membrane matrix.
11. The Ad used were: Ad GnRHR, Ad NFAT1c-EFP, Ad ERK2-GFP, Ad K52R ERK2-GFP, Ad NFAT-RE asRED, Ad Egr1zsGREEN, Ad Egr1-driven DUSP5 and Ad Egr1-driven R52/53A DUSP5. These were all engineered, bulked, stored and used as described [34, 35, 43, 53, 57–61].
12. Cell counting chamber or automatic cell counter.
13. ERK1 siRNA sequences:
    5′ -CGUCUAAUAUAUAAAUAUAdTdT-3′(sense)
    5′ -UAUAUUUAUAUAUUAGACGdGdG-3′(antisense)
    5′ -CCCUGACCCGUCUAAUAUAdTdT-3′(sense)
    5′ -UAUAUUAGACGGGUCAGGGdAdG-3 (antisense)
14. ERK2 siRNA sequences:
    5′ - CACUUGUCAAGAAGCGUUAdTdT-3′ (sense)
    5′ -UAACGCUUCUUGACAAGUGdTdT-3′ (antisense)
    5′ -CAUGGUAGUCACUAACAUAdTdT-3′ (sense)
    5′ -UAUGUUAGUGACUACCAUGdAdT-3′ (antisense)
15. Lipofectamine RNAiMAX transfection reagent.
16. Multichannel pipette.
17. Virkon disinfectant.
18. Polypropylene reagent reservoirs (60 mL volume) (autoclaved).

2.2 Agonist Stimulation

1. GnRH (stored in 10 mM acetic acid with 0.1% BSA as single-use aliquots at −20 °C).
2. Phorbol 12, 13 dibutyrate (PDBu) (stored in dimethyl sulfoxide (DMSO) as single use aliquots at −20 °C).
3. DMEM, 0.1% FBS, penicillin, and streptomycin.
4. 96-well clear plate ("working plate").
5. Multichannel pipette.
6. 4% paraformaldehyde (PFA) (stored in 20 mL aliquots at −20 °C).
7. Methanol (stored in 60 mL pots at −20 °C).
8. PBS.
9. Virkon disinfectant.

2.3 Staining for ppERK

1. 5% NGS in PBS.
2. PBS.
3. Multichannel pipette.

4. Platform rockers, one at room temperature and one in a cold room (4 °C).
5. Mouse anti-ppERK1/2 antibody (clone MAPK-YT, Sigma) at 1:200 in 1% NGS/PBS.
6. Alexa 546 or 488 labeled goat anti-mouse antibody at 1:200 in 1% NGS/PBS at 30 μL/well.
7. 4,6-diamidino-2-phenylindole (DAPI) in PBS (DAPI stock is stored in water at 4 °C and is diluted 1:5000 in PBS for working solution of 600 nM).
8. Polypropylene reagent reservoirs (60 mL volume) (not sterile for this step).

2.4 Image Acquisition and Analysis

1. InCell Analyzer 1000 (GE Healthcare, Amersham, UK) or similar system for automated capture of digital images of cells using a 10× or 20× objective and with three or four fluorophores.
2. InCell Investigator software or similar software for automated image segmentation. Routinely, definition of nuclear perimeters from images of DAPI-stained nuclei and definition of cell perimeters either from additional stains or as a collar of defined width around the nucleus.
3. Output files are raw image stacks (.tif files) and Microsoft Excel spreadsheets (.xls files) giving average fluorescence intensity (AFU) for all measured fluorophores in each segmented area for each individual cell and as population-averaged data for each field of view and for each well imaged.

2.5 Calculation of Mutual Information Values and Modeling

1. MatLab (version 2015b).

3 Methods

3.1 Cell Culture and Adenovirus Transduction

1. Culture HeLa cells and LβT2 cells as described [34, 35].
2. Trypsinize cells.
3. Count cells.
4. For LβT2 cells, all culture flasks and plates should be coated with Matrigel Basement Membrane Matrix. Stock Matrigel (10 mL) is thawed on ice/at 4 °C, aliquoted into 200 μL aliquots and stored at −20 °C. When required, a Matrigel aliquot is thawed on ice then diluted to 6 mL with the addition of unsupplemented DMEM. 3 mL of diluted Matrigel is used to coat a T75 cell culture flask at room temperature for 1 h (*see* **Note 1**).

5. Seed 4000 HeLa cells or 10,000 LβT2 cells per well (and, in DMEM, 10% FBS, 100 units/mL penicillin and 0.1 mg/mL streptomycin in a clear-bottomed, black-walled 96-well plates suitable for imaging, in 100 μL per well.
6. We typically use 2–4 replicate wells for each condition (i.e., for each stimulus concentration at each time point).
7. Approximately 16 h after plating, the cell culture media is removed and replaced with Ad for 4–6 h in DMEM, 2% FBS, penicillin and streptomycin. Concentrated Ad stocks are stored at −80 °C in 20 μL aliquots and are diluted directly into this transduction medium, which is pipetted directly onto the cells, typically at 80 μL per well.
8. Ad titer can be varied to vary transgene expression levels, but we routinely work at 1–5 plaque-forming units (pfu)/nl (*see* **Note 2**).
9. At the end of the Ad transduction period, remove the virus by aspiration.
10. Wash the cells once in 100 μL PBS per well.
11. Replace PBS with 100 μL per well DMEM, 0.1% FBS, penicillin and streptomycin.
12. Incubate overnight for serum starvation prior to stimulation.

3.2 siRNA Knockdown and ERK2-GFP Expression

1. Where ERK1 and ERK2 are knocked down (*see* **Note 3**), this is achieved by reverse transfection with the four siRNAs detailed below for human cells (i.e., HeLa) as previously described [53, 57, 62]. Briefly, each newly synthesized siRNA is reconstituted to 80 μM in siRNA universal. 120 μL of each siRNA solution is combined to give the final 20 μM working stock solution.
2. HeLa cells are trypsinized and counted.
3. The siRNA transfection complex is assembled according to the Lipofectamine RNAiMAX manufacturer's instructions, with some modifications. OptiMEM is warmed to 37 °C, and the appropriate amount for 20 μL per well of the 96-well plate is placed into a 15 mL falcon tube. To this, the ERK1/2 siRNA is added for the equivalent of 0.006 μL per well, to give 1 nM final siRNA concentration. Lipofectamine RNAiMAX is removed from the fridge, gently mixed, and the equivalent of 0.2 μL per well added to the liquid in the falcon tube. The tube contents are gently mixed and incubated at room temperature for 10–20 min.
4. Meanwhile, the HeLa cells are centrifuged at 120 × *g* for 5 min at room temperature. The cell pellet is resuspended in DMEM, 10% FBS without antibiotics, to a concentration of 4000 cells per well.

5. The siRNA mix is then distributed to each well of a black-walled 96-well plate, 20 μL per well using a repeat pipette.
6. To this, 100 μL cells are distributed to each well from a sterile reagent reservoir using a multichannel pipette.
7. The experimental plate(s) are placed into the incubator overnight.
8. The cells are then transduced with Ad ERK2-GFP to achieve ERK2-GFP expression levels approximating those of endogenous ERK in untransfected cells, as described above [53, 57, 62].
9. The nuclear translocation of ERK2-GFP provides a simple readout for ERK activation and facilitates use of modified reporters including catalytically inactive ERK2-GFP constructs, as described [59, 63] (*see* **Note 3**).

3.3 Agonist Stimulation

1. For stimulation, agonists are prepared at 5x concentrated in DMEM, 0.1% FBS, penicillin and streptomycin, in a 96-well clear "working plate" immediately before use.
2. For most experiments, we construct full concentration response curves (i.e., control and 10^{-12}–10^{-6} M GnRH) at multiple time points with replicate wells (typically 2–4 wells for each concentration at each time point) and collect images for 2–9 fields of view per well. This yields data for >10,000 individual cells (for each treatment in each experiment) (*see* **Note 4**).
3. Agonists are transferred from the appropriate well of the working plate to the appropriate well of the experimental plate (i.e., 25 μL of 5 × concentrated agonist is added to the 100 μL of medium in the experimental plate, to achieve the desired final concentration) using a multichannel pipette. This enables cell stimulation without a further medium change (*see* **Note 5**).
4. Stimulation times can be varied and for most experiments are between 5 min and 8 h (*see* **Note 6**).
5. Terminate stimulation by immediate removal of media from all wells. This can be done with an aspiration tube, a repeat pipette or simply by inverting the plate over a waste fluid collection tray. In each case the fluid removed should be treated with Virkon to destroy any remaining viable Ad.
6. If fluid is removed by inverting the plate, do not tap the plate to remove any fluid drops adhering, but instead remove them by touching the inverted plate on tissue paper.
7. Add 50 μL per well ice-cold 4% PFA and incubate in the fridge at 4 °C for 5 min
8. Replace PFA with 50 μL per well ice-cold methanol.

9. Incubate at −20 °C in the freezer for 5 min.
10. Replace methanol with 100 μL per well PBS.
11. The cells can be stained immediately as outlined below. Alternatively, replace the plate lid, seal the plate with Parafilm and/or aluminum foil to protect from light, and transfer the plate to the fridge before staining (*see* **Note 7**).

3.4 Staining for ppERK

1. Remove PBS (wash once with 100 μL PBS if staining is not being performed immediately) and replace with 30 μL per well 5% NGS in PBS to block nonspecific binding.
2. Rock plates at room temperature for 1–2 h.
3. Remove blocking solution by tipping into a waste tray and replace with mouse anti-ppERK1/2 antibody at 1:200 in 1% NGS/PBS (30 μL per well).
4. Rock plates overnight at 4 °C (preferably covered with aluminum foil).
5. Wash plates 3× with PBS, 100 μL per well.
6. Replace PBS with Alexa 546 or 488 labeled goat anti-mouse antibody at 1:200 in 1% NGS/PBS at 30 μL/well. From this point onward the plate must be protected from the light to prevent bleaching of the fluorophore.
7. Incubate for 90 min with rocking at room temperature.
8. Wash 3× with PBS, 100 μL per well.
9. Remove PBS and replace with 100 μL per well 600 nM DAPI in PBS (1:5000 dilution).
10. Incubate cells in DAPI solution for 20 min with rocking at room temperature. Wash twice in 100 μL PBS. Plates can then be stored in PBS at 4 °C until imaging.

3.5 Image Acquisition and Analysis

1. Our imaging methods have been described previously [34, 43, 53, 58, 59].
2. Briefly, digital images are acquired using, for example, an InCell Analyzer 1000 high content imaging platform with a 10× objective and filters for DAPI (blue channel), Alexa488, GFP, EFP and zsGREEN (green channel), or Alexa546 and asRED (red channel). Exposure time for each fluorophore is determined individually to obtain clear images whilst keeping exposure times short (typically <1 s).
3. Image analysis is used to determine whole cell or nuclear fluorophore intensities in arbitrary fluorescence units (AFU), as described [34, 35, 43, 53, 57, 59].
4. Use a "multi-target analysis" algorithm (InCell Investigator software) to define nuclear perimeters from the DAPI stain.

5. Set cell perimeters either as a collar around the nucleus (typically 2–4 μm) or by segmentation on green or red fluorophores (InCell Investigator software) (*see* **Note 8**).
6. Image analysis with InCell Investigator software generates Excel spreadsheets (.xls files) as output files. These contain fluorescence intensity measures (for each of the regions segmented and for each fluorophore) for each well, for each field of view and for each individual cell.
7. For population-averaged responses we typically pool data from all replicate wells in an experiment. We also routinely repeat experiments (typically 3–6 times) and pool data, reporting means and SEMs for "n" separate experiments [59, 62, 63].

3.6 Calculation of Mutual Information Values

1. Digital images (.tif files) are the raw data acquired in these experiments. The image analysis outlined above generates Excel spreadsheets (.xls files) including single cell measures for each segmented region and for each fluorophore. These single cell measures are imported into Matlab and used to calculate MI between measured outputs and the experimental inputs (i.e., hormone concentrations) as described [43, 48] (*see* **Note 4**).
2. Briefly, to estimate the MI between a continuous, possibly multivariate response (Z; e.g., ppERK readout), and a signal (S; e.g., GnRH concentration) which takes a finite number of possible values, we use the following expression:

$$I(Z;S) = h(Z) - h(Z|S) = h(Z) - E[h(Z|S = s).$$

   Here, h(Z) is the unconditional entropy of the response and h(Z|S) is the conditional entropy given the signal, which is obtained as the average of h(Z|S = s) taken over the distribution of the signal. We estimate each entropy term using the nearest-neighbor (knn) method of Kraskov et al. [64].
3. For the estimation of the conditional entropy estimate each h(Z|S = s), use all cell data (i.e., single cell measures of fluorescence intensity within segmented areas of the cells) obtained for the value of the signal s, and then take their weighted average based on the signal distribution (*see* **Note 9**).
4. To estimate the unconditional entropy, use an iterative scheme to sample from the marginal response distribution corresponding to the particular signal distribution (e.g., uniform). This should be done to ensure the maximum number of possible cells for each value.
5. At each iteration *i*, sample in total N_i cells across all of the empirical conditional populations. Cells should be sampled one at a time by drawing a signal value from the signal distribution and sampling (without replacement) a cell at random from the conditional population corresponding to that signal value.

6. Set N_i equal to the number of cells in the conditional population with the fewest cells at the start of iteration i (since sampling is without replacement, the total number of cells is reduced by each iteration).
7. Iterate until $N_i = 0$. Estimates can be obtained assuming a uniform distribution over the eight different levels of stimulation (between 0 and 10^{-6} M).
8. Our MatLab code for MI calculation is available through Open Research Exeter repository (http://hdl.handle.net/10871/27843).

3.7 Modeling

1. Since the dynamics of ERK responses are dictated largely by negative feedback loops [43, 65, 66], we developed a stochastic model to explore how fast and slow negative feedback loops might influence information transfer via ERK. We modeled ERK signaling as a dynamical system with the input signal driving the upstream effector E into a more active form E*, and with two negative feedback loops: a fast one from ppERK to E and E*, and a slow one via phosphatase ($pASE_2$) expression (Fig. 2, *see* [43] for parameters). The Stochastic Simulation Algorithm (SSA or Gillespie algorithm [67, 68]) is used for numerical simulations to study behavior of the model at different feedback strengths. Fast negative feedback is controlled by varying the binding rate of ppERK to E or E*, and slow negative feedback by varying the number of promoters expressing $pASE_2$.
2. The code used for the presented model is available through the Open Research Exeter repository (http:// hdl.handle.net/10871/27844).
3. MI is estimated from 2000 independent runs of the model with total ERK sampled from a $\log_{10}$-Normal distribution, with mean 3.9 and standard deviation 0.1.
4. The model is run to converge to steady state before the signal is introduced. The number of signal molecules (S) is randomly drawn from a uniform distribution on the logarithmic scale between 10 and 1000 molecules.
5. MI values are estimated from the simulated data using the following expression:

$$I(Z;S) = h(Z) + h(S) - h(Z;S),$$

where, h(Z) is the entropy of the response, h(S) is the entropy of the signal, and h(Z;S) is the joint entropy.

6. We estimate MI using the nearest-neighbor method of Kraskov et al. [64]. To use this method, convert simulated data from discrete (number of molecules) into continuous quantities by adding to each data point a random number uniformly distributed between 0 and 1.

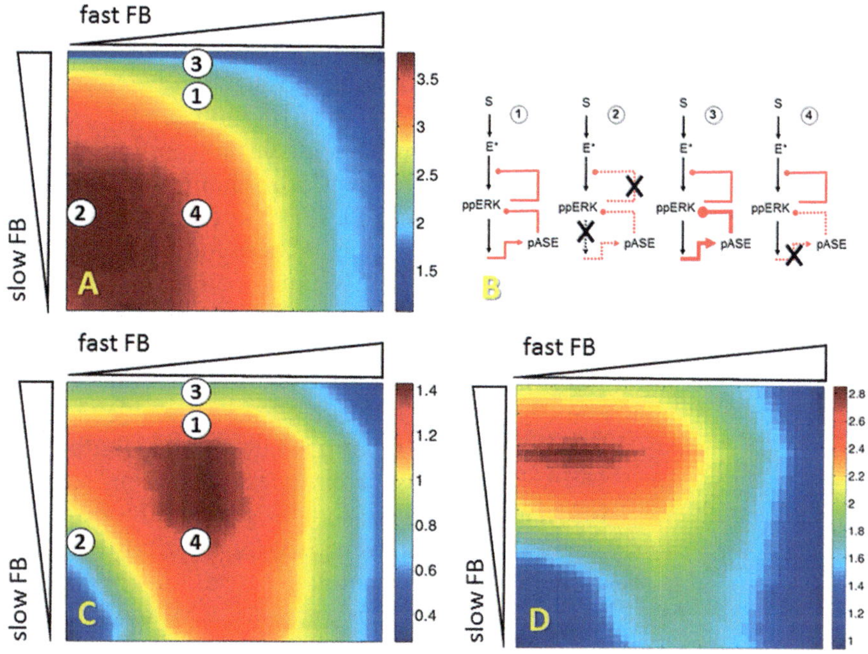

Fig. 2 Stochastic modeling of ERK signaling and summary of experimental manipulations. Panel A: heat map of population averaged S-stimulated ppERK responses (maximum-minimum) obtained from single cell responses with varied fast and slow feedback (FB). The data were generated by simulation as described in the text and in [43], with fast FB and slow FB at −0.5–3.5 and 0–4 ($\log_{10}$ scale), respectively. The simulations were run for 60 min and 10 min data are shown. Panel B: simplified model topologies for control cells (1), cells with catalytically inactive ERK (2), and cells where the slow (transcription-dependent) pathway is increased (3) or reduced (4). Panel C: heat map of predicted I(ppERK;S) for the 10 min snap-shot data corresponding to A. Panel D: heat map of predicted I(ppERK;S) with parameters identical to those for A and C, except that the simulations were run for 60 min and MI was calculated taking into account response trajectories. The numbers 1–4 on A and C correspond to the cartoons, showing possible positions of these topologies in the parameter space and illustrating consistency between the simulations and experimental manipulations. Adapted from [43]

3.8 Interpretation of Results

Figure 1 A shows representative images of HeLa cells transduced with Ad GnRHR before and after stimulation (with GnRH or PDBu) and staining for DAPI and ppERK. Image analysis was used to define perimeters of nuclei and of individual cells and fluorophore intensities within these areas. Figure 1b–e show data from a series of high-content imaging experiments, including population averaged nuclear ppERK measures for cells stimulated for the indicated periods with varied concentrations of GnRH (B) or PDBu (C) as well as for whole cell measures of (Egr1-driven) zsGREEN expression in cells stimulated for 6 h with the indicated concentrations of PDBu or GnRH. Figure 1d shows the MI between ppERK and the stimulus (either GnRH or PDBu) at different time-points after stimulation. A key observation here was that most information was actually lost through signaling, as in these experiments 3 Bits of information were available (i.e., 2^3

GnRH or PDBu concentrations) and <1 Bit of information was transferred for each experimental output considered (*see* **Note 10**).

It is well established that response kinetics and compartmentalization can dictate the effects of ERK on cell fate, largely because sustained activation can have much more pronounced effects on gene expression than transient activation [65, 66]. This is evident in the data shown, where PDBu causes a more sustained increase in ppERK than GnRH (Fig. 1b, c) and consequently caused a greater increase in zsGREEN expression (Fig. 1e). Note also that I(zsGREEN;GnRH) was much lower than I(zsGREEN;PDBu), showing that when a reporter sensitive to response dynamics is used, information transfer via ERK is also revealed to be sensitive to response dynamics (*see* **Note 11**).

ERK responses are modulated by multiple feedback loops [65, 66]. To explore effects of feedback on sensing we developed a stochastic model with slow and fast feedback loops. In this model, a stimulus (S) causes conversion of an upstream effector (E) into a more active form (E*) that in turn promotes phosphorylation of ERK to ppERK. We also incorporate ERK-mediated feedback; a fast loop paralleling ERK-mediated inhibitory phosphorylation of Raf, and a slow one paralleling a transcription-dependent increase in DUSP. Negative feedback can preserve sensing by opposing the "leak" due to basal activity in a generic protein phosphorylation and dephosphorylation pathway [48], so we included basal activity in the model. The basic model is shown in Fig. 2b (*see* also [43] for full system topology and parameters).

The key findings were that the negative feedback loops can both increase and reduce I(ppERK;S). For both rapid and slow loops, strong negative feedback impairs sensing by reducing the output dynamic range, whereas intermediate strength feedback can improve it by opposing noise and/or leak [43]. Consequently, the curves representing the relationships between fast feedback strength and sensing are bell-shaped, and this is most obvious when there is little or no slow feedback. The same relationship can be observed between slow feedback strength and sensing that is most evident when fast feedback is weak or intermediate. These features are illustrated in Fig. 2 where panel C shows I(ppERK;S) at different feedback levels and A shows the maximal substrate-driven increase in ppERK, each as heat maps generated from simulations with varied feedback intensities. The difference between these heat maps provides a striking demonstration of the fact that changes in population-averaged responses (panel A) do not necessarily parallel changes in information transfer (panel C). Indeed, the overriding effect of increasing fast or slow feedback is to reduce population-averaged ppERK values whereas for both fast and slow feedback, I(ppERK;S) is maximal at intermediate levels.

We also tested these model predictions experimentally [43]. First, we used siRNA to knockdown endogenous ERK1 and ERK2 and then transduced the cells with Ad for GnRHR expression and also for expression of either ERK2-GFP or K52R ERK2-GFP. Since the K52R ERK2-GFP mutant is catalytically inactive, both of the negative feedback paths are broken (2 in Fig. 2b). In a second experiment, we transduced cells with Ad for GnRHR expression and also for expression of a construct for Egr1-driven expression of either wild-type DUSP5 or R53A/54A mutant of DUSP5. While wild-type DUSP5 serves to increase DUSP-mediated negative feedback (3 in Fig. 2b), the R53A/54A mutant of DUSP5 has reduced affinity for interacting proteins and is used as a control. Third, we used cycloheximide to block protein synthesis in order to reduce negative feedback via nuclear-inducible DUSP expression (4 in Fig. 2b) (*see* **Note 12**). The experimental data were consistent with the simulations as illustrated by the numbers overlaid on Fig. 2a, c. Here (1) represents the situation where fast and slow feedback pathways are intact. In cells expressing K52R-ERK2, both feedback paths are broken and I(ppERK;S) is reduced, moving from (1) to (2). With Egr1-driven expression of WT DUSP5, slow feedback is increased and this reduces I(ppERK;S), moving from (1) to (3). Finally, with cycloheximide the slow feedback loop is broken and this increases I(ppERK;S), moving from (1) to (4). Overall, the simulations predict sensing to be dependent on the complex interplay of fast and slow feedback and that for broad ranges of feedback strength, information transfer is greatest at intermediate levels of fast and slow feedback. The experimental data are also consistent with these predictions from simulations, implying that intermediate levels of fast and slow feedback support hormone (GnRH) sensing [43].

Finally, Fig. 2a, c show values for 10 min snap-shot data, but the simulations were actually run for 60 min, enabling I(ppERK;S) calculation taking into account response trajectory and these values are shown in panel D (*see* **Note 11**). This analysis reenforces the general conclusion that information transfer is maximal at intermediate levels of fast and slow feedback [43]. However, the MI values are much higher (note the different scales for C and D), demonstrating the potential for cells to gain more information by dynamic sensing ERK trajectory and we are currently running live-cell imaging experiments to explore this possibility.

4 Conclusions and Outlook

Although our work focuses on GnRH signaling, NFAT and ERK are activated by a great number of different hormones making our methods applicable to many different cell types and signaling systems, pulsatile and nonpulsatile inputs, and may even be adapted for different signal outputs. Measuring hormone sensing by cells is fundamental to our investigations and we find that individual cells are not very good at sensing different concentrations of hormone in their environment (i.e., most information about GnRH concentration in the environment is actually lost through signaling) at least when, only one signaling pathway is considered. However, additional information may well be gained by sensing multiple pathways within the network simultaneously and/or by sensing response trajectories over time [43, 69]. Another particularly intriguing feature here is that GnRHR signaling networks are dependent on cellular context [4]. Consequently, information transfer via GnRHR to different effectors could also vary from one cell type to another (or even within one cell type over time) and more generally, our work illustrates an approach for exploring this relationship between network architecture and hormone sensing.

5 Notes

1. Matrigel used to coat a flask can be refrozen once and used to coat the wells of a 96-well plate (50 μL per well). A second refreezing is not advisable.
2. When developing in vitro reporter assays it is common practice to select cell subpopulations (often clonal lines) to minimize cell–cell variability. In contrast, for the data shown here we have used transient Ad-mediated transduction for reporter expression because we wish to exploit and understand the cell–cell variability rather than simply minimizing it. We assume that this introduces heterogeneity due to Ad uptake and reporter expression that could influence information transfer and lead to underestimation of MI. However, we routinely use ratiometric readouts (i.e., the proportion of ERK2-GFP or NFAT1c-EFP in the nucleus) that are likely to minimize the influence of expression level, and also avoid overexpression that would perturb localization. Moreover, we note that frequency distributions are similar for whole cell ERK measures in control cells and in cells subjected to siRNA-mediated knockdown of ERK and Ad-mediated add-back of ERK2-GFP (unpublished data). Accordingly, distributions obtained with Ad transduction may well be similar to those of the endogenous proteins and, consistent with this, we note that I(ppERK;GnRH) values are similar in cells expressing endogenous ERK and the Ad transduced ERK2-GFP reporter [43, 48]. As we do not have reliable antibodies against endogenous GnRHR we cannot compare frequency distributions for endogenous and exogenous receptors. However, I(ppERK;GnRH) values are similar for LβT2 cells (with endogenous GnRHR) and HeLa cells (transduced with GnRHR at comparable population averaged expression levels) and the simplest interpretation of this is that the frequency distributions for GnRHR expression

are comparable for both models. Such arguments cannot be made when Ad are used to express imaging reporters for transcriptional activation. In this case the whole cell measures of fluorophore would be expected to reflect the number of reporters expressed per cell and this additional variable would be expected to cause underestimation of MI. However, in preliminary (unpublished) experiments, we have cotransduced cells with Ad Egr1-zsGREEN and Ad for expression of a nuclear marker (histone 3 coupled to blue fluorescent protein, (BFP)). Since these Ads have identical tropism we anticipated similar uptake, as confirmed by excellent correlation between cellular levels of zsGREEN and BFP. Moreover, normalization of the zsGREEN signal to BFP (as a cell-specific expression control) did not increase measures of MI. Ideally this would be addressed with dual expression vectors but from the data available at present, we assume that these imaging reporters underestimate information transfer but suspect the underestimation to be small.

3. Imaging reporters have the potential to alter the pathways being monitored so validation and assay optimization is normally necessary to exclude or minimize such effects. For example, in unstimulated cells ERK is bound largely to cytoplasmic proteins (i.e., KSR and MEK). Upon activation it is liberated and typically translocates to the nucleus [11, 70]. This effect can be monitored with ERK-GFP reporters. However, when ERK is overexpressed ERK cytoplasmic anchors can be saturated. This causes nuclear accumulation in unstimulated cells which hinders the interpretation of ERK-GFP nuclear translocation assays. To circumvent this we typically use siRNA to knock down endogenous ERK and then use recombinant Ad to "add back" the ERK2-GFP reporter at a similar level to that of endogenous ERK in normal cells. Assay validation includes demonstration of the pronounced inhibition of ERK-driven transcription by knockdown, and recapitulation of near normal responses by Ad ERK2-GFP, as well as checking that the kinetics of ERK2-GFP translocation closely mirror those of changes in ppERK levels [53, 54, 57, 62]. Where single cell measures are available, an alternative strategy is to express ERK2-GFP at a broad range of concentrations and to bin cells for comparison of subpopulations with ERK expressed at levels similar to those of normal cells [59, 63]. In contrast, when we use the nuclear translocation of an NFAT1c-EFP reporter as a readout for GnRHR activation we find that the response is relatively insensitive to expression level [34, 71]. In this case, activation of the Ca^{2+}/CaM/Cn cascade causes dephosphorylation of the reporter, exposing a nuclear translocation sequence and we assume that there is simply too little reporter expressed to saturate the uptake mechanism. For

validation of the Egr1-driven DUSP5 constructs we varied Ad titer and established conditions for which GnRH caused comparable expression of wild-type DUSP5 and R52/53A DUSP5, and for which the wild-type phosphatase reduced ppERK responses and the mutant did not [43].

4. MI values are dependent on the stimulus distribution and the number of cells imaged. We typically expose cells with eight concentrations of stimulus (providing a $2^3 = 3$ Bit input) chosen to encompass the full dynamic range of responses in a typical sigmoid dose-response curve, and construct full concentration–response curves at each time point (as illustrated for GnRH in Fig. 1b). We routinely use 2–4 replicate wells (for each combination of stimulus concentration and time) and capture 2–4 fields per well. MI values are then calculated using the individual cell measures for the whole concentration response curve (typically $>10^4$ cells). We routinely use a 10× objective where whole cell measures are used (i.e., whole cell measures of ppERK or zsGREEN) or where nuclear–cytoplasmic ratios of reporters are calculated to measure nuclear translocation [34, 35, 53]. Where more detailed morphological readouts (such as receptor redistribution to endocytic vesicles) are monitored we use a 20× or 40× objective [71, 72]. This inevitably reduces the number of cells captured per field of view so the number of fields captured is increased (typically 4–16 fields). A related issue is that some cells grow unevenly over the bottom of the well (often concentrating toward the well edge or in the well center). We routinely collect multiple fields adjacent to one-another in the middle of the well but if necessary, define alternative locations (either randomly selected, or selected to avoid positions where cell density is atypically high or low). For any given experiment, locations of fields should be identical for all wells.

5. Pipette tips should be changed to avoid carryover of stimulus from one well to another, but this is typically minimized by the arrangement of treatment groups on the plate (i.e., when all 12 wells in a single row of a 96-well plate receive the same concentration of stimulus).

6. Stimulation times for most experiments are between 5 min and 8 h. During longer incubations the wells may begin to dry out, as evidenced by reduction in fluid volume that is most pronounced in the wells at the corner of the plates. To avoid this, plates should be returned to the water-saturated incubator during stimulations of 1 h or more. Similarly, a 5% CO_2 atmosphere is used for stimulations of 1 h or more to prevent pH changes. A related issue is that wells on the edge of plates have a tendency to dry out over prolonged periods in culture introducing artifactual "edge-effects." This problem

is mitigated by culturing cells in a humid environment, by changing medium regularly for long-term cultures and by simply confining experiments to the central 48 wells in a 96-well culture plate. We routinely retain cells in 96-well plates for only 2 days under a water-saturated atmosphere and rarely observe any such effects.

7. We routinely stain cells within 24 h of fixation but in some cases it may be possible to delay staining for up to a week by storing the experimental plate in the fridge covered with aluminum foil to prevent degradation of fluorophores by light and by placing 200 μL PBS into each well to prevent drying of the cells.

8. Measures of signaling in individual cells are essential for the approach to MI analysis outlined here. Obtaining such measures by automated fluorescence microscopy can be difficult where cells grow close to, or on top of, one another in culture. We routinely work with HeLa cells which are typically well-separated making segmentation simple. We also work with gonadotrope-derived LβT2 cells, which grow in tightly packed groups for which accurate delineation of cell perimeters is more difficult. To circumvent this, for LβT2 cells we often use a collar of 2–4 μm around the nucleus as the "cell perimeter" and in this way, define a region that can be assumed to be the cytoplasm of the cell for which the nucleus has been segmented (rather than attempting to identify all of the cytoplasm associated with it). Here, morphological measures are also very useful as objects defined by segmentation of the DAPI stain that are atypically large, elongated or irregularly shaped are almost invariably actually multiple overlapping nuclei. Thus, we would typically apply one or more filters to exclude nuclei with a too-large area or chord ratio. A related issue is that nuclear membrane breakdown during mitosis can influence reporter localization. However, dividing cells often have a characteristic appearance with dense DAPI staining over elongated nuclei. For rapidly dividing cells (such as HeLa cells), measures of DAPI stain intensity, nuclear area and nuclear chord ratio can also be used to exclude (or count) such cells [72].

9. MI values are not influenced by transformation of system inputs or outputs. Thus, although X-Y plots often show transformed data (i.e., $\log_{10}$ of stimulus concentration versus background subtracted fluorescence intensity), such transformations should have little or no influence on the estimated MI values.

10. Although MI is increasingly used to estimate information transfer through signaling pathways, the output is inevitably seen through the prism of the measurement method. What we really measure is the information about the cell's environment that is available to the experimenter, rather than that

which is available to the cell. This highlights the fact that the reliability of the measurement can impact the estimated reliability of sensing, so inadequate methodology can clearly lead to underestimation of MI. This is particularly problematic when one seeks to compare information transfer using different experimental readouts. Thus, one may wish to compare the amount of information that is transferred via a receptor to each of two effectors in a network, but if the reliability of the measurements differ, this could undermine the comparison. This would be the case if, for example, fluorescent staining with an antibody at optimal concentration provided a reliable measure for activation of one effector, whereas staining with another antibody at suboptimal conditions provided a less reliable estimate of the activation of another effector. However, this can be turned into an advantage, as the reliability with which the experimenter can sense the cell's environment is precisely the information needed for bioassay optimization. High content imaging experiments typically yield multiple measures (30–60 in most of our experiments) for each individual cell. The calculation of MI between stimulus and each of these responses provides an immediate indication of their value for environment sensing. Moreover, we find (for example) that I(ppERK;GnRH) values are reduced when primary or secondary antibody concentrations are too low, when image acquisition time is too low, when image segmentation is inadequate and when a 4x objective is used (similar MI values are obtained with 10×, 20× or 40× objectives). Accordingly, we use MI measures for assay optimization, selecting conditions for routine use that maximize information available to the experimentalist about the environment of the cells.

11. For most experiments, we measured MI at a single time-point or with a range of snap-shots taken over a time-course. However, this could lead to underestimation of information transfer when cells sense inputs (i.e., GnRH concentrations) from trajectories of outputs (i.e., ppERK levels) over time [49]. For example, we have found that I(ppERK;GnRH) is higher at 5 min than at 360 min [43] but this clearly does not mean that a cell obtains less information over 360 min than it does over 5 min. Instead, it shows that the 360 min snap-shot underestimates information transferred over the 360 min period. Measuring MI for ERK-driven transcription is an alternative approach that is sensitive to ppERK trajectory and stochastic modeling of the ERK cascade indicates the potential for additional information to be gained by trajectory sensing (*see* below and Fig. 2).

12. When using pharmacological inhibitors to explore cell signaling, off-target effects must always be considered. Here an

important advantage of the high content imaging approach is that although an experiment may be designed to monitor a single response (such as ERK activation), the image analysis can provide a great deal of additional information. For example, the nuclear stains used to identify nuclear perimeters also provide cell counts and can be used to estimate numbers of apoptotic and/or mitotic cells [73]. Thus, for example, we used cycloheximide to block protein synthesis in the study shown in Fig. 2. This would undoubtedly be expected to be cytotoxic over long exposure periods but had no measurable effect on cell number in the relatively brief (<2 h) experiments undertaken, although it did prevent the rapid induction of DUSP5 as required for this study [43].

References

1. Bliss SP, Navratil AM, Xie J, Roberson MS (2010) GnRH signaling, the gonadotrope and endocrine control of fertility. Front Neuroendocrinol 31(3):322–340. http://doi.org/10.1016/j.yfrne.2010.04.002
2. Ciccone NA, Kaiser UB (2009) The biology of gonadotroph regulation. Curr Opin Endocrinol Diabetes Obes 16(4):321–327. http://doi.org/10.1097/MED.0b013e32832d88fb
3. Conn PM, Huckle WR, Andrews WV, McArdle CA (1987) The molecular mechanism of action of gonadotropin releasing hormone (GnRH) in the pituitary. Recent Prog Horm Res 43:29–68
4. McArdle CA, Roberson MS (2015) Gonadotropes and gonadotropin-releasing hormone signaling. In: Plant TM (ed) Knobil and Neill's physiology of reproduction. Elsevier, Amsterdam, pp 335–397
5. Naor Z (2009) Signaling by G-protein-coupled receptor (GPCR): studies on the GnRH receptor. Front Neuroendocrinol 30(1):10–29. http://doi.org/10.1016/j.yfrne.2008.07.001
6. Stojilkovic SS, Catt KJ (1995) Novel aspects of GnRH-induced intracellular signaling and secretion in pituitary gonadotrophs. J Neuroendocrinol 7(10):739–757
7. Sealfon SC, Weinstein H, Millar RP (1997) Molecular mechanisms of ligand interaction with the gonadotropin-releasing hormone receptor. Endocr Rev 18(2):180–205
8. Hille B, Tse A, Tse FW, Almers W (1994) Calcium oscillations and exocytosis in pituitary gonadotropes. Ann N Y Acad Sci 710:261–270
9. Stojilkovic SS, Iida T, Merelli F, Torsello A, Krsmanovic LZ, Catt KJ (1991) Interactions between calcium and protein kinase C in the control of signaling and secretion in pituitary gonadotrophs. J Biol Chem 266(16):10377–10384
10. Zhu H, Hille B, Xu T (2002) Sensitization of regulated exocytosis by protein kinase C. Proc Natl Acad Sci U S A 99(26):17055–17059. http:// doi.org/10.1073/pnas.232588899
11. Caunt CJ, Finch AR, Sedgley KR, McArdle CA (2006) GnRH receptor signalling to ERK: kinetics and compartmentalization. Trends Endocrinol Metab 17(8):308–313. http://doi.org/10.1016/j.tem.2006.08.001
12. Burger LL, Haisenleder DJ, Aylor KW, Marshall JC (2009) Regulation of Lhb and Egr1 gene expression by GNRH pulses in rat pituitaries is both c-Jun N-terminal kinase (JNK)- and extracellular signal-regulated kinase (ERK)-dependent. Biol Reprod 81(6):1206–1215. http://doi.org/10.1095/biolreprod.109.079426
13. Coss D, Hand CM, Yaphockun KK, Ely HA, Mellon PL (2007) p38 mitogen-activated protein kinase is critical for synergistic induction of the FSH(beta) gene by gonadotropin-releasing hormone and activin through augmentation of c-Fos induction and Smad phosphorylation. Mol Endocrinol 21(12):3071–3086. http://doi.org/10.1210/me.2007-0247
14. Roberson MS, Zhang T, Li HL, Mulvaney JM (1999) Activation of the p38 mitogen-activated protein kinase pathway by

gonadotropin-releasing hormone. Endocrinology 140(3):1310–1318
15. Lim S, Pnueli L, Tan JH, Naor Z, Rajagopal G, Melamed P (2009) Negative feedback governs gonadotrope frequency-decoding of gonadotropin releasing hormone pulse-frequency. PLoS One 4(9):e7244. http:// doi.org/10.1371/journal.pone.0007244
16. Ruf F, Park MJ, Hayot F, Lin G, Roysam B, Ge Y, Sealfon SC (2006) Mixed analog/digital gonadotrope biosynthetic response to gonadotropin-releasing hormone. J Biol Chem 281(41):30967–30978. http:// doi.org/10.1074/jbc.M606486200
17. Yuen T, Ruf F, Chu T, Sealfon SC (2009) Microtranscriptome regulation by gonadotropin-releasing hormone. Mol Cell Endocrinol 302(1):12–17. http:// doi.org/10.1016/j.mce.2008.12.013
18. Burger LL, Haisenleder DJ, Dalkin AC, Marshall JC (2004) Regulation of gonadotropin subunit gene transcription. J Mol Endocrinol 33(3):559–584. http:// doi.org/10.1677/jme.1.01600
19. Haisenleder DJ, Dalkin AC, Ortolano GA, Marshall JC, Shupnik MA (1991) A pulsatile gonadotropin-releasing hormone stimulus is required to increase transcription of the gonadotropin subunit genes: evidence for differential regulation of transcription by pulse frequency in vivo. Endocrinology 128(1):509–517
20. Burger LL, Haisenleder DJ, Aylor KW, Marshall JC (2008) Regulation of intracellular signaling cascades by GNRH pulse frequency in the rat pituitary: roles for CaMK II, ERK, and JNK activation. Biol Reprod 79(5):947–953. http:// doi.org/10.1095/biolreprod.108.070987
21. Haisenleder DJ, Ferris HA, Shupnik MA (2003) The calcium component of gonadotropin-releasing hormone-stimulated luteinizing hormone subunit gene transcription is mediated by calcium/calmodulin-dependent protein kinase type II. Endocrinology 144(6):2409–2416
22. Clarke IJ, Cummins JT (1982) The temporal relationship between gonadotropin releasing hormone (GnRH) and luteinizing hormone (LH) secretion in ovariectomized ewes. Endocrinology 111(5):1737–1739. http:// doi.org/10.1210/endo-111-5-1737
23. Dierschke DJ, Bhattacharya AN, Atkinson LE, Knobil E (1970) Circhoral oscillations of plasma LH levels in the ovariectomized rhesus monkey. Endocrinology 87(5):850–853. http:// doi.org/10.1210/endo-87-5-850
24. Ciccone NA, Xu S, Lacza CT, Carroll RS, Kaiser UB (2010) Frequency-dependent regulation of follicle-stimulating hormone beta by pulsatile gonadotropin-releasing hormone is mediated by functional antagonism of bZIP transcription factors. Mol Cell Biol 30(4):1028–1040. http:// doi.org/10.1128/MCB.00848-09
25. Ferris HA, Shupnik MA (2006) Mechanisms for pulsatile regulation of the gonadotropin subunit genes by GNRH1. Biol Reprod 74(6):993–998. http:// doi.org/10.1095/biolreprod.105.049049
26. Marshall JC, Dalkin AC, Haisenleder DJ, Paul SJ, Ortolano GA, Kelch RP (1991) Gonadotropin-releasing hormone pulses: regulators of gonadotropin synthesis and ovulatory cycles. Recent Prog Horm Res 47:155–187 discussion 188-159
27. Thompson IR, Ciccone NA, Xu S, Zaytseva S, Carroll RS, Kaiser UB (2013) GnRH pulse frequency-dependent stimulation of FSHbeta transcription is mediated via activation of PKA and CREB. Mol Endocrinol 27(4):606–618. http:// doi.org/10.1210/me.2012-1281
28. Marshall JC, Dalkin AC, Haisenleder DJ, Griffin ML, Kelch RP (1993) GnRH pulses–the regulators of human reproduction. Trans Am Clin Climatol Assoc 104:31–46
29. Sisk CL, Foster DL (2004) The neural basis of puberty and adolescence. Nat Neurosci 7(10):1040–1047. http:// doi.org/10.1038/nn1326
30. McArdle CA, Davidson JS, Willars GB (1999) The tail of the gonadotrophin-releasing hormone receptor: desensitization at, and distal to, G protein-coupled receptors. Mol Cell Endocrinol 151(1–2):129–136
31. Pratap A, Garner KL, Voliotis M, Tsaneva-Atanasova K, McArdle CA (2017) Mathematical modeling of gonadotropin-releasing hormone signaling. Mol Cell Endocrinol 449:42–55. http:// doi.org/10.1016/j.mce.2016.08.022
32. Willars GB, McArdle CA, Nahorski SR (1998) Acute desensitization of phospholipase C-coupled muscarinic M3 receptors but not gonadotropin-releasing hormone receptors co-expressed in alphaT3-1 cells: implications for mechanisms of rapid desensitization. Biochem J 333(Pt 2):301–308

33. Tsaneva-Atanasova K, Mina P, Caunt CJ, Armstrong SP, McArdle CA (2012) Decoding GnRH neurohormone pulse frequency by convergent signalling modules. J R Soc Interface 9(66):170–182. http://doi.org/10.1098/rsif.2011.0215
34. Armstrong SP, Caunt CJ, Fowkes RC, Tsaneva-Atanasova K, McArdle CA (2009) Pulsatile and sustained gonadotropin-releasing hormone (GnRH) receptor signaling: does the Ca2+/NFAT signaling pathway decode GnRH pulse frequency? J Biol Chem 284(51):35746–35757. http://doi.org/10.1074/jbc.M109.063917
35. Armstrong SP, Caunt CJ, Fowkes RC, Tsaneva-Atanasova K, McArdle CA (2010) Pulsatile and sustained gonadotropin-releasing hormone (GnRH) receptor signaling: does the ERK signaling pathway decode GnRH pulse frequency? J Biol Chem 285(32):24360–24371. http://doi.org/10.1074/jbc.M110.115964
36. Bowsher CG, Swain PS (2014) Environmental sensing, information transfer, and cellular decision-making. Curr Opin Biotechnol 28:149–155. http://doi.org/10.1016/j.copbio.2014.04.010
37. Bowsher CG, Swain PS (2012) Identifying sources of variation and the flow of information in biochemical networks. Proc Natl Acad Sci U S A 109(20):E1320–E1328. http://doi.org/10.1073/pnas.1119407109
38. Bowsher CG, Voliotis M, Swain PS (2013) The fidelity of dynamic signaling by noisy biomolecular networks. PLoS Comput Biol 9(3):e1002965. http://doi.org/10.1371/journal.pcbi.1002965
39. Brennan MD, Cheong R, Levchenko A (2012) Systems biology. How information theory handles cell signaling and uncertainty. Science 338(6105):334–335. http://doi.org/10.1126/science.1227946
40. Cheong R, Rhee A, Wang CJ, Nemenman I, Levchenko A (2011) Information transduction capacity of noisy biochemical signaling networks. Science 334(6054):354–358. http://doi.org/10.1126/science.1204553
41. Rhee A, Cheong R, Levchenko A (2012) The application of information theory to biochemical signaling systems. Phys Biol 9(4):045011. http://doi.org/10.1088/1478-3975/9/4/045011
42. Caunt CJ, Perett RM, Fowkes RC, McArdle CA (2012) Mechanisms of GnRH-induced extracellular signal-regulated kinase nuclear localization. PLoS One 7(7):e40077. http://doi.org/10.1371/journal.pone.0040077
43. Garner KL, Perrett RM, Voliotis M, Bowsher C, Pope GR, Pham T, Caunt CJ, Tsaneva-Atanasova K, McArdle CA (2016) Information transfer in gonadotropin-releasing hormone (GnRH) signaling: EXTRACELLULAR SIGNAL-REGULATED KINASE (ERK)-MEDIATED FEEDBACK LOOPS CONTROL HORMONE SENSING. J Biol Chem 291(5):2246–2259. http://doi.org/10.1074/jbc.M115.686964
44. Lewis CE, Richards PS, Morris JF (1989) Heterogeneity of responses to LH-releasing hormone and phorbol ester among rat gonadotrophs: a study using a reverse haemolytic plaque assay for LH. J Mol Endocrinol 2(1):55–63
45. McArdle CA, Bunting R, Mason WT (1992) Dynamic video imaging of cystolic Ca(2+) in the alphaT3-1, gonadotrope-derived cell line. Mol Cell Neurosci 3(2):124–132
46. Ruf F, Hayot F, Park MJ, Ge Y, Lin G, Roysam B, Sealfon SC (2007) Noise propagation and scaling in regulation of gonadotrope biosynthesis. Biophys J 93(12):4474–4480. http://doi.org/10.1529/biophysj.107.115170
47. Perrett RM, Voliotis M, Armstrong SP, Fowkes RC, Pope GR, Tsaneva-Atanasova K, McArdle CA (2014) Pulsatile hormonal signaling to extracellular signal-regulated kinase: exploring system sensitivity to gonadotropin-releasing hormone pulse frequency and width. J Biol Chem. http://doi.org/10.1074/jbc.M113.532473
48. Voliotis M, Perrett RM, McWilliams C, McArdle CA, Bowsher CG (2014) Information transfer by leaky, heterogeneous, protein kinase signaling systems. Proc Natl Acad Sci U S A 111(3):E326–E333. http://doi.org/10.1073/pnas.1314446111
49. Selimkhanov J, Taylor B, Yao J, Pilko A, Albeck J, Hoffmann A, Tsimring L, Wollman R (2014) Systems biology. Accurate information transmission through dynamic biochemical signaling networks. Science 346(6215):1370–1373. http://doi.org/10.1126/science.1254933
50. Uda S, Saito TH, Kudo T, Kokaji T, Tsuchiya T, Kubota H, Komori Y, Ozaki Y, Kuroda S (2013) Robustness and compensation of information transmission of signaling pathways. Science 341(6145):558–561. http://doi.org/10.1126/science.1234511

51. Hislop JN, Everest HM, Flynn A, Harding T, Uney JB, Troskie BE, Millar RP, McArdle CA (2001) Differential internalization of mammalian and non-mammalian gonadotropin-releasing hormone receptors. Uncoupling of dynamin-dependent internalization from mitogen-activated protein kinase signaling. J Biol Chem 276(43):39685–39694. http://doi.org/10.1074/jbc.M104542200
52. Turgeon JL, Kimura Y, Waring DW, Mellon PL (1996) Steroid and pulsatile gonadotropin-releasing hormone (GnRH) regulation of luteinizing hormone and GnRH receptor in a novel gonadotrope cell line. Mol Endocrinol 10(4):439–450. http://doi.org/10.1210/mend.10.4.8721988
53. Armstrong SP, Caunt CJ, McArdle CA (2009) Gonadotropin-releasing hormone and protein kinase C signaling to ERK: spatiotemporal regulation of ERK by docking domains and dual-specificity phosphatases. Mol Endocrinol 23(4):510–519. http://doi.org/10.1210/me.2008-0333
54. Caunt CJ, Armstrong SP, McArdle CA (2010) Using high-content microscopy to study gonadotrophin-releasing hormone regulation of ERK. Methods Mol Biol 661:507–524. http://doi.org/10.1007/978-1-60761-795-2_32
55. Wurmbach E, Yuen T, Ebersole BJ, Sealfon SC (2001) Gonadotropin-releasing hormone receptor-coupled gene network organization. J Biol Chem 276(50):47195–47201. http://doi.org/10.1074/jbc.M108716200
56. Yuen T, Wurmbach E, Ebersole BJ, Ruf F, Pfeffer RL, Sealfon SC (2002) Coupling of GnRH concentration and the GnRH receptor-activated gene program. Mol Endocrinol 16(6):1145–1153. http://doi.org/10.1210/mend.16.6.0853
57. Caunt CJ, Armstrong SP, Rivers CA, Norman MR, McArdle CA (2008) Spatiotemporal regulation of ERK2 by dual specificity phosphatases. J Biol Chem 283(39):26612–26623. http://doi.org/10.1074/jbc.M801500200
58. Caunt CJ, Finch AR, Sedgley KR, Oakley L, Luttrell LM, McArdle CA (2006) Arrestin-mediated ERK activation by gonadotropin-releasing hormone receptors: receptor-specific activation mechanisms and compartmentalization. J Biol Chem 281(5):2701–2710. http://doi.org/10.1074/jbc.M507242200
59. Caunt CJ, McArdle CA (2010) Stimulus-induced uncoupling of extracellular signal-regulated kinase phosphorylation from nuclear localization is dependent on docking domain interactions. J Cell Sci 123(Pt 24):4310–4320. http:// doi.org/10.1242/jcs.076349
60. Hislop JN, Madziva MT, Everest HM, Harding T, Uney JB, Willars GB, Millar RP, Troskie BE, Davidson JS, McArdle CA (2000) Desensitization and internalization of human and xenopus gonadotropin-releasing hormone receptors expressed in alphaT4 pituitary cells using recombinant adenovirus. Endocrinology 141(12):4564–4575
61. Rushworth LK, Kidger AM, Delavaine L, Stewart G, van Schelven S, Davidson J, Bryant CJ, Caddye E, East P, Caunt CJ, Keyse SM (2014) Dual-specificity phosphatase 5 regulates nuclear ERK activity and suppresses skin cancer by inhibiting mutant Harvey-Ras (HRasQ61L)-driven SerpinB2 expression. Proc Natl Acad Sci U S A 111(51):18267–18272. http://doi.org/10.1073/pnas.1420159112
62. Caunt CJ, Rivers CA, Conway-Campbell BL, Norman MR, McArdle CA (2008) Epidermal growth factor receptor and protein kinase C signaling to ERK2: spatiotemporal regulation of ERK2 by dual specificity phosphatases. J Biol Chem 283(10):6241–6252. http://doi.org/10.1074/jbc.M706624200
63. Perrett RM, Fowkes RC, Caunt CJ, Tsaneva-Atanasova K, Bowsher CG, McArdle CA (2013) Signaling to extracellular signal-regulated kinase from ErbB1 kinase and protein kinase C: feedback, heterogeneity, and gating. J Biol Chem 288(29):21001–21014. http://doi.org/10.1074/jbc.M113.455345
64. Kraskov A, Stogbauer H, Grassberger P (2004) Estimating mutual information. Phys Rev E 69(6 Pt 2):066138
65. Kholodenko BN, Hancock JF, Kolch W (2010) Signalling ballet in space and time. Nat Rev Mol Cell Biol 11(6):414–426. http://doi.org/10.1038/nrm2901
66. Santos SD, Verveer PJ, Bastiaens PI (2007) Growth factor-induced MAPK network topology shapes Erk response determining PC-12 cell fate. Nat Cell Biol 9(3):324–330. http://doi.org/10.1038/ncb1543
67. Gillespie DT (1977) Exact stochastic simulation of coupled chemical reactions. J Phys Chem 81:2340–2361

68. Gillespie DT (1976) A general method for numerically simulating the stochastic time evolution of coupled chemical reactions. J Comp Phys 22:403–434
69. Garner KL, Perrett RM, Voliotis M, Pham T, Tsaneva-Atanasova K, McArdle CA (2017) Information transfer via gonadotropin-releasing hormone receptors to ERK and NFAT: sensing GnRH and sensing dynamics. J Endocr Soc 1:260–277
70. Caunt CJ, Finch AR, Sedgley KR, McArdle CA (2006) Seven-transmembrane receptor signalling and ERK compartmentalization. Trends Endocrinol Metab 17(7):276–283. http://doi.org/10.1016/j.tem.2006.07.008
71. Finch AR, Caunt CJ, Armstrong SP, McArdle CA (2009) Agonist-induced internalization and downregulation of gonadotropin-releasing hormone receptors. Am J Physiol Cell Physiol 297(3):C591–C600. http://doi.org/10.1152/ajpcell.00166.2009
72. Armstrong SP, Caunt CJ, Finch AR, McArdle CA (2011) Using automated imaging to interrogate gonadotrophin-releasing hormone receptor trafficking and function. Mol Cell Endocrinol 331(2):194–204. http://doi.org/10.1016/j.mce.2010.07.008
73. Finch AR, Sedgley KR, Armstrong SP, Caunt CJ, McArdle CA (2010) Trafficking and signalling of gonadotrophin-releasing hormone receptors: an automated imaging approach. Br J Pharmacol 159(4):751–760. http://doi.org/10.1111/j.1476-5381.2009.00413.x

INDEX

Louise von Stechow and Alberto Santos Delgado (eds.), *Computational Cell Biology: Methods and Protocols*, Methods in Molecular Biology, vol. 1819, https://doi.org/10.1007/978-1-4939-8618-7,

K

L

M

N

T

U

V

W

Y

Printed by Printforce, the Netherlands